Lecture Notes in Computer Science   2452

Edited by G. Goos, J. Hartmanis, and J. van Leeuwen

Springer
Berlin
Heidelberg
New York
Barcelona
Hong Kong
London
Milan
Paris
Tokyo

Roderic Guigó   Dan Gusfield (Eds.)

Algorithms in Bioinformatics

Second International Workshop, WABI 2002
Rome, Italy, September 17-21, 2002
Proceedings

 Springer

Series Editors

Gerhard Goos, Karlsruhe University, Germany
Juris Hartmanis, Cornell University, NY, USA
Jan van Leeuwen, Utrecht University, The Netherlands

Volume Editors

Roderic Guigó
IMIM-UPF-CRG, Dr. Aiguader 80, 08003 Barcelona, Spain
E-mail: rguigo@imim.es

Dan Gusfield
Department of Computer Science, University of California
Davis CA 95616, U.S.A.
E-mail: gusfield@cs.ucdavis.edu

Cataloging-in-Publication Data applied for

Die Deutsche Bibliothek - CIP-Einheitsaufnahme

Algorithms in bioinformatics : second international workshop ; proceedings /
WABI 2002, Rome, Italy, September 17 - 21, 2002. Roderic Guigó ; Dan
Gusfield (ed.). - Berlin ; Heidelberg ; New York ; Hong Kong ; London ;
Milan ; Paris ; Tokyo : Springer, 2002
    (Lecture notes in computer science ; Vol. 2452)
    ISBN 3-540-44211-1

CR Subject Classification (1998): F.1, F.2.2, E.1, G.1, G.2, G.3, J.3

ISSN 0302-9743
ISBN 3-540-44211-1 Springer-Verlag Berlin Heidelberg New York

Springer-Verlag Berlin Heidelberg New York
a member of BertelsmannSpringer Science+Business Media GmbH

http://www.springer.de

© Springer-Verlag Berlin Heidelberg 2002
Printed in Germany

Typesetting: Camera-ready by author, data conversion by Olgun Computergrafik
Printed on acid-free paper   SPIN: 10871128   06/3142   5 4 3 2 1 0

Preface

We are pleased to present the proceedings of the *Second Workshop on Algorithms in Bioinformatics (WABI 2002)*, which took place on September 17-21, 2002 in Rome, Italy. The WABI workshop was part of a three-conference meeting, which, in addition to WABI, included the ESA and APPROX 2002. The three conferences are jointly called ALGO 2002, and were hosted by the Faculty of Engineering, University of Rome "La Sapienza". See `http://www.dis.uniroma1.it/~algo02` for more details.

The Workshop on Algorithms in Bioinformatics covers research in all areas of algorithmic work in bioinformatics and computational biology. The emphasis is on discrete algorithms that address important problems in molecular biology, genomics, and genetics, that are founded on sound models, that are computationally efficient, and that have been implemented and tested in simulations and on real datasets. The goal is to present recent research results, including significant work in progress, and to identify and explore directions of future research.

Original research papers (including significant work in progress) or state-of-the-art surveys were solicited on all aspects of algorithms in bioinformatics, including, but not limited to: exact and approximate algorithms for genomics, genetics, sequence analysis, gene and signal recognition, alignment, molecular evolution, phylogenetics, structure determination or prediction, gene expression and gene networks, proteomics, functional genomics, and drug design.

We received 83 submissions in response to our call for papers, and were able to accept about half of the submissions. In addition, WABI hosted two invited, distinguished lectures, given to the entire ALGO 2002 conference, by Dr. Ehud Shapiro of the Weizmann Institute and Dr. Gene Myers of Celera Genomics. An abstract of Dr. Shapiro's lecture, and a full paper detailing Dr. Myers lecture, are included in these proceedings.

We would like to sincerely thank all the authors of submitted papers, and the participants of the workshop. We also thank the program committee for their hard work in reviewing and selecting the papers for the workshop. We were fortunate to have on the program committee the following distinguished group of researchers:

Pankaj Agarwal (GlaxoSmithKline Pharmaceuticals, King of Prussia)
Alberto Apostolico (Università di Padova and Purdue University, Lafayette)
Craig Benham (University of California, Davis)
Jean-Michel Claverie (CNRS-AVENTIS, Marseille)
Nir Friedman (Hebrew University, Jerusalem)
Olivier Gascuel (Université de Montpellier II and CNRS, Montpellier)
Misha Gelfand (IntegratedGenomics, Moscow)
Raffaele Giancarlo (Università di Palermo)

David Gilbert (University of Glasgow)
Roderic Guigo (Institut Municipal d'Investigacions Mèdiques,
    Barcelona, co-chair)
Dan Gusfield (University of California, Davis, co-chair)
Jotun Hein (University of Oxford)
Inge Jonassen (Universitetet i Bergen)
Giuseppe Lancia (Università di Padova)
Bernard M.E. Moret (University of New Mexico, Albuquerque)
Gene Myers (Celera Genomics, Rockville)
Christos Ouzonis (European Bioinformatics Institute, Hinxton Hall)
Lior Pachter (University of California, Berkeley)
Knut Reinert (Celera Genomics, Rockville)
Marie-France Sagot (Université Claude Bernard, Lyon)
David Sankoff (Université de Montréal)
Steve Skiena (State University of New York, Stony Brook)
Gary Stormo (Washington University, St. Louis)
Jens Stoye (Universität Bielefeld)
Martin Tompa (University of Washington, Seattle)
Alfonso Valencia (Centro Nacional de Biotecnología, Madrid)
Martin Vingron (Max-Planck-Institut für Molekulare Genetik, Berlin)
Lusheng Wang (City University of Hong Kong)
Tandy Warnow (University of Texas, Austin)

We also would like to thank the WABI steering committee, Olivier Gascuel, Jotun Hein, Raffaele Giancarlo, Erik Meineche-Schmidt, and Bernard Moret, for inviting us to co-chair this program committee, and for their help in carrying out that task.

We are particularly indebted to Terri Knight of the University of California, Davis, Robert Castelo of the Universitat Pompeu Fabra, Barcelona, and Bernard Moret of the University of New Mexico, Albuquerque, for the extensive technical and advisory help they gave us. We could not have managed the reviewing process and the preparation of the proceedings without their help and advice.

Thanks again to everyone who helped to make WABI 2002 a success. We hope to see everyone again at WABI 2003.

July, 2002   Roderic Guigó and Dan Gusfield

Table of Contents

Simultaneous Relevant Feature Identification
and Classification in High-Dimensional Spaces 1
> L.R. Grate (Lawrence Berkeley National Laboratory), C. Bhattacharyya,
> M.I. Jordan, and I.S. Mian (University of California Berkeley)

Pooled Genomic Indexing (PGI): Mathematical Analysis
and Experiment Design 10
> M. Csűrös (Université de Montréal), and A. Milosavljevic (Human
> Genome Sequencing Center)

Practical Algorithms and Fixed-Parameter Tractability
for the Single Individual SNP Haplotyping Problem 29
> R. Rizzi (Università di Trento), V. Bafna, S. Istrail (Celera Genomics),
> and G. Lancia (Università di Padova)

Methods for Inferring Block-Wise Ancestral History
from Haploid Sequences.. 44
> R. Schwartz, A.G. Clark (Celera Genomics), and S. Istrail (Celera
> Genomics)

Finding Signal Peptides in Human Protein Sequences
Using Recurrent Neural Networks 60
> M. Reczko (Synaptic Ltd.), P. Fiziev, E. Staub (metaGen Pharmaceu-
> ticals GmbH), and A. Hatzigeorgiou (University of Pennsylvania)

Generating Peptide Candidates from Amino-Acid Sequence Databases
for Protein Identification via Mass Spectrometry 68
> N. Edwards and R. Lippert (Celera Genomics)

Improved Approximation Algorithms for NMR Spectral Peak Assignment . 82
> Z.-Z. Chen (Tokyo Denki University), T. Jiang (University of Califor-
> nia, Riverside), G. Lin (University of Alberta), J. Wen (University of
> California, Riverside), D. Xu, and Y. Xu (Oak Ridge National Labo-
> ratory)

Efficient Methods for Inferring Tandem Duplication History 97
> L. Zhang (Nat. University of Singapore), B. Ma (University of Western
> Ontario), and L. Wang (City University of Hong Kong)

Genome Rearrangement Phylogeny Using Weighbor 112
> L.-S. Wang (University of Texas at Austin)

Segment Match Refinement and Applications 126
    A.L. Halpern (Celera Genomics), D.H. Huson (Tübingen University),
    and K. Reinert (Celera Genomics)

Extracting Common Motifs under the Levenshtein Measure:
Theory and Experimentation 140
    E.F. Adebiyi and M. Kaufmann (Universität Tübingen)

Fast Algorithms for Finding Maximum-Density Segments
of a Sequence with Applications to Bioinformatics 157
    M.H. Goldwasser (Loyola University Chicago), M.-Y. Kao (Northwest-
    ern University), and H.-I. Lu (Academia Sinica)

FAUST: An Algorithm for Extracting Functionally Relevant Templates
from Protein Structures 172
    M. Milik, S. Szalma, and K.A. Olszewski (Accelrys)

Efficient Unbound Docking of Rigid Molecules 185
    D. Duhovny, R. Nussinov, and H.J. Wolfson (Tel Aviv University)

A Method of Consolidating and Combining EST and mRNA Alignments
to a Genome to Enumerate Supported Splice Variants 201
    R. Wheeler (Affymetrix)

A Method to Improve the Performance of Translation Start Site Detection
and Its Application for Gene Finding 210
    M. Pertea and S.L. Salzberg (The Institute for Genomic Research)

Comparative Methods for Gene Structure Prediction
in Homologous Sequences 220
    C.N.S. Pedersen and T. Scharling (University of Aarhus)

MultiProt – A Multiple Protein Structural Alignment Algorithm 235
    M. Shatsky, R. Nussinov, and H.J. Wolfson (Tel Aviv University)

A Hybrid Scoring Function for Protein Multiple Alignment 251
    E. Rocke (University of Washington)

Functional Consequences in Metabolic Pathways
from Phylogenetic Profiles 263
    Y. Bilu and M. Linial (Hebrew University)

Finding Founder Sequences from a Set of Recombinants 277
    E. Ukkonen (University of Helsinki)

Estimating the Deviation from a Molecular Clock 287
    L. Nakhleh, U. Roshan (University of Texas at Austin), L. Vawter
    (Aventis Pharmaceuticals), and T. Warnow (University of Texas at
    Austin)

Exploring the Set of All Minimal Sequences of Reversals – An Application
to Test the Replication-Directed Reversal Hypothesis 300
    Y. Ajana, J.-F. Lefebvre (Université de Montréal), E.R.M. Tillier (University Health Network), and N. El-Mabrouk (Université de Montréal)

Approximating the Expected Number of Inversions
Given the Number of Breakpoints 316
    N. Eriksen (Royal Institute of Technology)

Invited Lecture – Accelerating Smith-Waterman Searches 331
    G. Myers (Celera Genomics) and R. Durbin (Sanger Centre)

Sequence-Length Requirements for Phylogenetic Methods 343
    B.M.E. Moret (University of New Mexico), U. Roshan, and T. Warnow (University of Texas at Austin)

Fast and Accurate Phylogeny Reconstruction Algorithms
Based on the Minimum-Evolution Principle 357
    R. Desper (National Library of Medicine, NIH) and O. Gascuel (LIRMM)

NeighborNet: An Agglomerative Method for the Construction
of Planar Phylogenetic Networks 375
    D. Bryant (McGill University) and V. Moulton (Uppsala University)

On the Control of Hybridization Noise
in DNA Sequencing-by-Hybridization 392
    H.-W. Leong (National University of Singapore), F.P. Preparata (Brown University), W.-K. Sung, and H. Willy (National University of Singapore)

Restricting SBH Ambiguity via Restriction Enzymes 404
    S. Skiena (SUNY Stony Brook) and S. Snir (Technion)

Invited Lecture – Molecule as Computation:
Towards an Abstraction of Biomolecular Systems 418
    E. Shapiro (Weizmann Institute)

Fast Optimal Genome Tiling with Applications to Microarray Design
and Homology Search 419
    P. Berman (Pennsylvania State University), P. Bertone (Yale University), B. DasGupta (University of Illinois at Chicago), M. Gerstein (Yale University), M.-Y. Kao (Northwestern University), and M. Snyder (Yale University)

Rapid Large-Scale Oligonucleotide Selection for Microarrays 434
    S. Rahmann (Max-Planck-Institute for Molecular Genetics)

Border Length Minimization in DNA Array Design.................... 435
   A.B. Kahng, I.I. Măndoiu, P.A. Pevzner, S. Reda (University of California at San Diego), and A.Z. Zelikovsky (Georgia State University)

The Enhanced Suffix Array and Its Applications to Genome Analysis..... 449
   M.I. Abouelhoda, S. Kurtz, and E. Ohlebusch (University of Bielefeld)

The Algorithmic of Gene Teams 464
   A. Bergeron (Université du Québec a Montreal), S. Corteel (CNRS -
   Université de Versailles), and M. Raffinot (CNRS - Laboratoire Génome
   et Informatique)

Combinatorial Use of Short Probes
for Differential Gene Expression Profiling 477
   L.L. Warren and B.H. Liu (North Carolina State University)

Designing Specific Oligonucleotide Probes
for the Entire S. cerevisiae Transcriptome 491
   D. Lipson (Technion), P. Webb, and Z. Yakhini (Agilent Laboratories)

K-ary Clustering with Optimal Leaf Ordering for Gene Expression Data .. 506
   Z. Bar-Joseph, E.D. Demaine, D.K. Gifford (MIT LCS), A.M. Hamel
   (Wilfrid Laurier University), T.S. Jaakkola (MIT AI Lab), and N. Sre-
   bro (MIT LCS)

Inversion Medians Outperform Breakpoint Medians
in Phylogeny Reconstruction from Gene-Order Data.................... 521
   B.M.E. Moret (University of New Mexico), A.C. Siepel (University of
   California at Santa Cruz), J. Tang, and T. Liu (University of New
   Mexico)

Modified Mincut Supertrees 537
   R.D.M. Page (University of Glasgow)

Author Index 553

Simultaneous Relevant Feature Identification and Classification in High-Dimensional Spaces

L.R. Grate[1], C. Bhattacharyya[2,3], M.I. Jordan[2,3], and I.S. Mian[1]

. Life Sciences Division, Lawrence Berkeley National Laboratory, Berkeley CA 94720
. Department of EECS, University of California Berkeley, Berkeley CA 94720
. Department of Statistics, University of California Berkeley, Berkeley CA 94720

Abstract. Molecular profiling technologies monitor thousands of transcripts, proteins, metabolites or other species concurrently in biological samples of interest. Given two-class, high-dimensional profiling data, nominal LIKNON [4] is a specific implementation of a methodology for performing simultaneous relevant feature identification and classification. It exploits the well-known property that minimizing an $l.$ norm (via linear programming) yields a sparse hyperplane [15,26,2,8,17]. This work (i) examines computational, software and practical issues required to realize nominal LIKNON, (ii) summarizes results from its application to five real world data sets, (iii) outlines heuristic solutions to problems posed by domain experts when interpreting the results and (iv) defines some future directions of the research.

1 Introduction

Biologists and clinicians are adopting high-throughput genomics, proteomics and related technologies to assist in interrogating normal and perturbed systems such as unaffected and tumor tissue specimens. Such investigations can generate data having the form $\mathcal{D} = \{(\mathbf{x}_n, y_n), n \in (1, \ldots, N)\}$ where $\mathbf{x}_n \in \mathbb{R}^P$ and, for two-class data, $y_n \in \{+1, -1\}$. Each element of a data point $\mathbf{x}_n$ is the absolute or relative abundance of a molecular species monitored. In transcript profiling, a data point represents transcript (gene) levels measured in a sample using cDNA, oligonucleotide or similar microarray technology. A data point from protein profiling can represent Mass/Charge (M/Z) values for low molecular weight molecules (proteins) measured in a sample using mass spectroscopy.

In cancer biology, profiling studies of different types of (tissue) specimens are motivated largely by a desire to create clinical decision support systems for accurate tumor classification and to identify robust and reliable targets, "biomarkers", for imaging, diagnosis, prognosis and therapeutic intervention [14,3,13,27,18,23,9,25,28,19,21,24]. Meeting these biological challenges includes addressing the general statistical problems of classification and prediction, and relevant feature identification.

Support Vector Machines (SVMs) [30,8] have been employed successfully for cancer classification based on transcript profiles [5,22,25,28]. Although mechanisms for reducing the number of features to more manageable numbers include

R. Guigó and D. Gusfield (Eds.): WABI 2002, LNCS 2452, pp. 1–9, 2002.

discarding those below a user-defined threshold, relevant feature identification is usually addressed via a filter-wrapper strategy [12,22,32]. The filter generates candidate feature subsets whilst the wrapper runs an induction algorithm to determine the discriminative ability of a subset. Although SVMs and the newly formulated Minimax Probability Machine (MPM) [20] are good wrappers [4], the choice of filtering statistic remains an open question.

Nominal LIKNON is a specific implementation of a strategy for performing simultaneous relevant feature identification and classification [4]. It exploits the well-known property that minimizing an l_1 norm (via linear programming) yields a sparse hyperplane [15,26,2,8,17]. The hyperplane constitutes the classifier whilst its sparsity, a weight vector with few non-zero elements, defines a small number of relevant features. Nominal LIKNON is computationally less demanding than the prevailing filter–(SVM/MPM) wrapper strategy which treats the problems of feature selection and classification as two independent tasks [4,16]. Biologically, nominal LIKNON performs well when applied to real world data generated not only by the ubiquitous transcript profiling technology, but also by the emergent protein profiling technology.

2   Simultaneous Relevant Feature Identification and Classification

Consider a data set $\mathcal{D} = \{(\mathbf{x}_n, y_n), n \in (1, \dots, N)\}$. Each of the N data points (profiling experiments) is a P-dimensional vector of features (gene or protein abundances) $\mathbf{x}_n \in \mathbb{R}^P$ (usually $N \sim 10^1 - 10^2; P \sim 10^3 - 10^4$). A data point n is assigned to one of two classes $y_n \in \{+1, -1\}$ such a normal or tumor tissue sample. Given such two-class high-dimensional data, the analytical goal is to estimate a sparse classifier, a model which distinguishes the two classes of data points (classification) *and* specifies a small subset of discriminatory features (relevant feature identification). Assume that the data $\mathcal{D}$ can be separated by a linear hyperplane in the P-dimensional input feature space. The learning task can be formulated as an attempt to estimate a hyperplane, parameterized in terms of a weight vector $\mathbf{w}$ and bias b, via a solution to the following N inequalities [30]:

$$y_n z_n = y_n(\mathbf{w}^T \mathbf{x}_n - b) \geq 0$$
$$\forall n = \{1, \dots, N\} \ . \tag{1}$$

The hyperplane satisfying $\mathbf{w}^T \mathbf{x} - b = 0$ is termed a classifier. A new data point $\mathbf{x}$ (abundances of P features in a new sample) is classified by computing $z = \mathbf{w}^T \mathbf{x} - b$. If $z > 0$, the data point is assigned to one class otherwise it belongs to the other class.

Enumerating relevant features at the same time as discovering a classifier can be addressed by finding a sparse hyperplane, a weight vector $\mathbf{w}$ in which most components are equal to zero. The rationale is that zero elements do not contribute to determining the value of z:

$$z = \sum_{p=1}^{P} w_p x_p - b \ .$$

If $w_p = 0$, feature p is "irrelevant" with regards to deciding the class. Since only non-zero elements $w_p \neq 0$ influence the value of z, they can be regarded as "relevant" features.

The task of defining a small number of relevant features can be equated with that of finding a small set of non-zero elements. This can be formulated as an optimization problem; namely that of minimizing the l_0 norm $\|\mathbf{w}\|_0$, where $\|\mathbf{w}\|_0 =$ number of $\{p : w_p \neq 0\}$, the number of non-zero elements of $\mathbf{w}$. Thus we obtain:

$$\min_{\mathbf{w},b} \quad \|\mathbf{w}\|_0$$
$$\text{subject to } y_n(\mathbf{w}^T \mathbf{x}_n - b) \geq 0$$
$$\forall n = \{1, \ldots, N\} \ . \tag{2}$$

Unfortunately, problem (2) is NP-hard [10]. A tractable, convex approximation to this problem can be obtained by replacing the l_0 norm with the l_1 norm $\|\mathbf{w}\|_1$, where $\|\mathbf{w}\|_1 = \sum_{p=1}^{P} |w_p|$, the sum of the absolute magnitudes of the elements of a vector [10]:

$$\min_{\mathbf{w},b} \|\mathbf{w}\|_1 = \sum_{p=1}^{P} |w_p|$$
$$\text{subject to } y_n(\mathbf{w}^T \mathbf{x}_n - b) \geq 0$$
$$\forall n = \{1, \ldots, N\} \ . \tag{3}$$

A solution to (3) yields the desired sparse weight vector $\mathbf{w}$.

Optimization problem (3) can be solved via linear programming [11]. The ensuing formulation requires the imposition of constraints on the allowed ranges of variables. The introduction of new variables $u_p, v_p \in \mathbb{R}^P$ such that $|w_p| = u_p + v_p$ and $w_p = u_p - v_p$ ensures non-negativity. The range of $w_p = u_p - v_p$ is unconstrained (positive or negative) whilst u_p and v_p remain non-negative. u_p and v_p are designated the "positive" and "negative" parts respectively. Similarly, the bias b is split into positive and negative components $b = b_+ - b_-$. Given a solution to problem (3), either u_p or v_p will be non-zero for feature p [11]:

$$\min_{\mathbf{u},\mathbf{v},b_+,b_-} \quad \sum_{p=1}^{P}(u_p + v_p)$$
$$\text{subject to } \quad y_n((\mathbf{u} - \mathbf{v})^T \mathbf{x}_n - (b_+ - b_-)) \geq 1$$
$$u_p \geq 0; v_p \geq 0; b_+ \geq 0; b_- \geq 0$$
$$\forall n = \{1, \ldots, N\}; \forall p = \{1, \ldots, P\} \ . \tag{4}$$

A detailed description of the origins of the ≥ 1 constraint can be found elsewhere [30].

If the data $\mathcal{D}$ are not linearly separable, misclassifications (errors in the class labels y_n) can be accounted for by the introduction of slack variables ξ_n. Problem (4) can be recast yielding the final optimization problem,

$$\min_{\mathbf{u},\mathbf{v},b_+,b_-} \quad \sum_{p=1}^{P}(u_p + v_p) + C\sum_{n=1}^{N}\xi_n$$

$$\text{subject to } y_n((\mathbf{u} - \mathbf{v})^T\mathbf{x}_n - (b_+ - b_-)) \geq 1 - \xi_n$$

$$u_p \geq 0; v_p \geq 0; b_+ \geq 0; b_- \geq 0; \xi_n \geq 0$$

$$\forall n = \{1, \ldots, N\}; \forall p = \{1, \ldots, P\} \ . \tag{5}$$

C is an adjustable parameter weighing the contribution of misclassified data points. Larger values lead to fewer misclassifications being ignored: $C = 0$ corresponds to no outliers being ignored whereas $C \to \infty$ leads to the hard margin limit.

3 Computational, Software and Practical Issues

Learning the sparse classifier defined by optimization problem (5) involves minimizing a linear function subject to linear constraints. Efficient algorithms for solving such linear programming problems involving ~10,000 variables (N) and ~10,000 constraints (P) are well-known. Standalone open source codes include lp_solve[1] and PCx[2].

Nominal LIKNON is an implementation of the sparse classifier (5). It incorporates routines written in Matlab[3] and a system utilizing perl[4] and lp_solve. The code is available from the authors upon request. The input consists of a file containing an $N \times (P + 1)$ data matrix in which each row represents a single profiling experiment. The first P columns are the feature values, abundances of molecular species, whilst column $P + 1$ is the class label $y_n \in \{+1, -1\}$. The output comprises the non-zero values of the weight vector $\mathbf{w}$ (relevant features), the bias b and the number of non-zero slack variables ξ_n.

The adjustable parameter C in problem (5) can be set using cross validation techniques. The results described here were obtained by choosing $C = 0.5$ or $C = 1$.

4 Application of Nominal Liknon to Real World Data

Nominal LIKNON was applied to five data sets in the size range ($N = 19$, $P = 1,987$) to ($N = 200$, $P = 15,154$). A data set $\mathcal{D}$ yielded a sparse classifier, $\mathbf{w}$ and b, and a specification of the l relevant features ($P \gg l$). Since the profiling studies produced only a small number of data points ($N \ll P$), the generalization error of a nominal LIKNON classifier was determined by computing the leave-one-out error for l-dimensional data points. A classifier trained using $N - 1$ data points was used to predict the class of the withheld data point; the procedure repeated N times. The results are shown in Table 1.

Nominal LIKNON performs well in terms of simultaneous relevant feature identification and classification. In all five transcript and protein profiling data

---

[*] http://www.netlib.org/ampl/solvers/lpsolve/
[*] http://www-fp.mcs.anl.gov/otc/Tools/PCx/
[*] http://www.mathworks.com
[*] http://www.perl.org/

Table 1. Summary of published and unpublished investigations using nominal LIKNON [4,16].

Transcript profiles	Sporadic breast carcinoma tissue samples [29]
	inkjet microarrays; relative transcript levels
	`http://www.rii.com/publications/vantveer.htm`
Two-class data	46 patients with distant metastases < 5 years
	51 patients with no distant metastases ≥ 5 years
Relevant features	72 out of $P=5,192$
Leave-one-out error	1 out of $N=97$
Transcript profiles	Tumor tissue samples [1]
	custom cDNA microarrays; relative transcript levels
	`http://www.nhgri.nih.gov/DIR/Microarray/`
	`selected_publications.html`
Two-class data	13 *KIT*-mutation positive gastrointestinal stromal tumors
	6 spindle cell tumors from locations outside the gastrointestinal tract
Relevant features	6 out of $P=1,987$
Leave-one-out error	0 out of $N=19$
Transcript profiles	Small round blue cell tumor samples (EWS, RMS, NHL, NB) [19]
	custom cDNA microarrays; relative transcript levels
	`http://www.nhgri.nih.gov/DIR/Microarray/Supplement`
Two-class data	46 EWS/RMS tumor biopsies
	38 EWS/RMS/NHL/NB cell lines
Relevant features	23 out of $P=2,308$
Leave-one-out error	0 out of $N=84$
Transcript profiles	Prostate tissue samples [31]
	Affymetrix arrays; absolute transcript levels
	`http://carrier.gnf.org/welsh/prostate`
Two-class data	9 normal
	25 malignant
Relevant features	7 out of $P=12,626$
Leave-one-out error	0 out of $N=34$
Protein profiles	Serum samples [24]
	SELDI-TOF mass spectrometry; M/Z values (spectral amplitudes)
	`http://clinicalproteomics.steem.com`
Two-class data	100 unaffected
	100 ovarian cancer
Relevant features	51 out of $P=15,154$
Leave-one-out error	3 out of $N=200$

sets a hyperplane was found, the weight vector was sparse (< 100 or $< 2\%$ non-zero components) and the relevant features were of interest to domain experts (they generated novel biological hypotheses amenable to subsequent experimental or clinical validation). For the protein profiles, better results were obtained using normalized as opposed to raw values: when employed to predict the class of 16 independent non-cancer samples, the 51 relevant features had a test error of 0 out of 16.

On a powerful desktop computer, a > 1 GHz Intel-like machine, the time required to create a sparse classifier varied from 2 seconds to 20 minutes. For the larger problems, the main memory RAM requirement exceeded 500 MBytes.

5   Heuristic Solutions to Problems Posed by Domain Experts

Domain experts wish to postprocess nominal LIKNON results to assist in the design of subsequent experiments aimed at validating, verifying and extending any biological predictions. In lieu of a theoretically sound statistical framework, heuristics have been developed to prioritize, reduce or increase the number of relevant features.

In order to priorities features, assume that all P features are on the same scale. The l relevant features can be ranked according to the magnitude and/or sign of the non-zero elements of the weight vector $\mathbf{w}$ ($w_p \neq 0$). To reduce the number of relevant features to a "smaller, most interesting" set, a histogram of $w_p \neq 0$ values can be used to determine a threshold for pruning the set. In order to increase the number of features to a "larger, more interesting" set, nominal LIKNON can be run in an iterative manner. The l relevant features identified in one pass through the data are removed from the data points to be used as input for the next pass. Each successive round generates a new set of relevant features. The procedure is terminated either by the domain expert or by monitoring the leave-one-out error of the classifier associated with each set of relevant features.

Preliminary results from analysis of the gastrointestinal stromal tumor/spindle cell tumor transcript profiling data set indicate that these extensions are likely to be of utility to domain experts. The leave-one-out error of the relevant features identified by five iterations of nominal LIKNON was at most one. The details are: iteration 0 (number of relevant features = 6, leave-one-out error = 0), iteration 1 (5, 0), iteration 2 (5, 1), iteration 3 (9, 0), iteration 4 (13, 1), iteration 5 (11, 1).

Iterative LIKNON may prove useful during explorations of the (qualitative) association between relevant features and their behavior in the N data points. The gastrointestinal stromal tumor/spindle cell tumor transcript profiling data set has been the subject of probabilistic clustering [16]. A finite Gaussian mixture model as implemented by the program AutoClass [6] was estimated from P=1,987, N=19-dimensional unlabeled data points. The trained model was used to assign each feature (gene) to one of the resultant clusters. Five iterations of nominal LIKNON identified the majority of genes assigned to a small number of discriminative clusters. Furthermore, these genes constituted most of the important distinguishing genes defined by the original authors [1].

6   Discussion

Nominal LIKNON implements a mathematical technique for finding a sparse hyperplane. When applied to two-class high-dimensional real-world molecular profiling data, it identifies a small number of relevant features and creates a classifier that generalizes well. As discussed elsewhere [4,7], many subsets of relevant features are likely to exist. Although nominal LIKNON specifies but one set of discriminatory features, this "low-hanging fruit" approach does suggest genes of interest to experimentalists. Iterating the procedure provides a rapid mechanism for highlighting additional sets of relevant features that yield good classifiers. Since nominal LIKNON is a single-pass method, one disadvantage is that the learned parameters cannot be adjusted (improved) as would be possible with a more typical train/test methodology.

7 Future Directions

Computational biology and chemistry are generating high-dimensional data so sparse solutions for classification and regression problems are of widespread importance. A general purpose toolbox containing specific implementations of particular statistical techniques would be of considerable practical utility. Future plans include developing a suite of software modules to aid in performing tasks such as the following. A. Create high-dimensional input data. (i) Direct generation by high-throughput experimental technologies. (ii) Systematic formulation and extraction of large numbers of features from data that may be in the form of strings, images, and so on (*a priori*, features "relevant" for one problem may be "irrelevant" for another). B. Enunciate sparse solutions for classification and regression problems in high-dimensions. C. Construct and assess models. (i) Learn a variety of models by a grid search through the space of adjustable parameters. (ii) Evaluate the generalization error of each model. D. Combine best models to create a final decision function. E. Propose hypotheses for domain expert.

Acknowledgements

This work was supported by NSF grant IIS-9988642, the Director, Office of Energy Research, Office of Health and Environmental Research, Division of the U.S. Department of Energy under Contract No. DE-AC03-76F00098 and an LBNL/LDRD through U.S. Department of Energy Contract No. DE-AC03-76SF00098.

References

1. S.V. Allander, N.N. Nupponen, M. Ringner, G. Hostetter, G.W. Maher, N. Goldberger, Y. Chen, Carpten J., A.G. Elkahloun, and P.S. Meltzer. Gastrointestinal Stromal Tumors with KIT mutations exhibit a remarkably homogeneous gene expression profile. *Cancer Research*, 61:8624–8628, 2001.
2. K. Bennett and A. Demiriz. Semi-supervised support vector machines. In *Neural and Information Processing Systems*, volume 11. MIT Press, Cambridge MA, 1999.
3. A. Bhattacharjee, W.G. Richards, J. Staunton, C. Li, S. Monti, P. Vasa, C. Ladd, J. Beheshti, R. Bueno, M. Gillette, M. Loda, G. Weber, E.J. Mark, E.S. Lander, W. Wong, B.E. Johnson, T.R. Golub, D.J. Sugarbaker, and M. Meyerson. Classification of human lung carcinomas by mrna expression profiling reveals distinct adenocarcinoma subclasses. *Proc. Natl. Acad. Sci.*, 98:13790–13795, 2001.
4. C. Bhattacharyya, L.R. Grate, A. Rizki, D.C. Radisky, F.J. Molina, M.I. Jordan, M.J. Bissell, and I.S. Mian. Simultaneous relevant feature identification and classification in high-dimensional spaces: application to molecular profiling data. *Submitted, Signal Processing*, 2002.

5. M.P. Brown, W.N. Grundy, D. Lin, N. Cristianini, C.W. Sugnet, T.S. Furey, M. Ares, Jr, and D. Haussler. Knowledge-based analysis of microarray gene expression data by using support vector machines. *Proc. Natl. Acad. Sci.*, 97:262–267, 2000.

6. P. Cheeseman and J. Stutz. Bayesian Classification (AutoClass): Theory and Results. In U.M. Fayyad, G. Piatetsky-Shapiro, P. Smyth, and R. Uthurusamy, editors, *Advances in Knowledge Discovery and Data Mining*, pages 153–180. AAAI Press/MIT Press, 1995. The software is available at the URL http://www.gnu.org/directory/autoclass.html.

7. M.L. Chow, E.J. Moler, and I.S. Mian. Identifying marker genes in transcription profile data using a mixture of feature relevance experts. *Physiological Genomics*, 5:99–111, 2001.

8. N. Cristianini and J. Shawe-Taylor. *Support Vector Machines and other kernel-based learning methods*. Cambridge University Press, Cambridge, England, 2000.

9. S.M. Dhanasekaran, T.R. Barrette, R. Ghosh, D. Shah, S. Varambally, K. Kurachi, K.J. Pienta, M.J. Rubin, and A.M. Chinnaiyan. Delineation of prognostic biomarkers in prostate cancer. *Nature*, 432, 2001.

10. D.L. Donoho and X. Huo. Uncertainty principles and idea atomic decomposition. Technical Report, Statistics Department, Stanford University, 1999.

11. R. Fletcher. *Practical Methods in Optimization*. John Wiley & Sons, New York, 2000.

12. T. Furey, N. Cristianini, N. Duffy, D. Bednarski, M. Schummer, and D. Haussler. Support vector machine classification and validation of cancer tissue samples using microarray expression data. *Bioinformatics*, 16:906–914, 2000.

13. M.E. Garber, O.G. Troyanskaya, K. Schluens, S. Petersen, Z. Thaesler, M. Pacyana-Gengelbach, M. van de Rijn, G.D. Rosen, C.M. Perou, R.I. Whyte, R.B. Altman, P.O. Brown, D. Botstein, and I. Petersen. Diversity of gene expression in adenocarcinoma of the lung. *Proc. Natl. Acad. Sci.*, 98:13784–13789, 2001.

14. T.R. Golub, D.K. Slonim, P. Tamayo, C. Huard, M. Gaasenbeek, J. Mesirov, H. Coller, M.L. Loh, J.R. Downing, M.A. Caligiuri, C.D. Bloomfeld, and E.S. Lander. Molecular classification of cancer: Class discovery and class prediction by gene expression monitoring. *Science*, 286:531–537, 1999. The data are available at the URL waldo.wi.mit.edu/MPR/data_sets.html.

15. T. Graepel, B. Herbrich, R. Schölkopf, A.J. Smola, P. Bartlett, K. Müller, K. Obermayer, and R.C. Williamson. Classification on proximity data with lp-machines. In *Ninth International Conference on Artificial Neural Networks*, volume 470, pages 304–309. IEE, London, 1999.

16. L.R. Grate, C. Bhattacharyya, M.I. Jordan, and I.S. Mian. Integrated analysis of transcript profiling and protein sequence data. *In press, Mechanisms of Ageing and Development*, 2002.

17. T. Hastie, R. Tibshirani, , and Friedman J. *The Elements of Statistical Learning: Data Mining, Inference, and Prediction*. Springer-Verlag, New York, 2000.

18. I. Hedenfalk, D. Duggan, Y. Chen, M. Radmacher, M. Bittner, R. Simon, P. Meltzer, B. Gusterson, M. Esteller, M. Raffeld, Z. Yakhini, A. Ben-Dor, E. Dougherty, J. Kononen, L. Bubendorf, W. Fehrle, S. Pittaluga, S. Gruvberger, N. Loman, O. Johannsson, H. Olsson, B. Wilfond, G. Sauter, O.-P. Kallioniemi, A. Borg, and J. Trent. Gene-expression profiles in hereditary breast cancer. *New England Journal of Medicine*, 344:539–548, 2001.

19. J. Khan, J.S. Wei, M. Ringner, L.H. Saal, M. Ladanyi, F. Westermann, F. Berthold, M. Schwab, Antonescu C.R., Peterson C., and P.S. Meltzer. Classification and diagnostic prediction of cancers using gene expression profiling and artificial neural networks. *Nature Medicine*, 7:673–679, 2001.

20. G. Lanckerit, L. El Ghaoui, C. Bhattacharyya, and M.I. Jordan. Minimax probability machine. *Advances in Neural Processing systems*, 14, 2001.

21. L.A. Liotta, E.C. Kohn, and E.F. Perticoin. Clinical proteomics. personalized molecular medicine. *JAMA*, 14:2211–2214, 2001.

22. E.J. Moler, M.L. Chow, and I.S. Mian. Analysis of molecular profile data using generative and discriminative methods. *Physiological Genomics*, 4:109–126, 2000.

23. D.A. Notterman, U. Alon, A.J. Sierk, and A.J. Levine. Transcriptional gene expression profiles of colorectal adenoma, adenocarcinoma, and normal tissue examined by oligonucleotide arrays. *Cancer Research*, 61:3124–3130, 2001.

24. E.F. Petricoin III, A.M. Ardekani, B.A. Hitt, P.J. Levine, V.A. Fusaro, S.M. Steinberg, G.B Mills, C. Simone, D.A. Fishman, E.C. Kohn, and L.A. Liotta. Use of proteomic patterns in serum to identify ovarian cancer. *The Lancet*, 359:572–577, 2002.

25. S. Ramaswamy, P. Tamayo, R. Rifkin, S. Mukherjee, C.-H. Yeang, M. Angelo, C. Ladd, M. Reich, E. Latulippe, J.P. Mesirov, T. Poggio, W. Gerald, M. Loda, E.S. Lander, and T.R. Golub. Multiclass cancer diagnosis using tumor gene expression signatures. *Proc. Natl. Acad. Sci.*, 98:15149–15154, 2001. The data are available from http://www-genome.wi.mit.edu/mpr/GCM.html.

26. A. Smola, T.T. Friess, and B. Schölkopf. Semiparametric support vector and linear programming machines. In *Neural and Information Processing Systems*, volume 11. MIT Press, Cambridge MA, 1999.

27. T. Sorlie, C.M. Perou, R. Tibshirani, T. Aas, S. Geisler, H. Johnsen, T. Hastie, M.B. Eisen, M. van de Rijn, S.S. Jeffrey, T. Thorsen, H. Quist, J.C. Matese, P.O. Brown, D. Botstein, P.E. Lonning, and A.-L. Borresen-Dale. Gene expression patterns of breast carcinomas distinguish tumor subclasses with clinical implications. *Proc. Natl. Acad. Sci.*, 98:10869–10874, 2001.

28. A.I. Su, J.B. Welsh, L.M. Sapinoso, S.G. Kern, P. Dimitrov, H. Lapp, P.G. Schultz, S.M. Powell, C.A. Moskaluk, H.F. Frierson Jr, and G.M. Hampton. Molecular classification of human carcinomas by use of gene expression signatures. *Cancer Research*, 61:7388–7393, 2001.

29. L.J. van 't Veer, H. Dai, M.J. van de Vijver, Y.D. He, A.A. Hart, M. Mao, H.L. Peterse, van der Kooy K., M.J. Marton, A.T. Witteveen, G.J. Schreiber, R.M. Kerkhoven, C. Roberts, P.S. Linsley, R. Bernards, and S.H. Friend. Gene expression profiling predicts clinical outcome of breast cancer. *Nature*, 415:530–536, 2002.

30. V. Vapnik. *Statistical Learning Theory*. Wiley, New York, 1998.

31. J.B. Welsh, L.M. Sapinoso, A.I. Su, S.G. Kern, J. Wang-Rodriguez, C.A. Moskaluk, J.F. Frierson Jr, and G.M. Hampton. Analysis of gene expression identifies candidate markers and pharmacological targets in prostate cancer. *Cancer Research*, 61:5974–5978, 2001.

32. J. Weston, Mukherjee S., O. Chapelle, M. Pontil, T. Poggio, and V. Vapnik. Feature Selection for SVMs. In *Advances in Neural Information Processing Systems*, volume 13, 2000.

Pooled Genomic Indexing (PGI): Mathematical Analysis and Experiment Design

Miklós Csűrös[1,2,3] and Aleksandar Milosavljevic[2,3]

Département d'informatique et de recherche opérationnelle, Université de Montréal
CP 6128 succ. Centre-Ville, Montréal, Québec H3C 3J7, Canada
csuros@iro.umontreal.ca
Human Genome Sequencing Center, Department of Molecular and Human Genetics
Baylor College of Medicine
Bioinformatics Research Laboratory
Department of Molecular and Human Genetics
Baylor College of Medicine, Houston, Texas 77030, USA
amilosav@bcm.tmc.edu

Abstract. Pooled Genomic Indexing (PGI) is a novel method for physical mapping of clones onto known macromolecular sequences. PGI is carried out by pooling arrayed clones, generating shotgun sequence reads from pools and by comparing the reads against a reference sequence. If two reads from two different pools match the reference sequence at a close distance, they are both assigned (deconvoluted) to the clone at the intersection of the two pools and the clone is mapped onto the region of the reference sequence between the two matches. A probabilistic model for PGI is developed, and several pooling schemes are designed and analyzed. The probabilistic model and the pooling schemes are validated in simulated experiments where 625 rat BAC clones and 207 mouse BAC clones are mapped onto homologous human sequence.

1 Introduction

Pooled Genomic Indexing (PGI) is a novel method for physical mapping of clones onto known macromolecular sequences. PGI enables targeted comparative sequencing of homologous regions for the purpose of discovery of genes, gene structure, and conserved regulatory regions through comparative sequence analysis. An application of the basic PGI method to BAC[1] clone mapping is illustrated in Figure 1. PGI first pools arrayed BAC clones, then shotgun sequences the pools at an appropriate coverage, and uses this information to map individual BACs onto homologous sequences of a related organism. Specifically, shotgun reads from the pools provide a set of short (cca. 500 base pair long) random subsequences of the unknown clone sequences (100–200 thousand base pair long). The reads are then individually compared to reference sequences, using standard sequence alignment techniques [1] to find homologies. In a clone-by-clone sequencing strategy [2], the shotgun reads are collected for each clone

---

[1] Bacterial Artificial Chromosome

R. Guigó and D. Gusfield (Eds.): WABI 2002, LNCS 2452, pp. 10–28, 2002.
© Springer-Verlag Berlin Heidelberg 2002

separately. Because of the pooling in PGI, the individual shotgun reads are not associated with the clones, but detected homologies may be in certain cases. If two reads from two different pools match the reference sequence at a close distance, they are both assigned (deconvoluted) to the clone at the intersection of the two pools. Simultaneously, the clone is mapped onto the region of the reference sequence between the two matches. Subsequently, known genomic or transcribed reference sequences are turned into an index into the yet-to-be sequenced homologous clones across species. As we will see below, this basic pooling scheme is somewhat modified in practice in order to achieve correct and unambiguous mapping.

PGI constructs comparative BAC-based physical maps at a fraction (on the order of 1%) of the cost of full genome sequencing. PGI requires only minor changes in the BAC-based sequencing pipeline already established in sequencing laboratories, and thus it takes full advantage of existing economies of scale. The key to the economy of PGI is BAC pooling, which reduces the amount of BAC and shotgun library preparations down to the order of the square root of the number of BAC clones. The depth of shotgun sequencing of the pools is adjusted to fit the evolutionary distance of comparatively mapped organisms. Shotgun sequencing, which represents the bulk of the effort involved in a PGI project, provides useful information irrespective of the pooling scheme. In other words, pooling by itself does not represent a significant overhead, and yet produces a comprehensive and accurate comparative physical map.

Our reason for proposing PGI is motivated by recent advances in sequencing technology [3] that allow shotgun sequencing of BAC pools. The Clone-Array Pooled Shotgun Sequencing (CAPSS) method, described by [3], relies on clone-array pooling and shotgun sequencing of the pools. CAPSS detects overlaps between shotgun sequence reads are used by and assembles the overlapping reads into sequence contigs. PGI offers a different use for the shotgun read information obtained in the same laboratory process. PGI compares the reads against another sequence, typically the genomic sequence of a related species for the purpose of comparative physical mapping. CAPSS does not use a reference sequence to deconvolute the pools. Instead, CAPSS deconvolutes by detecting overlaps between reads: a column-pool read and a row-pool read that significantly overlap are deconvoluted to the BAC at the intersection of the row and the column. Despite the clear distinction between PGI and CAPSS, the methods are compatible and, in fact, can be used simultaneously on the same data set. Moreover, the advanced pooling schemes that we present here in the context of PGI are also applicable to and increase performance of CAPSS, indicating that improvements of one method are potentially applicable to the other.

In what follows, we propose a probabilistic model for the PGI method. We then discuss and analyze different pooling schemes, and propose algorithms for experiment design. Finally, we validate the method in two simulated PGI experiments, involving 207 mouse and 625 rat BACs.

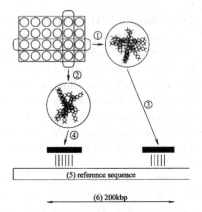

Fig. 1. The Pooled Genomic Indexing method maps arrayed clones of one species onto genomic sequence of another (5). Rows (1) and columns (2) are pooled and shotgun sequenced. If one row and one column fragment match the reference sequence (3 and 4 respectively) within a short distance (6), the two fragments are assigned (deconvoluted) to the clone at the intersection of the row and the column. The clone is simultaneously mapped onto the region between the matches, and the reference sequence is said to index the clone.

2   Probability of Successful Indexing

In order to study the efficiency of the PGI strategy formally, define the following values. Let N be the total number of clones on the array, and let m be the number of clones within a pool. For simplicity's sake, assume that every pool has the same number of clones, that clones within a pool are represented uniformly, and that every clone has the same length L. Let F be the total number of random shotgun reads, and let ℓ be the expected length of a read. The *shotgun coverage c* is defined by $c = \frac{F\ell}{NL}$.

Since reads are randomly distributed along the clones, it is not certain that homologies between reference sequences and clones are detected. However, with larger shotgun coverage, this probability increases rapidly. Consider the particular case of detecting homology between a given reference sequence and a clone. A random fragment of length λ from this clone is aligned locally to the reference sequence and if a significant alignment is found, the homology is detected. Such an alignment is called a *hit*. Let $M(\lambda)$ be the number of positions at which a fragment of length λ can begin and produce a significant alignment. The probability of a hit for a fixed length λ equals $M(\lambda)$ divided by the total number of possible start positions for the fragment, $(L - \lambda + 1)$.

When $L \gg \ell$, the expected probability of a random read aligning to the reference sequence equals

$$p_{\text{hit}} = \mathbb{E}\frac{M(\lambda)}{L - \lambda + 1} = \frac{M}{L},$$

(1)

where M is a shorthand notation for $\mathbb{E}M(\lambda)$. The value M measures the homology between the clone and the reference sequence. We call this value the *effective length* of the (possibly undetected) index between the clone and the reference sequence in question. For typical definitions of significant alignment, such as an identical region of a certain minimal length, $M(\lambda)$ is a linear function of λ, and thus $\mathbb{E}M(\lambda) = M(\ell)$. Example 1: let the reference sequence be a subsequence of length h of the clone, and define a hit as identity of at least o base pairs. Then $M = h + \ell - 2o$. Example 2: let the reference sequence be the transcribed sequence of total length g of a gene on the genomic clone, consisting of e exons, separated by long ($\gg \ell$) introns, and define a hit as identity of at least o base pairs. Then $M = g + e(\ell - 2o)$.

Assuming uniform coverage and a $m \times m$ array, the number of reads coming from a fixed pool equals $\frac{cmL}{2\ell}$. If there is an index between a reference sequence and a clone in the pool, then the number of hits in the pool is distributed binomially with expected value $\left(\frac{cmL}{2\ell}\right)\left(\frac{p_{\text{hit}}}{m}\right) = \frac{cM}{2\ell}$. Propositions 1, 2, and 3 rely on the properties of this distribution, using approximation techniques pioneered by [4] in the context of physical mapping.

Proposition 1. *Consider an index with effective length M between a clone and a reference sequence The probability that the index is detected equals approximately*

$$p_M \approx \left(1 - e^{-c\frac{M}{2\ell}}\right)^2.$$

(Proof in Appendix.)

Thus, by Proposition 1, the probability of false negatives decreases exponentially with the shotgun coverage level. The expected number of hits for a detected index can be calculated similarly.

Proposition 2. *The number of hits for a detected index of effective length M equals approximately*

$$E_M \approx c\frac{M}{\ell}\left(1 - e^{-c\frac{M}{2\ell}}\right)^{-1}.$$

(Proof in Appendix.)

3   Pooling Designs

3.1   Ambiguous Indexes

The success of indexing in the PGI method depends on the possibility of deconvoluting the local alignments. In the simplest case, homology between a clone and a reference sequence is recognized by finding alignments with fragments from one row and one column pool. It may happen, however, that more than one clone are homologous to the same region in a reference sequence (and therefore to each other). This is the case if the clones overlap, contain similar genes, or contain similar repeat sequences. Subsequently, close alignments may be found between

	$C.$	$C.$
$R.$	$B..$	$B..$
$R.$	$B..$	$B..$

Fig. 2. Ambiguity caused by overlap or homology between clones. If clones $B..$, $B..$, $B..$, and $B..$ are at the intersections of rows $R.$, $R.$ and columns $C.$, $C.$ as shown, then alignments from the pools for $R.$, $R.$, $C.$, $C.$ may originate from homologies in $B..$ and $B..$, or $B..$ and $B..$, or even $B..$, $B..$, $B..$, etc.

a reference sequence and fragments from more than two pools. If, for example, two rows and one column align to the same reference sequence, an index can be created to the *two* clones at the intersections simultaneously. However, alignments from two rows and two columns cannot be deconvoluted conclusively, as illustrated by Figure 2.

A simple clone layout on an array cannot remedy this problem, thus calling for more sophisticated pooling designs. The problem can be alleviated, for instance, by arranging the clones on more than one array, thereby reducing the chance of assigning overlapping clones to the same array. We propose other alternatives. One of them is based on construction of extremal graphs, while another uses reshuffling of the clones.

In addition to the problem of deconvolution, multiple homologies may also lead to incorrect indexing at low coverage levels. Referring again to the example of Figure 2, assume that the clones B_{11} and B_{22} contain a particular homology. If the shotgun coverage level is low, it may happen that the only homologies found are from row R_1 and column C_2. In that case, the clone B_{12} gets indexed erroneously. Such indexes are *false positives*. The probability of false positive indexing decreases rapidly with the coverage level as shown by the following result.

Proposition 3. *Consider an index between a reference sequence and two clones with the same effective length M. If the two clones are not in the same row or same column, the probability of false positive indexing equals approximately*

$$p_{\mathrm{FP}} \approx 2e^{-c\frac{M}{\ell}} \left(1 - e^{-c\frac{M}{2\ell}}\right)^2. \tag{2}$$

(Proof in Appendix.)

3.2 Sparse Arrays

The array layout of clones can be represented by an edge-labeled graph $\mathcal{G}$ in the following manner. Edges in $\mathcal{G}$ are bijectively labeled with the clones. We call such graphs *clone-labeled*. The pooling is defined by incidence, so that each vertex corresponds to a pool, and the incident edges define the clones in that pool. If $\mathcal{G}$ is bipartite, then it represents arrayed pooling with rows and columns corresponding to the two sets of vertices, and cells corresponding to edges. Notice

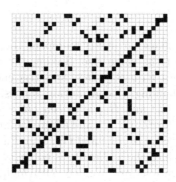

Fig. 3. Sparse array used in conjunction with the mouse experiments. This 39×39 array contains 207 clones, placed at the darkened cells. The array was obtained by randomly adding clones while preserving the sparse property, i.e., the property that for all choices of two rows and two columns, at most three out of the four cells at the intersections have clones assigned to them.

that every clone-labeled graph defines a pooling design, even if it is not bipartite. For instance, a clone-labeled full graph with $N = K(K-1)/2$ edges defines a pooling that minimizes the number K of pools and thus number of shotgun libraries, at the expense of increasing ambiguities.

Ambiguities originating from the existence of homologies and overlaps between exactly two clones correspond to cycles of length four in $\mathcal{G}$. If $\mathcal{G}$ is bipartite, and it contains no cycles of length four, then deconvolution is always possible for two clones. Such a graph represents an array in which cells are left empty systematically, so that for all choices of two rows and two columns, at most three out of the four cells at the intersections have clones assigned to them. An array with that property is called a *sparse array*. Figure 3 shows a sparse array. A sparse array is represented by a bipartite graph $\mathcal{G}^*$ with given size N and a small number K of vertices, which has no cycle of length four. This is a specific case of a well-studied problem in extremal graph theory [5] known as the problem of Zarankiewicz. It is known that if $N > 180K^{3/2}$, then $\mathcal{G}^*$ does contain a cycle of length four, hence $K > N^{2/3}/32$.

Let m be a prime power. We design a $m^2 \times m^2$ sparse array for placing $N = m^3$ clones, achieving the $K = \Theta(N^{2/3})$ density, by using an idea of Reiman [6]. The number m is the size of a pool, i.e., the number of clones in a row or a column. Number the rows as $R_{a,b}$ with $a, b \in \{0, 1, \ldots, m-1\}$. Similarly, number the columns as $C_{x,y}$ with $x, y \in \{0, 1, \ldots, m-1\}$. Place a clone in each cell $(R_{a,b}, C_{x,y})$ for which $ax + b = y$, where the arithmetic is carried out over the finite field $\mathbb{F}_m$. This design results in a sparse array by the following reasoning. Considering the affine plane of order m, rows correspond to lines, and columns correspond to points. A cell contains a clone only if the column's point lies on the row's line. Since there are no two distinct lines going through the same two points, for all choices of two rows and two columns, at least one of the cells at the intersections is empty.

3.3 Double Shuffling

An alternative to using sparse arrays is to repeat the pooling with the clones reshuffled on an array of the same size. Taking the example of Figure 2, it is unlikely that the clones B_{11}, B_{12}, B_{21}, and B_{22} end up again in the same configuration after the shuffling. Deconvolution is possible if after the shuffling, clone B_{11} is in cell (R'_1, C'_1) and B_{22} is in cell (R'_2, C'_2), and alignments with fragments from the pools for R_i, C_i, R'_i, and C'_i are found, except if the clones B_{12} and B_{21} got assigned to cells (R'_1, C'_2) and (R'_2, C'_1). Figure 4 shows the possible placements of clones in which the ambiguity is not resolved despite the shuffling. We prove that such situations are avoided with high probability for random shufflings.

Define the following notions. A *rectangle* is formed by the four clones at the intersections of two arbitrary rows and two arbitrary columns. A rectangle is *preserved* after a shuffling if the same four clones are at the intersections of exactly two rows and two columns on the reshuffled array, and the diagonals contain the same two clone pairs as before (see Figure 4).

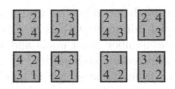

Fig. 4. Possible placements of four clones (1–4) within two rows and two columns forming a rectangle, which give rise to the same ambiguity.

Theorem 1. *Let $R(m)$ be the number of preserved rectangles on a $m \times m$ array after a random shuffling. Then, for the expected value $\mathbb{E}R(m)$,*

$$\mathbb{E}R(m) = \frac{1}{2} - \frac{2}{m}\Big(1 + o(1)\Big).$$

Moreover, for all $m > 2$, $\mathbb{E}R(m+1) > \mathbb{E}R(m)$.

(Proof in Appendix.)

Consequently, the expected number of preserved rectangles equals $\frac{1}{2}$ asymptotically. Theorem 1 also implies that with significant probability, a random shuffling produces an array with at most one preserved rectangle. Specifically, by Markov's inequality, $\mathbb{P}\big\{R(m) \geq 1\big\} \leq \mathbb{E}R(m)$, and thus

$$\mathbb{P}\big\{R(m) = 0\big\} \geq \frac{1}{2} + \frac{2}{m}\Big(1 + o(1)\Big).$$

In particular, $\mathbb{P}\big\{R(m) = 0\big\} > \frac{1}{2}$ holds for all $m > 2$. Therefore, a random shuffling will preserve no rectangles with at least $\frac{1}{2}$ probability. This allows us to use a random algorithm: pick a random shuffling, and count the number of preserved rectangles. If that number is greater than zero, repeat the step. The algorithm finishes in at most two steps on average, and takes more than one hundred steps with less than 10^{-33} probability.

Remark. Theorem 1 can be extended to non-square arrays without much difficulty. Let $R(m_r, m_c)$ be the number of preserved rectangles on a $m_r \times m_c$ array after a random shuffling. Then, for the expected value $\mathbb{E}R(m_r, m_c)$,

$$\mathbb{E}R(m_r, m_c) = \frac{1}{2} - \frac{m_r + m_c}{m_r m_c}\Big(1 + o(1)\Big).$$

Consequently, random shuffling can be used to produce arrays with no preserved rectangles even in case of non-square arrays (such as 12×8, for example).

3.4   Pooling Designs in General

Let $\mathcal{B} = \{B_1, B_2, \ldots, B_N\}$ be the set of clones, and $\mathcal{P} = \{P_1, P_2, \ldots, P_K\}$ be the set of pools. A general pooling design is described by an incidence structure that is represented by an $N \times K$ 0-1 matrix $\mathbf{M}$. The entry $\mathbf{M}[i, j]$ equals one if clone B_i is included in P_j, otherwise it is 0. The *signature* $\mathbf{c}(B_i)$ of clone B_i is the i-th row vector, a binary vector of length K. In general, the signature of a subset $\mathcal{S} \subseteq \mathcal{B}$ of clones is the binary vector of length K, defined by $\mathbf{c}(\mathcal{S}) = \vee_{B \in \mathcal{S}} \mathbf{c}(B)$, where $\vee$ denotes the bitwise OR operation. In order to assign an index to a set of clones, one first calculates the signature $\mathbf{x}$ of the index defined as a binary vector of length K, in which the j-th bit is 1 if and only if there is a hit coming from pool P_j. For all binary vectors $\mathbf{x}$ and $\mathbf{c}$ of length K, define

$$\Delta(\mathbf{x}, \mathbf{c}) = \begin{cases} \sum_{j=1}^{K}(\mathbf{c}_j - \mathbf{x}_j) & \text{if } \forall j = 1 \ldots K : \mathbf{x}_j \leq \mathbf{c}_j; \\ \infty & \text{otherwise;} \end{cases} \quad (3a)$$

and let

$$\Delta(\mathbf{x}, \mathcal{B}) = \min_{\mathcal{S} \subseteq \mathcal{B}} \Delta\Big(\mathbf{x}, \mathbf{c}(\mathcal{S})\Big). \quad (3b)$$

An index with signature $\mathbf{x}$ can be deconvoluted unambiguously if and only if the minimum in Equation (3b) is unique. The *weight* $w(\mathbf{x})$ of a binary vector $\mathbf{x}$ is the number of coordinates that equal one, i.e. $w(\mathbf{x}) = \sum_{j=1}^{K} \mathbf{x}_j$. Equation (3a) implies that if $\Delta(\mathbf{x}, \mathbf{c}) < \infty$, then $\Delta(\mathbf{x}, \mathbf{c}) = w(\mathbf{c}) - w(\mathbf{x})$.

Similar problems to PGI pool design have been considered in other applications of combinatorial group testing [7], and pooling designs have often been used for clone library screening [8]. The design of sparse arrays based on combinatorial geometries is a basic design method in combinatorics (eg., [9]). Reshuffled array designs are sometimes called transversal designs. Instead of preserved rectangles, [10] consider *collinear* clones, i.e., clone pairs in the same row or column, and propose designs with the unique collinearity condition, in which clone pairs are collinear at most once on the reshuffled arrays. Such a condition is more restrictive than ours and leads to incidence structures obtained by more complicated combinatorial methods than our random algorithm. We describe here a construction of arrays satisfying the unique collinearity condition. Let q be a prime power. Based on results of design theory [9], the following method can be used for producing up to $q/2$ reshuffled arrays, each of size $q \times q$, for pooling q^2 clones. Let $\mathbb{F}_q$ be a finite field of size q. Pools are indexed with elements of $\mathbb{F}_q^2$ as $P_{x,y}\colon x, y \in \mathbb{F}_q$. Define pool set $\mathcal{P}_i = \{P_{i,y}\colon y \in \mathbb{F}_q\}$ for all i and $\mathcal{P} = \cup_i \mathcal{P}_i$. Index the clones as $B_{a,b}\colon a, b \in \mathbb{F}_q$. Pool $P_{x,y}$ contains clone $B_{a,b}$ if and only if $y = a + bx$ holds.

Proposition 4. *We claim that the pooling design described above has the following properties.*

1. *each pool belongs to exactly one pool set;*
2. *each clone is included in exactly one pool from each pool set;*
3. *for every pair of pools that are not in the same pool set there is exactly one clone that is included in both pools.*

(Proof in Appendix.)

Let $d \le q/2$. This pooling design can be used for arranging the clones on d reshuffled arrays, each one of size $q \times q$. Select $2d$ pool sets, and pair them arbitrarily. By Properties 1–3 of the design, every pool set pair can define an array layout, by setting one pool set as the row pools and the other pool set as the column pools. Moreover, this set of reshuffled arrays gives a pooling satisfying the unique collinearity condition by Property 3 since two clones are in the same row or column on at most one array.

Similar questions also arise in coding theory, in the context of superimposed codes [11,12]. Based on the idea of [11], consider the following pooling design method using error-correcting block codes. Let $\mathcal{C}$ be a code of block length n over the finite field $\mathbb{F}_q$. In other words, let $\mathcal{C}$ be a set of length-n vectors over $\mathbb{F}_q$. A corresponding binary code is constructed by replacing the elements of $\mathbb{F}_q$ in the codewords with binary vectors of length q. The substitution uses binary vectors of weight one, i.e., vectors in which exactly one coordinate is 1, using the simple rule that the z-th element of $\mathbb{F}_q$ is replaced by a binary vector in which the z-th coordinate is 1. The resulting binary code $\mathcal{C}'$ has length qn, and each binary codeword has weight n. Using the binary code vectors as clone signatures, the binary code defines a pooling design with $K = qn$ pools and $N = |\mathcal{C}|$ clones, for which each clone is included in n pools. If d is the the minimum distance of the original code $\mathcal{C}$, i.e., if two codewords differ in at least d coordinates, then $\mathcal{C}'$ has minimum distance $2d$. In order to formalize this procedure, define ϕ as the operator of "binary vector substitution," mapping elements of $\mathbb{F}_q$ onto column vectors of length q: $\phi(0) = [1, 0, \ldots, 0]$, $\phi(1) = [0, 1, 0, \ldots, 0]$, $\ldots$, $\phi(q-1) = [0, \ldots, 0, 1]$. Furthermore, for every codeword $\mathbf{c}$ represented as a row vector of length n, let $\phi(\mathbf{c})$ denote the $q \times n$ array, in which the j-th column vector equals $\phi(\mathbf{c}_j)$ for all j. Enumerating the entries of $\phi(\mathbf{c})$ in any fixed order gives a binary vector, giving the signature for the clone corresponding to $\mathbf{c}$. Let $f\colon \mathbb{F}_q^n \mapsto \mathbb{F}_2^{qn}$ denote the mapping of the original codewords onto binary vectors defined by ϕ and the enumeration of the matrix entries.

Designs from Linear Codes. A linear code of dimension k is defined by a $k \times n$ generator matrix $\mathbf{G}$ with entries over $\mathbb{F}_q$ in the following manner. For each message $\mathbf{u}$ that is a row vector of length k over $\mathbb{F}_q$, a codeword $\mathbf{c}$ is generated by calculating $\mathbf{c} = \mathbf{u}\mathbf{G}$. The code $\mathcal{C}_{\mathbf{G}}$ is the set of all codewords obtained in this way. Such a linear code with minimum distance d is called a $[n, k, d]$ code. It is assumed that the rows of $\mathbf{G}$ are linearly independent, and thus the number of codewords equals q^k. Linear codes can lead to designs with balanced pool sizes as shown by the next lemma.

Lemma 1. *Let $\mathcal{C}$ be a $[n, k, d]$ code over $\mathbb{F}_q$ with generator matrix $\mathbf{G}$, and let $f \colon \mathbb{F}_q^n \mapsto \mathbb{F}_2^{qn}$ denote the mapping of codewords onto binary vectors as defined above. Let $\mathbf{u}^{(1)}, \mathbf{u}^{(2)}, \ldots, \mathbf{u}^{(q^k)}$ be the lexicographic enumeration of length k vectors over $\mathbb{F}_q$. For an arbitrary $1 \le N \le q^k$, let the incidence matrix $\mathbf{M}$ of the pooling be defined by the mapping f of the first N codewords $\{\mathbf{c}^{(i)} = \mathbf{u}^{(i)}\mathbf{G} : i = 1, \ldots, N\}$ onto binary vectors. If the last row of $\mathbf{G}$ has no zero entries, then every column of $\mathbf{M}$ has $\left\lceil \frac{N}{q} \right\rceil$ or $\left\lfloor \frac{N}{q} \right\rfloor$ ones, i.e., every pool contains about $m = \frac{N}{q}$ clones.*

(Proof in Appendix.)

Designs from MDS Codes. A $[n, k, d]$ code is *maximum distance separable* (MDS) if it has minimum distance $d = n-k+1$. (The inequality $d \le n-k+1$ holds for all linear codes, so MDS codes achieve maximum distance for fixed code length and dimension, hence the name.) The Reed-Solomon codes are MDS codes over $\mathbb{F}_q$, and are defined as follows. Let $n = q - 1$, and $\alpha_0, \alpha_2, \ldots, \alpha_{n-1}$ be different non-zero elements of $\mathbb{F}_q$. The generator matrix $\mathbf{G}$ of the RS(n, k) code is

$$
\mathbf{G} = \begin{pmatrix}
1 & 1 & \ldots & 1 \\
\alpha_0 & \alpha_1 & \ldots & \alpha_{n-1} \\
\alpha_0^2 & \alpha_1^2 & \ldots & \alpha_{n-1}^2 \\
\cdots & \cdots & \cdots & \cdots \\
\alpha_0^{k-1} & \alpha_1^{k-1} & \ldots & \alpha_{n-1}^{k-1}
\end{pmatrix}.
$$

Using a mapping $f \colon \mathbb{F}_q^n \mapsto \mathbb{F}_2^{qn}$ as before, for the first $N \le q^k$ codewords, a pooling design is obtained with N clones and K pools. This design has many advantageous properties. Kautz and Singleton [11] prove that if $t = \lfloor \frac{n-1}{k-1} \rfloor$, then the signature of any t-set of clones is unique. Since $\alpha_i^{k-1} \ne 0$ in $\mathbf{G}$, Lemma 1 applies and thus each pool has about the same size $m = \frac{N}{q}$.

Proposition 5. *Suppose that the pooling design with $N = q^k$ is based on a RS(n, k) code and $\mathbf{x}$ is a binary vector of positive weight and length K. If there is a clone B such that $\Delta(\mathbf{x}, \mathbf{c}(B)) < \infty$, then $\Delta(\mathbf{x}, \mathcal{B}) = n - w(\mathbf{x})$, and the following holds. If $w(\mathbf{x}) \ge k$, then the minimum in Equation (3b) is unique,* *and is attained for the singleton set containing B. Conversely, if $w(\mathbf{x}) < k$, then the minimum in Equation (3b) is attained for $q^{k-w(\mathbf{x})}$ choices of singleton clone sets.*

(Proof in Appendix.)
For instance, a design based on the RS(6, 3) code has the following properties.

- 343 clones are pooled in 42 pools;
- each clone is included in 6 pools, if at least 3 of those are included in an index to the clone, the index can be deconvoluted unambiguously;
- each pool contains 49 clones;
- signatures of 2-sets of clones are unique and have weights 10–12;
- signatures of 3-sets of clones have weights between 12 and 18; if the weight of a 3-sets' signature is less than 14, than the signature is unique (determined by a computer program).

The success of indexing in the general case is shown by the following claim.

Proposition 6. *Consider an index with effective length M between a clone and a reference sequence. If the clone appears in n pools, and a set of at least $n_{\min} \leq n$ of these pools uniquely determines the clone, then the probability that the index is detected equals*

$$p_M = \sum_{t=n_{\min}}^{n} \binom{n}{t} (1 - p_0)^t p_0^{n-t}$$

where $p_0 \approx e^{-c\frac{M}{n\ell}}$.

(Proof in Appendix.)
In simple arrayed pooling $n_{\min} = n = 2$. In the pooling based on the RS(6, 3) code, $n_{\min} = 3$, $n = 6$. When is the probability of success larger with the latter indexing method, at a fixed coverage? The probabilities can be compared using the following analogy. Let 6 balls be colored with red and green independently, each ball is colored randomly to green with probability p. Let $\mathcal{X}$ denote the event that at least one of the first three balls, and at least one of the last three balls are red: $\mathbb{P}\mathcal{X} = (1 - p^3)^2$. Let $\mathcal{Y}$ denote the event that at least three balls are red: $\mathbb{P}\mathcal{Y} = \sum_{t=3}^{6} \binom{6}{t}(1 - p)^t p^{n-t}$. $\mathbb{P}\mathcal{X}$ equals the probability of successful indexing in simple arrayed pooling (with $p_0 = p^3$). $\mathbb{P}\mathcal{Y}$ equals the probability of successful indexing in RS(6,3) pooling (with $p_0 = p$). We claim that $\mathbb{P}\mathcal{Y} > \mathbb{P}\mathcal{X}$ if $p < 2/11$. Consider the event $\mathcal{X} - \mathcal{Y}$: when exactly one of the first three balls is red and exactly one of the second three balls is red. $\mathbb{P}(\mathcal{X}-\mathcal{Y}) = (3(1-p)p^2)^2$. Consider the event $\mathcal{Y} - \mathcal{X}$: when the first three or the second three balls are red, and the others green. $\mathbb{P}(\mathcal{Y}-\mathcal{X}) = 2(1-p)^3 p^3$. Now, $\mathbb{P}\mathcal{Y} > \mathbb{P}\mathcal{X}$ if and only if $\mathbb{P}(\mathcal{Y}-\mathcal{X}) > \mathbb{P}(\mathcal{X}-\mathcal{Y})$, i.e., when $p < 2/11$. The inequality holds if $M > \ell\left(-6c\ln(2/11)\right) \approx 10c\ell$. Thus, for longer homologous regions (cca. 5000 bp effective length if $c = 1$), the RS(6,3) pooling is predicted to have better success. As c grows, simple arrayed pooling becomes better for the same fixed index. Furthermore, at $p < 2/11$, even simple arrayed pooling gives around 99% or higher success probability, thus

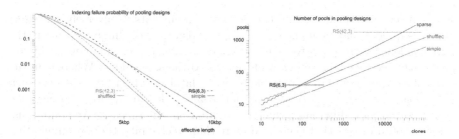

Fig. 5. The graphs on the left-hand side show the probabilities for failing to find an index as a function of the effective length. The graphs are plotted for coverage $c = 1$ and expected shotgun read length $\ell = 500$. The values are calculated from Proposition 6 for simple arraying and double shuffling with unique collinearity, as well as for designs based on the $RS(6, 3)$ and $RS(42, 3)$ codes. Notice that in case of a simple index, the failure probabilities for the sparse array design are the same as for the simple array. The graphs on the right-hand side compare the costs of different pooling designs by plotting how the number of pools depends on the total number of clones in different designs.

the improvements are marginal, while the difference between the probabilities is significantly larger when p is large. On the other hand, a similar argument shows that double shuffling with unique collinearity ($n = 4$, $n_{min} = 2$) has always higher success probability than simple arrayed pooling. Figure 5 compares the probabilities of successful indexing for some pooling designs.

4   Experiments

We tested the efficiency of the PGI method for indexing mouse and rat clones by human reference sequences in simulated experiments. The reference databases included the public human genome draft sequence [2], the Human Transcript Database (HTDB) [13], and the Unigene database of human transcripts [14]. Local alignments were computed using BLASTN [1] with default search parameters (word size=11, gap open cost=5, gap extension cost=2, mismatch penalty=2). A hit was defined as a local alignment with an E-value less than 10^{-5}, a length of at most 40 bases, and a score of at least 60.

In the case of transcribed reference sequences, hits on the same human reference sequence were grouped together to form the indexes. In the case of genomic sequences, we grouped close hits together within the same reference sequence to form the indexes. In particular, indexes were defined as maximal sets of hits on the same reference sequence, with a certain threshold on the maximum distance between consecutive hits, called the *resolution*. After experimenting with resolution values between 1kbp and 200kbp, we decided to use 2kbp resolution throughout the experiments. A particular difficulty we encountered in the experiments was the abundance of repetitive elements in eukaryotic DNA. If part of a shotgun read has many homologies in the human sequences, as is the case with common repeat sequences, the read generates many hits. Conversely, the same human sequence may be homologous to many reads. Accordingly, repeats in the arrayed clones correspond to highly ambiguous indexes, and human-specific repeats may produce large number of indexes to the same clone. Whereas in the cases of rat and mouse, it is possible to use a database of repeat sequences such as Repbase [15], such information is not available for many other species. We thus resorted to a different technique for filtering out repeats. We simply discarded shotgun reads that generated more than twelve hits, thereby eliminating almost all repeats without using a database of repeated elements.

For the deconvolution, we set the maximum number of clones to three, that is, each index was assigned to one, two, or three clones, or was declared ambiguous.

Finally, due to the fact that pooling was simulated and that in all experiments the original clone for each read was known enabled a straightforward test of accuracy of indexing: an index was considered accurate if both reads deconvoluted to the correct original clone.

Table 1. Experimental results for simulated indexing of 207 mouse clones with coverage level 2. The table gives the results for four pooling designs (simple, shuffled, sparse array, and RS(6,3)), and three databases of reference sequences: Unigene (UG), HTDB, and human genome draft (HS). The RS(6,3) design was used with the UG database only.

	Pools	Number of correctly indexed clones			Number of correct indexes / false positives		
		UG	HTDB	HS	UG	HTDB	HS
simple	29	159 (77%)	139 (67%)	172 (83%)	723 / 248	488 / 108	1472 / 69
RS(6,3)	42	172 (83%)	-	-	611 / 150	-	-
shuffled	58	172 (83%)	150 (72%)	180 (87%)	756 / 76	514 / 18	1549 / 238
sparse	78	175 (85%)	152 (74%)	185 (88%)	823 / 22	569 / 11	1634 / 69

4.1 Mouse Experiments

In one set of experiments we studied the efficiency of indexing mouse clones with human sequences. We selected 207 phase 3 sequences in the mouse sequencing project with lengths greater than 50k bases. We used four pooling designs: two 14×15 arrays for double shuffling, a 39×39 sparse array, shown in Figure 3, and a design based on the RS(6,3) code with 42 pools.

Random shotgun reads were produced by simulation. Each random read from a fixed pool was obtained in the following manner. A clone was selected randomly with probabilities proportional to clone lengths. A read length was picked using a Poisson distribution with mean $\ell = 550$, truncated at 1000. The random position of the read was picked uniformly from the possible places for the read length on the selected clone. Each position of the read was corrupted independently with probability 0.01. The procedure was repeated to attain the desired coverage within each pool. Fragments were generated for the different pooling designs with $c = 2$.

The results of the experiments are summarized in Table 1. The main finding is that 67%–88% of the clones have at least one correct index, and 500–1600 correct indexes are created, depending on the array design and the reference database. Note that double shuffling and sparse array methods significantly reduce the number of false positives, as expected based on the discussion above. One of the reasons for the excellent performance of the sparse array method is the fact that BAC redundancy in the data set does not exceed 2X. In the course of the experiments, between 7% and 10% of a total of approximately 121,400 simulated reads gave alignments against human sequences that were informative for the deconvolution, a percentage that is consistent with the expected overall level of genomic sequence conservation between mouse and human.

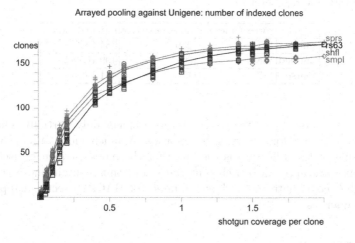

Fig. 6. Number of correctly mapped clones is indicated as a function of shotgun coverage for three different pooling schemes. PGI was tested in a simulated experiment involving 207 publicly available mouse genomic sequences of length between 50kbp and 300kbp. Pooling and shotgun sequencing were then simulated. For each coverage level, reads for $c = 2$ were resampled ten times. Graphs go through the median values for four poling designs: simple (smpl), shuffled (shfl), and sparse (sprs), and Reed-Solomon (rs63). Deconvolution was performed using human Unigene database. Notice that the curves level off at approximately $c = 1.0$, indicating limited benefit of much greater shotgun coverages.

In order to explore the effect of shotgun coverage on the success of indexing, we repeated the indexing with lower coverage levels. We resampled the simulated reads by selecting appropriate portions randomly from those produced with coverage 2. Figure 6 plots the number of indexed clones as a function of coverage. It is worth noting that even for $c = 0.5$, about 2/3 of the clones get indexed by at least one human sequence. In addition, the curves level off at about $c = 1.0$, and higher coverage levels yield limited benefits.

4.2 Rat Experiments

In contrast to the mouse experiments, PGI simulation on rat was performed using publicly available real shotgun reads from individual BACs being sequenced as part of the rat genome sequencing project. The only simulated aspect was BAC pooling, for which we pooled reads from individual BACs computationally. We selected a total of 625 rat BACs, each with more than 570 publicly available reads from the rat sequencing project. An average of 285 random reads per clone were used in each pool — corresponding to an approximate $c = 1.5$ coverage. The results are summarized in Table 2.

Table 2. Experimental results on simulated indexing of 625 rat clones by Unigene sequences with coverage level 1.5.

	correct indexes	false positives	indexed clones
simple	1418	236	384 (61%)
shuffled	1383	30	409 (65%)
sparse	1574	17	451 (72%)

The lower percentage of correctly mapped clones only partially reflects the lower coverage (1.5 for rat vs. 2 for mouse). A much more important factor contributing to the difference is the fact that the mouse sequences used in our experiments are apparently richer in genes and thus contain more conserved regions that contribute to the larger number of BACs correctly mapped onto human sequences.

5 Discussion

PGI is a novel method for physical mapping of clones onto known macromolecular sequences. It employs available sequences of humans and model organisms to index genomes of new organisms at a fraction of full genome sequencings cost. The key idea of PGI is the pooled shotgun library construction, which reduces the amount of library preparations down to the order of the square root of the number of BAC clones. In addition to setting priorities for targeted sequencing, PGI has the advantage that the libraries and reads it needs can be reused in the sequencing phase. Consequently, it is ideally suited for a two-staged approach to comparative genome explorations yielding maximum biological information for given amounts of sequencing efforts.

We presented a probabilistic analysis of indexing success, and described pooling designs that increase the efficiency of *in silico* deconvolution of pooled shotgun reads. Using publicly available mouse and rat sequences, we demonstrated the power of the PGI method in simulated experiments. In particular, we showed that using relatively few shotgun reads corresponding to 0.5-2.0 coverage of the clones, 60-90% of the clones can be indexed with human genomic or transcribed sequences.

Due to the low level of chromosomal rearrangements across mammals, the order of BACs in a comparative physical map should provide an almost correct ordering of BACs along the genome of a newly indexed mammal. Such information should be very useful for whole-genome sequencing of such organisms. Moreover, the already assembled reference sequences of model organisms onto which the BACs are mapped may guide the sequence assembly of the homologous sequence of a newly sequenced organism[2] [16].

Comparative physical maps will allow efficient, targeted, cross-species sequencing for the purpose of comparative annotation of genomic regions in model organisms that are of particular biomedical importance. PGI is not limited to the mapping of BAC clones. Other applications currently include the mapping of arrayed cDNA clones onto genomic or known full or partial cDNA sequences within and across species, and the mapping of bacterial genomic clones across different bacterial strains. Sampling efficiency of PGI is in practice increased by an order of magnitude by sequencing short sequence fragments. This is accomplished by breaking the pooled DNA into short fragments, selecting them by size, forming concatenamers by ligation, and then sequencing the concatenamers [17,18,19]. Assuming a sequence read of 600bp and tag size of 20–200bp, a total of 3–30 different clones may be sampled in a single sequencing reaction. This technique is particularly useful when the clone sequences and reference sequences are highly similar, e.g., cDNA mapping against the genome of the same species, bacterial genome mapping across similar strains, and mapping of primate genomic BACs against human sequence.

Acknowledgements

The authors are grateful to Richard Gibbs and George Weinstock for sharing pre-publication information on CAPSS and for useful comments, and to Paul Havlak and David Wheeler for contributing database access and computational resources at HGSC.

---

* Claim 1 of US Patent 6,001,562 [16] reads as follows: A method for detecting sequence similarity between at least two nucleic acids, comprising the steps of:

(a) identifying a plurality of putative subsequences from a first nucleic acid;
(b) comparing said subsequences with at least a second nucleic acid sequence; and
(c) aligning said subsequences using said second nucleic acid sequence in order to simultaneously maximize
  (i) matching between said subsequences and said second nucleic acid sequence and
  (ii) mutual overlap between said subsequences,

whereby said aligning predicts a subsequence that occurs within both said first and said second nucleic acids.

References

1. Altschul, S.F., Madden, T.L., Schäffer, A.A., Zhang, J., Zhang, Z., Miller, W., Lipman, D.J.: Gapped BLAST and PSI-BLAST: a new generation of protein database search programs. Nucleic Acids Res. **25** (1997) 3389–3402

2. IHGSC: Initial sequencing and analysis of the human genome. Nature **609** (2001) 860–921

3. Cai, W.W., Chen, R., Gibbs, R.A., Bradley, A.: A clone-array pooled strategy for sequencing large genomes. Genome Res. **11** (2001) 1619–1623

4. Lander, E.S., Waterman, M.S.: Genomic mapping by fingerprinting random clones: a mathematical analysis. Genomics **2** (1988) 231–239

5. Bollobás, B.: Extremal graph theory. In Graham, R.L., Grötschel, M., Lovász, L., eds.: Handbook of Combinatorics. Volume II. Elsevier, Amsterdam (1995) 1231–1292

6. Reiman, I.: Über ein Problem von K. Zarankiewicz. Acta Math. Sci. Hung. **9** (1958) 269–279

7. Du, D.Z., Hwang, F.K.: Combinatorial Group Testing and Its Applications. 2nd edn. World Scientific, Singapore (2000)

8. Bruno, W.J., Knill, E., Balding, D.J., Bruce, D.C., Doggett, N.A., Sawhill, W.W., Stallings, R.L., Whittaker, C.C., Torney, D.C.: Efficient pooling designs for library screening. Genomics **26** (1995) 21–30

9. Beth, T., Jungnickel, D., Lenz, H.: Design Theory. 2nd edn. Cambridge University Press, UK (1999)

10. Barillot, E., Lacroix, B., Cohen, D.: Theoretical analysis of library screening using an n-dimensional strategy. Nucleic Acids Res. **19** (1991) 6241–6247

11. Kautz, W.H., Singleton, R.C.: Nonrandom binary superimposed codes. IEEE Trans. Inform. Theory **IT-10** (1964) 363–377

12. D'yachkov, A.G., Macula, Jr., A.J., Rykov, V.V.: New constructions of superimposed codes. IEEE Trans. Inform. Theory **IT-46** (2000) 284–290

13. Bouck, J., McLeod, M.P., Worley, K., Gibbs, R.A.: The Human Transcript Database: a catalogue of full length cDNA inserts. Bioinformatics **16** (2000) 176–177 ⟨http://www.hgsc.bcm.tmc.edu/HTDB/⟩.

14. Schuler, G.D.: Pieces of the puzzle: expressed sequence tags and the catalog of human genes. J. Mol. Med. **75** (1997) 694–698 ⟨http://www.ncbi.nlm.nih.gov/Unigene/⟩.

15. Jurka, J.: Repbase update: a database and an electronic journal of repetitive elements. Trends Genet. **16** (2000) 418–420 ⟨http://www.girinst.org/⟩.

16. Milosavljevic, A.: DNA sequence similarity recognition by hybridization to short oligomers (1999) U. S. patent 6,001,562.

17. Andersson, B., Lu, J., Shen, Y., Wentland, M.A., Gibbs, R.A.: Simultaneous shotgun sequencing of multiple cDNA clones. DNA Seq. **7** (1997) 63–70

18. Yu, W., Andersson, B., Worley, K.C., Muzny, D.M., Ding, Y., Liu, W., Ricafrente, J.Y., Wentland, M.A., Lennon, G., Gibbs, R.A.: Large-scale concatenation cDNA sequencing. Genome Res. **7** (1997) 353–358

19. Velculescu, V.E., Vogelstein, B., Kinzler, K.W.: Analysing uncharted transcriptomes with SAGE. Trends Genet. **16** (2000) 423–425

Appendix

Proof (Proposition 1). The number of random shotgun reads from the row pool associated with the clone equals $\frac{cmL}{2\ell}$. By Equation (1), the probability that at least one of them aligns with the reference sequence equals

$$p_{\geq 1} = 1 - \left(1 - \frac{p_{\text{hit}}}{m}\right)^{\frac{cmL}{2\ell}} = 1 - \left(1 - \frac{M}{mL}\right)^{\frac{cmL}{2\ell}} \approx 1 - e^{-c\frac{M}{2\ell}}. \qquad (4)$$

The probability that the row and column pools both generate reads aligning to the reference sequence equals $p_{\geq 1}^2$, as claimed.

Proof (Proposition 2). The number of hits for an index coming from a fixed row or column pool is distributed binomially with parameters $n = \frac{cmL}{2\ell}$ and $p = \frac{M}{mL}$. Let ξ_r denote the number of hits coming from the clone's row pool, and let ξ_c denote the number of hits coming from the clone's column pool. Then

$$E_M = \mathbb{E}\Big[\xi_r + \xi_c \ \Big|\ \xi_r > 0, \xi_c > 0\Big] = \mathbb{E}\Big[\xi_r \ \Big|\ \xi_r > 0\Big] + \mathbb{E}\Big[\xi_c \ \Big|\ \xi_c > 0\Big].$$

In order to calculate the conditional expectations on the right-hand side, notice that if ξ is a non-negative random variable, then $\mathbb{E}\xi = \mathbb{E}\Big[\xi \ \Big|\ \xi > 0\Big]\mathbb{P}\{\xi > 0\}$. Since $\mathbb{P}\{\xi_c > 0\} = \mathbb{P}\{\xi_r > 0\} = p_{\geq 1}$,

$$E_M = 2\mathbb{E}\Big[\xi_r \ \Big|\ \xi_r > 0\Big] = \frac{2np}{p_{\geq 1}}$$

Resubstituting p and n, the proposition follows from Equation (4).

Proof (Proposition 3). Let $n = \frac{cmL}{2\ell}$ be the number of reads from a pool and $p = \frac{M}{\ell}$ the probability of a hit within a pool containing one of the clones. Then a false positive occurs if there are no hits in the row pool for one clone and the column pool for the other, and there is at least one hit in each of the other two pools containing the clones. The probability of that event equals

$$2\Big((1-p)^n\Big)^2\Big(1 - (1-p)^n\Big)^2.$$

Using the same approximation technique as in Proposition 1 leads to Equation (2).

Proof (Theorem 1). Let $\mathcal{R}$ be an arbitrary rectangle in the array. We calculate the probability that $\mathcal{R}$ is preserved after a random shuffling by counting the number of shufflings in which it is preserved. There are $\binom{m}{2}$ choices for selecting two rows for the position of the preserved rectangle. Similarly, there are $\binom{m}{2}$ ways to select two columns. There are 8 ways of placing the clones of the rectangle in the cells at the four intersections in such a way that the diagonals are preserved (see Figure 4). There are $(m^2 - 4)!$ ways of placing the remaining clones on the array. Thus, the total number of shufflings in which $\mathcal{R}$ is preserved equals

$$8\binom{m}{2}^2(m^2 - 4)! = 8\left(\frac{m(m-1)}{2}\right)^2(m^2 - 4)!.$$

Dividing this number by the number of possible shufflings gives the probability of preserving $\mathcal{R}$:

$$p = \frac{8\binom{m}{2}^2}{(m^2)(m^2-1)(m^2-2)(m^2-3)}. \tag{5}$$

For every rectangle $\mathcal{R}$, define the indicator variable $I(\mathcal{R})$ for the event that it is preserved. Obviously, $\mathbb{E}I(\mathcal{R}) = p$. Using the linearity of expectations and Equation (5),

$$\mathbb{E}R(m) = \sum_{\mathcal{R}} \mathbb{E}I(\mathcal{R}) = \binom{m}{2}^2 p = \frac{1}{2} \cdot \frac{m^4(m-1)^4}{m^2(m^2-1)(m^2-2)(m^2-3)}. \tag{6}$$

Thus,

$$\mathbb{E}R(m) = \frac{1}{2}\left(1 - \frac{4}{m}\left(1 + o(1)\right)\right),$$

proving the theorem. Equation (6) also implies that $\frac{\mathbb{E}R(m+1)}{\mathbb{E}R(m)} > 1$ for $m > 2$, and thus $\mathbb{E}R(m)$ is increasing monotonically.

Proof (Proposition 4). Property 1 is trivial. For Property 2, notice that clone $B_{a,b}$ is included in pool $P_{i,a+ib}$ in every $\mathcal{P}_i$, and in no other pools. For Property 3, let P_{x_1,y_1} and P_{x_2,y_2} be two arbitrary pools. Each clone $B_{a,b}$ is included in both pools if and only if

$$y_1 = a + bx_1 \quad \text{and} \quad y_2 = a + bx_2.$$

If $x_1 \neq x_2$, then there is exactly one solution for (a,b) that satisfies both equalities.

Proof (Lemma 1). Fix the first $(k-1)$ coordinates of $\mathbf{u}$ and let the last one vary from 0 to $(q-1)$. The i-th coordinate of the corresponding codeword $\mathbf{c} = \mathbf{uG}$ takes all values of $\mathbb{F}_q$ if the entry $\mathbf{G}[k,i]$ is not 0.

Proof (Proposition 5). The first part of the claim for the case $w(\mathbf{x}) \geq k$ is a consequence of the error-correcting properties of the code. The second part of the claim follows from the MDS property.

Proof (Proposition 6). This proof generalizes that of Proposition 1. The number of random shotgun reads from one pool associated with the clone equals $\frac{cmL}{n\ell}$. By Equation (1), the probability that at least one of them aligns with the reference sequence equals

$$p_{\geq 1} = 1 - \left(1 - \frac{p_{\text{hit}}}{m}\right)^{\frac{cmL}{n\ell}} \approx 1 - e^{-c\frac{M}{n\ell}}.$$

The probability that at least $n_{\min}$ pools generate reads aligning to the reference sequence equals

$$\sum_{t=n_{\min}}^{n} \binom{n}{t} p_{\geq 1}^t (1 - p_{\geq 1})^{n-t}$$

proving the claim with $p_0 = 1 - p_{\geq 1}$.

Practical Algorithms and Fixed-Parameter Tractability for the Single Individual SNP Haplotyping Problem

Romeo Rizzi[1,*], Vineet Bafna[2], Sorin Istrail[2], and Giuseppe Lancia[3]

[1] Math. Dept., Università di Trento, 38050 Povo (Tn), Italy
rrizzi@science.unitn.it
[2] Celera Genomics, Rockville MD, USA,
{Vineet.Bafna,Sorin.Istrail}@celera.com
[3] D.E.I., Università di Padova, 35100 Padova, Italy
lancia@dei.unipd.it

Abstract. Single nucleotide polymorphisms (SNPs) are the most frequent form of human genetic variation, of foremost importance for a variety of applications including medical diagnostic, phylogenies and drug design.

The complete SNPs sequence information from each of the two copies of a given chromosome in a diploid genome is called a *haplotype*. The Haplotyping Problem for a single individual is as follows: Given a set of fragments from one individual's DNA, find a maximally consistent pair of SNPs haplotypes (one per chromosome copy) by removing data "errors" related to sequencing errors, repeats, and paralogous recruitment. Two versions of the problem, i.e. the Minimum Fragment Removal (MFR) and the Minimum SNP Removal (MSR), are considered.

The Haplotyping Problem was introduced in [8], where it was proved that both MSR and MFR are polynomially solvable when each fragment covers a set of consecutive SNPs (i.e., it is a *gapless* fragment), and NP-hard in general. The original algorithms of [8] are of theoretical interest, but by no means practical. In fact, one relies on finding the maximum stable set in a perfect graph, and the other is a reduction to a network flow problem. Furthermore, the reduction does not work when there are fragments completely included in others, and neither algorithm can be generalized to deal with a bounded total number of holes in the data.

In this paper, we give the first practical algorithms for the Haplotyping Problem, based on Dynamic Programming. Our algorithms do not require the fragments to not include each other, and are polynomial for each constant k bounding the total number of holes in the data. For m SNPs and n fragments, we give an $O(mn^{2k+2})$ algorithm for the MSR problem, and an $O(2^{2k}m^2n + 2^{3k}m^3)$ algorithm for the MFR problem, when each fragment has at most k holes. In particular, we obtain an $O(mn^2)$ algorithm for MSR and an $O(m^2n + m^3)$ algorithm for MFR on gapless fragments.

Finally, we prove that both MFR and MSR are APX-hard in general.

---

* Research partially done while enjoying hospitality at BRICS, Department of Computer Science, University of Aarhus, Denmark.

1   Introduction

With the sequencing of the human genome [12,7] has come the confirmation that all humans are almost identical at DNA level (99% and greater identity). Hence, small regions of differences must be responsible for the observed diversities at phenotype level. The smallest possible variation is at a single nucleotide, and is called *Single Nucleotide Polymorphism*, or SNP (pronounced "snip"). Broadly speaking, a polymorphism is a trait, common to everybody, whose value can be different but drawn in a limited range of possibilities, called *alleles*. A SNP is a specific nucleotide, placed in the middle of a DNA region which is otherwise identical for all of us, whose value varies within a population. In particular, each SNP shows a variability of only two alleles. These alleles can be different for different SNPs.

Recent studies have shown that SNPs are the predominant form of human variation [2] occurring, on average, every thousand bases. Their importance cannot be overestimated for therapeutic, diagnostic and forensic applications. Nowadays, there is a large amount of research going on in determining SNP sites in humans as well as other species, with a SNP consortium founded with the aim of designing a detailed SNP map for the human genome [11,6].

Since DNA of *diploid* organisms is organized in pairs of chromosomes, for each SNP one can either be *homozygous* (same allele on both chromosomes) or *heterozygous* (different alleles). The values of a set of SNPs on a particular chromosome copy define a *haplotype*. *Haplotyping* an individual consists in determining a pair of haplotypes, one for each copy of a given chromosome. The pair provides full information of the SNP fingerprint for that individual at the specific chromosome.

There exist different combinatorial versions of haplotyping problems. In particular, the problem of haplotyping a population (i.e., a set of individuals) has been extensively studied, under many objective functions [4,5,3], while haplotyping for a single individual has been studied in [8] and in [9].

Given complete DNA sequence, haplotyping an individual would consist of a trivial check of the value of some nucleotides. However, the complete DNA sequence is obtained by the assembly of smaller fragments, each of which can contain errors, and is very sensitive to repeats. Therefore, it is better to define the haplotyping problem considering as input data the fragments instead of the fully assembled sequence.

Computationally, the haplotyping problem calls for determining the "best" pair of haplotypes, which can be inferred from data which is possibly inconsistent and contradictory. The problem was formally defined in [8], where conditions are derived under which it results solvable in polynomial time and others for which it is NP-hard. We remark that both situations are likely to occur in real-life contexts, depending on the type of data available and the methodology used for sequencing. In this paper we improve on both the polynomial and hardness results of [8]. In particular, we describe practical effective algorithms based on Dynamic Programming, which are low–degree polynomial in the number of SNPs and fragments, and remain polynomial even if the fragments are allowed to skip

Table 1. A chromosome and the two haplotypes

```
Chrom. c, paternal: ataggtccCtatttccaggcgcCgtatacttcgacgggActata
Chrom. c, maternal: ataggtccGtatttccaggcgcCgtatacttcgacgggTctata
```

Haplotype 1 →   C   C   A
Haplotype 2 →   G   C   T

some SNPs (up to a fixed maximum). As for the complexity results, we show that the problems are not just NP-hard, but in fact APX-hard.

Despite the biological origin of the problem, the model turns out to be purely combinatorial, and has many nice mathematical properties. The basic biological motivations behind the problem are provided in Section 2. The mathematical model of the problems, together with the notation and two useful reductions, are described in Section 3. In Section 4, we introduce a suitable bipartite labeled graph, used to characterize the problems and to give an APX-hardness result for their general versions. In Section 5 we describe Dynamic Programming-based polynomial algorithms for the gapless versions of the problem, while in Section 6 we extend these results to bounded-length gaps.

2   SNPs and Haplotypes

The process of passing from the sequence of nucleotides in a DNA molecule to a string over the DNA alphabet is called *sequencing*. A *sequencer* is a machine that is fed some DNA and whose output is a string of As, Ts, Cs and Gs. To each letter, the sequencer attaches a value (*confidence level*) which represents, essentially, the probability that the letter has been correctly read.

The main problem with sequencing is that the technology is not powerful enough to sequence a long DNA molecule, which must therefore first be cloned into many copies, and then be broken, at random, into several pieces (called *fragments*), of a few hundred nucleotides each, that are individually fed to a sequencer. The cloning phase is necessary so that the fragments can have nonempty overlap. From the overlap of two fragments one infers a longer fragment, and so on, until the original DNA sequence has been reconstructed. This is, in essence, the principle of *Shotgun Sequencing* [13], in which the fragments are *assembled* back into the original sequence by using sophisticated algorithms. The assembly phase is complicated from the fact that in a genome there exist many regions with identical content (*repeats*) scattered all around, which may fool the assembler into thinking that they are all copies of a same, unique, region. The situation is complicated further from the fact that *diploid* genomes are organized into pairs of chromosomes (a paternal and a maternal copy) which may have identical or nearly identical content, a situation that makes the assembly process even harder.

To partly overcome these difficulties, the fragments used in shotgun sequencing sometimes have some extra information attached. In fact, they are obtained via a process that generates pairs (called *mate pairs*) of fragments instead than individual ones, with a fairly precise estimate of the distance between them. These pairs are guaranteed to come from the same copy of a chromosome, and there is a good chance that, even if one of them comes from a repeat region, the other does not.

A *Single Nucleotide Polymorphism*, or SNP, is a position in a genome at which, within a population, some individuals have a certain base while the others have a different one. In this sense, that nucleotide is polymorphic, from which the name. For each SNP, an individual is *homozygous* if the SNP has the same allele on both chromosome copies, and otherwise the individual is *heterozygous*. The values of a set of SNPs on a particular chromosome copy define a *haplotype*. In Figure 1 we give a simplistic example of a chromosome with three SNP sites. The individual is heterozygous at SNPs 1 and 3 and homozygous at SNP 2. The haplotypes are CCA and GCT.

The *Haplotyping Problem* consists in determining a pair of haplotypes, one for each copy of a given chromosome, from some input genomic data. Given the assembly output (i.e., a fully sequenced genome) haplotyping would simply consist in checking the value of some specific sites. However, there are unavoidable errors, some due to the assembler, some to the sequencer, that complicate the problem and make it necessary to proceed in a different way. One problem is due to repeat regions and "paralogous recruitment" [9]. In practice, fragments with high similarity are merged together, even if they really come from different chromosome copies, and the assembler tends to reconstruct a single copy of each chromosome. Note that in these cases, heterozygous SNP sites could be used to correct the assembly and segregate two distinct copies of the similar regions. Another problem is related to the quality of the reads. For these reasons, the haplotyping problem has been recently formalized as a combinatorial problem, defined not over the assembly output, but over the original set of fragments. The framework for this problem was introduced in [8]. The data consists of small, overlapping fragments, which can come from either one of two chromosome copies. Further, e.g. in shotgun sequencing, there may be pairs of fragments known to come from the same chromosome copy and to have a given distance between them. Because of unavoidable errors, under a general parsimony principle, the basic problem is the following:

– *Given a set of fragments obtained by DNA sequencing from the two copies of a chromosome, find the smallest number of errors so that there exist two haplotypes compatible with all the (corrected) fragments observed.*

Depending on the errors considered, different combinatorial problems have been defined in the literature. "Bad" fragments can be due either to contaminants (i.e. DNA coming from a different organism than the actual target) or to read errors. An alternative point of view assigns the errors to the SNPs, i.e. a "bad" SNP is a SNP for which some fragments contain read errors. Correspondingly, we have the following optimization problems: *"Find the minimum number of fragments to ignore"* or *"Find the minimum number of SNPs to ignore"*, such that *"the (corrected) data is consistent with the existence of two haplotypes. Find such haplotypes."*

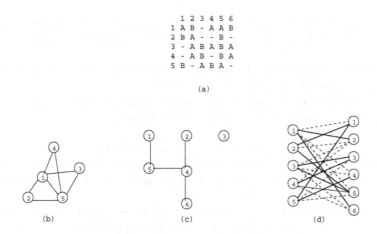

```
        1 2 3 4 5 6
1 A B - A A B
2 B A - - B -
3 - A B A B A
4 - A B - B A
5 B - A B A -
```

(a)

(b)   (c)   (d)

Fig. 1. (a) A SNP matrix. (b) The Fragment conflict graph $G_{\mathcal{F}}(M)$. (c) The SNP conflict graph $G_{\mathcal{S}}(M)$. (d) The labeled bipartite graph (G, ℓ)

3 Terminology and Notation

Let $\mathcal{S} = \{1, \ldots, n\}$ be a set of SNPs and $\mathcal{F} = \{1, \ldots, m\}$ be a set of fragments (where, for each fragment, only the nucleotides at positions corresponding to some SNP are considered). Each SNP is covered by some of the fragments, and can take only two values. The actual values (nucleotides) are irrelevant to the combinatorics of the problem and hence we will denote, for each SNP, by A and B the two values it can take. Given any ordering of the SNPs (e.g., the natural one, induced by their physical location on the chromosome), the data can also be represented by an $m \times n$ matrix over the alphabet $\{A, B, -\}$, which we call the *SNP matrix* (read "snip matrix"), defined in the obvious way. The symbol $-$ appears in all cells $M[f, s]$ for which a fragment f does not cover a SNP s, and it is called a *hole*.

For a SNP s, two fragments f and g are said to *conflict on* s if $M[f, s] = A$ and $M[g, s] = B$ or vice-versa. Two fragments f and g are said to *conflict* if there exists a SNP s such that they conflict on s, otherwise f and g are said to *agree*. A SNP matrix M is called *error-free* if we can partition the rows (fragments) into two classes of non-conflicting fragments.

Given a SNP matrix M, the *fragment conflict graph* is the graph $G_{\mathcal{F}}(M) = (\mathcal{F}, E_{\mathcal{F}})$ with an edge for each pair of conflicting fragments (see Figure 1(a) and (b)). Note that if M is error-free, $G_{\mathcal{F}}(M)$ is a bipartite graph, since each haplotype defines a shore of $G_{\mathcal{F}}(M)$, made of all the fragments coming from that haplotype. Conversely, if $G_{\mathcal{F}}(M)$ is bipartite, with shores H_1 and H_2, all the fragments in H_1 can be merged into one haplotype and similarly for H_2. Hence, M *is error-free if and only if* $G_{\mathcal{F}}(M)$ *is bipartite*.

The fundamental underlying problem in SNP haplotyping is determining an optimal set of changes to M (e.g., row and/or column- deletion) so that M becomes error-free. Given a matrix M, and where X is any set of rows or columns of M, we denote by $M \setminus X$ the matrix obtained from M by dropping the rows or columns in X. In this work, we will consider the following problems.

MSR **Minimum SNP Removal** - Find a minimum number of columns (SNPs) whose removal makes M error-free;

MFR **Minimum Fragment Removal** - Find a minimum number of rows (fragments) whose removal makes M error-free.

For better readability, from now on we will refer to "a matrix M" instead of "a SNP matrix M", unless a possible confusion arises with another type of matrix. For a problem $\Pi \in \{MSR, MFR\}$ on input a matrix M, we will denote by $\Pi(M)$ the value of an optimal solution.

The following two reductions can be used to remove redundant data from the input, and hence to clean the structure of the problems.

We start by considering the minimum number of columns (SNPs) whose removal makes M error-free. We have the following proposition:

Proposition 1 (S-reduction) *Let M' be the matrix obtained from M by dropping those columns where no A's or no B's occur. Clearly, $\mathrm{MSR}(M') \leq \mathrm{MSR}(M)$. Let X be any set of SNPs such that $M' \setminus X$ is error-free. Then also $M \setminus X$ is error-free.*

Essentially, Proposition 1 says that when solving the problem we can simply concentrate our attention to M', the other columns being inessential. Matrix M' so obtained is called S-*reduced*. When M is error-free, then we say that two fragments f and g are *allies* (*enemies*) when they must be in the same class (in separate classes) for every partition of the rows of M into two classes of non-conflicting fragments.

Now, on an S-reduced M', we have the following structure for solutions.

Lemma 2 ("o con noi o contro di noi")[1] *Let M be an S-reduced matrix. Let X be any set of SNPs whose removal makes M error-free. Let $f, g \in \mathcal{F}$ be any two fragments. Consider any SNP $s \in S \setminus X$ such that $\{M[f, s], M[g, s]\} \subseteq \{A, B\}$. Then, if $M[f, s] \neq M[g, s]$ then f and g are enemies, otherwise, they are allies.*

Proof: The first part is obvious. As for the second, assume, e.g., $M[f, s] = M[g, s] = A$. Then , since M is S-reduced, there exists a third fragment h such that $M[h, s] = B$. □

We also have a similar reduction which applies to rows.

Proposition 3 ($\mathcal{F}$-reduction) *Let M' be the matrix obtained from M by dropping those rows which conflict with at most one other row. Clearly, $\mathrm{MSR}(M') \leq \mathrm{MSR}(M)$. Let X be any set of SNPs whose removal makes M' error-free. Then the removal of X makes also M error-free.*

---

[1] Literal translation: either with us or against us. Meaning: you have to choose, either be friend or enemy

We can hence assume that every row conflicts with at least two other rows, simply by dropping those rows which conflict with at most one row. Matrix M' so obtained is called $\mathcal{F}$-*reduced*.

We now proceed to check that the two reductions just introduced for MSR are also valid for MFR (they are valid for MER as well). So, consider the minimum number of rows (fragments) whose removal makes M error-free. The following propositions are easy to prove

Proposition 4 ($\mathcal{S}$-reduction, again) *Let M' be the matrix obtained from M by dropping those columns where only A's or only B's occur. Clearly,* $\mathrm{MFR}(M') \leq \mathrm{MFR}(M)$. *Let X be any set of rows whose removal makes M' error-free. Then also $M \setminus X$ is error-free.*

Proposition 5 ($\mathcal{F}$-reduction, again) *Let M' be the matrix obtained from M by dropping those rows which conflict with at most one single other row. Clearly,* $\mathrm{MFR}(M') \leq \mathrm{MFR}(M)$. *Let X be any set of rows whose removal makes M' error-free. Then also $M \setminus X$ is error-free.*

3.1   Old and New Results

We define a *gapless* fragment i as one for which the As and Bs appear consecutively, with no - s between them, in row i of M. In general, a *gap* is a maximal run of consecutive holes between two non-hole symbols. As an example, --ABABBBA----, is a gapless fragment, while there are 2 gaps in --AB---B--AB-. The *length of a gap* is the number of holes it contains (so, the above second example, has a total gap length of $5 = 3+2$). The *body* of a fragment extends from the leftmost non-hole to the rightmost non-hole (e.g., the body of ---ABB-B--AB- is ABB-B--AB).

In haplotyping, there can be gaps for mainly two reasons:

1. *thresholding of low-quality reads.* This happens if the sequencer cannot call a SNP A or B with enough confidence; then, no call is made, and the position is marked with -.
2. *mate-pairing in shotgun sequencing.* One pair of mates are two fragments coming from the same chromosome copy, with a given distance between them. Hence, they are logically equivalent to a single fragment with one gap.

We call a matrix *gapless* if all fragments are gapless. We say that a matrix *has the consecutive-1 property* (is C1P) if the columns can be rearranged so as to obtain a gapless matrix. Note that determining if such a rearrangement exists is a well-known polynomial problem [1]. As a consequence, all polynomial results for gapless matrices can be readily extended to polynomial results for C1P matrices. We note also that the consecutive-1 property is conserved under both $\mathcal{S}$-reduction and $\mathcal{F}$-reduction.

We now briefly recall the main results that were obtained in [8] for gapless matrices and in Sections 5.1 and 5.2 we go on to improve them to more practical and effective algorithms. In a nutshell, in [8] it is shown that for gapless matrices the problems are polynomial, while, in general, they are NP-hard. Note that these cases are both likely to occur in real-life applications. For instance, an EST (Expressed Tagged Sequence) is a short DNA fragment with no gaps, fully sequenced. When the input consists only of ESTs, the matrix M is C1P (i.e., has the consecutive 1 property). On the other hand, when also mate pairs are used, the matrix is not necessarily C1P.

Let M be an $\mathcal{S}$-reduced matrix. Two SNPs s and t are said to *be in conflict* if there exists fragments f and g such that $M[f,a], M[g,a], M[f,c], M[g,c] \neq -$ and the boolean value $(M[f,a] = M[g,a])$ is the negation of $(M[f,c] = M[g,c])$. In other words, the 2×2 submatrix of M defined by rows f and g and columns s and t has 3 symbols of one type (A or B) and one of the opposite (B or A respectively). Given a matrix M, the *SNP conflict graph* is the graph $G_{\mathcal{S}}(M) = (\mathcal{S}, E_{\mathcal{S}})$, with an edge for each pair of SNPs in conflict (see Figure 1(c)).

The following theorems are proved in [8]

Theorem 6 *Let M be a gapless matrix. Then M is error-free if and only if $G_{\mathcal{S}}(M)$ is a stable set.*

Theorem 7 *If M is a gapless matrix, then $G_{\mathcal{S}}(M)$ is a weakly triangulated graph.*

From theorems 6 and 7 it followed that MSR is polynomial for a gapless matrix, since it amounts to finding the largest stable set in a perfect graph. The main result in [8] was to show that the problem is polynomial, and little attention was paid to the fact that the running times of algorithms based on the perfectness of graphs are bounded by too high-degree polynomials for practical applications (let apart the problem of actually coding these algorithms). Similarly, in [8] a point was made that, for gapless data, also MFR is polynomial, which was shown via an expensive reduction to a network flow problem. Furthermore, the result is only valid when no fragment is contained in another. In the next two sections, we set up for quicker and much simpler algorithms for both MFR and MSR. We also manage to get rid of the assumption that no fragment is contained in another. The main results we establish in this paper are the following:

Theorem 8 *There is an $O(mn^{2k+2})$ polynomial time algorithm for MSR problem on matrices in which each fragment has total gap length at most k.*

Corollary 9 *There is an $O(mn^2)$ polynomial time algorithm for MSR problem on matrices which have the consecutive-1 property.*

Theorem 10 *There is an $O(2^{2k}m^2n + 2^{3k}m^3)$ polynomial time algorithm for MFR problem on matrices in which each fragment has total gap length at most k.*

Corollary 11 *There is an $O(m^2n + m^3)$ polynomial time algorithm for MFR problem on matrices which have the consecutive-1 property.*

Theorem 12 *The problems MFR and MSR are, in general, APX-hard.*

4   The Data as Labeled Bipartite Graph

We assume that we are working on an S-reduced matrix M. To M we associate a labeled graph as follows. Let $G = (\mathcal{F}, \mathcal{S}; E)$ be a bipartite graph on color classes $\mathcal{F}$ and $\mathcal{S}$ and with edge set $E := \{sf : s \in \mathcal{S}, f \in \mathcal{F}, M[s, f] \neq -\}$. When $M[s, f] = \mathtt{A}$ we label sf *even*, and set $\ell(sf) = 0$. When $M[s, f] = \mathtt{B}$ we label sf *odd*, and set $\ell(sf) = 1$ (see Figure 1(d), where even edges are dashed and odd edges are thick). An edge set $F \subseteq E$ is called ℓ-*odd* if it contains an odd number of odd edges, and ℓ-*even* otherwise. Indeed, we can extend our labeling to edge sets as $\ell(F) := \left(\sum_{f \in F} \ell(f)\right)_{\bmod 2}$. We say that the pair (G, ℓ) is ℓ-*bipartite* if and only if it contains no ℓ-odd cycle (regard a cycle as an edge set). By the "o con noi o contro di noi" Lemma 1, we have the following consequence:

Proposition 13 *The matrix M is error-free if and only if (G, ℓ) is ℓ-bipartite.*

We now give a formal proof of the above proposition in a more general form.

Lemma 14 *Let X be a set of columns. If M is S-reduced and $M \setminus X$ is error-free, then $(G \setminus X, \ell)$ is ℓ-bipartite.*

Proof: Indeed, let $C = f_1, s_1, \ldots, f_k, s_k$ be any cycle in $G \setminus X$. Define $\ell(\delta_C(s_i)) = \ell(f_i s_i) + \ell(s_i f_{i+1})$, so that $\ell(C) = \sum_{i=1}^{k} \ell(\delta_C(s_i))$. Consider a partition of $\mathcal{F}$ into two haplotypes. By Lemma 1, whenever $\ell(\delta_C(s_i))$ is even, f_i and f_{i+1} are in in the same haplotype, while when it is odd, they are in different haplotypes. But, along the cycle $f_1, f_2, \ldots, f_k$, the number of haplotype switches must be even (like jumping forth and back between the two sides of a river but eventually returning to the starting side). □
   Now we go for the converse.

Lemma 15 *If (G, ℓ) is ℓ-bipartite then M (not necessarily reduced) is error-free.*

Proof: By definition, the fragment conflict graph $G_{\mathcal{F}}(M) = (\mathcal{F}, E_{\mathcal{F}})$ is such that $uv \in E_{\mathcal{F}}$ if there is an s such that us and sv are in (G, ℓ) and $\ell(us) + \ell(sv) = 1$. Hence, to each cycle C in $G_{\mathcal{F}}(M)$ corresponds a cycle C' in (G, ℓ), and C has an even number of edges if and only if C' is ℓ-even. So, $G_{\mathcal{F}}(M)$ is bipartite and hence M is error-free. □
   We finally prove Theorem 12 of Section 3.1.

Theorem 16 *The problems MFR and MSR are APX-hard.*

Proof: Given a graph G, the problem (MINEDGEBIPARTIZER) of finding a minimum cardinality set of edges F such that $G \setminus F$ is bipartite, and the problem (MINNODEBIPARTIZER) of finding a minimum cardinality set of nodes Z such that $G \setminus Z$ is bipartite, are known to be APX-hard [10]. Moreover, the same problems are not known to be in APX. We give simple L-reductions from these two problems to MFR and MSR, hence showing the APX-hardness of MFR and

MSR, but also that finding a constant ratio approximation algorithm for any of these problems in general is somewhat of a challenge.

Given an input graph G as instance for either MINEDGEBIPARTIZER or MINN-ODEBIPARTIZER, we subdivide each edge into two edges in series. (The resulting graph G' is clearly bipartite, we call SNPs the new nodes and fragments the old ones). Now, of the two edges incident with every SNP, we declare one to be odd and one to be even. Clearly, solving or approximating MSR on (G', ℓ) amount to solving or approximating (within exactly the same approximation) MINEDGEBIPARTIZER on G. Moreover, solving or approximating MFR on (G', ℓ) amounts to solving or approximating (within exactly the same approximation) MINNODEBIPARTIZER on G. $\qquad\square$

5   Polynomial Algorithms for the Gapless Case

In this section we prove Corollaries 9 and 11 of Section 3.1. Although they would follow from the more general theorems for matrices with bounded length gaps (described in Section 6) we prove them here directly because it is didactically better to do so. In fact, the results of Section 6 will be obtained as generalizations of the ideas described in this section.

5.1   MSR: A Dynamic Programming $O(mn^2)$ Algorithm

In this section we propose a dynamic programming approach for the solution of MSR. The resulting algorithm can be coded as to take $O(mn^2)$ time. In the following, we assume M to be $\mathcal{S}$-reduced.

It is easier to understand our dynamic program if we state it for the complementary problem of MSR, i.e., find the maximum number of SNPs that can be kept so that M becomes error-free. Clearly, if k is the largest number of SNPs that we can keep, then $n - k$ is the smallest number of SNPs to remove.

For $j \le n$ (with $j \ge 0$), we define $K[j]$ as the maximum number of columns that can be kept to make M error-free, under the condition that j is the rightmost column kept (if all columns are removed then $j = 0$, and $K[0] = 0$).

Once all the $K[j]$ are known, the solution to the problem is given by

$$\max_{j \in \{0, \ldots, n\}} K[j].$$

For every j, we define $\mathrm{OK}(j)$ as the set of those i with $i < j$ such that columns i and j do not conflict. We assume that 0 belongs to $\mathrm{OK}(j)$ for every j. Now, for every j,

$$K[j] := 1 + \max_{i \in \mathrm{OK}(j)} K[i] \tag{1}$$

where Equation (1) is correct by the following easily proven fact.

Lemma 17 *Let M be a gapless $\mathcal{S}$-reduced matrix. Consider columns $a < b < c \in \mathcal{S}$. If a is not in conflict with b and b is not in conflict with c, then a is not in conflict with c.*

Proof: Assume SNPs a and c to be conflicting, that is, there exist fragments f and g such that $M[f,a]$, $M[g,a]$, $M[f,c]$, $M[g,c] \neq -$ and the boolean value $(M[f,a] = M[g,a])$ is the negation of $(M[f,c] = M[g,c])$. Since $a < b < c$, then $M[f,b], M[g,b] \neq -$ since M is gapless. Therefore, $(M[f,b] = M[g,b])$ is either the negation of $(M[f,a] = M[g,a])$ or the negation of $(M[f,c] = M[g,c])$. That is, b is either conflicting with a or with c. □

Note that for computing the entries $K[j]$ we only need to know the sets $OK(j)$. Note that $OK(j)$ is the set of those $i < j$ which are neighbors of j in the SNP-conflict graph $G_S(M)$. Since determining if two SNPs are in conflict can be done in time $O(m)$, the cost of creating the $OK(j)$ is $O(mn^2)$. This dominates the cost $O(n^2)$ of solving all equations (1).

5.2   MFR: A Dynamic Programming $O(m^2n + m^3)$ Algorithm

In this section we propose a dynamic programming approach for the solution of MFR. We remark that, contrary to the approach suggested in [8], nested fragments will not be a problem. The resulting algorithm can be coded as to take $O(m^2n + m^3)$ time.

Given a row f of M we denote by $l(f)$ the index of the leftmost SNP s such that $M[f,s] \neq -$ and by $r(f)$ the index of the rightmost SNP s such that $M[f,s] \neq -$. In other words, the body of the fragment f is all contained between the SNPs $l(f)$ and $r(f)$. We assume that the rows of M are ordered so that $l(i) \leq l(j)$ whenever $i < j$. For every $i \in \{1, \ldots, m\}$, let M_i be the matrix made up by the first i rows of M. For $h, k \leq i$ (with $h, k \geq -1$) such that $r(h) \leq r(k)$, we define $D[h, k; i]$ as the minimum number of rows to remove to make M_i error-free, under the condition that

- row k is not removed, and among the non-removed rows maximizes $r(k)$;
- row h is not removed and goes into the opposite haplotype as k, and among such rows maximizes $r(h)$.

(If all rows are removed then $h = -1, k = 0$, and $D[-1, 0; i] := i$. Rows -1 and 0 are all $-$, that is, empty).

Once all the $D[h, k; i]$ are known, the solution to the MSR problem is given by

$$\min_{h,k\,:\,r(h)\,\leq\,r(k)} D[h, k; m].$$

Clearly, for every i, and for every $h, k < i$ with $r(h) \leq r(k)$,

$$D[h, k; i] := \begin{cases} D[h, k; i-1] & \text{if } r(i) \leq r(k) \text{ and rows } i \text{ and } k \text{ agree} \\ D[h, k; i-1] & \text{if } r(i) \leq r(h) \text{ and rows } i \text{ and } h \text{ agree} \\ D[h, k; i-1] + 1 & \text{otherwise,} \end{cases} \quad (2)$$

where Equation (2), as well as Equation (3) and Equation (4), finds explanation into the following easily proven fact.

Lemma 18 *Consider rows $a, b, c \in \mathcal{F}$. Assume $a, b < c$ and $r(a) \leq r(b)$. If a agrees with b and b agrees with c, then a agrees with c.*

For every i, we define $OK(i)$ as the set of those j with $j < i$ such that rows i and j agree. We assume that $0, -1$ belong to $OK(i)$ for every i. (Just think to append a pair of all "$-$" rows at the top of M).

Now, for every i, and for every $h < i$ with $r(h) \leq r(i)$,

$$D[h, i; i] := \min_{j \in OK(i),\, j \neq h,\, r(j) \leq r(i)} \begin{cases} D[j, h; i-1] \text{ if } r(h) \geq r(j) \\ D[h, j; i-1] \text{ if } r(h) < r(j) \end{cases} \tag{3}$$

Finally, for every i, and for every $k < i$ with $r(k) \geq r(i)$,

$$D[i, k; i] := \min_{j \in OK(i),\, j \neq k,\, r(j) \leq r(i)} D[j, k; i-1]. \tag{4}$$

Note that for computing the entries $D[h, k; i]$ we only need to know the sets $OK(i)$. The cost of creating the $OK(i)$ data structure (done in a first phase) is $O(m^2 n)$. The cost of computing the entries $D[h, k; i]$ (done in a second phase). is $O(m^3)$, since it can be seen as the cost of computing the $O(m^3)$ entries $D[h, k; i]$ by using Equation (2) (costs $O(1)$ each) plus the cost of computing the $O(m^2)$ entries $D[h, i; i]$ and $D[i, k; i]$ by using Equations (3) and (4) (costs $O(m)$ each).

6   Dealing with Gaps

In this section we propose a practical approach to deal with any number of gaps, when the body of each fragment does not contain many holes. For the remainder of this section, let k be a constant such that the body of each fragment in the input instance contains at most k holes. We will derive dynamic programming-based polynomial (for a constant k) algorithms for both MFR and MSR, proving Theorems 8 and 10 of Section 3.

6.1   MSR: An $O(mn^{2k+2})$ Algorithm

We modify the basic dynamic programming approach for MSR introduced in Section 5.1. More precisely, the new dynamic programming algorithm for MSR will now consider $2k$ columns of history. We remind the reader that we show how to determine the largest number of SNPs that can be kept to make M error-free, which is equivalent to solving MSR. The resulting algorithm can be coded as to take $O(mn^{2k+2})$ time. In the following, we assume M to be $\mathcal{S}$-reduced.

We say that SNPs $s_1, \ldots, s_t$ are *consistent* when no two of them conflict. For consistent $j_1 < \ldots < j_{2k+1} \leq i$ (with $j_1 \geq -2k$), we define $K[j_1, \ldots, j_{2k+1}]$ as the maximum number of columns to keep to make M error-free, under the condition that $j_1, \ldots, j_{2k+1}$ are the $2k+1$ rightmost columns kept (if all columns are removed then $j_{2k+1} = 0$, and $K[-2k, \ldots, -1, 0] = 0$).

Once all the $K[\ldots]$ are known the solution to the problem is given by

$$\max_{-2k < j_1 < \ldots < j_{2k+1} \leq n} K[j_1, \ldots, j_{2k+1}].$$

For $i \leq n$ and consistent $j_1 < \ldots < j_{2k} < i$ we define $OK(j_1, \ldots, j_{2k}, i)$ as the set of those $j < j_1$ such that columns $j, j_1 < \ldots < j_{2k}, i$ are consistent.

Now, for every consistent $j_1 < \ldots < j_{2k} < i$,

$$K[j_1, \ldots, j_{2k}, i] := 1 + \max_{j \in OK(j_1, \ldots, j_{2k}, i)} K[j, j_1, \ldots, j_{2k}] \tag{5}$$

where Equation (5) is correct by the following easily proven fact.

Lemma 19 *Let M be an S-reduced matrix where each fragment contains at most k holes. Consider columns $a < j_1, \ldots, j_{2k} < c \in S$. Assume columns $a, j_1, \ldots, j_{2k}$ are consistent. Assume further columns $j_1, \ldots, j_{2k}, c$ are consistent. Then, columns $a, j_1, \ldots, j_{2k}, c$ are consistent.*

Proof: Assume on the contrary that a and c conflict. Let f and g be fragments such that $M[f, a], M[g, a], M[f, c], M[g, c] \neq -$ and the boolean value $(M[f, a] = M[g, a])$ is the negation of $(M[f, c] = M[g, c])$.

Since each row has at most k holes, as a consequence, on $2k + 1$ columns, any two rows must have a common non-hole or one of the considered columns is out of the body of one of the two rows. Since a and c are both in the body of f then $j_1, \ldots, j_{2k}$ are all in the body of f. Similarly, $j_1, \ldots, j_{2k}$ are all in the body of g. Hence, let $b \in \{j_1, \ldots, j_{2k}\}$ such that $M[f, b], M[g, b] \neq -$. Since $a < b < c$, then $M[f, b], M[g, b] \neq -$ since M is gapless. Therefore, $(M[f, b] = M[g, b])$ is either the negation of $(M[f, a] = M[g, a])$ or the negation of $(M[f, c] = M[g, c])$. That is, b is either conflicting with a or with c. □

Note that for computing the entries $K[\ldots; i]$ we only need to know the sets $OK(j_1, \ldots, j_{2k}, i)$. The cost of creating the $OK(j_1, \ldots, j_{2k}, i)$ (done in a first phase) is $O(mn^{2k+2})$, which dominates the cost $O(n^{2k+2})$ of solving all equations (5).

6.2 MFR: An $O(2^{2k}nm^2 + 2^{3k}m^3)$ Algorithm

We show how to extend the dynamic programming approach given in Section 5.2 to solve gapless problem instances with holes in $O(2^{2k}nm^2 + 2^{3k}m^3)$ time. We point out that the form of the necessarily-exponential dependence on k is a very good one, i.e., it shows the fixed-parameter tractability of the problem. The memory requirement is $2^{2k}m^3$.

Let f be a fragment and let $x \in \{A, B\}^k$. We denote by $f[x]$ the fragment obtained from f by filling in the holes one by one, using the first characters in x. Since we assumed that the body of each fragment in our input instance contains at most k holes, the characters in x will always suffice to fill in all the holes of f. Given $x_1, x_2 \in \{A, B\}^k$ and two rows f_1, f_2 of matrix M, we denote by $M[f_1[x_1], f_2[x_2]]$ the matrix obtained from M by substituting f_1 with $f_1[x_1]$ and f_2 with $f_2[x_2]$. The following consideration suggests how to extend the dynamic programming algorithm given in Section 5.2 to the case with holes:

Proposition 20 *Let $F_1 = \{f_1^1, \ldots, f_1^p\}, F_2 = \{f_2^1, \ldots, f_2^q\}$ be sets of fragments in M such that any two fragments in F_i ($i = 1, 2$) agree. Then, for every $i \leq p$*

and $j \leq q$ we can give $x_1^i, x_2^j \in \{A, B\}^k$ such that $F_1' = \{f_1^1, \ldots, f_i^i[x_1^i], \ldots, f_1^p\}$ and $F_2' = \{f_2^1, \ldots, f_2^j[x_2^j], \ldots, f_2^q\}$ would still be both without conflicts.

We assume that the rows of M are ordered so that $l(i) \leq l(j)$ whenever $i < j$. For every $i \in \{1, \ldots, m\}$, let M_i be the matrix made up by the first i rows of M. For $h, k \leq i$ (with $h, k \geq -1$) such that $r(h) \leq r(k)$, and for $x, y \in \{A, B\}^k$ we define $D[h, x; k, y; i]$ as the minimum number of rows to remove to make $M_i[h[x], k[y]]$ error-free, under the condition that

- row $k[y]$ is not removed, and among the non-removed rows maximizes $r(k)$;
- row $h[x]$ is not removed and goes into the opposite haplotype as $k[y]$, and among such rows maximizes $r(h)$.

(If all rows are removed then $h = -1, k = 0$, and $D[-1, x; 0, y; i] = i$ for all $x, y \in \{A, B\}^k$.

Once all the $D[h, x; k, y; i]$ are known, the solution to the problem is given by

$$\min_{x, y \in \{A, B\}^k; h, k : r(h) \leq r(k)} D[h, x; k, y; m].$$

Clearly, for every i, and for every $h, k < i$ with $r(h) \leq r(k)$, and for every $x, y \in \{A, B\}^k$,

$$D[h, x; k, y; i] := \begin{cases} D[h, x; k, y; i-1] & \text{if } r(i) \leq r(k) \text{ and rows } i \\ & \text{and } k[y] \text{ agree} \\ D[h, x; k, y; i-1] & \text{if } r(i) \leq r(h) \text{ and rows } i \\ & \text{and } h[x] \text{ agree} \\ D[h, x; k, y; i-1] + 1 & \text{otherwise} \end{cases} \quad (6)$$

For every fragment i and for every $x \in \{A, B\}^k$ we define $OK(i, x)$ as the set of those pairs (j, y) such that j is a fragment with $j < i$ and $y \in \{A, B\}^k$ such that rows $i[x]$ and $j[y]$ agree. Now, for every i, and for every $h < i$ with $r(h) \leq r(i)$, and for every $x_i, x_h \in \{A, B\}^k$,

$$D[h, x_h; i, x_i; i] := \min_{(j, x_j) \in OK(i, x_i), j \neq h, r(j) \leq r(i)} \begin{cases} D[j, x_j; h, x_h; i-1] \\ \quad \text{if } r(h) \geq r(j) \\ D[h, x_h; j, x_j; i-1] \\ \quad \text{if } r(h) < r(j) \end{cases} \quad (7)$$

Finally, for every i, and for every $k < i$ with $r(k) \geq r(i)$, and for every $x_i, x_k \in \{A, B\}^k$,

$$D[i, x_i; k, x_k; i] := \min_{(j, x_j) \in OK(i, x_i), j \neq k, r(j) \leq r(i)} D[j, x_j; k, x_k; i-1]. \quad (8)$$

Note that for computing the entries $D[h, x; k, y; i]$ we only need to know the sets $OK(i)$. The cost of creating the $OK(i, x)$ data structure (done in a first phase) is $O(2^{2k}nm^2)$. The cost of computing the entries $D[h, x; k, y; i]$ (done in a second phase) is $O(2^{3k}m^3)$, since it can be seen as the cost of computing the $O(2^{2k}m^3)$ entries $D[h, x; k, y; i]$ with $h < k < i$ by using Equation (6) (costs $O(1)$ each) plus the cost of computing the $O(2^{2k}m^2)$ entries $D[h, x; i, y; i]$ by using Equations (7) and (8) (costs $O(2^k m)$ each).

References

1. K. S. Booth and G. S. Lueker. Testing for the consecutive ones property, intervals graphs and graph planarity testing using PQ-tree algorithms. *J. Comput. System Sci.*, 13:335–379, 1976.
2. A. Chakravarti. It's raining SNP, hallelujah? *Nature Genetics*, 19:216–217, 1998.
3. A. Clark. Inference of haplotypes from PCR–amplified samples of diploid populations. *Molecular Biology Evolution*, 7:111–122, 1990.
4. D. Gusfield. A practical algorithm for optimal inference of haplotypes from diploid populations. In R. Altman, T.L. Bailey, P. Bourne, M. Gribskov, T. Lengauer, I.N. Shindyalov, L.F. Ten Eyck, and H. Weissig, editors, *Proceedings of the Eighth International Conference on Intelligent Systems for Molecular Biology*, pages 183–189, Menlo Park, CA, 2000. AAAI Press.
5. D. Gusfield. Haplotyping as perfect phylogeny: Conceptual framework and efficient solutions. In G. Myers, S. Hannenhalli, S. Istrail, P. Pevzner, and M. Watermand, editors, *Proceedings of the Sixth Annual International Conference on Computational Biology*, pages 166–175, New York, NY, 2002. ACM Press.
6. L. Helmuth. Genome research: Map of the human genome 3.0. *Science*, 293(5530):583–585, 2001.
7. International Human Genome Sequencing Consortium. Initial sequencing and analysis of the human genome. *Nature*, 409:860–921, 2001.
8. G. Lancia, V. Bafna, S. Istrail, R. Lippert, and R. Schwartz. SNPs problems, complexity and algorithms. In *Proceedings of Annual European Symposium on Algorithms (ESA)*, volume 2161 of *Lecture Notes in Computer Science*, pages 182–193. Springer, 2001.
9. R. Lippert, R. Schwartz, G. Lancia, and S. Istrail. Algorithmic strategies for the SNPs haplotype assembly problem. *Briefings in Bioinformatics*, 3(1):23–31, 2002.
10. C. Lund and M. Yannakakis. The approximation of maximum subgraph problems. In *Proceedings of 20th Int. Colloqium on Automata, Languages and Programming*, pages 40–51. Springer-Verlag, 1994.
11. E. Marshall. Drug firms to create public database of genetic mutations. *Science Magazine*, 284(5413):406–407, 1999.
12. J.C. Venter *et al.* The sequence of the human genome. *Science*, 291:1304–1351, 2001.
13. J. Weber and E. Myers. Human whole genome shotgun sequencing. *Genome Research*, 7:401–409, 1997.

Methods for Inferring Block-Wise Ancestral History from Haploid Sequences
The Haplotype Coloring Problem

Russell Schwartz[1], Andrew G. Clark[1,2], and Sorin Istrail[1,*]

[1] Celera Genomics, Inc.
45 West Gude Dr.
Rockville, MD 20850 USA
Phone: (240) 453-3668, Fax: (240) 453-3324
{Russell.Schwartz,Sorin.Istrail}@celera.com
[2] Department of Biology
Pennsylvania State University
University Park, PA 16802 USA
c92@psu.edu

Abstract. Recent evidence for a "blocky" haplotype structure to the human genome and for its importance to disease inference studies has created a pressing need for tools that identify patterns of past recombination in sequences of samples of human genes and gene regions. We present two new approaches to the reconstruction of likely recombination patterns from a set of haploid sequences which each combine combinatorial optimization techniques with statistically motivated recombination models. The first breaks the problem into two discrete steps: finding recombination sites then coloring sequences to signify the likely ancestry of each segment. The second poses the problem as optimizing a single probability function for parsing a sequence in terms of ancestral haplotypes. We explain the motivation for each method, present algorithms, show their correctness, and analyze their complexity. We illustrate and analyze the methods with results on real, contrived, and simulated datasets.

1 Introduction

The sequencing of the human genome [12,25] has created a tremendous opportunity for medical advances through the discovery of genetic predictors of disease. So far, though, catalogs of the genetic differences between individuals have proven difficult to apply. Examined in isolation, these differences - which occur predominantly in the form of isolated changes called single nucleotide polymorphisms (SNPs) - may fail to distinguish real relationships from the background noise millions of SNPs produce. Recent analysis of the structure of the human genome [14] has given hope that greater success will be achieved through studies of haplotypes, sets of alleles from all SNPs in a region that tend to travel together

---

[*] To whom correspondence should be addressed

R. Guigó and D. Gusfield (Eds.): WABI 2002, LNCS 2452, pp. 44–59, 2002.

A ACGATCGATCATGAT B ACGATCG|ATCAT|GAT C
   GGTGATTGCATCGAT   GGTGATT|GCATC|GAT   ---A---|G---C|-C-
   ACGATCGGGCTTCCG   ACGATCG|GGCTT|CCG   ---G---|A---T|-A-
   ACGATCGGCATCCCG   ACGATCG|GCATC|CCG   ---G---|G---T|-C-
   GGTGATTATCATGAT   GGTGATT|ATCAT|GAT   ---A---|A---T|-A-
   GGTGATTGGCTTGAT   GGTGATT|GGCTT|GAT

Fig. 1. An illustration of the value of haplotype blocks. A: A hypothetical population sample of a set of polymorphic sites. B: A pattern of haplotype blocks inferred from the population sample. C: The results of a hypothetical assay conducted on additional individuals based on the block patterns. If the original sample adequately captured the full population variability, then typing four sites per individual would be sufficient to determine their block patterns, allowing inference of their untyped sites.

through evolutionary history. Several recent studies [3,13,20] have suggested that the human genome consists largely of blocks of common SNPs organized in haplotypes separated by recombination sites, such that most human chromosome segments have one of a few possible sets of variations. It may therefore be possible to classify most human genetic variation in terms of a small number of SNPs identifying the common haplotype blocks. If so, then determining the genome's haplotype structure and defining reduced SNP sets characterizing common haplotype variants could greatly reduce the time and cost of performing disease association studies without significantly reducing their power to find disease-related genes and genetic predictors of disease phenotypes. Figure 1 illustrates haplotype blocks and their potential value with a contrived example.

There is considerable prior work on detecting ancestral recombination events from a sample of gene sequences of known haplotype phase (i.e. haploid sequences). Some methods, such as those of Sawyer [22], Maynard Smith [17], and Maynard Smith and Smith [18], detect whether any recombination has occurred in a set of sequences. Others, such as the methods of Hudson and Kaplan [11] and Weiler [27] further attempt to find the locations of the recombination events. More difficult is assigning haplotypes and recombination patterns to individual sequences in a sample — as was done for example by Daly et al. [3] and Zhang et al. [29] — which provides the information that would be necessary for haplotype-based LD mapping and associated inference studies. The ultimate goal of such methods would be reconstruction of the ancestral recombination graph from a set of sequences, a problem addressed by the methods of Hein [9], Kececioglu and Gusfield [15], and Wang et al. [26]. Simulation studies of recombination detection methods [28,21,6] suggest some room for improvement. One suggestion of these studies is that more adaptable methods and methods suited to special cases of recombination might have greater power in detecting recombination.

There is also a need for better integration of the statistical theory of recombination with the theory of algorithmic optimization methods. With the notable exception of the work of Kececioglu and Gusfield, there has been little interaction between these fields; it seems likely that progress will be best achieved at the intersection of the best models and the best methods for solving for them.

Our hope is to suggest methods that may help in better combining statistically motivated recombination models with combinatorial optimization methods.

We specifically address the problem of inferring from a set of haploid sequences the recombination patterns and regions of shared recent ancestry between sequences in the sample, similar to the computational problem approached by Daly et al. [3]. We formulate the task as two variants of what we call the "haplotype coloring problem," the goal of which is to assign colors to regions of haploid sequences to indicate common ancestry. Thus, shared colors between sequences at a given site would indicate descent from a common haplotypes at that site. Shared colors between sites in a single sequence would indicate descent of those sites from a common ancestral sequence undisrupted by recombination. The first method treats the problem as two discrete steps: locating haplotype blocks and coloring sequences to indicate likely ancestry of haplotypes. The second method performs all aspects of haplotype coloring as the optimization of a unified objective function. We also describe an iterative expectation maximization (EM) algorithm based on the second method to allow simultaneous inference of population haplotype frequencies and assignment of haplotypes to individual haploid sequences within a sample. In the remainder of this paper, we formalize the methods as computational problems, describe algorithms, prove their correctness, and analyze their efficiency. We also describe applications of the methods to contrived, real, and simulated data. Finally, we discuss implications of the methods and prospects for future work.

2   The Block-Color Method

Inferring ancestry in the presence of recombination can be decomposed into two distinct stages: infer the block pattern of the haplotypes and color haplotypes in individual sequences in the way that is most likely to reflect the ancestry of the observed sequences. Related problems are commonly understood in terms of the neutral infinite sites model [16], which assumes that any site mutates at most once in the ancestral history of a genomic region. Under the assumptions of the infinite sites model, the only way that gametic types AB, Ab, aB, and ab can all be present in a sample is if recombination had generated the fourth gametic type from the others. Recurrent or back mutations could also produce the fourth gamete in reality, but are assumed not to occur under the infinite sites model. Thus, any pair of sites for which all 4 gametes are found can be inferred to have incurred a recombination between them at some time in the past [11]. Tallying all pairs of sites having four gametes allows one to infer a minimum number of recombination events necessary, giving rise to a two-step method for the haplotype coloring problem. First, identify maximal blocks of sites for which no pair has all four gametes and separate sequences within blocks into distinct haplotypes. Second, color sequences within each block to minimize the number of color changes across blocks over all sequences. When the infinite sites model is not obeyed, as is often the case with real data, recurrent mutation at a site also generates site-pairs with all four gametes, but the pattern of haplotypes is quite different. We would therefore like the method to be insensitive to minor changes between haplotypes generated by such recurrent mutation. Very rare variants and mistyping errors are also difficult to distinguish, and past recombination events may be obscured by subsequent mutation events. We thus perform a pre-filtering step that removes from consideration all polymorphic loci for which the minor variant occurs in less than a user-specified fraction of the total population.

The two-step method is meant to apply the well-understood four-gamete test to the broader problem of inferring the ancestral history of the set of sequences. The additional coloring stage provides a method for inferring which sequences were likely to have been formed through recombinations at identified recombination sites, allowing us to infer a history of the sequence sample and reduce the amount of information needed to specify an individual's haplotypes.

2.1 Identifying Blocks

Hudson and Kaplan [11] developed a method for finding the minimum number of blocks in a set of phased sequences such that the four-gamete constraint is satisfied within each block. Their method finds a minimum-size set of blocks for any set of phased sequences such that all blocks satisfy the four-gamete constraint. Gusfield [8] developed a method for testing for the equivalent perfect phylogeny constraint in unphased sequences. Gusfield's method can be applied straightforwardly to simultaneously infer haplotypes and a block structure by locating the minimal size set of blocks consistent with a perfect phylogeny for each block. Within each block, haplotypes can be inferred efficiently using an earlier algorithm of Gusfield [7] for inferring phylogenetic trees. Either the Gusfield method for unphased data or the Hudson and Kaplan method for phased data can therefore be used to derive sets of blocks associated with haploid sequences, which provides the input necessary for the coloring stage of our block-color method.

Jeffreys et al. [13] suggested that within the hotspots of high recombination, the recombination events might localize to a few nearby markers rather than occurring at a single point. It may therefore be valuable to develop methods for enumerating or sampling optimal or near-optimal solutions or finding common substructures among them, rather than finding a single optimum. We might also want to consider other tests for block viability than the four-gamete constraint that are more robust to the number of sequences or more liberal or conservative in selecting which regions may be considered blocks. We have therefore also developed a slower but more general dynamic programming algorithm for the block-identification problem, similar that used by Zhang et al. [29], which takes an arbitrary set of pair-wise constraints and constructs bottom-up a minimum-size partition into blocks such that every constraint spans a block boundary. Due to space limitations, we omit a detailed description of this algorithm.

2.2 Coloring

Our goal in block coloring is to assign colors to blocks such that each distinct haplotype on a given sequence region is assigned a different color so as to mini-

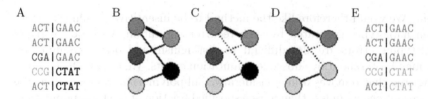

A
```
ACT I GAAC
ACT I GAAC
CGA I GAAC
CCG I CTAT
ACT I CTAT
```

E
```
ACT I GAAC
ACT I GAAC
CGA I GAAC
CCG I CTAT
ACT I CTAT
```

Fig. 2. An illustration of one step in the coloring algorithm. A: Two consecutive blocks in an example set of sequences. The sequences in the first block are assumed to have already been colored. B: The graph construction, with each node on the left representing one of the three haplotypes in the first block and each node on the right representing one haplotype in the second block. The uppermost edge (between the red and gray nodes) is thicker to represent the fact that two examples of that haplotype pair occur and the edge therefore has double weight. C: The solution to the maximum matching. D: The coloring of the right nodes implied by the matching. E: The translation of the coloring back into the sequences yielding a coloring of the haplotypes in the second block that minimizes color changes between the blocks.

mize the total number of color changes between haplotypes in our sequence set. Intuitively, this procedure is meant to explain the data with as few recombinations as possible, providing a maximally parsimonious solution to the coloring problem given the assumption that recombination events are relatively rare. For the purposes of this analysis, we will decompose each sequence into a string of haplotypes arranged in blocks, with all sequences sharing the same block structure. We informally define a block to be an interval of polymorphic sites and the set of haplotypes that occur on that interval. Within each block, distinct haplotypes are assigned distinct colors and identical haplotypes are assigned identical colors. Where two consecutive haplotypes in a sequence are given the same color, it is implied that they are likely to have come from a common ancestral sequence. Where colors change between two haplotypes in a sequence, the implication is that a recombination event was likely involved in forming that sequence from two different ancestral sequences that were sources of the two haplotypes. In expressing the coloring stage as a computational problem, we more formally define the input as the following:

B, a sequence $b_1, \ldots, b_k$ of blocks where each b_i is a set of haplotypes in a given interval of polymorphic sites. Element x of b_i is denoted $b_{i,x}$.

S, a set of block-decomposed sequences $s_1, \ldots, s_n$ where each sequence $s_i \in b_1 \times \cdots \times b_k$ has a multiplicity n_i. Element j of s_i is denoted $s_{i,j}$. Let C be a set of positive integer colors. Then our output is a set of assignments $X = \chi_1, .., \chi_k$ where each χ_i is a function from b_i to C such that $b_{i,x} = b_{i,y} \Leftrightarrow \chi_i(b_{i,x}) = \chi_i(b_{i,y})$, for which we minimize the following objective function (expressing the sum of color changes):

$$G = \sum_{j=1}^{k-1} \sum_{i=1}^{n} n_i ID(\chi_j(s_{i,j}) \neq \chi_{j+1}(s_{i,j+1})), \text{where} ID(b) = \left\{ \begin{array}{l} 1 \text{ if } b \text{ is true} \\ 0 \text{ if } b \text{ is false} \end{array} \right\}.$$

The function ID tests whether for a given i and j there is a color change in sequence i between blocks j and $j + 1$. G thus counts the total number of color changes between consecutive blocks over all sequences. We will show two key properties of this problem that allow us to solve it efficiently. First, greedily coloring each block optimally in terms of the previous block (minimizing the number of color changes given the previous block's coloring) yields a globally optimal solution. Second, coloring each block is an instance of the weighted bipartite maximum matching problem, for which there exist efficient algorithms [5]. Figure 2 illustrates the resulting method, which we now examine.

Lemma 1. *Any set X such that χ_{j+1} minimizes $G_j(X) = \sum_{i=1}^n ID(\chi_j(s_{i,j}) \neq \chi_{j+1}(s_{i,j+1}))$ given χ_j $\forall j \in [1, k-1]$ will minimize the objective function G.*

Proof. Assume we have a mapping X such that χ_{j+1} minimizes $G_j(X)$ given χ_j for all j in $[1, k-1]$. Assume further for the purposes of contradiction that there exists another solution X' such that $\sum_{j=1}^{k-1} G_j(X') < \sum_{j=1}^{k-1} G_j(X)$. Then there must be some smallest j such that $G_j(X') < G_j(X)$. We can then create a new X'' such that X'' is identical to X up to position j and $\chi''_{j+1}(c) = \chi'_j(c)$ if and only if $\chi'_{j+1}(c) = \chi'_j(c)$. Then X'' must have the same cost as X' for the transition from region j to $j + 1$, which is strictly lower than that for X on that transition. Thus, X could have chosen a better solution for the transition from j to $j + 1$ given its solutions for all previous transitions, contradicting the assumption that X minimizes $G_j(X)$ given χ_j for all j. Thus, X' cannot exist and X must minimize G.

Lemma 2. *Finding the optimal χ_{j+1} given χ_j can be expressed as weighted maximum matching.*

Proof. We prove the lemma by construction of the instance of weighted maximum matching. We first rewrite G as $(k-1)\sum_{i=1}^n n_i - \sum_{j=1}^{k-1} \sum_{i=1}^n n_i ID(\chi_j(s_{i,j}) = \chi_{j+1}(s_{i,j+1}))$. Since $(k-1)\sum_{i=1}^n n_i$ does not depend on X, minimizing our original objective function is equivalent to maximizing $\sum_{j=1}^{k-1} \sum_{i=1}^n n_i ID(\chi_j(s_{i,j}) = \chi_{j+1}(s_{i,j+1}))$. We create a bipartite graph B in which each node u_i in the first part corresponds to a haplotype c_{ji} in block j and each node $v_{i'}$ in the second part corresponds to a haplotype $c_{j+1,i'}$ in block $j + 1$. If any sequence has haplotypes c_{ji} and $c_{j+1,i'}$, then we create an edge $(u_i, v_{i'})$ with weight

$$\sum_{i=1}^n n_i HAS_{j,i,i'}(s_i), \text{ where } HAS_{j,i,i'}(s) = \begin{cases} 1 \text{ if } s_j = c_{j,i} \text{ and } s_{j+1} = c_{j+1,i'} \\ 0 \qquad\qquad \text{otherwise} \end{cases}.$$

A matching in B corresponds to a set of edges pairing haplotypes in block j with haplotypes in block $j + 1$. We construct X so that it assigns the same color to $c_{j+1,i'}$ as was assigned to $c_{j,i}$ if and only if the matching of B selects the edge between the nodes corresponding to those two haplotypes. Any given matching will have a weight equal to the sums of the frequencies of sequences sharing both of each pair of haplotypes whose corresponding nodes are connected by the matching. Thus the coloring corresponding to a maximum matching will maximize $\sum_{i=1}^{n} n_i \sum_{j=1}^{k-1} ID(\chi_j(s_{i,j}) = \chi_{j+1}(s_{i,j+1}))$, which yields an optimal assignment of χ_{j+1} given χ_j.

Lemmas 1 and 2 imply the following algorithm for optimal coloring, which we refer to as Algorithm 1:

create an arbitrary assignment χ_1 of haplotypes to distinct colors in block 1
for $i = 2$ to m
    construct an instance of weighted maximum matching as described above
    solve the instance with the algorithm of Edmonds [5]
    for each pair $(c_{i-1,j}, c_{i,k})$ joined in the matching assign $\chi_i(c_{i,k}) = \chi_{i-1}(c_{i-1,j})$
    for each $c_{i,k}$ that is unmatched assign an arbitrary unused color to $\chi_i(c_{i,k})$

Theorem 1. *Algorithm 1 produces a coloring of haplotype blocks minimizing the number of color changes across sequences in time $O(mn^2 \log n)$.*

Proof. The proof of correctness follows directly from Lemmas 1 and 2. Creating an instance of weighted maximum matching and assigning colors requires $O(n)$ time. The run time for each iteration of i is therefore dominated by the $O(n^2 \log n)$ run time of the maximum matching algorithm for this type of dataset (where the number of non-zero edges of the graph is bounded by n). There are $O(m)$ rounds of computation, yielding a total run time of $O(mn^2 \log n)$.

Using the Hudson and Kaplan algorithm for block assignment gives the block-color method an overall complexity of $O(nm^2 + mn^2 \log n)$ for n sequences and m polymorphic sites. In practice, the input data would typically come from resequencing some number of individuals in one sequenced gene - with a bound on m and n typically on the order of 100 - yielding run times well within what can be handled by standard desktop computers.

3   The Alignment Method

Although solving the problem in well-defined stages has advantages, it may also be fruitful to find a unified global solution to all aspects of the problem. Our hope is that such an approach will help in finding more biologically meaningful probabilistic models that capture the essential features of the system but are computationally tractable. We therefore developed a second approach based on techniques from sequence alignment. Sequence alignment can be viewed as assigning to each position of the target sequence a frame of a reference sequence, with changes of frame representing gaps. We can analogously "align" a single target to multiple references, but instead of shifting frames to model gaps, shift reference sequences to model recombination. Figure 3 illustrates this analogy. This approach is similar to the Recombination Cost problem of Kececioglu and

A   ACTAGCTAGCATCG   B   TGAGGCATGTACGA

    ACTAGCTAGCATCG   ACTAGCTAGCATCG
    ACT   CATCG   ACT   GTATGA

Fig. 3. An illustration of the analogy between sequence alignment and haplotype coloring as variants of the problem of "parsing" a query sequence in terms of a set of reference sequences. Each sub-figure shows a query sequence (bottom) aligned to three reference sequences (top). A: sequence alignment as parsing of a sequence in terms of a set of identical reference sequences in distinct frames; B: haplotype coloring as parsing a sequence in terms of a set of distinct reference sequences in identical frames.

Gusfield [15] and the "jumping alignments" of Spang et al. [23]. We, however, assume we have a population sequenced for specific polymorphic sites and therefore need not consider insertion/deletion costs. Deletion polymorphisms can be treated simply as additional allowed symbols. Dealing with missing data is more complicated, but can be handled through the use of a special symbol representing an undefined site, which can match any other symbol. Bayesian methods, similar to that of Stephens et al. [24], might also be used to impute missing data.

A major advantage of this technique over our other method is that it does not assume that there are recombination hotspots or a block structure. It may therefore be better suited to testing that assumption or examining genomes or genome regions in which it proves to be inapplicable. It should also allow us to distinguish recent recombination sites affecting a small fraction of the sequenced population from more ancient or frequently recombining sites and is easily parameterized to be more or less sensitive to recent mutations in identifying haplotypes. Among its disadvantages are that it requires an additive objective function and therefore uses a probability model different from those traditionally used in LD studies and that it is parameterized by values that may be unknown and difficult to estimate. In addition, the function being optimized is harder to understand intuitively than those used in the block-color method, making the alignment method harder to judge and improve upon.

Our probability model parses a sequence as a string of haplotype identifiers describing the ancestral source of each of its polymorphic sites. A given sequence chooses its first value from a distribution of haplotypes at the first polymorphic position. Each subsequent polymorphic site may follow a recombination event with some probability ρ. If there is no recombination event then the sequence continues with the same haplotype as it had at the prior site. Otherwise, the sequence samples among all available haplotypes according to a site-specific distribution for the new site. There is also a mutation probability μ that any given site will be mutated from that of its ancestral haplotype. This model leads to the following formalization of the inputs:

   m, the log probability of a mutation event at any one site of a sequence
   r, the log probability of a recombination event between any two sites
   $S = s_1, \ldots, s_n$, a set of n reference sequences. Each s_i is a sequence of l polymorphic sites $s_{i1}, \ldots, s_{il}$

F, an $n \times l$ matrix of log frequencies in which f_{ij} specifies the probability of choosing a given haplotype i at each site j following a recombination immediately prior to that site

$\Sigma = \sigma_1, \ldots, \sigma_t$, a set of t target sequences. Each σ_i is a sequence of l polymorphic values, $\sigma_{i1}, \ldots, \sigma_{il}$

Our goal is to produce a $t \times l$ matrix H, where h_{ij} specifies which haplotype from the set $[1, n]$ has been assigned to position j of target sequence σ_i, maximizing the following objective function:

$$G(H) = \sum_{i=1}^{t} f_{h_{i1}1} + \sum_{i=1}^{t} \sum_{j=2}^{l} (f_{h_{ij}j} + r) D(h_{i,j}, h_{i,j-1})$$

$$+ \sum_{i=1}^{t} \sum_{j=2}^{l} \log((1 - e^r) + e^{r+f_{h_{ij}j}})(1 - D(h_{i,j}, h_{i,j-1})) + \sum_{i=1}^{t} \sum_{j=1}^{l} M(s_{h_{i,j},j}, \sigma_{i,j})$$

$$\text{where} \quad D(a, b) = \begin{Bmatrix} 0, & a = b \\ 1, & a \neq b \end{Bmatrix} \quad \text{and} \quad M(a, b) = \begin{Bmatrix} 0, & a = b \\ m, & a \neq b \end{Bmatrix}.$$

$G(H)$ gives a normalized log probability of the assignment H given the sequences and haplotype frequencies. The first sum reflects the probabilities of choosing different starting haplotypes. The next two sums reflect the contributions of respectively choosing to recombine at each site or choosing not to recombine. The final sum gives the contribution to the probability of mismatches.

The function implies some assumptions about the independence of different events whose validity must be considered. The assumptions that probabilities of recombination and mutation are independent and identical at all sites are imperfect but may be reasonable *a priori* in the absence of additional information. It is less clearly reasonable to assume that the selection of recombination positions and the choices of haplotypes between them can be considered independent, although this too may be reasonable absent additional information.

It is important to note that there is no unique "correct" pair of r and m parameters for a given genome or gene region. The right parameters depend on how much tolerance is desired in allowing slightly different sequences to be considered identical haplotypes, analogous to the need for different sensitivities in sequence similarity searches conducted for different purposes. We can thus propose that the value of m should be determined by the particular application of the method, with r then being an unknown that depends on both m and the nature of the genome region under examination.

3.1   The Alignment Algorithm

We maximize $G(H)$ with a dynamic programming algorithm. The following formula describes the basic dynamic programming recursion for assigning haplotypes to a single sequence σ:

$$C_\sigma(i, j) = \max_k \left\{ \begin{matrix} C_\sigma(i, j-1) + \log(1 - e^r + e^{r+f_{ij}}) \\ C_\sigma(k, j-1) + r + f_{ij} \end{matrix} \right\} + \left\{ \begin{matrix} 0, & \sigma_j = s_{ij} \\ m, & \sigma_j \neq s_{ij} \end{matrix} \right\}.$$

$C_\sigma(i, j)$ represents the cost of the optimal parse of sequence σ up to position j ending with an assignment to haplotype i at position j. The right-hand term accounts for the mismatch penalty, if any, between σ and haplotype i at position j. The full algorithm follows directly from the recurrence. The following pseudocode describes the complete algorithm, which we call Algorithm 2:

> for $i = 1$ to n: if $(\sigma_1 = s_{i1})$ then $C_\sigma[i, 1] \leftarrow f_{i1}$ else $C_\sigma[i, 1] \leftarrow f_{i1} + m$
> for $j = 2$ to l: for $i = 1$ to n:
> $\quad$ $best \leftarrow C_\sigma[i, j-1] + \log(1 - e^r + e^{r+f_{ij}})$; $argbest \leftarrow i$
> $\quad$ for $k = 1$ to n, $k \neq i$:
> $\quad\quad$ if $(C_\sigma[k, j-1] + r + f_{ij} < best)$
> $\quad\quad\quad$ $best \leftarrow C_\sigma[k, j-1] + r + f_{ij}$; $argbest \leftarrow k$
> $\quad$ if $(\sigma_j = s_{ij})$ then $C\sigma[i, j] \leftarrow best$ else $C\sigma[i, j] \leftarrow best + m$
> $\quad$ $P_\sigma[i, j] \leftarrow argbest$
> $best \leftarrow -\infty$
> for $i = 1$ to n: if $C_\sigma[i, l] > best$ then $best \leftarrow C_\sigma[i, l]$; $H_\sigma[l] \leftarrow i$
> for $j = l\text{-}1$ downto 1: $H_\sigma[j] \leftarrow P_\sigma[H_\sigma[j+1], j+1]$

Lemma 3. $C_\sigma[i, j]$ *is the optimal cost of any assignment of positions 1 through j of σ to haplotypes in S such that position j of σ is assigned to haplotype i.*

Proof. We prove the statement by induction on j. For the base case of $j = 1$, each $C_\sigma[i, j]$ is uniquely determined by the log frequency f_{i1} plus a mutation penalty m if σ_1 and s_{i1} do not match. The first for loop sets each $C_\sigma[i, 1]$ accordingly, satisfying the inductive hypothesis. Now assume the lemma is true for $j - 1$. We can decompose $C_\sigma[i, j]$ into two terms, $A_\sigma[i, j] + B_\sigma[i, j]$, where

$$A_\sigma[i, j] = \sum_{i=1}^{t} f_{h_{i1}1} + \sum_{i=1}^{t}\sum_{j'=2}^{j-1} (f_{h_{ij'}j'} + r)D(h_{i,j'}, h_{i,j'-1})$$

$$+ \sum_{i=1}^{t}\sum_{j'=2}^{j-1} \log((1 - e^r) + e^{r+f_{h_{ij'}j'}})(1 - D(h_{i,j'}, h_{i,j'-1}))$$

$$+ \sum_{i=1}^{t}\sum_{j'=1}^{j-1} M(s_{h_{i,j'},j'}, \sigma_{i,j'})$$

and

$$B_\sigma[i, j] = \sum_{i=1}^{t} (f_{h_{ij}j} + r)D(h_{i,j}, h_{i,j-1})$$

$$+ \sum_{i=1}^{t} \log((1 - e^r) + e^{r+f_{h_{ijj}}})(1 - D(h_{i,j}, h_{i,j-1})) + \sum_{i=1}^{t} M(s_{h_{i,j},j}, \sigma_{i,j}).$$

A_σ is exactly the function optimized by $C_\sigma[i, j-1]$. B_σ depends only on assignments at positions j and $j - 1$; it is therefore optimized for a given assignment i at position j by maximizing it over all assignments at position $j - 1$, which the algorithm does in deriving $C_\sigma[i,j]$ in the second loop. $C_\sigma[i,j]$ is thus the cost of the optimal assignment of positions 1 through j ending on haplotype i.

Theorem 2. *Algorithm 2 will find an H maximizing G(H) for sets of n reference sequences S and t target sequences Σ with length l in $O(n^2lt)$ time.*

Proof. It follows from lemma 3 that $C_\sigma[i,l]$ will be the cost of the optimal solution to the global problem terminating in haplotype i. Finding an i maximizing $C_\sigma[i,l]$, as is done by the third outer loop, therefore yields the global optimum to the problem. The final loop performs backtracking, reconstructing the optimal solution H. Run time is dominated by an inner loop requiring constant time for each of $O(n^2l)$ iterations, for a total of $O(n^2lt)$ run time when run on t targets.

The definition of the problem solved by the alignment method requires that we know in advance the frequencies from which haplotypes are sampled following a recombination event. As this information may not be available, we would like a way to estimate it from measured frequencies of full-length haploid sequences. We therefore developed an iterative method to optimize this probability by successively performing a discrete optimization of the coloring given the haplotype frequencies followed by a continuous optimization of the frequencies given the coloring, which we perform through a steepest-descent search. The algorithm is a form of generalized expectation maximization (EM) algorithm [1,4] that treats the frequencies as hidden variables, repeatedly finding a maximum aposteriori probability (MAP) coloring H given the frequencies F maximizing $\Pr(H|F)$ by Algorithm 2 then improving $\Pr(H|F)$ in terms of F by steepest descent. The number of rounds required for the resulting method to converge might theoretically be large, although we have found convergence in practice to occur reliably within ten iterations on real gene region data sets.

4   Results

Both methods were implemented in C++. All tests were run on four-processor 500 MHz Compaq Alpha machines, although the code itself is serial.

We used a contrived data set, shown in Figure 4, to illustrate the strengths and weaknesses of the methods. We strongly penalized mutations in order to simplify the illustration. The block-color method correctly detects the recombination site, although the coloring appears suboptimal due to the constraint that identical haplotypes must have identical colors. The alignment method correctly identifies the recombinants, but only with an appropriate choice of parameters.

In order to demonstrate the methods and explore the parameter space, we further applied the methods to a real data set: the apolipoprotein E (APOE) gene region core sample of Nickerson et al. [19], a set of 72 individuals typed on 22 polymorphic sites, with full-length haplotypes determined by a combination of the Clark haplotype inference method [2] and allele-specific PCR to verify phases. Figure 5 demonstrates the effects on the block-color algorithm of varying

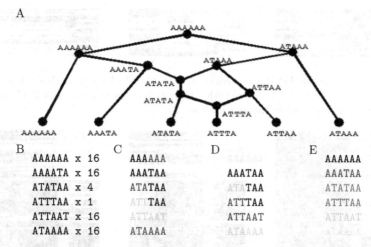

A

B
AAAAAA	x 16
AAAATA	x 16
ATATAA	x 4
ATTTAA	x 1
ATTAAT	x 16
ATAAAA	x 16

C
AAAAAA
AAATAA
ATATAA
ATTTAA
ATTAAT
ATAAAA

D
AAATAA
ATATAA
ATTTAA
ATTAAT
ATAAAA

E
AAAAAA
AAATAA
ATATAA
ATTTAA
ATTAAT

Fig. 4. A contrived sample problem. Different colors represent different predicted ancestral sequences. A: An ancestral recombination graph showing a proposed sequence ancestry; B: The extant sequences with frequencies chosen to make the recombinant and double recombinant relatively rare; C: The output of the block-color method screening out sites with minor allele frequencies below 0.1; D: The output of the alignment method with parameters $r = -1$ and $m = -100$; E: The output of the alignment method with parameters $r = -2$ and $m = -100$.

tolerance to infrequent SNPs. Moving from considering all SNPs in Figure 5A to considering only those with minor frequencies above 10% in Figure 5B then 25% in Figure 5C, leads to progressively simpler block structures, with fewer blocks and less variability within them. Figure 6 illustrates the effects of varying recombination and mutation penalties for the alignment algorithm with the EM extension on the same dataset. Higher recombination penalties generally lead to greater numbers of haplotypes being assigned while higher mutation penalties reveal more subtle regions of variation. We note that inferred haplotype regions do not necessarily line up at "recombination hotspots," suggesting that the data might be more parsimoniously explained by not assuming the existence of discrete haplotype blocks with the same structure across all sequences.

While a real dataset can demonstrate the methods, it cannot rigorously validate them, as we do not definitively know the recombination history of any real gene region. We therefore resorted to simulated data to perform a partial test of the methods. Simulations were generated through Hudson's coalescent simulator [10], using populations of 50 individuals with 70 segregating sites and allowing recombination between any pair of sites. A simulated data set was generated for each recombination parameter ρ in the set $\{0,1,2,5,10,20,30\}$. We then calculated for the block-color method how many recombination sites were predicted, screening out sites with minor allele frequency below 0.1. We further calculated for the alignment method, with parameters $m = -1.5$ and $r = -1.0$, at how many sites at least one recombination event was predicted. Figure 7 shows the

Fig. 5. Coloring of the APOE gene region [19] by the block-color method. A: coloring using all sites. B: coloring using sites with minor allele frequency above 10%. C: coloring using sites with minor allele frequency above 25%.

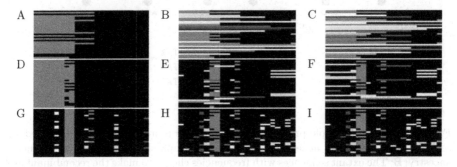

Fig. 6. Coloring of the APOE gene region [19] by the alignment-color method. Parameter values are A: $r = -1.0$ $m = -2.0$; B: $r = -1.0$ $m = -4.0$; C: $r = -1.0$ $m = -6.0$; D: $r = -0.5$ $m = -2.0$; E: $r = -0.5$ $m = -4.0$; F: $r = -0.5$ $m = -6.0$; G: $r = 0.0$ $m = -2.0$; H: $r = 0.0$ $m = -4.0$; I: $r = 0.0$ $m = -6.0$;

resulting plot. As the plot indicates, the number of detected recombination sites generally increases with increasing recombination rate, although the correlation is imperfect and grows more slowly than the increase in ρ. Of course the process of simulating such data involves a high degree of stochasiticity, so one does not expect a perfect correlation. For the block-color method, this result is consistent with the analogous experiments performed by Hudson and Kaplan [11].

5   Discussion

We have presented two new methods for detecting recombination patterns in sets of haploid sequences, phrased in terms of the problem of "coloring" sequences to reflect their ancestral histories. The first method uses parsimony principles to separately infer a minimum number of recombination sites capable of explaining the data and then color sequences between sites to denote likely ancestry. The second uses a discrete optimization method similar to those used in sequence alignment to find the most probable parse of a set of sequences in terms of haplotypes. We have also incorporated that technique into an iterative method using alternating discrete and continuous optimization to simultaneously infer haplotype frequencies and color sequences optimally given the inferred frequencies.

Each method has strengths that might make it more appropriate for certain cases. The block-color method creates a general algorithmic framework in which optimal solutions can be found efficiently for a range of possible tests of

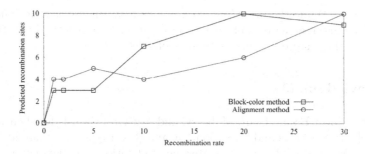

Fig. 7. Predicted recombination sites versus coalescent parameter ρ for simulated data.

compatibility of pairs of sites with the assumption of no recombination. It thus might provide a useful general method for the evaluation and application of new statistical tests for recombination. The alignment method solves optimally for a single objective function, making it potentially more useful in testing and applying a unified probabilistic model of sequence evolution. It also does not rely on a prior assumption that haplotypes have a block structure and therefore might be more useful for testing hypotheses about the existence of such a block structure, finding instances in which it is not conserved, or processing data from organisms or gene regions that do not exhibit haplotype blocks. The EM algorithm may be independently useful in estimating haplotype block frequencies.

We can consider possible generalizations of the methods described above. It may be worthwhile to try other tests for deriving pair-wise constraints for the block-color method. As Hudson and Kaplan [11] note, the number of recombination events detected by the four-gamete constraint may be substantially smaller than the actual number of recombination sites, especially for small population sizes; their method finds a maximally parsimonious explanation for the data and will therefore miss instances in which recombination has occurred but has not yielded a violation of the four-gamete test or in which the population examined does not contain sufficient examples to demonstrate such a violation. The Gusfield [8] method for diploid data can be expected to be similarly conservative, suggesting the value of pursuing more sensitive tests of recombination. The alignment model could be extended to handle position-specific mutation weights or recombination probabilities when an empirical basis is available for choosing them. It might also be possible to adapt the EM algorithm to infer mutation or recombination probabilities at the same time as it infers haplotype frequencies. In addition to providing greater versatility and accuracy, automating inference of the mutation and recombination rates might substantially improve ease-of-use.

Making the best use of sequencing data for understanding human diversity and applying that understanding to association studies will require first developing a more complete picture of the processes involved; second, building and validating statistical and probabilistic models that reliably capture that picture; and third, developing methods that can best interpret the available data given those models. While the preceding work is meant to suggest avenues and techniques for pursuing the overall goal of applying human genetic diversity data, all three of those steps remain far from resolved.

Acknowledgments

We thank Vineet Bafna, Clark Mobarry, and Nathan Edwards for their insights into block identification and Hagit Shatkay and Ross Lippert for their advice and guidance. We also thank Michael Waterman and various anonymous referees for their advice on the preparation of this manuscript.

References

1. Baum, L.E., Petrie, T., Soules, G., and Weiss, N. A maximization technique occurring in the statistical analysis of probabilistic functions of Markov chains. Annals Math. Stat., 41, 164–171, 1970.
2. Clark, A. G. Inference of Haplotypes from PCR-amplified samples of diploid populations. Mol. Biol. Evol., 7, 111–122, 1990.
3. Daly, M.J., Rioux, J.D., Schaffner, S.F., Hudson, T.J., and Lander, E.S. High-resolution haplotype structure in the human genome. Nature Gen., 29, 229–232, 2001.
4. Dempster, A.P., Laird, N.M., and Rubin, D.B. Maximum likelihood from incomplete data via the EM algorithm. J. Royal Stat. Soc. B, 39, 1–38, 1977.
5. Edmonds, J. Paths, trees, and flowers. Canad. J. Math., 17, 449–467, 1965.
6. Fearnhead, P. and Donnelly, P. Estimating recombination rates from population genetic data. Genetics, 159:1299–1318, 2001.
7. Gusfield, D. Efficient algorithms for inferring evolutionary history. Networks, 21:19–28, 1991.
8. Gusfield, D. Haplotyping as perfect phylogeny: Conceptual framework and efficient solutions. In Proc. 6th Intl. Conf. Comp. Biol., RECOMB'02, 166–175, 2002.
9. Hein, J. A heuristic method to reconstruct the history of sequences subject to recombination. J. Mol. Evol., 20, 402–411, 1993.
10. Hudson, R.R. Properties of the neutral allele model with intergenic recombination. Theoret. Pop. Biol., 23, 183–201, 1983.
11. Hudson, R.R. and Kaplan, N.L. Statistical properties of the number of recombination events in the history of a sample of DNA sequences. Genetics, 111, 147–164, 1985.
12. International Human Genome Sequencing Consortium. Initial sequencing and analysis of the human genome. Nature, 409, 860–921, 2001.
13. Jeffreys, A.J., Kauppi, L., and Neumann, R. Intensely punctate meiotic recombination in the class II region of the major histocompatibility complex. Nature Gen., 29, 217–222, 2001.
14. Johnson, G.C.L., Esposito, L., Barratt, B.J., Smith, A.N., Heward, J., Di Genova, G., Ueda, H., Cordell, H.J., Eaves, I.A., Dudbridge, F., Twells, R.C.J., Payne, F., Hughes, W., Nutland, S., Stevens, H., Carr, P., Tuomilehto-Wolf, E., Tuomilehto, J., Gough, S.C.L., Clayton, D.G., and Todd, J.A. Haplotype tagging for the identification of common disease genes. Nature Gen., 29, 233–237, 2001.

15. Kececioglu, J. and Gusfield, D. Reconstructing a history of recombinations from a set of sequences. Disc. Appl. Math., 88, 239–260, 1998.
16. Kimura, M. Theoretical foundations of population genetics at the molecular level. Theoret. Pop. Biol., 2, 174–208, 1971.
17. Maynard Smith, J. Analyzing the mosaic structure of genes. J. Mol. Evol., 34, 126–129, 1992.
18. Maynard Smith, J. and Smith, N.H. Detecting recombination from gene trees. Mol. Biol. Evol., 15, 590–599, 1998.
19. Nickerson, D. A., Taylor, S. L., Fullerton, S. M., Weiss, K. M., Clark, A. G., Stengrd, J. H., Salomaa, V., Boerwinkle, E., and Sing, C. F. Sequence diversity and large-scale typing of SNPs in the human apolipoprotein E gene. Gen. Res., 10, 1532–1545, 2000.
20. Patil, N., Berno, A.J., Hinds, D.A., Barrett, W.A., Doshi, J.M., Hacker, C.R., Kautzer, C.R., Lee, D.H., Marjoribanks, C., McDonough, D.P., Nguyen, B.T.N., Norris, M.C., Sheehan, J.B., Shen, N., Stern, D., Stokowski, R.P., Thomas, D.J., Trulson, M.O., Vyas, K.R., Frazer, K.A., Fodor, S.P.A., and Cox, D.R. Blocks of limited haplotype diversity revealed by high-resolution scanning of human chromosome 21. Science, 294, 1719–1723, 2001.
21. Posada, D., and Crandall, K.A. Evaluation of methods for detecting recombination from DNA sequences: computer simulations. Proc. Natl. Acad. Sci. USA, 98, 13757–13762, 2001.
22. Sawyer, S. Statistical tests for detecting gene conversion. Mol. Biol. Evol., 6, 526–536, 1989.
23. Spang, R., Rehmsmeier, M., and Stoye, J. Sequence database search using jumping alignments. In Proc. Intel. Sys. Mol. Biol., ISMB'00, 367–375, 2000.
24. Stephens, M., Smith, N.J., and Donnelly, P. A new statistical method for haplotype reconstruction from population data. Am. J. Hum. Gen., 68, 978–989, 2001.
25. Venter, J.C., Adams, M.D., Myers, E.W., et al. The sequence of the human genome. Science, 291, 1304–1351, 2001.
26. Wang, L. Zhang, K., and Zhang, L. Perfect phylogentic networks with recombination. J. Comp. Biol., 8, 69–78, 2001.
27. Weiler, G. F. Phylogenetic profiles: a graphical method for detecting genetic recombinations in homologous sequences. Mol. Biol. Evol., 15, 326–335, 1998.
28. Wiuf, C., Christensen, T., and Hein, J. A simulation study of the reliability of recombination detection methods. Mol. Biol. Evol., 18,1929–1939, 2001.
29. Zhang, K., Deng, M., Chen, T., Waterman, M. S., and Sun, F. A dynamic programming algorithm for haplotype block partition. Proc. Natl. Acad. Sci. USA, 99, 7335–7339, 2002.

Finding Signal Peptides in Human Protein Sequences Using Recurrent Neural Networks

Martin Reczko[1], Petko Fiziev[2], Eike Staub[2], and Artemis Hatzigeorgiou[3]

[1] Synaptic Ltd., Science and Technology Park of Crete
P.O. Box 1447, 711 10 Voutes Heraklion, Greece
[2] metaGen Pharmaceuticals GmbH, Oudenader Str.16, D-13347 Berlin, Germany
[3] Department of Genetics, University of Pennsylvania, School of Medicine
Philadelphia, PA 19104-6145, USA

Abstract. A new approach called Sigfind for the prediction of signal peptides in human protein sequences is introduced. The method is based on the bidirectional recurrent neural network architecture. The modifications to this architecture and a better learning algorithm result in a very accurate identification of signal peptides (99.5% correct in fivefold cross-validation). The Sigfind system is available on the WWW for predictions (http://www.stepc.gr/ synaptic/sigfind.html).

1 Introduction

In the last few years the complete or nearly the complete sequence for a large number of genomes including also the human genome has been determined. One of the expected consequences from sequencing these genomes is the discovery of new drugs. The sorting of subcellular proteins is a fundamental aspect of finding such drugs, because such proteins have the potential to leave the cell and move through the organism.

In humans secretory proteins account for about one-tenth of the human proteome. As many functional characteristics of proteins can be correlated with more or less well-defined linear motifs in their aminoacid sequences, also in this case sorting depends on *signals* that can already be identified in the primary structure of a protein.

Signal peptides are N-terminal extensions on the mature polypeptide chain. They are cleaved from the mature part of the protein at the *cleavage site*. The basic design of signal peptides is constituted by three distinct regions: a short positively charged N-terminal domain (N-region) a central hydrophobic stretch of 7-15 residues (H-region) and a more polar C-terminal part of 3-7 residues that defines a processing site for the signal peptidase enzyme (C-region) [18].

Prediction of signal peptides has a long history starting back in early eighties by simple algorithms based on hydrophobicity calculations [8] and the later use of a positional weight matrix for the cleavage site [17]. The application of more sophisticated methods such as rule-based systems [7], neural networks (NNs) [13], hidden Markov models (HMMs) [14] and linear programming [9] has increased the levels of accuracy considerably. A more detailed review can be found in [12].

R. Guigó and D. Gusfield (Eds.): WABI 2002, LNCS 2452, pp. 60–67, 2002.

In this paper a new method for defining signal peptides is introduced. The method is based on a novel neural network architecture called bidirectional recurrent neural nets (BRNN) that was recently introduced by [2]. The BRNN architecture is used for sequential learning problems with sequences of finite lengths. Other recurrent NNs can only model causal dynamical system, where the output of the system at a certain time does not depend on future inputs. The BRNN model has a symmetric memory for storing influences of past and future inputs to an output at time t.

This non causal behavior is useful for the prediction of properties of biosequences, since most properties are influenced by elements of those sequences both upstream and downstream of that property. The concept of time does not apply to this class of sequential learning problems, so we know about the 'future' sequence following a sequence position t and the system is not violating any physical causality.

We modify the architecture and introduce the combination with a more efficient learning algorithm that produces networks with better generalization performance in less computational time. In the results section we present a comparison with the methods available on the Internet and we are documenting the achieved improvements.

2 Material and Methods

The data used here is the same data as was used for the first SignalP prediction system [13] that was made available by the authors[1]. That data was derived from SWISS-PROT version 29 [1]. For the removal of redundancy in their data set, Nielsen et. al. [13] eliminated any pair of sequences with more than 17 identical residues in a local alignment.

From this non-redundant data set we use only the human sequences which contain 416 signal peptides and the N-terminal part of 97 cytoplasmic and 154 nuclear proteins as negative examples.

2.1 Neural Network Architecture

The neural network architecture employed for the task is a modified version of the BRNN as described in [2]. The BRNN architecture is summarized here and the modifications are introduced.

The state vectors are defined as F_t and B_t in $\mathbb{R}^n$ and $\mathbb{R}^m$ and contain the context information while reading forward and backwards, respectively. They are calculated as:

$$F_t = \phi(F_{t-1}, U_t) \qquad \text{and} \qquad (1)$$

$$B_t = \beta(B_{t+1}, U_t) \qquad (2)$$

---

[1] The SignalP datasets are available at: ftp://virus.cbs.dtu.dk/pub/signalp

where $\phi()$ and $\beta()$ are nonlinear transition functions realized by the multilayer perceptrons (MLPs) $\aleph_\phi$ and $\aleph_\beta$ and the vector $U_t \in \mathbb{R}^k$ is the input at time $t \in [1, T]$. The state vectors are initialized with $F_0 = B_{T+1} = 0$.

The output $Y_t \in \mathbb{R}^s$ is calculated after the calculation of F_t and B_t has finished using

$$Y_t = \eta(F_t, B_t, U_t) \tag{3}$$

where $\eta()$ is again realized by a MLP $\aleph_\eta$. In the realization of [2], the MLP $\aleph_\eta$ contains one hidden layer that is connected to U_t and the state units in F_t and B_t are connected directly to the output layer.

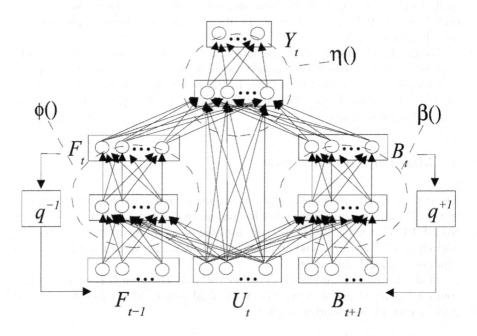

Fig. 1. The modified bidirectional recurrent neural network architecture. The input at time t is applied at U_t. Y_t contains the output vector. F_t is the forward state vector, and B_t the backward state vector. The shift operator q^{-1} copies the last F_t state vector while reading the input sequence forwards while the operator q^{+1} copies the last B_t state vector while reading the input backwards.

In our modified BRNN, the state units are not connected directly to the output, but to the hidden layer. It is reasonable to assume that several activation patterns occurring on the combination of the forward and backward state units are not linearly separable and thus cannot be distinguished without the additional transformation performed by this hidden layer. There is an implicit asymmetry in the amount of context information accumulated in the state vectors. At the start of the sequence ($t = 1$), the forward state vector F_1 contains information about only one input vector U_1 whereas the backward state vector B_1 has accumulated all information of the complete sequence. At the end of the sequence ($t = T$), the opposite situation occurs. The processing of the state vectors with a hidden layer can help detecting these asymmetric situations and process the more relevant activations.

The architecture is shown in figure 1 where the shift operator q^{-1} to copy the forward state vector is defined as $X_{t-1} = q^{-1}X_t$ and the inverse shift operator q^{+1} to copy the backward state vector is defined as $X_{t+1} = q^{+1}X_t$.

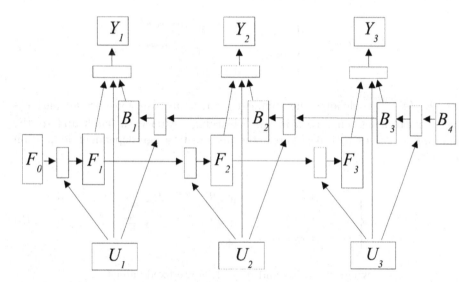

Fig. 2. The unfolded feedforward structure of a BRNN processing an input sequence U_t with length $T = 3$. First, the forward states $F_0, \ldots, F_3$ are calculated sequentially from left to right, while the backward states $B_4, \ldots, B_1$ are calculated beginning with the end of the sequence and continuing from right to left. After all F_t and B_t are processed, the output sequence Y_t can be calculated.

2.2   Learning Algorithm

To adapt the weight parameters in the MLPs $\aleph_\eta, \aleph_\phi$ and $\aleph_\beta$ the recurrent NN is unfolded in time and the backpropagation algorithm [16] is used to calculate the gradient for all weights. The unfolding is illustrated in figure 2. This transformation of a recurrent NN into a equivalent feedforward NN is called backpropagation through time and was first described in [11] and the application of backpropagation learning to these networks was introduced in [16].

The gradient for a weight is summed up for all training sequences and each shared position within each sequence.

One well known obstacle when training recurrent NNs is the problem of the vanishing gradients [3]. In most cases, the gradient decays exponentially with the number of layers in a NN. Since the unfolded network has at least as many layers as the length of the input sequences, there is a limit for storing features that might be relevant for the prediction of a property and are separated by longer sequences from that property.

A learning algorithm that performs well without depending on the magnitude of weight gradients is the resilient backpropagation (RPROP) algorithm [15]. Only the sign of the derivative is considered to indicate the direction of the weight update. The size of the weight change is determined by a weight-specific update-value $\triangle_{ij}^{(epoch)}$ and defined as

$$\triangle w_{ij}^{(epoch)} = \begin{cases} -\triangle_{ij}^{(epoch)} & , \quad \text{if } \frac{\partial E}{\partial w_{ij}}^{(epoch)} > 0 \\ +\triangle_{ij}^{(epoch)} & , \quad \text{if } \frac{\partial E}{\partial w_{ij}}^{(epoch)} < 0 \\ 0 & , \quad \text{else} \end{cases} \tag{4}$$

where $\frac{\partial E}{\partial w_{ij}}^{(epoch)}$ denotes the summed gradient information over all patterns of the pattern set (one *epoch* in batch learning). After each batch update, the new update-values $\triangle_{ij}(epoch)$ are determined using a sign-dependent adaptation process.

$$\triangle_{ij}^{(epoch)} = \begin{cases} \eta^+ * \triangle_{ij}^{(epoch-1)} & , \quad \text{if } \frac{\partial E}{\partial w_{ij}}^{(epoch-1)} * \frac{\partial E}{\partial w_{ij}}^{(epoch)} > 0 \\ \eta^- * \triangle_{ij}^{(epoch-1)} & , \quad \text{if } \frac{\partial E}{\partial w_{ij}}^{(epoch-1)} * \frac{\partial E}{\partial w_{ij}}^{(epoch)} < 0 \\ \triangle_{ij}^{(epoch-1)} & , \quad \text{else} \end{cases} \tag{5}$$

where $\eta^- = 0.5$ and $\eta^+ = 1.05$ are fixed values.

Every time the partial derivative of the corresponding weight w_{ij} changes its sign, which indicates that the last update was too big and the algorithm has jumped over a local minimum, the update-value $\triangle_{ij}^{(epoch)}$ is decreased by the factor η^-. If the derivative retains its sign, the update-value is slightly increased in order to accelerate convergence in shallow regions.

Using this scheme the magnitude of the update-values is completely decoupled from the magnitude of the gradients. It is possible that a weight with a very small gradient has a very large weight change and vice versa, if it just is in accordance with the topology of the error landscape. This learning algorithm converges substantially faster than standard backpropagation and allows all BRNN simulations to be run on standard PC hardware.

3   Results

A sequence of the first 45 residues of the N-terminal is used for the input vectors $U_t \in [0, 1]^{20}$ where each vector codes the residue in the standard one out of 20 coding. Each element of output sequence $Y_t \in [0, 1]$ is set to 1, if the residue at position t is part of the signal peptide before the cleavage site and 0 otherwise.

The set of the signal peptide sequences and the two sets of non-secretory sequences are each split into 5 subsets of equal size. Fivefold crossvalidation is used by training 5 networks on 3/5 of each set, using 1/5 of each set as a validation set to stop the training and 1/5 of each set as a test set to measure the performance. The sizes of the different layers in the BRNN are determined experimentally. The number of units in the forward and backward state vectors is 5 or 6, the number of hidden units feeding into the state vectors is between 10 and 13 and the number of hidden units feeding into the output layer is between 2 and 6.

The criterion to discriminate the signal peptides is optimized using the validation set. If a segment of more than 7 consecutive output activations meets the condition $Y_t > 0.75, t \in [1, 45]$, the input sequence is classified as a signal peptide.

The performance is measured by calculating the correlation coefficient, [10] the sensitivity *sens* and specificity *spec* defined as

$$sens = cp/(cp + fn) \tag{6}$$
$$spec = cp/(cp + fp), \tag{7}$$

where cp is the number of correct positive examples, fn is the number of false negative and fp is the number of false positive examples. The results are given in table 1. The average percentage of correctly predicted sequences is 99.55%

Table 1. Performance on the test sets in fivefold crossvalidation.

dataset	sensitivity	specificity	correlation
1	100.0	100.0	1.0
2	100.0	98.8	0.984
3	100.0	98.8	0.984
4	100.0	98.8	0.984
5	100.0	100.0	1.0
avg	100.0	99.3	0.99

Another test set is created by extracting human protein sequences from the release 39.25 of SWISS-PROT that have the feature keyword SIGNAL and that are not contained in the used training set. The non signal peptide proteins used in this set are full-length mRNAs with the N-terminus experimentally verified. Since the data set for training the SignalP2 system is not available, it is not guaranteed that all of these sequences are not novel to that system. For the performance measurements shown in table 2 and the general use of the Sigfind system, the predictions of the 5 networks are combined using a majority vote of the jury of these networks.

For scanning targets in complete genomes, the execution time of an algorithm should not be neglected. As shown in table 3 the Sigfind system compares very favorably with SignalP2.

Table 2. Performances on an independent testset.

system	sensitivity	specificity	correlation
signalp2 with NN and HMM	88.28	96.55	0.851
signalp2 with NN	93.24	92.83	0.856
signalp2 with HMM	90.99	94.83	0.857
signalp2 with NN or HMM	95.94	91.42	0.867
sigfind	98.64	93.59	0.918

Table 3. Execution times on a Sun Sparc.

system	222 signal peptides	210 non signal peptides
signalp2	300 seconds	251 seconds
sigfind	1.98 seconds	1.94 seconds

4 Availability

The sigfind system can be used for predictions using a WWW interface with the
URL http://www.stepc.gr/ synaptic/sigfind.html .

5 Discussion

The approach introduced here is shown to be very accurate for the identification of signal peptides in human protein sequences. The identification of signal
peptides has become of major importance in the analysis of sequenced data.
Programs like SignalP2 and Psort [7] are widely used for this purpose and produce in combination quite reliable predictions. The algorithm that we developed
here can be used as an computationally efficient, additional tool to find signal
peptides.

The method is designed to determine if the N-terminal part of a protein is
a signal peptide or not. However in many cases the complete N-terminal part
of the sequence is not experimentally verified but predicted in silico. Particularly in protein sequences derived from genomic data it is possible that the real
startcodon is not predicted correctly. The same occurs also for protein sequences
derived out of expressed sequence tags (ESTs). The use of Sigfind on the whole
protein might produce false positive predictions mostly on the location of trans-
membrane regions which has a similar structure as the signal peptides. For these
reasons we are planning to combine the Sigfind predictions with reliable start-
codon [4] and coding region predictions. For ESTs it would also be useful to
apply a frame shift correction software [6,5]. The modified BRNN algorithm described has very great potential for these tasks as well as for other biosequence
analysis problems.

References

1. Bairoch, A., Boeckmann, B.: The swiss-prot protein sequence data bank: current status. Nucleic Acids Res. 22 (1994) 3578–3580
2. Baldi, P., Brunak, S., Frasconi, P., Pollastri, G., Soda, G.: Bidirectional dynamics for protein secondary structure prediction. In: Sun, R., Giles, L. (eds.): Sequence Learning: Paradigms, Algorithms, and Applications. Springer Verlag (2000)
3. Bengio, Y., P.Simard, Frasconi, P.: Learning long-term dependencies with gradient descent is difficult. IEEE Trans. on Neural Networks, 5 (1994) 157–166
4. Hatzigeorgiou, A.: Translation initiation site prediction in human cDNAs with high accuracy. Bioinformatics, 18 (2002) 343–350
5. Hatzigeorgiou, A., Fizief, P., Reczko, M. Diana-est: A statistical analysis. Bioinformatics, 17 (2001) 913–919
6. Hatzigeorgiou, A., Papanikolaou, H., Reczko, M.: Finding the reading frame in protein coding regions on dna sequences: a combination of statistical and neural network methods. In: Computational Intelligence: Neural Networks & Advanced Control Strategies. IOS Press, Vienna (1999) 148–153
7. Horton, P., Nakai, K.: Better prediction of protein cellular localization sites with the k nearest neighbors classifier. In: ISMB (1997) 147–152
8. Kyte, J., Doolittle, R.: A simple method dor displaying the hydrophatic character of a protein. J. Mol. Biol., 157 (1982) 105–132
9. Ladunga, I.: Large-scale predictions of secretory proteins from mammalian genomic and est sequences. Curr. Opin. in Biotechnolgy, 11 (2000) 13–18
10. Mathews, B. W.: Comparison of the predicted and observed secondary structure of T4 phage lysozyme. Biochem. Biophys. Acta, Vol. 405 (1975) 442–451
11. Minsky, M., Papert, S.: Perceptrons: An Introduction to Computational Geometry. The MIT Press, Cambridge, Massachusetts (1969) 145
12. Nielsen, H., Brunak, S., von Heijne, G. Machine learning approaches for the prediction of signal peptides and other protein sorting signals. Protein Engineering, 12 (1999) 3–9
13. Nielsen, H., Engelbrecht, J., S.Brunak, von Heijne, G.: Identification of prokaryotic and eukaryotic signal peptides and prediction of their cleavage sites. Protein Engineering, 10 (1997) 1–6
14. Nielsen, H., Krogh, A.: Prediction of signal peptides and signal anchors by a hidden markov model. In: ISMB (1998) 122–130
15. Riedmiller, M., Braun, H.: A direct adaptive method for faster backpropagation learning: The RPROP algorithm. In: Ruspini, H., (ed.): Proceedings of the IEEE International Conference on Neural Networks (ICNN 93). IEEE, San Francisco (1993) 586–591
16. Rumelhart, D. E., Hinton, G. E., Williams, R. J.: Learning internal representations by error propagation. In: Rumelhart, D. E., McClelland, J. L. (eds.): Parallel Distributed Processing: Explorations in the microstructure of cognition; Vol. 1: Foundations. The MIT Press, Cambridge, Massachusetts (1986)
17. v. Heijne, G.: A new method for predicting signal sequence cleavage sites. Nucleid Acids Res., 14 (1986) 4683–4690
18. von Heijne, G.: Computer-assisted identification of protein sorting signals and prediction of membrane protein topology and structure. In: Advances in Computational Biology, volume 2, Jai Press Inc. (1996) 1–14

Generating Peptide Candidates from Amino-Acid Sequence Databases for Protein Identification via Mass Spectrometry

Nathan Edwards and Ross Lippert

Celera Genomics, 45 West Gude Drive, Rockville, MD
{Nathan.Edwards,Ross.Lippert}@Celera.Com

Abstract. Protein identification via mass spectrometry forms the foundation of high-throughput proteomics. Tandem mass spectrometry, when applied to a complex mixture of peptides, selects and fragments each peptide to reveal its amino-acid sequence structure. The successful analysis of such an experiment typically relies on amino-acid sequence databases to provide a set of biologically relevant peptides to examine. A key subproblem, then, for amino-acid sequence database search engines that analyze tandem mass spectra is to efficiently generate all the peptide candidates from a sequence database with mass equal to one of a large set of observed peptide masses. We demonstrate that to solve the problem efficiently, we must deal with substring redundancy in the amino-acid sequence database and focus our attention on looking up the observed peptide masses quickly. We show that it is possible, with some preprocessing and memory overhead, to solve the peptide candidate generation problem in time asymptotically proportional to the size of the sequence database and the number of peptide candidates output.

1   Introduction

Reliable methods for identifying proteins form the foundation of proteomics, the large scale study of expressed proteins. Of the available technologies for protein identification, mass spectrometry is the most suitable for high-throughput automation and analysis.

Tandem mass spectrometry, in which peptides are selected and fragmented, reveals peptides' amino-acid sequence structure. In a typical high-throughput setting, a complex mixture of unknown proteins is cut into peptides using a digestion enzyme such as trypsin; fractionated into reduced complexity samples on the basis of some physical or chemical property such as hydrophobicity; and then a tandem mass spectrum is taken for all the observed peptides in each fraction. The end result of such an experiment is a set of a few hundred to a few thousand tandem mass spectra, each of which represents a peptide of about 6-20 amino acid residues. Typically, amino-acid sequences of 8-10 residues carry sufficient information content to determine the protein from which the peptide is derived, at least up to significant homology. This experimental protocol can

R. Guigó and D. Gusfield (Eds.): WABI 2002, LNCS 2452, pp. 68–81, 2002.

reliably identify hundreds of proteins from a complex mixture in a couple of hours of instrument time.

The traditional approach to automated protein identification using mass spectrometry is not nearly as suited to a high-throughput environment. Peptide mass fingerprinting, see Pappin, Hojrup, and Bleasby [11], James et. al. [8], Cottrell and Sutton [4], and Pappin [10], considers the mass spectrum generated by the (typically tryptic) peptides of a single protein. The mass of each peptide, by itself, carries little information content about the original protein, it is only by the simultaneous observation of the masses of many peptides from a protein that reliable protein identification can be carried out. Furthermore, if peptides from many different proteins are observed in the spectrum, the task of reliably determining which peptide ions belong to the same protein becomes near impossible. This means that significant wet lab work must be done in order to reduce the number of proteins in each fraction to as near to one as possible, which ultimately makes peptide mass fingerprinting unsuitable for high-throughput protein identification.

The analysis of each tandem mass spectrum can be done in a variety of ways. Given a good quality spectrum and a peptide that fragments nicely, the amino-acid sequence of the peptide can often be determined *de novo*, merely by looking at the mass differences between peaks in the spectrum. Many algorithmic approaches have been proposed for *de novo* tandem mass spectrum interpretation, see Taylor and Johnson [14], Dancik et. al. [5], Pevzner, Dancik and Tang [13], and Chen et. al. [2] for some examples. In reality, however, in a high-throughput setting, the number of spectra that can be reliably analyzed *de novo* is small. The most reliable approach for analyzing high-throughput tandem mass spectra for protein identification uses amino-acid sequence databases to suggest peptide candidates which are then ranked according to how well they explain the observed spectrum. A number of papers have been published describing tandem mass spectra database search engines and peptide candidate ranking schemes. Some of these approaches have even been commercialized. The first widely adopted program for tandem mass spectra identification via sequence database search was SEQUEST, based on work by Eng, McCormack and Yates [6]. Another successful commercial product, Mascot, is based on work by Perkins et. al. [12]. Bafna and Edwards [1] published a probability based model for scoring and ranking peptide candidates that could be tuned to the peptide fragmentation propensities of different mass spectrometry technologies. Unfortunately, all of this work focuses primarily on determining the correct peptide candidate from among the possibilities suggested by the sequence database. In this paper, we address the problem at the core of all tandem mass spectrum identification sequence search engines, enumerating all appropriate peptide candidates from an amino-acid sequence database efficiently. For each tandem mass spectrum, we require peptide candidates that match the mass of the peptide that was selected for fragmentation.

2   Generating Peptide Candidates

2.1   Problem Formulation and Notation

Peptide Candidate Generation with Integer Weights [PCG(Int)].
Given a string σ of length n over an alphabet $\mathcal{A}$ of size m; a positive integer
mass $\mu(a)$ for each $a \in \mathcal{A}$; and k integer mass queries $M_1, \ldots, M_k$, enumerate
all (distinct) pairs (i, ω), where $1 \leq i \leq k$ and ω is a substring of σ, such that

$$\sum_{j=1}^{|\omega|} \mu(\omega_j) = M_i.$$

For convenience, we denote the maximum relevant substring mass by $M_{\max} = \max_i M_i$ and the minimum relevant substring mass by $M_{\min} = \min_i M_i$. Further,
since the set of query masses may contain repeated values, we define $R_{\max}$ to be
the maximum number of repeats.

While we will concentrate primarily on the integer mass version of the peptide
candidate generation problem, we must not lose sight of the real mass version
which we must eventually implement. The real mass version of the problem
defines real masses for each alphabet symbol and supports real query masses
with lower and upper mass tolerances.

Peptide Candidate Generation with Real Weights [PCG(Real)]. Given
a string σ of length n over an alphabet $\mathcal{A}$ of size m; a positive real mass $\mu(a)$ for
each $a \in \mathcal{A}$; and k positive real mass queries with positive lower and upper tolerances $(M_1, l_1, u_1), \ldots, (M_k, l_k, u_k)$, enumerate all (distinct) pairs (i, ω), where
$1 \leq i \leq k$ and ω is a substring of σ, such that

$$M_i - l_i \leq \sum_{j=1}^{|\omega|} \mu(\omega_j) \leq M_i + u_i.$$

We define $M_{\max}$ and $M_{\min}$ to be the minimum and maximum relevant substring
mass, $M_{\max} = \max_i (M_i + u_i)$ and $M_{\min} = \min_i (M_i - l_i)$ and define $O_{\max}$ to be
the maximum number of $[M_i - l_i, M_i + u_i]$ intervals to overlap any mass.

2.2   Application to Peptide Identification
via Tandem Mass Spectrometry

In order to focus our analysis of different algorithms for generating peptide candidates, we first describe some of the issues that need be resolved before a peptide
candidate generation technique can be used as part of a peptide identification
amino-acid sequence search engine.

First, most amino-acid sequence databases do not consist of a single sequence
of amino-acids, they instead contain many distinct amino-acid sequences, representing many proteins. Our peptide candidates should not straddle amino-acid

sequence boundaries. This is easily dealt with by introducing an addition symbol to the alphabet $\mathcal{A}$ with a mass guaranteed to be larger than the biggest query.

We usually require significantly more information about each peptide candidate than is represented by its sequence alone. A peptide candidate's *protein context* minimally consists of a reference, such as an accession number, to the sequence database entry containing it, and its position within the sequence. Often, the protein context also contains flanking residues, for checking enzymatic digest consistency quickly, and taxonomy information, for restricting candidates to a species of interest.

Since our experimental protocol cuts each protein with a digestion enzyme, we usually require our peptide candidates have one or both ends consistent with the putative digest cut sites. Trypsin, very commonly used in mass spectrometry applications, cuts proteins immediately after a lysine (K) or arginine (R) unless the next residue is a proline (P). In order to determine whether a peptide candidate is consistent with trypsin, for example, we must use the protein context to examine the symbols before and after a peptide candidate.

If we will always require that both ends of the peptide candidate be consistent with the enzymatic digest, we can improve the performance of any algorithm for the peptide candidate generation problem significantly. However, in practice, we have observed that a significant number of the tandem mass spectra generated in the high throughput setting do not have enzymatic digest consistent endpoints, and requiring this property of our peptide candidates significantly affects our ability to successfully interpret many spectra. As such, we currently do not impose this constraint at the algorithmic level, instead we filter out candidates without digestion consistent endpoints at candidate generation time. Section 6 deals with this issue further.

Peptide redundancy must be carefully considered too. It does us no good to score the same peptide candidate against a spectrum multiple times, but if a peptide scores well, we must be able list all the proteins that it occurs in. If our algorithm does not explicitly eliminate redundant peptide candidates, we can choose to score them multiple times, or store all scored candidates in some data-structure and only score candidates the first time they are observed. Section 4 deals with this issue further.

Post-translational modifications provide yet another consideration that affects algorithmic implementation decisions. If we permit the online specification of the amino-acid mass table in order to support the generation of peptide candidates with a particular post-translational modification, we must discard any algorithm which precomputes peptide candidate masses, unless we can somehow correct for amino-acids with modified masses.

Modeling post-translational modifications becomes more difficult when we must generate peptide candidates in which some amino-acid symbols have multiple masses. In this setting, we can still use the peptide candidate generation formulation above but we must use additional query masses and do additional checks on the generated candidates. For each tandem mass spectrum, we generate a set of mass queries that correct for the presence of a certain number of residues being translated to a particular modified amino-acid, and check each returned candidate-query mass pair to ensure that the required number of ambiguous amino-acid symbols is present.

2.3   Implications for Peptide Candidate Generation

We consider amino-acid sequence databases of a few megabytes to a few gigabytes of sequence, over an alphabet of size 20. The alphabet is small enough that we can look up the mass table for each symbol in the alphabet in constant time.

A typical set of tandem mass spectra from a single high throughput run consists of hundreds to thousands of spectra, so we expect hundreds to tens of thousands of distinct query masses. Furthermore, these query masses are typically constrained between 600-3000 Daltons with a 2 Dalton tolerance on acceptable peptide candidates. A 2 Dalton tolerance is typically necessary since mass spectrometers select ions for further fragmentation with much less accuracy than they measure the mass of the same ions, and we must consider any peptide candidate that might have been selected, not only those that have been measured. For our peptide candidate generation problem, this implies that the query masses occupy much of the feasible mass range. The importance of this consideration will become clear in Section 5.

We consider the amino-acid sequence database to be provided offline, and that it may be pre-processed at will, the cost amortized over many identification searches. On the other hand, we assume that the alphabet mass function and the query masses are provided only at runtime and any preprocessing based on weights must be accounted for in the algorithm run time.

3   Simple Algorithms

3.1   Linear Scan

Suppose initially that we have a single query mass, M. Finding all substrings of σ with weight M involves a simple linear scan. The algorithm maintains indices b and e for the beginning and end of the current substring and accumulates the current substring mass in $\widehat{M}$. If the current mass is less than M, e is incremented and $\widehat{M}$ increased. If the current mass is greater than $\widehat{M}$, b is incremented and $\widehat{M}$ decreased. If the current mass equals M, then $\sigma_{b...e}$ is output. See Algorithm 1 for pseudo-code.

This algorithm outputs all peptide candidates that match the query mass in $O(n)$ time. **Linear Scan** requires no preprocessing and requires only the string itself be stored in memory. The recent work of Cieliebak, Erlebach, Lipták, Stoye, and Welzl [3] demonstrates how to generalize **Linear Scan** by creating blocks of contiguous alphabet symbols and changing the pointers b and e in block size increments.

3.2   Sequential Linear Scan

The **Sequential Linear Scan** algorithm for the peptide candidate generation problem merely calls **Linear Scan** for each query mass M_i, $i = 1, \ldots, k$. **Sequential Linear Scan** requires no preprocessing of the string and requires

---

Algorithm 1 Linear Scan

$b \leftarrow 1$, $e \leftarrow 0$, $\widehat{M} \leftarrow 0$.
  while $e < n$ **or** $\widehat{M} \geq M$ **do**
    if $\widehat{M} = M$ **then**
      Output $\sigma_{b...e}$.
    if $\widehat{M} < M$ **and** $e < n$ **then**
      $e \leftarrow e + 1$, $\widehat{M} \leftarrow \widehat{M} + \mu(\sigma_e)$.
    else $\{\widehat{M} \geq M\}$
      $\widehat{M} \leftarrow \widehat{M} - \mu(\sigma_b)$, $b \leftarrow b + 1$.

---

memory only to store the sequence data and query masses. The algorithm takes $O(nk)$ time to enumerate all peptide candidates that match the k query masses. The algorithm not only outputs redundant peptide candidates, it computes their mass from scratch every time it encounters them. On the other hand, the algorithm can easily keep track of protein context information as it scans the sequence database.

3.3   Simultaneous Linear Scan

Unlike **Sequential Linear Scan**, **Simultaneous Linear Scan** solves the peptide candidate generation problem with a single linear scan of the sequence database. For every peptide candidate start position, we generate all candidates with mass between the smallest and largest query masses, and check, for each one, whether or not there is a query mass that it matches. Assuming positive symbol masses and query masses bounded above by some constant, there are at most L candidates that must be considered at each start position. If we pre-sort the query masses, collapsing identical queries into a list associated with a single table entry, we can look up all queries corresponding to a particular substring mass in $O(R_{\max} \log k)$ time. Therefore, we can crudely estimate the running time of **Simultaneous Linear Scan** as $O(k \log k + nLR_{\max} \log k)$.

Theoretically, while it is possible to construct an input for the peptide candidate generation problem that requires the consideration of this many candidates, such an input will never occur in practice. Assuming typical amino-acid frequencies and a 3000 Dalton maximum query, the average number of candidates that must be considered at each start position, or L_{ave}, is about 27, much less than the worst case L of 54. Table 1(a) provides L_{ave} for a variety of publicly available sequence databases. For the case when enzymatic digest consistent endpoints are required, many fewer candidates need to be generated. Table 1(b) gives L_{ave} for tryptic peptide candidates.

In order to take advantage of this, **Simultaneous Linear Scan** must examine only the peptide candidates with masses up to the maximum query mass $M_{\max}$. Algorithm 2 implements **Simultaneous Linear Scan**, achieving a running time bound of $O(k \log k + nL_{\text{ave}}R_{\max} \log k)$. Clearly, as k gets large, this approach will do much better than **Sequential Linear Scan**.

Table 1. Average (Maximum) number of (a) candidates (b) tryptic candidates per start position for various values of $M_{\max}$.

	Sequence Database	500 Da	1000 Da	2000 Da	3000 Da
	SwissPROT	4.491(8)	9.008(17)	18.007(32)	27.010(48)
(a)	TrEMBL	4.492(8)	9.004(17)	18.006(33)	27.010(49)
	GenPept	4.491(8)	9.004(17)	18.005(33)	27.006(49)
	SwissPROT	1.443(4)	2.046(7)	3.137(12)	4.204(18)
(b)	TrEMBL	1.418(4)	2.022(7)	3.066(14)	4.189(19)
	GenPept	1.427(4)	2.036(7)	3.101(14)	4.195(22)

Algorithm 2 Simultaneous Linear Scan

Sort the query masses $M_1, \ldots, M_k$.
$b \leftarrow 1$, $e \leftarrow 0$, $\widehat{M} \leftarrow 0$.
while $b \leq n$ **do**
  while $\widehat{M} \leq M_{\max}$ **and** $e < n$ **do**
    $e \leftarrow e + 1$, $\widehat{M} \leftarrow \widehat{M} + \mu(\sigma_e)$.
  $Q \leftarrow$ **QueryLookup**$(\widehat{M})$.
  Output $(i, \sigma_{b\ldots e})$ for each $i \in Q$.
  $b \leftarrow b + 1$, $e \leftarrow b - 1$, $\widehat{M} \leftarrow 0$.

Like **Sequential Linear Scan**, **Simultaneous Linear Scan** does no preprocessing of the amino-acid sequence database and requires no additional memory above that required for the sequence and the queries. **Simultaneous Linear Scan** generates redundant peptide candidates and can easily provide protein context for each peptide candidates.

4   Redundant Candidate Elimination

Typical amino-acid sequence databases, particularly those that combine sequences from different sources, contain some entries with identical sequences. Fortunately, these redundant entries can be efficiently eliminated by computing a suitable hash on each sequence. Substring redundancy, however, is not as easy to deal with. When all peptide candidates of mass between 600 and 4000 Daltons are enumerated from the publicly available GenPept amino-acid sequence database, over 60% of the peptide candidates output have the same sequence as a candidate already seen. We will denote the compression in the number of candidates that must be considered when this redundancy is eliminated by the substring density, ρ. Figure 1 plots the substring density of various publicly available amino-acid sequence databases for candidates (and tryptic candidates) with mass between 600 and $M_{\max}$, where $M_{\max}$ varies from 610 to 4000 Daltons. Notice that ρ is quite flat for large $M_{\max}$, suggesting substring density significantly less than 1 is not merely an artifact of the length of the substrings considered.

Notice also that the carefully curated SwissPROT sequence database exhibits much greater substring density than the others. This probably reflects

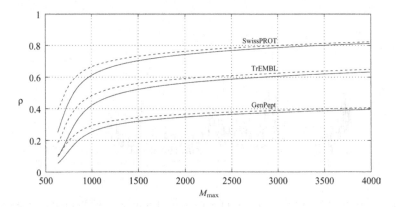

Fig. 1. Substring density of all candidates (solid line) and tryptic candidates (dashed line) with mass between 600 and $M_{\max}$ Daltons.

the curation policy of SwissPROT to collapse polymorphic and variant forms of the same protein to a single entry with feature annotations. If each variant form of a protein was a distinct entry in the database, then its substring density would decrease significantly. On the other hand, GenPept, which contains protein sequence data from many sources and is not hand curated, has much lower substring density. We claim that substring density is not merely inversely proportional to database size, but that it is increasingly difficult to find appropriate ways to collapse similar, but not identical, entries in ever larger sequence databases.

4.1   Suffix Tree Traversal

Suffix trees provide a compact representation of all distinct substrings of a string, effectively eliminating redundant peptide candidates. We discuss the salient properties of suffix trees here, see Gusfield [7] for a comprehensive introduction. Suffix trees can be built in time and space linear in the string length. Once built, all distinct substrings are represented by some path from the root of the suffix tree. Having constructed a suffix tree representation of our sequence database, a depth first traversal of the suffix tree that examines substrings on paths from the root with mass at most the maximum query mass generates all sequence distinct peptide candidates of our sequence database.

As the depth first traversal of the suffix tree proceeds, each candidate substring must be checked to see whether it corresponds to a query mass. As before, we bound the worst case performance of query mass lookup by $O(R_{\max} \log k)$. This results in a running time bound of $O(k \log k + \rho n L_{\mathrm{ave}} R_{\max} \log k)$.

The additional memory overhead of the suffix tree is the only downside of this approach. A naive implementation of the suffix tree data-structure can require as much as $24n$ additional memory for a string of length n. However, as suffix trees have found application in almost every aspect of exact string matching, a great

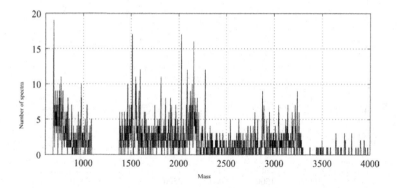

Fig. 2. Overlap plot for 1493 query masses from Finnigan LCQ tandem mass spectrometry experiment.

deal of effort has been made to find more compact representations. For example, Kurtz [9] claims a memory overhead of approximately $8n$ for various biological databases. This degree of compression can also be achieved, in our experience, by using a suffix array data-structure, which represents the same information as a suffix tree, but in a typically more compact form. Again, see Gusfield [7] for an introduction. In theory, a depth first traversal of a suffix array should be quite a bit more expensive than for a suffix tree, but in our experience, the linear scan through the suffix array that is necessary to implement the depth first traversal costs little more than pointer following in the suffix tree.

We note that using a suffix tree traversal for peptide candidate generation can make the protein context of peptide candidates more expensive to compute, particularly for context upstream of the candidate, necessary for determining if the peptide candidate is consistent with an enzymatic digest.

5   Query Lookup

Whether we consider the **Simultaneous Linear Scan** or **Suffix Tree Traversal** algorithm for peptide candidate generation, it should be clear that as the number of distinct query masses k gets large, our running time depends critically on the query mass lookup time. Once k gets too big for the **Sequential Linear Scan** algorithm to be appropriate, effective algorithms for the peptide candidate generation problem must no longer focus on the string scanning subproblem, but instead on the query mass lookup subproblem. Figure 2 plots the degree of overlap of a typical set of query mass intervals derived from a Finnigan LCQ mass spectrometer run containing 1493 tandem mass spectra. Notice the wide range of candidate masses that must be considered by the candidate generation algorithm.

We can obtain an $O(1)$ query lookup time per output peptide candidate by constructing a lookup table, indexed by the set of all possible candidate masses, containing the corresponding query mass indices. This is feasible for

---

Algorithm 3 Query Mass Lookup Table: Construction: Real Case

$\delta \leftarrow \min_i (l_i + u_i)$.
allocate and **initialize** a lookup table of size $N = \left\lfloor \frac{M_{\max}}{\delta} \right\rfloor + 1$.
for all query masses i **do**
    for all $j = \left\lfloor \frac{M_i - l_i}{\delta} \right\rfloor, \ldots, \left\lfloor \frac{M_i + u_i}{\delta} \right\rfloor$ **do**
        if $j\delta \leq M_i - l_i < (j+1)\delta$ **then**
            $T[j].L \leftarrow i$
        else if $j\delta \leq M_i + u_i < (j+1)\delta$ **then**
            $T[j].U \leftarrow i$
        else $\{M_i - l_i < j\delta$ **and** $M_i + u_i \geq (j+1)\delta\}$
            $T[j].O \leftarrow i$
for all $j = 0, \ldots, N - 1$ **do**
    sort the elements i of $T[j].L$ in increasing $M_i - l_i$ order.
    sort the elements i of $T[j].U$ in decreasing $M_i + u_i$ order.

Algorithm 4 Query Mass Lookup Table: Lookup: Real Case

---

input query mass $\widehat{M}$.
$j \leftarrow \left\lfloor \frac{\widehat{M}}{\delta} \right\rfloor$.
output i for each $i \in T[j].O$.
output i in order from $T[j].L$ until $M_i - l_i > \widehat{M}$.
output i in order from $T[j].U$ until $M_i + u_i < \widehat{M}$.

---

suitably small integer query masses, as the resulting data-structure will have size $O(M_{\max} + k)$.

We can readily adapt this approach to the real mass case. Algorithm 3 describes the procedure for building the lookup table. Having selected the discretization factor δ for the real masses to match the smallest query mass tolerance and allocating a table of the appropriate size, we iterate through the query masses to populate the table. For each query mass, we determine the relevant table entries that represent intervals that intersect with the query mass tolerance interval. The query mass intervals that intersect the interval $J = [j\delta, (j+1)\delta)$ represented by a table entry j do so in one of three ways: the lower endpoint falls inside J, in which case the query mass index is stored in $T[j].L$; the upper endpoint falls inside J, in which case the query mass index is stored in $T[j].U$; or the query mass interval completely overlaps J, in which case the query mass index is stored in $T[j].O$. The choice of δ ensures that no query mass interval is properly contained in J. Once all the table entries are populated, then for all table entries j, the query mass indices $i \in T[j].L$ are sorted in increasing $M_i - l_i$ order and similarly, the query mass indices $i \in T[j].U$ are sorted in decreasing $M_i + u_i$ order. The resulting lookup table has size bounded above by $O(M_{\max}/\delta + k \max_i (l_i + u_i)/\delta)$ and can be constructed in time bounded by $O(M_{\max}/\delta + k \max_i (l_i + u_i)/\delta + (M_{\max}/\delta)O_{\max} \log O_{\max})$.

For typical values, building this table is quite feasible. $M_{\max}$ is usually between 2000 and 4000 Daltons, while δ is usually about 2 Daltons. Even tens of thousands of query masses spread over 1000-2000 bins will not cause significant memory or construction time problems. Once built, the cost in memory and construction time is quickly amortized by the $O(1)$ query lookup time.

The bottom line is that it is possible to solve the peptide candidate generation problem (with integer masses) in time $O(M_{max} + k + \rho n L_{ave} R_{max})$ and that it is practical to implement this approach for real query masses. Note that the number of peptide candidates output by any correct algorithm is bounded above by $\rho n L_{ave} R_{max}$. This bound is quite tight, particularly when R_{max} is close to its lower bound of k/M_{max}. Ultimately, this means that we cannot expect to do much better than this running time bound when building the query lookup table is feasible.

6   Enzymatic Digest Considerations

Enzymatic digest considerations offer both a challenge and an opportunity for peptide candidate generation. Enzymatic digest considerations present a challenge, depending on the underlying scan or traversal, because obtaining the necessary protein context in order to check whether a peptide candidate is consistent with a particular enzymatic digest may not be trivial, particularly during a suffix tree based traversal. On the other hand, enzymatic digest considerations present an opportunity, because digest constraints significantly reduce the number of peptide candidates that need to be examined. With careful consideration for the underlying data-structures, it is possible to enumerate peptide candidates that satisfy enzymatic digest constraints at no additional cost, which means that the reduction in the number of candidates output directly impacts the algorithm run-time.

First, we consider how to manage enzymatic digest constraints for the linear scan algorithms. Notice first that since the motifs for enzymatic digestion are typically simple and short, we can find all putative digestion sites in $O(n)$ time. If both ends of the peptide candidates are constrained to be consistent with the enzymatic digest, then the begin and end pointers of Algorithms 1 and 2 can skip from one putative site to the next. Further, if we consider the case when only a small number of missed digestion sites are permitted, we can avoid the generation of candidates with moderate mass, but many missed digestion sites. If only one end of the peptide candidate is required to be consistent with the digest, we must use two passes through the sequence. The first pass constrains the left pointer to putative digest sites, and the second pass constrains the right pointer to putative digest sites and the left pointer to non-digest sites.

Next, we consider how to manage enzymatic digest consistency for the suffix tree based algorithms. Unlike the linear scan algorithms, peptide candidate generation is asymmetric with respect to the ends of the candidate. In fact, there is little we can do within the suffix tree traversal to reduce the amount of work that needs to be done to generate a new peptide candidate in which the right endpoint is consistent with the enzymatic digest. We must still explore all the children of any node we visit, as long as the current peptide candidate mass is

Table 2. total time (s)/number of calls ($\times 10^9$)/time per call ($\times 10^{-6}s$) of the traversal subroutines.

(μsec)	generation	interval search	string scan
Suffix tree	237/1.62/0.146	548/2.90/0.189	115/1.33/0.086
Sim. linear scan	228/2.02/0.112	844/3.53/0.239	87/1.55/0.056

less than the maximum query mass. We can avoid query mass lookups, just as we did for the linear scans, by checking each subsequence with a valid mass for an digest consistent right endpoint before we lookup the query masses. On the other hand, constraining the left endpoint of all candidates to be consistent with the digest can restrict the suffix tree traversal algorithms to a subtree of the suffix tree. In order to accomplish this, we must consider a peptide candidate to consist of the protein context to the left of the candidate prepended to the peptide candidate proper. The non-candidate symbols prepended to the candidate do not contribute to the mass of the candidate, but permit us to look only at those subtrees of the root that begin with a digest consistent motif.

7   Computational Observations

We implemented solutions with a suffix tree and a simultaneous linear scan as described in Algorithm 2. Profiles of these implementations give an estimate to the relative constant factors associated with string traversal versus interval search costs. We ran our implementations against the 1493 sample query masses on the SwissPROT database.

Although we have shown that $O(1)$ lookup of intervals is possible, we have used a skip list to store the intervals in both examples. As the average interval coverage is not large in this example, we thought this would suffice.

Our implementations were compiled in DEC Alpha EV6.7 processors (500 MHz) with the native C++ compiler, CXX (with the option -O4). Timing was done with gprof. We have aggregated the timings for the interval traversal routines, the string traversal routines, and the candidate generation function which calls them alternately.

Table 2 demonstrates that the largest fraction of the cost (by number of calls or time) is spent in the interval lookup, examining about 2 intervals per string search, as Figure 2 suggests. The extra time cost from the redundant string traversals in the simultaneous linear scan implementation is compensated by the speed per-call coming from a simpler scanning routine. The extra interval search time for the simultaneous linear scan is likely due to the increased time spent re-traversing short intervals.

8   Conclusion

We have identified and formulated a key subproblem, the peptide candidate generation problem, that must be solved in order to identify proteins via tandem mass spectrometry and amino-acid sequence database search. We have outlined the context in which any solution to the peptide candidate generation problem must operate and carefully examined how this context influences the classic algorithmic tradeoffs of time and space.

We have identified a key property of amino-acid sequence databases, substring density, that quantifies the unnecessary peptide candidates output by a linear scan of the sequence database. We have further proposed the use of the suffix array as a compact representation of all substrings of the sequence database that eliminates candidate redundancy.

We have also demonstrated that as the number of query masses increases, query mass lookup time becomes more significant than sequence database scan time. We proposed a constant time query lookup algorithm based on preprocessing the query masses to form a lookup table, and showed that this technique is quite feasible in practice.

Our results for the peptide candidate generation problem depend heavily on our empirical observations about the nature of the problem instances we see every day. We need to obtain better worst case bounds for this problem to make our performance less sensitive to less common extreme instances. We also need to do a comprehensive empirical study to more carefully determine the scenarios in which each of the algorithms identified is the method of choice.

References

1. V. Bafna and N. Edwards. Scope: A probabilistic model for scoring tandem mass spectra against a peptide database. *Bioinformatics*, 17(Suppl. 1):S13–S21, 2001.
2. T. Chen, M. Kao, M. Tepel, J. Rush, and G. Church. A dynamic programming approach to de novo peptide sequencing via tandem mass spectrometry. In *ACM-SIAM Symposium on Discrete Algorithms*, 2000.
3. M. Cieliebak, T. Erlebach, S. Lipták, J. Stoye, and E. Welzl. Algorithmic complexity of protein identification: Combinatorics of weighted strings. Submitted to Discrete Applied Mathematics special issue on Combinatorics of Searching, Sorting, and Coding., 2002.
4. J. Cottrell and C. Sutton. The identification of electrophoretically separated proteins by peptide mass fingerprinting. *Methods in Molecular Biology*, 61:67–82, 1996.
5. V. Dancik, T. Addona, K. Clauser, J. Vath, and P. Pevzner. De novo peptide sequencing via tandem mass spectrometry. *Journal of Computational Biology*, 6:327–342, 1999.
6. J. Eng, A. McCormack, and J. Yates. An approach to correlate tandem mass spectral data of peptides with amino acid sequences in a protein database. *Journal of American Society of Mass Spectrometry*, 5:976–989, 1994.
7. D. Gusfield. *Algorithms on Strings, Trees, and Sequences: Computer Science and Computational Biology*. Cambridge University Press, 1997.
8. P. James, M. Quadroni, E. Carafoli, and G. Gonnet. Protein identification in dna databases by peptide mass fingerprinting. *Protein Science*, 3(8):1347–1350, 1994.
9. S. Kurtz. Reducing the space requirement of suffix trees. *Software–Practice and Experience*, 29(13):1149–1171, 1999.

10. D. Pappin. Peptide mass fingerprinting using maldi-tof mass spectrometry. *Methods in Molecular Biology*, 64:165–173, 1997.

11. D. Pappin, P. Hojrup, and A. Bleasby. Rapid identification of proteins by peptide-mass fingerprinting. *Currents in Biology*, 3(6):327–332, 1993.

12. D. Perkins, D. Pappin, D. Creasy, and J. Cottrell. Probability-based protein identification by searching sequence databases using mass spectrometry data. *Electrophoresis*, 20(18):3551–3567, 1997.

13. P. Pevzner, V. Dancik, and C. Tang. Mutation-tolerant protein identification by mass-spectrometry. In R. Shamir, S. Miyano, S. Istrail, P. Pevzner, and M. Waterman, editors, *International Conference on Computational Molecular Biology (RECOMB)*, pages 231–236. ACM Press, 2000.

14. J. Taylor and R. Johnson. Sequence database searches via *de novo* peptide sequencing by mass spectrometry. *Rapid Communications in Mass Spectrometry*, 11:1067–1075, 1997.

Improved Approximation Algorithms for NMR Spectral Peak Assignment

Zhi-Zhong Chen[1,*], Tao Jiang[2,**], Guohui Lin[3,***], Jianjun Wen[2,†], Dong Xu[4,‡], and Ying Xu[4,‡]

[1] Dept. of Math. Sci., Tokyo Denki Univ., Hatoyama, Saitama 350-0394, Japan
chen@r.dendai.ac.jp
[2] Dept. of Comput. Sci., Univ. of California, Riverside, CA 92521
{jiang,wjianju}@cs.ucr.edu
[3] Dept. of Comput. Sci., Univ. of Alberta, Edmonton, Alberta T6G 2E8, Canada
ghlin@cs.ualberta.ca
[4] Protein Informatics Group, Life Sciences Division
Oak Ridge National Lab. Oak Ridge, TN 37831
{xud,xyn}@ornl.gov

Abstract. We study a constrained bipartite matching problem where the input is a weighted bipartite graph $G = (U, V, E)$, U is a set of vertices following a sequential order, V is another set of vertices partitioned into a collection of disjoint subsets, each following a sequential order, and E is a set of edges between U and V with non-negative weights. The objective is to find a matching in G with the maximum weight that satisfies the given sequential orders on both U and V, *i.e.*, if u_{i+1} follows u_i in U and if v_{j+1} follows v_j in V, then u_i is matched with v_j if and only if u_{i+1} is matched with v_{j+1}. The problem has recently been formulated as a crucial step in an algorithmic approach for interpreting NMR spectral data [15]. The interpretation of NMR spectral data is known as a key problem in protein structure determination via NMR spectroscopy. Unfortunately, the constrained bipartite matching problem is NP-hard [15]. We first propose a 2-approximation algorithm for the problem, which follows directly from the recent result of Bar-Noy *et al.* [2] on interval scheduling. However, our extensive experimental results on real NMR spectral data illustrate that the algorithm performs poorly in terms of recovering the target-matching (*i.e.* correct) edges. We then propose another approximation algorithm that tries to take advantage

* Supported in part by the Grant-in-Aid for Scientific Research of the Ministry of Education, Science, Sports and Culture of Japan, under Grant No. 12780241. Part of the work done while visiting at UC Riverside.
** Supported in part by a UCR startup grant and NSF Grants CCR-9988353 and ITR-0085910.
*** Supported in part by NSERC grants RGPIN249633 and A008599, and Startup Grant REE-P5-01-02-Sci from the University of Alberta.
† Supported by NSF Grant CCR-9988353.
‡ Supported by the Office of Biological and Environmental Research, U.S. Department of Energy, under Contract DE-AC05-00OR22725, managed by UT-Battelle, LLC.

R. Guigó and D. Gusfield (Eds.): WABI 2002, LNCS 2452, pp. 82–96, 2002.

of the "density" of the sequential order information in V. Although we are only able to prove an approximation ratio of $3 \log_2 D$ for this algorithm, where D is the length of a longest string in V, the experimental results demonstrate that this new algorithm performs much better on real data, *i.e.* it is able to recover a large fraction of the target-matching edges and the weight of its output matching is often in fact close to the maximum. We also prove that the problem is MAX SNP-hard, even if the input bipartite graph is unweighted. We further present an approximation algorithm for a nontrivial special case that breaks the ratio 2 barrier.

1 Introduction

The Human Genome Project [1] has led to the identification of a vast majority of protein-encoding genes in the human genome. To facilitate a systematic study of the biological functions of these proteins, the US National Institutes of Health (NIH) has recently launched another ambitious project, the Structural Genomics Project [9]. Its main goal is to solve about 100,000 protein structures within the next ten years, through the development and application of significantly improved experimental and computational technologies. Along with *X-ray crystallography, nuclear magnetic resonance* (NMR) spectroscopy has been one of the two main experimental methods for solving protein structures. Among the seven pilot Structural Genomics Centers set up by NIH, one center is devoted to protein structure determination via NMR.

Protein structure determination via NMR generally involves the following three key steps:

- NMR spectral data generation, which produces
  - resonance peaks corresponding to amino acids in the target protein sequence. Peaks corresponding to a common amino acid are grouped into a *spin system*;
  - certain geometric relationships (*e.g.* distances and angles) between the spin systems;
- NMR data interpretation, which involves relating the spin systems to the amino acids in the target protein sequence, providing both inter- and intra-amino acid distance and angle information;
- NMR structure calculation, which calculates the target protein structure through molecular dynamics (MD) and energy minimization (EM) under the constraints of the identified geometric relationships.

It typically takes several months to a year to solve a single protein structure by NMR, and a major part of that time is used for NMR data interpretation. Up until very recently, NMR data interpretation has been done mainly using manual procedures. Though a number of computer programs [4,6,7,14,16] have recently been developed to assist the data interpretation, most NMR labs are still doing the peak assignments manually or semi-manually for quality reasons. With the recent progress in NMR technologies for speeding up the data production rate, we expect that NMR data interpretation will soon become the sole bottleneck in a high-throughput NMR structure determination process.

Two key pieces of information form the foundation of NMR peak assignment:

- Each amino acid has a somewhat "unique" spin system[1];
- The sequential adjacency information between spin systems in a protein sequence is often inferable from the spectral data. However, this type of information is generally incomplete, *i.e.* we may often be able to obtain the adjacency relationship between some of the spin systems but not all.

In a recently developed computational framework [15], the NMR peak assignment problem has been formulated as a constrained bipartite matching problem. In this framework, each amino acid (also called *residue*) is represented as a vertex of U and each spin system is represented as a vertex of V (and thus generally $|U| = |V|$). A pair $(u_i, v_j) \in U \times V$ of vertices that represents a potential assignment has a non-negative weight $w_{i,j} = w(u_i, v_j)$, which scores the preference of assigning spin system v_j to amino acid u_i. Let E denote the set of all potential assignments. Clearly $G = (U, V, E \subseteq U \times V)$ is a bipartite graph. In general, the edges in E have different weights and G is said *weighted*. In the special case that the edges have equal weight, G is said *unweighted*. For more detailed information about the weighting scheme, we refer the reader to [15]. The MAXIMUM WEIGHT BIPARTITE MATCHING [5] provides a natural framework for the study of the NMR peak assignment problem. Nonetheless, some resonance peaks from a single NMR experiment are known to belong to atoms from consecutive amino acids and thus their host spin systems should be mapped to consecutive amino acids. Such spin systems that should be mapped consecutively are said to be *adjacent* and their corresponding vertices in V are required to follow a sequential order. For convenience, we number the amino acids consecutively in the order that they appear in the protein sequence, and number the spin systems in such a way that adjacent spin systems have consecutive indices. In this formulation, a *feasible* matching M in G is one such that if v_j and v_{j+1} are sequentially adjacent, then edge $(u_i, v_j) \in M$ if and only if edge $(u_{i+1}, v_{j+1}) \in M$. The CONSTRAINED BIPARTITE MATCHING (CBM) problem is to find a feasible matching in G achieving the maximum weight.

We call a maximal set of vertices in V that are consecutively adjacent a *string*. Thus, the set V is partitioned into a collection of strings. The CBM problem in which the maximum length of strings in V is D is called the D-STRING CBM problem. Without loss of generality, assume $D > 1$. In the practice of NMR peak assignment, D is usually between 4 and 10. One may notice that the standard MAXIMUM WEIGHT BIPARTITE MATCHING problem is simply the 1-STRING CBM problem, and it is known to be solvable in polynomial time [5]. Unfortunately, the D-STRING CBM problem is intractable even when it is unweighted and $D = 2$.

---

[1] This information alone is not sufficient for a correct assignment since a protein sequence typically contains multiple copies of the same amino acid. Additional information is necessary in order to tell if a particular spin system corresponds to, for example, an alanine at a particular sequence position.

Theorem 1. [15] *The unweighted* 2-STRING CBM *is NP-hard.*

A *two-layer* algorithm for D-STRING CBM has been proposed in [15] that attempts to fix likely assignments and filter out unlikely assignments for *long* strings (*i.e.* at least 3 spin systems) in the first layer of computation. In the second layer, it tries *all* possible combinations of assignments for long strings and extends them to perfect matchings (recall that $|U| = |V|$) by exhaustive enumeration. A perfect matching with the maximum weight generated in this way is output as the result. The current implementation of the algorithm runs efficiently for cases where the number of long strings is relatively small and most of the long strings consist of at least 4 or 5 spin systems. Its running time goes up quickly (*i.e.* exponentially) when the instance has many strings consisting of 2 or 3 spin systems.

In this paper, we first propose a simple 2-approximation algorithm for D-STRING CBM that directly follows from the recent result of Bar-Noy *et al.* [2] on interval scheduling. However, our experimental results on 126 instances of NMR spectral data derived from 14 proteins illustrate that the algorithm performs poorly in terms of recovering the *target-matching edges* (*i.e.* matching edges that assign spin systems to their correct amino acids)[2]. One explanation is that the algorithm looks for matching edges by scanning U from left to right, hence giving preference to edges close to the beginning of U. Consequently, it may miss many target-matching edges. We thus propose a second approximation algorithm that attempts to take advantage of the "density" of the spin system adjacency information in V. Although we are only able to prove an approximation ratio of $3\log_2 D$ for this algorithm, the experimental results demonstrate that this new algorithm performs much better than the 2-approximation algorithm on real data. In fact, it often recovers as many target-matching edges as the two-layer algorithm [15] (of exhaustive search nature) and the weight of its output matching is often close to the maximum. We then prove that unweighted 2-STRING CBM is MAX SNP-hard, implying that the problem has no polynomial-time approximation scheme (PTAS) unless $P = NP$. The proof extends to all constants $D \geq 2$. Although ratio 2 seems to be a barrier to polynomial-time approximation algorithms for D-STRING CBM, we show that this barrier can be broken for unweighted 2-STRING CBM, by presenting a $\frac{5}{3}$-approximation algorithm. We remark that unweighted D-STRING CBM could be interesting because it is simpler and is useful in NMR peak assignment when the edge weights fall into a small range. Moreover, since long strings in V are usually associated with good quality spectral data, algorithms that attempt to solve unweighted D-STRING CBM could yield reasonably good NMR peak assignment since they tend to favor long strings. We expect that the techniques developed in this work, in conjunction with the work of [15], will lead to a significantly improved capability for NMR data interpretation, providing a highly effective tool for high-throughput protein structure determination.

---

[2] Note that, target-matching edges are known in the simulations from BioMagRes-Bank [13] and they were used to generate the adjacency data.

The paper is organized as follows. Section 2 describes the 2-approximation and the $3\log_2 D$-approximation algorithms for D-STRING CBM, and compares their performances (as well as that of the two-layer algorithm) on 126 real NMR spectral data derived from 14 proteins. It also gives a proof of the MAX SNP-hardness of unweighted 2-STRING CBM. Section 3 presents an improved approximation algorithm for unweighted 2-STRING CBM. Section 4 concludes the paper with some future research directions.

2   Weighted Constrained Bipartite Matching

We first present two approximation algorithms for D-STRING CBM. Consider an instance of D-STRING CBM: $G = (U, V, E)$, where $U = \{u_1, u_2, \ldots, u_{n_1}\}$, $V = \{v_1 \cdots v_{i_1}, v_{i_1+1} \cdots v_{i_2}, \ldots, v_{i_p} \cdots v_{n_2}\}$, and $E \subseteq U \times V$ is the set of edges. Here, $v_{i_{j-1}+1} \cdots v_{i_j}$ in V denotes a string of consecutively adjacent spin systems. We may assume that for every substring $v_j v_{j+1}$ of a string in V, $(u_i, v_j) \in E$ if and only if $(u_{i+1}, v_{j+1}) \in E$, because otherwise (u_i, v_j) cannot be in any feasible matching and thus can be eliminated without further consideration. Based on $G = (U, V, E)$, we construct a new edge-weighted bipartite graph $G' = (U, V, E')$ as follows: For each $u_i \in U$ and each string $v_j v_{j+1} \cdots v_k \in V$ such that $(u_i, v_j) \in E$, let (u_i, v_j) be an edge in E' and its weight be the total weight of edges $\{(u_{i+x}, v_{j+x}) \mid 0 \le x \le k - j\}$ in E. For convenience, we call the subset $\{(u_{i+x}, v_{j+x}) \mid 0 \le x \le k - j\}$ of E the *expanded matching* of edge (u_i, v_j) of E'.

We say that two edges of E' are *conflicting* if the union of their expanded matchings is not a feasible matching in G. Note that a set of non-conflicting edges in E' is always a matching in G' but the reverse is not necessarily true. A matching in G' is *feasible* if it consists of non-conflicting edges. There is an obvious one-to-one correspondence between feasible matchings in G and feasible matchings in G'. Namely, the feasible matching M in G corresponding to a feasible matching M' in G' is the union of the expanded matchings of edges in M'. Note that the weight of M in G is the same as that of M' in G'. Thus, it remains to show how to compute a feasible approximate matching in G'.

Define an *innermost edge* of G' to be an edge (u_i, v_j) in G' satisfying the following condition:

- G' has no edge $(u_{i'}, v_{j'})$ other than (u_i, v_j) such that $i \le i' \le i' + s' - 1 \le i + s - 1$, where s (respectively, s') is the size of the expanded matching of (u_i, v_j) (respectively, $(u_{i'}, v_{j'})$).

Note that for every $u_i \in U$, G' has at most one innermost edge incident to u_i (*i.e.*, there cannot exist $v_{j_1} \in V$ and $v_{j_2} \in V$ with $j_1 \ne j_2$ such that both (u_i, v_{j_1}) and (u_i, v_{j_2}) are innermost edges of G'). Define a *leading innermost edge* of G' to be an innermost edge (u_i, v_j) such that index i is minimized. The crucial point is that for every leading innermost edge (u_i, v_j) of G' and every feasible matching M' in G', at most two edges of M' conflict with (u_i, v_j). To see this, let $(u_{i'}, v_{j'})$ be an edge in M' that conflicts with (u_i, v_j). Let s (respectively, s') be the size of the expanded matching of (u_i, v_j) (respectively, $(u_{i'}, v_{j'})$). Since (u_i, v_j) is an innermost edge of G', at least one of the following conditions holds:

1. $j' = j$.
2. $i' \le i \le i + s - 1 \le i' + s' - 1$.
3. $i < i' \le i + s - 1 < i' + s' - 1$.
4. $i' < i \le i' + s' - 1 < i + s - 1$.

For each of these conditions, M' contains at most one edge $(u_{i'}, v_{j'})$ satisfying the condition because M' is a feasible matching in G'. Moreover, if M' contains an edge $(u_{i'}, v_{j'})$ satisfying Condition 2, then it contains no edge satisfying Condition 3 or 4. Furthermore, M' contains no edge $(u_{i'}, v_{j'})$ satisfying Condition 4 or else there would be an innermost edge $(u_{i''}, v_{i''})$ in G' with $i' \le i'' < i \le i'' + s'' - 1 \le i' + s' - 1$ (where s'' is the size of the expanded matching of $(u_{i''}, v_{j''})$), contradicting the assumption that (u_i, v_j) is a leading innermost edge in G'. Thus, at most two edges of M' conflict with (u_i, v_j).

Using the above fact (that at most two edges of M' conflict with a leading innermost edge) and the *local ratio* technique in [3], we can construct a recursive algorithm to find a (heavy) feasible matching in G' as shown in Figure 1. The algorithm in fact, as we were informed very recently, follows directly from the recent result of Bar-Noy *et al.* [2] on interval scheduling.

2-APPROXIMATION on G':
    1. **if** $(E(G') = \emptyset)$
          output the empty set and halt;
    2. find a leading innermost edge e in G';
    3. $\Gamma = \{e\} \cup \{e' \mid e' \in E(G'), e' \text{ conflicts with } e\}$;
    4. find the minimum weight c of an edge of Γ in G';
    5. **for** (every edge $f \in \Gamma$)
          subtract c from the weight of f;
    6. $F = \{e \mid e \in \Gamma, e \text{ has weight } 0\}$;
    7. $G'' = G' - F$;
    8. recursively call 2-APPROXIMATION on G'' and output M_1';
    9. find a maximal $M_2' \subseteq F$ such that $M_1' \cup M_2'$ is a feasible matching in G';
   10. output $M_1' \cup M_2'$ and halt.

Fig. 1. A recursive algorithm for finding a feasible matching in G'.

Theorem 2. [2] *The algorithm described in* Figure 1 *outputs a feasible matching of the graph* $G' = (U, V, E')$ *with weight at least half of the optimum.*

We have implemented the algorithm and tested it on a set of 14 proteins from BioMagResBank [13]. For each protein, we randomly generated 9 instances of spin-system adjacency by adding links (getting from the correct adjacency from BioMagResBank) between neighboring spin systems. If the spin systems are connected by the links, they will map to the sequence as a string together.

We increased the number of links from 10% of the sequence length to 90% of the sequence length. In other words, the algorithm was tested on 126 bipartite graphs with positive edge weights and adjacency constraints. The test results are summarized in Table 1 (Columns 5–7). In the tests, the *target* assignments are matchings consisting of edges of form (u_i, v_i) that assign spin systems to correct amino acids from BioMagResBank. Although these target assignments do not always have the maximum weights, their weights are not far from the maxima. As can be seen from the table, although the algorithm did very well in terms of maximizing the weight of its output matching, it recovered very few target-matching edges and is thus almost useless in practice. A possible explanation of the poor performance of the algorithm in this experiment is that the algorithm looks for edges by scanning amino acids in U from left to right, hence giving preference to edges close to the beginning of U. As a consequence, it may miss many target-matching edges. Another reason of the poor performance is probably due to the scoring function that was used. The goal of the scoring function is to force the correct assignment to have the maximum score. However, given the statistical nature of the scoring function, this goal can not be achieved completely currently. That is why even the "two-layer" algorithm [15] (briefly described in the Introduction section. For the detail description of the algorithm, please refer to [15].) recovers small numbers of correct assignments (Table 1, Column 4) in many cases, although as the number of links between adjacent spin systems increases, the performance improves. The development of the scoring function, which we are working on, will not be addressed in this paper. As the scoring function improves, the correct assignment should get closer to the maximum score, especially when the number of links between adjacent spin systems is large.

In trying to improve the performance on recovering the target-matching edges, we next present a second approximation algorithm that tries to take advantage of the presence of many long strings in the instance, as described in Figure 2. Basically, the algorithm partitions the strings in V into groups of strings of approximately the same length, greedily finds a maximal feasible matching in each group, and then greedily extends the matching to a maximal feasible matching in G'. It outputs the heaviest one among the matchings found for all groups.

Theorem 3. *The algorithm described in* Figure 2 *outputs a feasible matching in* G' *with weight at least* $\frac{1}{3\max\{1,\log_2 r\}}$ *of the maximum weight in* $\widetilde{O}(|U||V|)$ *(i.e. quadratic up to a poly-logarithmic factor) time, where* r *is as defined in* Figure 2. *It is thus an approximation algorithm for* D-STRING CBM *with performance ratio* $3\log_2 D$.

Proof. For each $i \in \{1, 2, \ldots, g\}$, consider the bipartite graph $G'_i = (U, V_i, E_i)$. Let M^*_i denote an optimal feasible matching for graph G'_i. Right before the execution of Step 3.4 of the algorithm, M'_i is clearly a feasible matching for graph G'_i, and its weight is at least $\frac{1}{6}$ of that of M^*_i because we can claim that each execution of Step 3.3.2 only rules out at most 6 edges of M^*_i from further

Table 1. Summary on the performances of the 2-approximation and $3\log_2 D$-approximation algorithms on 126 instances of NMR peak assignment. The number after the underscore symbol in the name of each instance indicates the number of adjacent pairs of spin system in the instance (more precisely, _5 means that the number of adjacent pairs of spin systems is 50% of the total number of residues). M^* represents the target assignment that we want to recover, and M_1 (M_2, or M_3) is the assignment computed by the two-layer (2-approximation, or $3\log_2 D$-approximation, respectively) algorithm. The parameters $R_1 = |M^* \cap M_1|$, $R_2 = |M^* \cap M_2|$, and $R_3 = |M^* \cap M_3|$ show how many target-matching edges were recovered by the two-layer, 2-approximation, and $3\log_2 D$-approximation algorithms, respectively, on each instance. Each N/A indicates that the computation of the two-layer algorithm was taking too long (> 2 days on a supercomputer) and had to be killed before any result could be obtained.

	$|M^*|$	$w(M^*)$	R_1	$|M_2|$	$w(M_2)$	R_2	$|M_3|$	$w(M_3)$	R_3		$|M^*|$	$w(M^*)$	R_1	$|M_2|$	$w(M_2)$	R_2	$|M_3|$	$w(M_3)$	R_3
bmr4027_1	158	1896284	11	158	1871519	28	158	1931099	4	bmr4144_1	78	949170	10	78	936144	10	78	845578	5
bmr4027_2			18	157	1849500	7	158	1927193	2	bmr4144_2			8	78	928175	4	78	869229	1
bmr4027_3			18	158	1841683	8	158	1930119	23	bmr4144_3			10	77	917197	5	78	881665	1
bmr4027_4			43	158	1829367	11	158	1925237	36	bmr4144_4			0	78	907130	10	78	886147	6
bmr4027_5			33	156	1827498	3	158	1923556	37	bmr4144_5			14	77	921816	17	78	914564	14
bmr4027_6			36	157	1818131	8	158	1916814	48	bmr4144_6			30	77	897500	11	76	876005	3
bmr4027_7			79	155	1784027	44	158	1885779	90	bmr4144_7			34	76	842073	2	78	888087	6
bmr4027_8			19	154	1671475	113	158	1875058	117	bmr4144_8			67	77	804531	5	78	896088	22
bmr4027_9			155	155	1652859	60	158	1896606	156	bmr4144_9			75	76	837519	35	78	949844	76
bmr4288_1	105	1249465	9	105	1238612	8	105	1208142	6	bmr4302_1	115	1298321	9	115	1305677	0	115	1316209	8
bmr4288_2			9	105	1220481	8	105	1194198	9	bmr4302_2			12	115	1273146	0	115	1324173	8
bmr4288_3			16	103	1206095	17	105	1199374	17	bmr4302_3			7	114	1276372	8	115	1313288	8
bmr4288_4			33	105	1185685	5	105	1214237	21	bmr4302_4			16	114	1246952	4	115	1307472	10
bmr4288_5			38	103	1169907	6	105	1211226	34	bmr4302_5			34	113	1219920	11	115	1295035	24
bmr4288_6			52	102	1179110	15	105	1217006	52	bmr4302_6			44	114	1174564	0	115	1255172	66
bmr4288_7			55	103	1112288	22	105	1230117	62	bmr4302_7			65	112	1181267	8	115	1294044	78
bmr4288_8			N/A	101	1133554	35	105	1232331	66	bmr4302_8			N/A	113	1152323	27	113	1283268	99
bmr4288_9			105	100	1051817	48	105	1249465	105	bmr4302_9			111	115	1293954	107	115	1298321	111
bmr4309_1	178	2048987	6	178	2066506	4	178	2118482	4	bmr4316_1	89	1029827	4	89	997300	13	89	1011408	7
bmr4309_2			10	178	2023648	9	178	2108291	4	bmr4316_2			15	89	976270	2	89	1019640	7
bmr4309_3			33	177	2013099	9	178	2115356	22	bmr4316_3			21	88	972224	0	89	1020190	9
bmr4309_4			34	176	2024268	14	178	2107417	18	bmr4316_4			20	87	936852	5	89	1028608	31
bmr4309_5			46	174	1954955	13	178	2090346	31	bmr4316_5			42	86	890944	2	89	1007619	43
bmr4309_6			59	177	1924727	12	178	2074540	55	bmr4316_6			60	84	863207	13	89	1012008	48
bmr4309_7			122	174	1885986	24	178	2078322	114	bmr4316_7			79	87	882818	9	87	1004449	67
bmr4309_8			106	173	1868338	55	178	2026479	112	bmr4316_8			87	87	957378	62	89	1029827	89
bmr4309_9			176	170	1796864	95	175	1999734	153	bmr4316_9			89	85	984774	85	89	1029827	89
bmr4318_1	215	2390881	8	215	2418440	17	215	2495022	2	bmr4353_1	126	1498891	6	126	1482821	20	126	1492927	7
bmr4318_2			5	215	2398412	0	215	2481997	6	bmr4353_2			8	126	1473982	9	126	1499720	7
bmr4318_3			N/A	214	2409316	17	215	2481867	10	bmr4353_3			4	125	1455084	6	126	1491933	7
bmr4318_4			23	213	2394682	3	215	2481099	12	bmr4353_4			20	126	1441162	9	126	1511112	14
bmr4318_5			38	215	2355926	2	215	2473707	27	bmr4353_5			17	125	1417351	8	126	1502628	21
bmr4318_6			38	214	2312260	13	215	2440684	31	bmr4353_6			35	125	1421633	18	126	1514294	11
bmr4318_7			87	210	2259377	52	215	2421426	70	bmr4353_7			29	125	1370235	14	126	1499010	58
bmr4318_8			113	212	2214174	63	209	2326045	91	bmr4353_8			N/A	123	1337329	9	122	1443144	81
bmr4318_9			N/A	207	2158223	122	215	2390651	197	bmr4353_9			126	122	1273988	15	126	1498891	126
bmr4391_1	66	710914	5	66	723525	8	66	750059	5	bmr4393_1	156	1850868	6	156	1826257	10	156	1876203	5
bmr4391_2			8	66	720589	6	66	755718	3	bmr4393_2			14	156	1805561	3	156	1873989	6
bmr4391_3			7	66	724102	8	66	749505	5	bmr4393_3			N/A	156	1782350	5	156	1859924	4
bmr4391_4			6	65	681286	9	66	745159	5	bmr4393_4			22	156	1778165	3	156	1868573	12
bmr4391_5			13	64	688400	5	66	741824	0	bmr4393_5			30	155	1742954	3	156	1862071	42
bmr4391_6			10	66	699066	8	66	739778	0	bmr4393_6			45	155	1772955	42	156	1857579	67
bmr4391_7			0	66	684953	37	66	717888	21	bmr4393_7			74	154	1722026	22	151	1794248	94
bmr4391_8			18	64	663147	30	66	705513	20	bmr4393_8			128	156	1640682	15	154	1830609	136
bmr4391_9			N/A	66	687290	45	66	652235	45	bmr4393_9			143	152	1527885	3	156	1851298	152
bmr4579_1	86	950173	7	86	931328	12	86	967574	5	bmr4670_1	120	1391055	8	120	1378876	27	120	1434117	5
bmr4579_2			12	86	933035	7	86	977013	9	bmr4670_2			10	120	1366541	14	120	1437469	5
bmr4579_3			11	85	923916	4	86	973431	14	bmr4670_3			20	120	1370848	6	120	1437484	16
bmr4579_4			16	86	935901	6	86	961214	11	bmr4670_4			32	119	1341300	6	120	1423323	28
bmr4579_5			13	85	894084	2	86	968378	21	bmr4670_5			35	117	1309727	11	120	1393428	28
bmr4579_6			15	86	911564	8	86	945148	21	bmr4670_6			48	118	1290812	13	120	1394903	40
bmr4579_7			42	86	873884	17	86	952794	45	bmr4670_7			45	118	1239001	6	120	1377578	45
bmr4579_8			49	83	877556	26	86	950136	78	bmr4670_8			N/A	120	1236726	19	114	1370011	101
bmr4579_9			86	83	760356	0	86	950173	86	bmr4670_9			N/A	113	1237614	60	114	1319698	94
bmr4752_1	68	882755	8	68	862523	20	68	889083	9	bmr4929_1	114	1477704	7	114	1432825	5	114	1502375	3
bmr4752_2			12	68	848225	16	68	886989	11	bmr4929_2			10	114	1424433	5	114	1500838	7
bmr4752_3			13	68	834299	2	68	886910	18	bmr4929_3			16	113	1417722	7	114	1499302	18
bmr4752_4			20	67	820207	2	68	892854	16	bmr4929_4			20	113	1411387	7	114	1497361	27
bmr4752_5			28	67	796019	8	68	878244	29	bmr4929_5			24	114	1408112	4	114	1487741	26
bmr4752_6			28	67	824289	6	68	879380	35	bmr4929_6			24	112	1385673	12	114	1480528	31
bmr4752_7			43	66	752633	3	68	868981	40	bmr4929_7			65	112	1378166	30	114	1449648	55
bmr4752_8			N/A	65	730276	17	68	860366	42	bmr4929_8			86	114	1424433	5	114	1471279	87
bmr4752_9			68	67	812950	44	68	882755	68	bmr4929_9			112	107	1178499	20	114	1477704	114

$3 \log_2 D$-APPROXIMATION on G':

    1. compute ratio $r = \frac{\ell_{max}}{\ell_{min}}$, where ℓ_{max} (respectively, ℓ_{min}) is the maximum (respectively, minimum) length of strings in V;

    2. partition V into $g = \max\{1, \log_4 r\}$ subsets $V_1, V_2, \ldots, V_g$ such that a string s is included in subset V_i if and only if $4^{i-1} \leq \frac{|s|}{\ell_{min}} \leq 4^i$; (Note: V_{i-1} and V_i may not be disjoint.)

    3. **for** (every $i \in \{1, 2, \ldots, g\}$)

        3.1 compute the set E_i of edges of G' incident to strings in V_i;

        3.2 initialize $M'_i = \emptyset$;

        3.3 **while** ($E_i \neq \emptyset$)

            3.3.1 find an edge $e \in E_i$ of maximum weight;

            3.3.2 add e to M'_i, and delete e and all edges conflicting with e from E_i;

        3.4 greedily extend M'_i to a maximal feasible matching of G';

    4. output the heaviest one among $M'_1, M'_2, \ldots, M'_g$ and halt.

Fig. 2. A new algorithm for finding a feasible matching in G'.

consideration. To see the claim, consider an edge $e = (u_x, v_y)$ added to M'_i in Step 3.3.2. Let $e' = (u_{x'}, v_{y'})$ be an edge conflicting with e. Let s (respectively, s') be the size of the expanded matching of e (respectively, e'). Then, at least one of the following conditions 1 through 6 holds:

1. $y' = y$.
2. $x' = x$ and $s' = s$.
3. $x' < x \leq x' + s' - 1 < x + s - 1$.
4. $x < x' \leq x + s - 1 < x' + s' - 1$.
5. $x' < x \leq x + s - 1 \leq x' + s' - 1$ or $x' \leq x \leq x + s - 1 < x' + s' - 1$.
6. $x < x' \leq x' + s' - 1 \leq x + s - 1$ or $x \leq x' \leq x' + s' - 1 < x + s - 1$.

Since M_i^* is a feasible matching of G'_i, M_i^* may contain at most one edge satisfying Condition 1, at most one edge satisfying Condition 2, at most one edge satisfying Condition 3, at most one edge satisfying Condition 4, at most one edge satisfying Condition 5, and at most four edges satisfying Condition 6 (because of the construction of V_i). Due to the same reason, if M_i^* contains an edge satisfying Condition 2 (respectively, 5), then M_i^* contains no edge satisfying Condition 6. Similarly, if M_i^* contains an edge satisfying Condition 3 or 4, then M_i^* contains at most three edges satisfying Condition 6 (because of the construction of V_i). So, in the worse case (where M_i^* contains the largest number of edges conflicting with e), M_i^* may contain one edge satisfying Condition 1, one edge satisfying Condition 3, one edge satisfying Condition 4, and three edges satisfying Condition 6. This proves the claim.

Let M' denote the output matching of the algorithm. Let $\bar{M}^*$ denote an optimal feasible matching for graph G', and $\bar{M}_i^*$ be the sub-matching of $\bar{M}^*$ in edge set E_i. Suppose without loss of generality that $\bar{M}_j^*$ is the heaviest one among $\bar{M}_1^*, \bar{M}_2^*, \ldots, \bar{M}_g^*$. Clearly, we have $w(\bar{M}_j^*) \geq \frac{1}{g}w(\bar{M}^*)$. Thus, $w(M') \geq \frac{1}{6}w(\bar{M}_j^*) \geq \frac{1}{6g}w(\bar{M}^*)$. The time complexity analysis is straightforward. $\qquad\square$

The above $3 \log_2 D$-approximation has been implemented and tested on the same set of 126 instances of NMR peak assignment. The test results are also summarized in Table 1 (Columns 8–10). It is quite clear that this algorithm is much more superior to the 2-approximation algorithm both in terms of maximizing the weight of the output matching and in terms of maximizing the number of target-matching edges recovered. In fact, on over half of the instances (more precisely, 65 out of the 126 instances), the $3 \log_2 D$-approximation algorithm recovered at least as many target-matching edges as the (exhaustive) two-layer algorithm. Because the $3 \log_2 D$-approximation algorithm is much more efficient than the two-layer algorithm, it will be very useful in NMR peak assignment.

Observe that the (feasible) matchings found by the approximation algorithms have weights greater than that of the target assignments on quite a few instances, especially when the adjacency information is sparse. This implies that the weighting scheme as formulated in [15] may not work very well when the adjacency information is sparse, and more work on weighting scheme is needed in the future.

A natural question is if D-STRING CBM admits a ρ-approximation algorithm for some constant $\rho < 2$. Our next theorem shows that there is a constant $\rho > 1$ such that D-STRING CBM does not admit a ρ-approximation algorithm for every $D \geq 2$, unless $P = NP$, even if the input bipartite graph is unweighted.

Theorem 4. *For all $D \geq 2$, unweighted D-STRING CBM is MAX SNP-hard.*

Proof. We prove the theorem for $D = 2$ by a simple L-reduction from MAXIMUM 3-DIMENSIONAL MATCHING (3DM), which is known to be MAX SNP-complete [8]. The proof can be easily extended to any constant $D \geq 2$.

MAXIMUM BOUNDED 3-DIMENSIONAL MATCHING (MB3DM): Given a universal set $\mathcal{U} = \{1, 2, \ldots, m\}$ and a collection of subsets $S_1, S_2, \ldots, S_n$, where $S_i \subseteq \mathcal{U}$, $|S_i| = 3$, and every element $u \in \mathcal{U}$ is contained in at most 3 subsets, find a largest subcollection of pairwise disjoint subsets.

Given an instance of MB3DM, without loss of generality, suppose that $m = 3q$ and $n \geq q$. Observe that $n \leq m$, because every element of $\mathcal{U}$ appears in at most 3 subsets. For each subset S_i, construct 7 vertices $a_{i,1}, a_{i,2}, \ldots, a_{i,7}$ in set U and for each element $i \in \mathcal{U}$ construct a 2-vertex string $b_{i,1}b_{i,2}$ in set V. We will also have in V q 1-vertex strings $f_1, f_2, \ldots, f_q$ and $3n$ 2-vertex strings $c_{1,1}c_{1,2}$, $c_{1,3}c_{1,4}$, $c_{1,5}c_{1,6}$, $\cdots$, $c_{n,1}c_{n,2}$, $c_{n,3}c_{n,4}$, $c_{n,5}c_{n,6}$. Finally, for every $i = 1, 2, \ldots, m$, we connect string $b_{i,1}b_{i,2}$ to $a_{j,2k}a_{j,2k+1}$ (*i.e.* connect vertex $b_{i,1}$ to vertex $a_{j,2k}$ and vertex $b_{i,2}$ to vertex $a_{j,2k+1}$), for each $1 \leq k \leq 3$, if $i \in S_j$; for every $i = 1, 2, \ldots, q$ and every $j = 1, 2, \ldots, n$, connect string f_i to $a_{j,1}$; and for every $i = 1, 2, \ldots, n$ and every $j = 1, 2, \ldots, n$, connect string $c_{i,2k-1}c_{i,2k}$ to $a_{j,2k-1}a_{j,2k}$, for each $1 \leq k \leq 3$. All the edges have the unit weight. This forms an instance of unweighted 2-STRING CBM: $G = (U, V, E)$, where $|U| = 7n$, $|V| = 7q + 6n$.

We claim that the above construction is an L-reduction [10] from MB3DM to unweighted 2-STRING CBM. It is straightforward to see that each subcollection of p (where $p \leq q$) disjoint subsets implies a constrained matching in G of weight

$7p + 6(n - p) = 6n + p$. To complete the proof of the claim, we only need to observe that, for any given constrained matching in the above bipartite graph, we can always rearrange it without decreasing the weight so that each group of vertices $a_{i,1}, a_{i,2}, \ldots, a_{i,7}$ are matched either with three c-type strings or with a combination of one f-type string and three b-type strings, due to the special construction of the edges. This completes the L-reduction.   □

3   Unweighted Constrained Bipartite Matching

As noted in the last section, a natural question is to ask if D-STRING CBM admits an approximation algorithm with ratio less than 2. In this section, we answer the question affirmatively for a special case, namely, unweighted 2-STRING CBM. More specifically, we will give a $\frac{5}{3}$-approximation algorithm for unweighted 2-STRING CBM.

Consider an instance of unweighted D-STRING CBM: $G = (U, V, E)$, where $U = \{u_1, u_2, \ldots, u_{n_1}\}$, $V = \{v_1, v_2, \cdots v_k, v_{k+1}v_{k+2}, v_{k+3}v_{k+4}, \ldots, v_{k+2\ell-1}v_{k+2\ell}\}$, and $E \subseteq U \times V$ is the set of edges in G. Here, v_j alone forms a 1-string in V for each $1 \leq j \leq k$, while $v_{k+2j-1}v_{k+2j}$ is a 2-string in V for each $1 \leq j \leq \ell$. Note that $k \geq 0$ and $\ell \geq 0$. Let $n_2 = k + 2\ell$ and $m = |E|$. We may assume that for every $u_i \in U$ and every 2-string $v_{k+2j-1}v_{k+2j}$ in V, $(u_i, v_{k+2j-1}) \in E$ if and only if $(u_{i+1}, v_{k+2j}) \in E$, because otherwise (u_i, v_{k+2j-1}) or (u_i, v_{k+2j}) cannot be in any feasible matching and thus can be eliminated without further consideration.

Fix a maximum size feasible matching M^* in G. Let m_1^* be the number of edges $(u_i, v_j) \in M^*$ with $1 \leq j \leq k$. Similarly, let m_2^* be the number of edges $(u_i, v_{k+2j-1}) \in M^*$ with $1 \leq j \leq \ell$. Then, $|M^*| = m_1^* + 2m_2^*$.

Lemma 1. *A feasible matching in G can be found in $O(m\sqrt{n_1 n_2})$ time, whose size is at least $m_1^* + m_2^*$.*

Proof. Construct a new bipartite graph $G' = (U, V, E_1 \cup E_2)$, where $E_1 = \{(u_i, v_j) \in E \mid 1 \leq j \leq k\}$ and $E_2 = \{(u_i, v_{k+2j-1}) \in E \mid 1 \leq j \leq \ell\}$. Let M' be a maximum matching in G'. Obviously, we can obtain a matching in G' from M^* by deleting all edges $(u_i, v_{k+2j}) \in M^*$ with $1 \leq i \leq n_1$ and $1 \leq j \leq \ell$. So, $|M'| \geq m_1^* + m_2^*$. To obtain a feasible matching M of G from M', we perform the following steps:

1. Initialize $M = \emptyset$.
2. Construct an auxiliary graph H as follows. The vertex set of H is M'. The edge set of H consists of all (e_1, e_2) such that $e_1 \in M'$, $e_2 \in M'$, and e_1 conflicts with e_2. [*Comment:* Each connected component of H is a path P (possibly consisting of a single vertex); if P contains two or more vertices, then there exist integers $i, j_1, \ldots, j_h$ ($h \geq 2$) such that the vertices of P are $(u_i, v_{j_1}), (u_{i+1}, v_{j_2}), \ldots, (u_{i+h-1}, v_{j_h})$, and each of v_{j_1} through $v_{j_{h-1}}$ is the leading vertex of a 2-string in V.]
3. For each connected component of H formed by only one vertex $(u_i, v_j) \in M'$, if v_j is a 1-string in V, then add edge (u_i, v_j) to M; otherwise, add edges (u_i, v_j) and (u_{i+1}, v_{j+1}) to M.

4. For each connected component P of H formed by two or more vertices, perform the following three substeps:
   (a) Let the vertices of P be $(u_i, v_{j_1}), (u_{i+1}, v_{j_2}), \ldots, (u_{i+h-1}, v_{j_h}) \in M'$.
   (b) If h is even or v_{j_h} is the leading vertex of a 2-string in V, then for each $1 \le x \le \lceil \frac{h}{2} \rceil$, add edges $(u_{i+2x-2}, v_{j_{2x-1}})$ and $(u_{i+2x-1}, v_{j_{2x-1}+1})$ to M.
   (c) If h is odd and v_{j_h} alone forms a 1-string in V, then for each $1 \le x \le \frac{h-1}{2}$, add edges $(u_{i+2x-2}, v_{j_{2x-1}})$ and $(u_{i+2x-1}, v_{j_{2x-1}+1})$ to M; further add edge (u_{i+h-1}, v_{j_h}) to M.

It is clear that for each connected component P of H, we add at least as many edges to M as the number of vertices in P. Thus, $|M| \ge |M'| \ge m_1^* + m_2^*$. □

Lemma 2. *A feasible matching in G can be found in $O(m\sqrt{n_1 n_2})$ time, whose size is at least $\frac{m_1^*}{3} + \frac{4m_2^*}{3}$.*

Proof. For each index $i \in \{0, 1, 2\}$, let G_i be the edge-weighted bipartite graph obtained from G as follows:

1. For every 2-string $v_j v_{j+1}$ in V, merge the two vertices in the string into a single super-vertex $s_{j,j+1}$ (with all resulting multiple edges deleted).
2. For all j such that $i + 1 \le j \le n_1 - 2$ and $j - 1 \equiv i \pmod{3}$, perform the following three substeps:
   (a) Merge u_j, u_{j+1}, and u_{j+2} into a single super-vertex $t_{j,j+1,j+2}$ (with all resulting multiple edges deleted).
   (b) For every 1-string v_h that is a neighbor of $t_{j,j+1,j+2}$, if edge $\{u_{j+1}, v_h\}$ is not in the original input graph, then delete the edge between $t_{j,j+1,j+2}$ and v_h; otherwise, assign a weight of 1 to the edge between $t_{j,j+1,j+2}$ and v_h.
   (c) For every 2-string $v_h v_{h+1}$ such that $s_{h,h+1}$ is a neighbor of $t_{j,j+1,j+2}$, if neither $\{(u_j, v_h), (u_{j+1}, v_{h+1})\}$ nor $\{(u_{j+1}, v_h), (u_{j+2}, v_{h+1})\}$ is a matching in the original input graph, then delete the edge between $t_{j,j+1,j+2}$ and $s_{h,h+1}$; otherwise, assign a weight of 2 to the edge between $t_{j,j+1,j+2}$ and $s_{h,h+1}$.
3. If neither u_1 nor u_2 was merged in Step 2a, then perform the following three substeps:
   (a) Merge u_1 and u_2 into a single super-vertex $t_{1,2}$ (with all resulting multiple edges deleted).
   (b) For every 1-string v_h that is a neighbor of $t_{1,2}$, if edge $\{u_1, v_h\}$ is not in the original input graph, then delete the edge between $t_{1,2}$ and v_h; otherwise, assign a weight of 1 to the edge between $t_{1,2}$ and v_h.
   (c) For every 2-string $v_h v_{h+1}$ such that $s_{h,h+1}$ is a neighbor of $t_{1,2}$, if $\{(u_1, v_h), (u_2, v_{h+1})\}$ is not a matching in the original input graph, then delete the edge between $t_{1,2}$ and $s_{h,h+1}$; otherwise, assign a weight of 2 to the edge between $t_{1,2}$ and $s_{h,h+1}$.
4. If neither u_{n_1-1} nor u_{n_1} was merged in Step 2a, then perform the following three substeps:

(a) Merge u_{n_1-1} and u_{n_1} into a single super-vertex t_{n_1-1,n_1} (with all result-ing multiple edges deleted).

(b) For every 1-string v_h that is a neighbor of t_{n_1-1,n_1}, if edge $\{u_{n_1}, v_h\}$ is not in the original input graph, then delete the edge between t_{n_1-1,n_1} and v_h; otherwise, assign a weight of 1 to the edge between t_{n_1-1,n_1} and v_h.

(c) For every 2-string $v_h v_{h+1}$ such that $s_{h,h+1}$ is a neighbor of t_{n_1-1,n_1}, if $\{(u_{n_1-1}, v_h), (u_{n_1}, v_{h+1})\}$ is not a matching in the original input graph, then delete the edge between t_{n_1-1,n_1} and $s_{h,h+1}$; otherwise, assign a weight of 2 to the edge between t_{n_1-1,n_1} and $s_{h,h+1}$.

For each $i \in \{0, 1, 2\}$, let M_i be a maximum-weighted matching in G_i. From each M_i, we can obtain a feasible matching $\bar{M}_i$ in the original input graph by performing the following steps in turn:

- Initialize $\bar{M}_i = \emptyset$.
- For each edge $(u_j, v_h) \in M_i$, add (u_j, v_h) to $\bar{M}_i$.
- For each edge $(t_{j,j+1,j+2}, v_h) \in M_i$, add (u_{j+1}, v_h) to $\bar{M}_i$.
- For each edge $(t_{1,2}, v_h) \in M_i$, add (u_1, v_h) to $\bar{M}_i$.
- For each edge $(t_{n_1-1,n_1}, v_h) \in M_i$, add (u_{n_1}, v_h) to $\bar{M}_i$.
- For each edge $(t_{j,j+1,j+2}, s_{h,h+1}) \in M_i$, if $\{(u_j, v_h), (u_{j+1}, v_{h+1})\}$ is a match-ing in the original input graph, then add edges (u_j, v_h) and (u_{j+1}, v_{h+1}) to $\bar{M}_i$; otherwise, add edges (u_{j+1}, v_h) and (u_{j+2}, v_{h+1}) to $\bar{M}_i$.
- For each edge $(t_{1,2}, s_{h,h+1}) \in M_i$, add edges (u_1, v_h) and (u_2, v_{h+1}) to $\bar{M}_i$.
- For each edge $(t_{n_1-1,n_1}, s_{h,h+1}) \in M_i$, add edges (u_{n_1-1}, v_h) and (u_{n_1}, v_{h+1}) to $\bar{M}_i$.

Note that the total weight of edges in M_i is exactly $|\bar{M}_i|$. Let $\bar{M}$ be the maximum size one among $\bar{M}_0, \bar{M}_1, \bar{M}_2$. We claim that $|\bar{M}| \geq \frac{m_1^*}{3} + \frac{4m_2^*}{3}$. To see this, for each $i \in \{0, 1, 2\}$, let M_i^* be the union of the set $\{(u_j, v_h) \in M^* \mid j+1 \equiv i \pmod 3$ and v_h is a 1-string in $V\}$ and the set $\{(u_j, v_h), (u_{j+1}, v_{h+1}) \in M^* \mid u_j$ and u_{j+1} belong to the same super-vertex in G_i and $v_h v_{h+1}$ is a 2-string in $V\}$. It holds that $\sum_{i=0}^{2} |M_i^*| = m_1^* + 4m_2^*$, because each edge in M^* incident to a 1-string belongs to exactly one of M_0^*, M_1^*, M_2^* while each edge in M^* incident to a 2-string belongs to exactly two of M_0^*, M_1^*, M_2^*. This implies that the maximum size one among M_0^*, M_1^*, M_2^* has size at least $\frac{m_1^*}{3} + \frac{4m_2^*}{3}$. On the other hand, for each $i \in \{0, 1, 2\}$, we can obtain a matching $\tilde{M}_i$ in G_i by modifying M_i^* as follows:

- For each edge $(u_j, v_h) \in M_i^*$ such that v_h is a 1-string in V, replace u_j by the super-vertex of G_i to which u_j belongs.
- For each pair of edges $(u_j, v_h), (u_{j+1}, v_{h+1}) \in M_i^*$ such that $v_h v_{h+1}$ is a 2-string in V, replace the two edges by a single edge between $s_{h,h+1}$ and the super-vertex in G_i to which both u_j and u_{j+1} belong.

The total weight of edges in $\tilde{M}_i$ is exactly $|M_i^*|$. On the other hand, the total weight of edges in M_i is larger than or equal to that of edges in $\tilde{M}_i$, because M_i

is a maximum-weighted matching in G_i. Thus, the total weight of edges in M_i is at least $|M_i^*|$. In turn, $|\bar{M}_i| \geq |M_i^*|$. Therefore,

$$|\bar{M}| \geq \frac{1}{3} \sum_{i=0}^{2} |\bar{M}_i| \geq \frac{1}{3} \sum_{i=0}^{2} |M_i^*| = \frac{1}{3}(m_1^* + 4m_2^*).$$

This completes the proof of the claim and hence that of the lemma. □

Combining Lemmas 1 and 2, we now have:

Theorem 5. *We can compute a feasible matching in G whose size is at least $\frac{3}{5}|M^*|$, in $O(m\sqrt{n_1 n_2})$ time. Consequently, there is a $\frac{5}{3}$-approximation algorithm for the unweighted 2-STRING CBM problem; it runs in $O(m\sqrt{n_1 n_2})$ time.*

4 Concluding Remarks

It would be interesting to test if the $\frac{5}{3}$-approximation algorithm works well in practice. An obvious open question is if D-STRING CBM admits a ρ-approximation algorithm for some constant $\rho < 2$.

In the real NMR spectral peak assignment, we want to assign every amino acid to a spin system. Therefore, the desired output matchings are perfect matchings. So far our theoretical analysis on the approximation algorithms does not involve this requirement, although during the implementation we did put priority on perfect matchings. Designing algorithms that guarantee to output perfect matchings (given that the real data is a complete weighted bipartite graph) is a practical consideration. On the other hand, not putting the perfect requirement on approximation algorithms could be one of the reasons that they produce matchings with better weights than the correct assignments.

Our current framework provides a proof-of-principle for automated backbone spectral assignment. In real NMR experiments, it can easily incorporate other NMR data to further improve the assignment accuracy. For example, one can use residual dipolar couplings, which are easy and fast to obtain. Residual dipolar couplings have been used to help peak assignment and structure determination [12,11]. Under our framework, we can use residual dipolar coupling data to provide more constraint information for spin systems. We believe further studies along the line could lead to a software package that moves closer to the goal of fully automated NMR peak assignments.

References

1. International Human Genome Sequencing Consortium. Initial Sequencing and Analysis of the Human Genome. *Nature*, 409:860-921, 2001.
2. A. Bar-Noy, R. Bar-Yehuda, A. Freund, J. Naor, and B. Schieber. A unified approach to approximating resource allocation and scheduling. *J. ACM*, 48:1069–1090, 2001.
3. R. Bar-Yehuda and S. Even. A local-ratio theorem for approximating the weighted vertex cover problem. *Annals of Discrete Mathematics*, 25:27–46, 1985.

4. C. Bartels, P. Güntert, M. Billeter, and K. Wüthrich. GARANT-A general algorithm for resonance assignment of multidimensional nuclear magnetic resonance spectra. *Journal of Computational Chemistry*, 18:139–149, 1996.
5. T. Cormen, C. Leiserson, and R. Rivest. *Introduction to Algorithms*. The MIT Press, 1990.
6. P. Güntert, M. Salzmann, D. Braun, and K. Wüthrich. Sequence-specific NMR assignment of proteins by global fragment mapping with the program mapper. *Journal of Biomolecular NMR*, 18:129–137, 2000.
7. K. Huang, M. Andrec, S. Heald, P. Blake, and J.H. Prestegard. Performance of a neural-network-based determination of amino acid class and secondary structure from ^{1}H-^{15}N NMR data. *Journal of Biomolecular NMR*, 10:45–52, 1997.
8. V. Kann. Maximum bounded 3-dimensional matching is MAX SNP-complete. *Information Processing Letters*, 37:27–35, 1991.
9. National Institute of General Medical Sciences. Pilot projects for the protein structure initiative (structural genomics). June 1999. Web page at http://www.nih.gov/grants/guide/rfa-files/RFA-GM-99-009.html.
10. C.H. Papadimitriou and M. Yannakakis. Optimization, approximation, and complexity classes. *Journal of Computer and System Sciences*, 43:425–440, 1991.
11. C. A. Rohl and D. Baker. De novo determination of protein backbone structure from residual dipolar couplings using Rosetta. *Journal of The American Chemical Society*, 124:2723–2729, 2002.
12. F. Tian, H. Valafar, and J. H. Prestegard. A dipolar coupling based strategy for simultaneous resonance assignment and structure determination of protein backbones. *Journal of The American Chemical Society*, 123:11791–11796, 2001.
13. University of Wisconsin. BioMagResBank. http://www.bmrb.wisc.edu. 2001.
14. J. Xu, S.K. Straus, B.C. Sanctuary, and L. Trimble. Use of fuzzy mathematics for complete automated assignment of peptide ^{1}H 2D NMR spectra. *Journal of Magnetic Resonance*, 103:53–58, 1994.
15. Y. Xu, D. Xu, D. Kim, V. Olman, J. Razumovskaya, and T. Jiang. Automated assignment of backbone NMR peaks using constrained bipartite matching. *IEEE Computing in Science & Engineering*, 4:50–62, 2002.
16. D.E. Zimmerman, C.A. Kulikowski, Y. Huang, W.F.M. Tashiro, S. Shimotakahara, C. Chien, R. Powers, and G.T. Montelione. Automated analysis of protein NMR assignments using methods from artificial intelligence. *Journal of Molecular Biology*, 269:592–610, 1997.

Efficient Methods for Inferring Tandem Duplication History

Louxin Zhang[1], Bin Ma[2], and Lusheng Wang[3]

[1] Labs for Infor. Tech. & Dept. of Mathematics
Nat. University of Singapore, Singapore 117543
matzlx@nus.edu.sg
[2] Department of Computer Science
University of Western Ontario, Canada N6A 5B8
bma@cs.uwo.ca
[3] Department of Computer Science
City University of Hong Kong, Hong Kong
lwang@cs.cityu.edu.hk

Abstract. In this paper, we study the problem of inferring duplication history of a tandem gene family using the duplication model proposed by Tang *et al.*. We provide an efficient algorithm for inferring a duplication model for a given set of (gene) sequences by combining a linear-time algorithm, which is for determining whether a rooted tree is associated with a duplication model, with the nearest neighbor interchange operation. Finally, using our proposed method, we derive duplication hypotheses for an exon of a mucin gene MUC5B, a ZNF gene family, and a OR gene family.

Keywords: Algorithm, tandem duplication model, phylogeny, Mucin genes, ZNF genes, olfactory receptors.

1 Introduction

Tandem duplication is a less well understood mutational process for DNA molecules in which a short stretch of DNA is transformed into several adjacent copies. For example, in the following tandem duplication,

$$\cdots \text{TT}\underline{\text{CGGA}}\text{GA}\cdots \rightarrow \cdots \text{TT}\underline{\text{CGGA}}\ \underline{\text{CGGA}}\ \underline{\text{CGGA}}\text{GA}\cdots$$

the segment CGGA is transformed into three adjacent copies. Since individual copies in a tandem repeat may undergo additional mutations later on, approximate repeats are usually present in a genome.

This process is a primary mechanism for generating gene family clusters on a chromosome. Thirty-one percent of known and predicted genes (340 genes) in human chromosome 19 (HSA19) are arranged in tandem arrays [11]. It was reported that 262 distinct C2H2 zinc-finger (ZNF) loci are distributed in 11 different clusters in three cytogenetic bands 19p12, 19p13 and 19q13 and 49 olfactory receptors (OR) loci in 4 different clusters [3]. Gene families organized

R. Guigó and D. Gusfield (Eds.): WABI 2002, LNCS 2452, pp. 97–111, 2002.

in tandem arrays have also been described in *C. elegans* [18], *Drosophila* [1], and *Arabidopsis* [17].

Genome analysis suggests tandem duplication is an important mode of evolutionary novelty by permitting one copy of each gene to drift and potentially to acquire a new function. Hence, evolution study of a tandem gene family together with other analyzes may yield valuable insights into predicting their functions and resolve important species-specific biological questions. With these motivations in mind, researchers have begun to study tandem gene families and their duplication history (see for example [2,6,16]).

In this paper, we study the problem of systematically inferring duplication history of a tandem gene family. Previous and related works [2,6,10,16] are summarized in the next section. Here, we attack this problem using a different approach. A *duplication model* (DM) $\mathcal{M}$ for a given family of (gene) sequences is a directed graph that contains *nodes, edges* and *blocks* as shown in Figure 1. If only the parent-child relations are considered in a duplication model $\mathcal{M}$, then, the resulting structure is just a rooted binary tree $T_{\mathcal{M}}$ which is unique and called the *associated phylogeny* for $\mathcal{M}$. We present an optimal linear-time algorithm which, given a phylogeny for a set of (gene) sequences, reconstruct the unique duplication model $\mathcal{M}$ such that $T = T_{\mathcal{M}}$ if it exists. Most importantly, we propose an efficient algorithm for inferring a duplication model for a set of (gene) sequences by combining this linear-time algorithm with the nearest neighbor interchange (NNI) operation (see Section 4 for its definition). The algorithm was used to infer the history of tandem repeats within a large exon of a mucin gene MUC5B [6] and performed better than Benson and Dong's algorithm; it was also used to derive the duplication history of tandem gene families ZNF45 and OLF3 reported in [3]. Our algorithm outperformed the Window method on ZNF45 family according to the parsimony principle.

2   Tandem Duplication Models

Let $F = \{s_1, s_2, \cdots, s_n\}$ be a set of DNA (or protein) sequences of a fixed length l over an alphabet A. Let A^l denote the set of all the sequences of length l over A and let $d(,)$ be a pairwise distance function on sequences in A^l. Let T be a phylogenetic tree (or phylogeny) on F. Then, it is a rooted binary tree in which each leaf is labeled by a sequence in F and each internal node v a sequence $s(v) \in A^l$. We measure T using the following cost $C(T)$:

$$C(T) = \sum_{(u,v)\in E(T)} d(s(u), s(v)),$$

where $E(T)$ denotes the set of directed edges in T.

An *ordered phylogenetic tree* on F is a rooted phylogenetic tree in which the left-to-right order of the leaves is $s_1, s_2, \cdots, s_n$. Obviously, ordered phylogenetic tree form a proper subclass of phylogenetic trees on F. In the final version of this paper, we prove that there are $\binom{2(n-1)}{n-1}/n$ ordered phylogenetic trees on F.

Benson and Dong described a greedy algorithm for the tandem repeat history (TRHIST) problem stated as [2]

TRHIST problem
Input: A set $F = \{s_1, s_2, \cdots, s_n\}$ of gene sequences of length l and a distance function $d(,)$ on sequences.
Output: An ordered phylogenetic tree T on F such that the cost of T is minimum.

Later, Jaitly *et al.* and Tang *et al.* proved that TRHIST problem is NP-hard; they also discovered a polynomial-time approximation scheme (PTAS) for it [10,16]. Such a PTAS is obtained from combining the lifting technique [20,19] with dynamic programming. A PTAS for a minimization problem is an algorithm which, for every $\epsilon > 0$, returns a solution whose score is at most $(1 + \epsilon)$ times the optimal score, and runs in polynomial time (depending on ϵ and the input size).

Furthermore, Tang *et al.* extended the study in this aspect by proposing the so-called duplication model [16] that captures evolutionary history, and the observed left-to-right order of (gene) sequences appearing on a chromosome. They then presented a heuristic method called *the Window Method* for inferring the tandem duplication history.

A *duplication model* (DM) $\mathcal{M}$ for a given family of (gene) sequences is a directed graph that contains *nodes, edges* and *blocks* [16] as shown in Figure 1. A node in $\mathcal{M}$ represents a sequence. A directed edge (u, v) from u to v indicates that v is a child of u. A node s is an *ancestor* of a node t if there is a directed path from s to t. A node that has no outgoing edges is called a *leaf;* each leaf

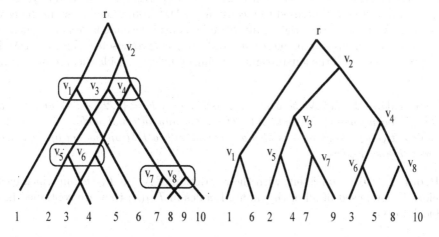

Fig. 1. A DM on a gene (sequence) family $\{1, 2, \cdots, n\}$, in which each directed edge is from top to bottom, and its associated phylogeny. There are five blocks $[r], [v_2], [v_1, v_3, v_4], [v_5, v_6], [v_7, v_8]$ in the DM; blocks with more than one nodes are represented by boxes.

is labeled with a unique sequence in the family. Each non-leaf node is called an *internal* node; it has a left and right child. There is a special node r called the *root* that has only outgoing edges. The root represents the ancestral sequence of all the given sequences.

Each *block* in $\mathcal{M}$ represents a duplication event. Each internal node appears in a unique block; no node is an ancestor of another in a block. Assume that a block B contains k nodes $v_1, v_2, \cdots, v_k$ in left-to-right order. Let $lc(v_i)$ and $rc(v_i)$ denote the left and right child of v_i respectively for each i from 1 to k. Then,

$$lc(v_1), lc(v_2), \cdots, lc(v_k), rc(v_1), rc(v_2), \cdots, rc(v_k)$$

are placed in left-to-right order in the model. Therefore, for any i and j, $1 \leq i < j \leq k$, the edges $(v_i, rc(v_i))$ and $(v_j, lc(v_j))$ cross each other. However, no other edges in the model cross. For simplicity, we will only draw a box for a block with more than one internal nodes.

Each leaf is labeled by a unique sequence of the given family. The left-to-right order of leaves in the duplication model is identical to the order of the given tandem sequences appearing on a chromosome.

A duplication model captures not only evolutionary relationship, but also the observed order of sequences appearing on a chromosome. If we consider only the evolutionary relationships defined in a duplication model $\mathcal{M}$, the underlying structure $T_{\mathcal{M}}$ is just a rooted binary tree, called the *associated phylogeny* of $\mathcal{M}$. Clearly, $T_{\mathcal{M}}$ is unique. Obviously, any ordered phylogenetic tree is the associated phylogeny of a duplication model in which all the duplication blocks contains only an internal node. Note that Figure 1 gives an associated phylogeny that is not ordered. Counting associated phylogenies on a set of n genes seems to be an interesting problem.

In the rest of this section, we list two basic properties of associated phylogeny. These properties will be used to reconstruct a DM from a phylogeny in the next section. We assume that the family F under consideration contains n sequences $1, 2, \cdots, n$, and appear in increasing order on a chromosome. Thus, a DM is *consistent* with F if the left-to-right ordering of its leaves is identical to increasing order.

Proposition 1. *Let $\mathcal{M}$ be a DM consistent with $F = \{1, 2, \cdots, n\}$ and $T_{\mathcal{M}}$ be the phylogeny associated with $\mathcal{M}$. Then, for any internal node u in $T_{\mathcal{M}}$, the subtree $T_{\mathcal{M}}(u)$ rooted at u is also associated with a DM on the set of sequences appearing in $T_{\mathcal{M}}(u)$.*

Proof. $T_{\mathcal{M}}(u)$ is consistent with the restriction of $\mathcal{M}$ on $T_{\mathcal{M}}(u)$, in which each block is composed of all of the internal nodes in $T_{\mathcal{M}}(u)$ that are contained in a block in $\mathcal{M}$.

Recall that, for an internal node u in a DM $\mathcal{M}$, we use $lc(u)$ and $rc(u)$ to denote its left and right child in $\mathcal{M}$ or $T_{\mathcal{M}}$ respectively. Similarly, we use $l^*c(u)$ and $r^*c(u)$ to denote the sequence labels of the leftmost and rightmost leaf in the subtree $T_{\mathcal{M}}(u)$ rooted at u respectively.

Proposition 2. *Let $\mathcal{M}$ be a DM consistent with $F = \{1, 2, \cdots, n\}$ and $T_{\mathcal{M}}$ be the phylogeny associated with $\mathcal{M}$. Then, $T_{\mathcal{M}}$ satisfies the following two properties:*

(i) For each internal node u in $T_{\mathcal{M}}$, $r^*c(u) > r^*c(lc(u))$ and $l^*c(u) < l^*c(rc(u))$; equivalently,

(ii) For each internal node u in $T_{\mathcal{M}}$, $l^*c(u)$ and $r^*c(u)$ are the smallest and largest sequence labels in the subtree $T_{\mathcal{M}}(u)$ respectively.

Proof. Both (i) and (ii) can be proved by induction. Their full proof is omitted here.

Proposition 2 can be illustrated by the model given in Figure 1. For v_2 in the model, $lc(v_2) = v_3$ and $r^*c(v_2) = 10 > r^*c(v_3) = 9$.

3   Reconstructing a Duplication Model from a Phylogeny

We have known that each DM $\mathcal{M}$ has a unique associated phylogeny $T_{\mathcal{M}}$. In this section, we present an optimal algorithm for the following reverse problem:

Given a phylogeny T, reconstruct a DM $\mathcal{M}$ such that $T = T_{\mathcal{M}}$ if it exists.

Let $\mathcal{M}$ be a DM consistent with the sequence family $F = \{1, 2, \cdots, n\}$. A duplication block is *single* if it contains only one node; it is *double* if it contains two nodes. A DM is said to be *double* if every duplication block of it is either single or double. In Section 3.1, we shall present a linear-time algorithm for reconstructing double DMs from phylogenies. Then, we generalize the algorithm to arbitrary DMs in Section 3.2. The best known algorithm for this problem takes quadratic time [16].

Note that, to specify a DM, we need only to list all non-single blocks on the associated phylogeny. Hence, the output from our algorithm is just a list of non-single duplication blocks.

3.1   Double Duplication Models

Let T be a phylogeny. We associate a pair (L_v, R_v) of leaf indices to each node v in T as follows [16]: The pair of a leaf labeled by gene i is (i, i); $(L_v, R_v) = (l^*c(v), r^*c(v))$ for an internal node v. Finally, we define $P(T) = \{(L_v, R_v) \mid v \in T\}$.

Proposition 3. *For a phylogeny T of n leaves, the set $P(T)$ can be computed in $O(n)$ time.*

Proof. Note that $(L_v, R_v) = (L_{lc(v)}, R_{rc(v)})$, where $lc(v)$ and $rc(v)$ are the left and right child of v. The set $P(T)$ can be computed using the dynamic programming approach in a bottom up fashion: $(L_v, R_v) = (i, i)$ for a leaf v labeled with sequence i; for a non-leaf node v, we calculate (L_v, R_v) once (L_u, R_u) is known for $u = lc(v)$ and $u = rc(v)$. Since T contains $2n - 1$ nodes, the above procedure takes $O(n)$ times.

The set $P(T)$ will play a critical role in designing algorithms for reconstructing a DM from a phylogeny. To use $P(T)$ efficiently, we store each pair (L_v, R_v) at the node v by introducing an extra pointer. The following fact is straightforward.

Proposition 4. *The properties in Proposition 2 can be verified in $O(n)$ time for a phylogeny T of n nodes using the set $P(T)$.*

By Proposition 4, we may assume that a given phylogeny T satisfies the properties stated in Proposition 2. Otherwise, T cannot be associated with a DM. Note that the leftmost and rightmost leaf in T are labeled with 1 and n respectively. Where does the leaf labeled with 2 locate? Let

$$v_0 = r, v_1, v_2, \cdots, v_{p-1}, v_p = 1 \tag{1}$$

be the path from the root r to the leftmost leaf 1. Such a path is said to be the *leftmost path* of T. Since $l^*c(v) < l^*c(rc(v))$ for every v in T (Proposition 2), 2 can only appear as the leftmost leaf of a subtree rooted at some $rc(v_i)$, $0 \le i \le p$ (see Figure 2 (a)). Let

$$u_1 = rc(v_i), u_2, \cdots, u_{q-1}, u_q = 2 \tag{2}$$

be the path from $rc(v_i)$ to the leaf 2, where $q \ge p - i$ as we will see later.

Proposition 5. *Suppose that T is associated with a double DM $\mathcal{M}$ that is consistent with $F = \{1, 2, \cdots, n\}$. Then,*

(1) $\mathcal{M}$ contains a unique series of $p - i - 1$ double duplication blocks of the following form:

$$[v_{i+1}, u_{j_1}], [v_{i+2}, u_{j_2}], \cdots, [v_{p-1}, u_{j_{p-i-1}}],$$

where $1 \le j_1 < j_2 < \cdots < j_{p-i-1} \le q - 1$. Hence, $q \ge p - i$.

(2) Such a series of $p - i - 1$ blocks can be determined by using at most $2q$ comparisons, where q is the length of the left path leading to label 2 given in (2).

Proof. (1) Let k be an integer between 1 and $p - i - 1$. If v_{i+k} does not belong to a double block in $\mathcal{M}$, 2 cannot be placed before the leftmost leaf in the subtree rooted at $rc(v_{i+k})$, contradicting the fact that 2 is right next to 1 in $\mathcal{M}$. (A tedious full proof of this statement will appear in the final version of this paper.) Hence, v_{i+k} must appear in a double duplication block for each k, $1 \le k \le p - i - 1$.

Since the subtree rooted at $rc(v_{i+1})$ is left to the subtree rooted at u_1 in T, v_{i+1} forms a block with some u_{j_1}, $1 \le j_1 \le q - 1$. By Proposition 2, $R_{u_1} > R_{u_2} > \cdots > R_{u_{q-1}} > R_{u_q}$. Since the left-to-right ordering of the sequences is identical to increasing order in $\mathcal{M}$, v_{i+1} and u_{j_1} form a duplication block only if $R_{u_{j_1}} > R_{v_{i+1}} > R_{u_{j_1+1}}$. Hence, u_{j_1} is unique. Similarly, we can prove that u_{j_k} is unique for any $k > 1$.

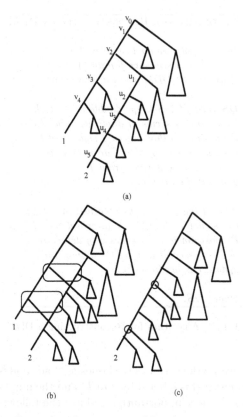

Fig. 2. *(a) Example illustrating the position of 2, where $p = 5$, $i = 2$ and $q = 6$. (b) The configuration after duplication blocks $[v_3, u_2]$ and $[v_4, u_4]$ are placed. (c) The tree T' derived after the duplication blocks are eliminated. The 'bad' nodes are marked with circles.*

(2) Notice that $R_{u_1} > R_{u_2} > \cdots > R_{u_{q-1}} > R_{u_q}$ and $R_{v_{i+1}} > R_{v_{i+2}} > \cdots > R_{v_{p-1}}$. By merging these two sequences in $p - i + q \leq 2q$ comparisons, we are able to determine all the u_{j_k}'s, $1 \leq k \leq p - i - 1$, by just scanning the merged sequences. This is because $R_{v_{i+k}}$ appears between $R_{u_{j_k}}$ and $R_{u_{j_k+1}}$ in the merged sequences. To implement this procedure efficiently, we need not only a pointer from a parent node to its child, but also a pointer from a child node to its parent.

Assume that all u_{j_k}'s have been determined for all $k = 1, 2, \cdots, p - i - 1$. After all the duplication blocks $[v_{i+k}, u_{j_k}]$ are placed on T, the leaf 2 should be right next to the leaf 1 (see Figure 2 (b)). We derive a rooted binary tree T'' from the subtree $T(u_1)$ by inserting a new node v'_k in the edge (u_{j_k}, u_{j_k+1}) for each $1 \leq k \leq p - i - 1$, and assigning the subtree $T(rc(v_{i+k}))$ rooted at $rc(v_{i+k})$ as the right subtree of v'_k. Notice that the left child of v'_k is u_{j_k+1}. Then, we form a new phylogeny T' from T by replacing subtree $T(v_i)$ by T'' as illustrated in Figure 2 (c). To distinguish the inserted nodes v'_k, $1 \leq k \leq p - i - 1$, from the

---

ALGORITHM FOR RECONSTRUCTING A DDM

Input: A phylogeny T with root r
on a sequence family $\{1, 2, \cdots, n\}$.
Output: The DDM derived from T if it exists.

$DS = \phi$; /* DS keeps double dup. blocks */
Determine a series of duplication blocks in
Proposition 5 that places 2 right next to 1;
if such a series of blocks do not exist, then
no DM can be derived from T
and exit the procedure.

Add all the dup. blocks obtained into DS;
Construct T' from T as described before
Proposition 6, and recursively determine
if T' is associated to a DM or not and
update DS accordingly.

Ouput DS.

Fig. 3. Algorithm for Reconstructing a DDM.

original ones in T, we mark these inserted ones as 'bad' nodes. It is not difficult to see that the leftmost leaf is labeled by 2 in T' and the number of nodes in T' is 1 less than those of T. Most importantly, we have the following straightforward fact.

Proposition 6. T *is associated with a double DM $\mathcal{M}$ only if T' is associated with the double DM $\mathcal{M}'$ that is the restriction of $\mathcal{M}$ on the nodes of T', in which the 'bad' nodes cannot participate double duplication blocks.*

Using Proposition 6, we obtain the algorithm in Figure 3 for reconstructing a double DM (DDM) from a phylogeny. By Proposition 5, we could charge the number of comparisons taken in different recursive steps to disjoint left paths in the input tree T. Since the total number of nodes in all the disjoint left paths is at most $2n$, the whole algorithm takes at most $2 \times 2n$ comparisons for determining all the duplication blocks. Therefore, we have

Theorem 1. *A double DM can be reconstructed from a given phylogenetic tree T of n leaves in $O(n)$ time if it exists.*

3.2 Generalization to Arbitrary Duplication Models

In this subsection, we generalize the algorithm presented above to arbitrary DMs. Such a generalization is based on the following proposition. Again, we assume the leftmost paths leading to sequence 1 and sequence 2 in T are given as (1) and (2) in the previous subsection.

Proposition 7. *Assume a phylogeny T is associated with a DM $\mathcal{M}$. Then, there exists a unique series of $p - i - 1$ double duplication blocks $[v_{i+k}, u_{j_k}]$ such that, after these duplications are placed in T, sequence 2 is right next to sequence 1. Notice that these double duplication blocks may not be in $\mathcal{M}$.*

Proof. Let k be an integer such that $1 \leq k \leq p - i - 1$. If v_{i+k} and u_j are not involved in a duplication block for any $j \leq q - 1$ in the model $\mathcal{M}$, sequence 2 cannot be moved across the leftmost leaf $l^* c(rc(v_{i+k}))$ in the subtree rooted at the right child of v_{i+k}, contradicting to the hypothesis that 2 is placed next to 1. Therefore, for any v_{i+k}, $1 \leq k \leq p - i - 1$, there exists a duplication block in $\mathcal{M}$ that contains v_{i+k} and u_j for some j. Restricting all the duplications in $\mathcal{M}$ on the leftmost paths specified in (1) and (2) gives a desired series of $p - i - 1$ double duplication blocks. ∎

We construct a new phylogeny T' from T in the same way as described before Proposition 6 in the previous subsection. Recall that there are two types of nodes on the leftmost path of T'. Some nodes are original ones in the input tree T; some are inserted due to duplication blocks we have considered. To extend the existing duplication blocks into large ones, we associated a flag to each original node on the leftmost path of T', which indicates whether the node is in an existing block or not. Let x be an original node on the leftmost path of T' involving in a duplication $[x_1, x_2, \cdots, x_t, x]$ of size $t + 1$ so far, then, there are t inserted nodes x_i' below x that corresponds to x_i for $i = 1, 2, \cdots, t$. To decide whether x is in a double duplication block in T', we need to consider x and all the x_i's ($1 \leq i \leq t$) simultaneously. We call such a double block a *hyper-double block*. If x involves in a hyper-double block $[x, y]$ in T', the block $[x_1, x_2, \cdots, x_t, x]$ is extended into block $[x_1, x_2, \cdots, x_t, x, y]$ of size $t + 2$ in the original tree T. With these notations in mind, we have

Proposition 8. *A phylogeny T is associated with a DM if and only if*
(1) There exists a unique series of $p-i-1$ double duplication blocks $[v_{i+k}, u_{j_k}]$ in T such that, after these duplication blocks are placed in T, sequence 2 is right next to sequence 1, and
(2) T' constructed above is associated to a DM.

Hence, the algorithm given in Subsection 3.1 can be used to reconstruct a DM from a phylogeny. To make the algorithm run in linear time, we assign a pair (R_x', R_x'') of indices to a node x on the leftmost path of T' in each recursive step: if x is in a duplication block $[x_1, x_2, \cdot, x_t, x]$ in the current stage, we let $R_x' = R_{x_1}$ and $R_x'' = R_x$, which are defined in the beginning of Subsection 3.1. Since $R_x' < R_{x_i} < R_x''$ for any $2 \leq i \leq t$, only R_x' and R_x'' will be examined for determining if x is in a hyper-double block in next step. In this way, we can reconstruct a DM from a phylogeny in linear time. Take the tree in Figure 4 as an example. To arrange 2 next to 1 in step 1, we obtain 3 hyper-double duplications $[v_1, v_2]$, $[v_3, v_5]$ and $[v_8, v_6]$. After this step, sequences $1, 2, 3, 4$ have been arranged in increasing order. To arrange 5 next to 4 in step 2, we obtain a hyper-double block $[v_5, v_4]$ and hence we extend the double duplication $[v_3, v_5]$ into $[v_3, v_5, v_4]$;

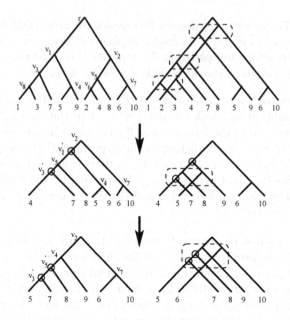

Fig. 4. Step-by-step demonstration of reconstructing a DM from a rooted tree. Here, we use a box drawn in break lines to denote a hyper-double block; we mark the inserted nodes with circles.

finally, to arrange 6 next to 5 in step 3, we derive a hyper-double block $[v_4, v_7]$ and further extend the duplication $[v_3, v_5, v_4]$ into $[v_3, v_5, v_4, v_7]$. After step 3, all sequences have been arranged in increasing order and the algorithm terminates successfully by outputting 3 duplications $[v_1, v_2]$, $[v_3, v_5, v_4, v_7]$ and $[v_8.v_6]$.

4   Inferring Duplication Models from Sequence Data

Let $F = \{s_1, s_2, \cdots, s_n\}$ be a set of tandem sequences of a fixed length l over an alphabet A and let $\mathcal{M}$ be a DM consistent with F, in which each node v is assigned with a sequence $s(v) \in A^l$, where $s(v)$ is in F if v is a leaf. We measure $\mathcal{M}$ using the following cost $C(\mathcal{M})$:

$$C(\mathcal{M}) = \sum_{(u,v) \in E(\mathcal{M})} dist(s(u), s(v)),$$

where $E(\mathcal{M})$ denotes the set of directed edges in $\mathcal{M}$.

By the parsimony principle, inferring a DM on F is to find a DM consistent with F and of minimum cost. Such an inference problem is NP-hard as proved in [10]. Therefore, there is unlikely a polynomial time algorithm for it. Here, we introduce an efficient heuristic for inferring a DM from sequence data using the algorithm presented in Section 3 and the *Nearest Neighbor Interchange* (NNI) operation for tree transformation (see Figure 5). Every edge, e, in a unrooted tree

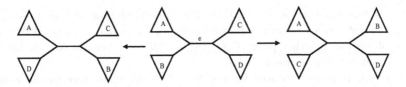

Fig. 5. *By a NNI operation on e, we can swap two subtrees at the different ends of e.*

partitions the set of leafs. We may consider alternate hypothesis associated with e by applying a NNI on e. This rearrangement generates alternate bipartition for e while leaving all the other parts of the tree unchanged (see [15] for more information).

Our method is analogous to the Feng-Doolittle algorithm for multiple alignments. First, a guided tree is constructed for the given set of sequences using the parsimony method or other efficient distance-based methods such as Neighbor-Joining (see [15]). Then, the NNI's are applied to the guided tree T for searching a DM that is nearest to T in terms of NNI operations. During the search process, we use the algorithm in section 3 for determining whether the resulting tree is associated with a DM or not. Since a phylogeny with n leaves can always be transformed into another by applying at most $n \log_2 n + O(n)$ NNI's [12], our algorithm will terminate quickly. For more efficiency, we can even use the following greedy approach to obtain a DM.

> **Greedy NNI Search:** We root the given guided tree at a fixed edge on the path from the leaf labeled with s_1 to the leaf labeled with s_n. For each $k = 2, 3, \cdots, n-1$, we repeatedly arrange s_k next to s_{k-1} using duplications as described in Section 3. If we cannot arrange s_k next to s_{k-1}, we use the NNI's to move the leaf labeled with s_k to the branch having s_{k-1} as an end so that these two leaves become siblings.

Greedy NNI Search is proposed based on the hypothesis that a tandem gene family has evolved mainly through single gene duplication events [2,16]. Since we can determine whether a rooted tree is associated with a duplication model in linear time, the Greedy NNI Search takes only quadratic time.

5   Analysis of Tandem Repeats

5.1   Mucin Gene MUC5B

Mucus consists mainly of mucin proteins, which are heterogeneous, highly glycosylated and produced from epithelial cells [8]. Four mucin genes MUC6, MUC2, MUC5AC and MUC5B have been identified within a 400kb genomic segment in human 11p15.5 [14]. All mucin genes contains a central part of tandem repeats. In the central part of MUC5B, a single large exon of about 11K bp [4], three Cyc-subdomains are followed by four repeats. Each repeat contains a R-subdomain

(11 or 17 tandem repeats of a motif of 29 amino acids), an R-End subdomain, and a Cys-subdomain. Another R-subdomain of 23 tandem repeats of 29 amino acids follows the fourth repeat. This suggests that the central part of MUC5B arose through tandem duplications.

The amino acid sequences for the five R-subdomains were taken from [6] and aligned using an alignment program at Michigan Tech [9]. We chose the following parameters - substitution matrix: Blosum62, mismatch score: -15, gap open penalty 10, gap extension penalty 5. The alignment was used to produce a guided tree by using the parsimony method in [7,13], which is rooted on the edge that partitions the set of leafs into $\{RI, RII, RIV\}$ and $\{RIII, RV\}$. This guided tree was then given as input to the Greedy NNI Searching algorithm described in last section. The resulting tandem DM for the central part of MUC5B agrees with the evolutionary history described verbally by Desseyn et al. in their original papers. The central repeat (Cyc/R/R-End subdomains) of an ancestral gene duplicated into two repeats. This event was then followed by a further duplication of a region containing all the two repeats. After that, the first repeat duplicated.

We also derive a tandem duplication hypothesis of the forth repeat segment RIV. It composed of 17 tandem repeats of an irregular motif of 29 amino acids rich in Ser and Thr. We obtained a guided parsimony tree from the multiple alignment of the 17 tandem repeats (RIV-1 to RIV-17) (see Figure6 (a)). Then, a duplication model was derived after 3 local NNI arrangements on the guided tree (see Figure 6 (b)). This model has the substitution score 126, which is better than the one with score 133 that we produced using Benson's greedy algorithm [2]. This hypothesis is also consistent with the conclusion drawn in [6].

5.2   ZNF and OLF Gene Families

Human genome contains roughly 700 *Krüppel*-type (C2H2) zinc-finger (ZNF) genes, which encode putative transcription factors, and 900 olfactory receptors (OR), which encode proteins that recognize distinct types of olfactants and that function in different regions of the olfactory epithelium. Because of their importance in species-specific aspects of biology, Dehal et al. analyzed the evolutionary relationship between C2H2 ZNF genes and OR genes in human chromosome 19 (HSA19) [3]. Two hundred sixty two C2H2 ZNF genes were identified in the HSA19 and were grouped into in 11 different locations (Z1 to Z11); forty nine OR genes were grouped into 4 location clusters (OLF1 to OLF4) (Table C and Table G in the supplementary material of [3]). *In situ* tandem duplication events are likely to have given rise to these features.

Because of their locations, we choose gene clusters ZNF45 and OLF3 and obtained gene-duplication history of them using our heuristic algorithm described in last section. The gene family ZNF45 contains 16 genes and OLF3 14 genes in HSA19q13.2. DNA or protein sequences for each gene family were aligned using the program at Michigan Tech [9]. The alignment of each gene family was used to produce a matrix of pairwise distances between the family members and then to produce a guided tree by using the parsimony method [7,13]. This guided tree was given as input to the Greedy NNI Searching algorithm described in last

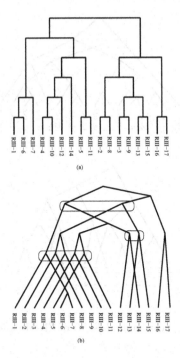

Fig. 6. The duplication history of the third repeat RIII in the MUC5B central part.

section. The resulting DMs for these two gene families are shown in Figure 7. Our algorithm performed better than the window method on the gene family ZNF45 according to parsimony principle. Our model for ZNF45 has substitution score 4690 while the one in [16] has score 4800, which were constructed from same alignment data.

These hypotheses in conjunction with other sequence analyzes could provide valuable clues into their functions at various stages of development.

6   Conclusion

In this paper, we present an optimal linear-time algorithm for reconstructing a DM from a phylogeny. Using this algorithm, we propose an efficient heuristic algorithm for inferring a DM from sequence data. Such an algorithm performed well on three test data sets - MUC5B [6], gene families ZNF45 and OLF3 [3]. Currently, we are conducting the simulation test for evaluating the algorithm further. Its performance on simulation data will be reported in the journal version of this paper.

Genome analysis suggests that sequence inversion and gene loss occurs during the course of tandem duplications. As a future research topic, the authors shall study how to incorporate gene sequence inversion and loss events into tandem duplication model and inferring algorithms.

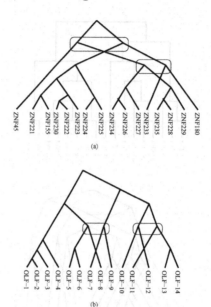

Fig. 7. The duplication models (a) of ZNF45 and (b) of OLF3.

Acknowledgments

Ma Bin would like to thank Chengzhi Liang for providing us the references on mucin genes and valuable discussions; he would also like to thank Mengxiang Tang for sharing their data. L. Zhang is financially supported in part by a NSTB grant LS/99/001. B. Ma is supported in part by NSERC RGP0238748. L. Wang is supported in part by a RGC grant CityU/1130/99E.

References

1. M.D. Adams *et al.* The genome sequence of *Drosophila melanogaster. Science* 287 (2000), 2185-2195.
2. G. Benson and L. Dong. Reconstructing the duplication history of a tandem repeat. In *Proceedings of the 7th ISMB*, 44-53, 1999.
3. P. Dehal *at al..* Human Chromosome 19 and related regions in Mouse: Conservative and lineage-specific evolution. *Science* 293 (2001), July, 104-111.
4. J. L. Desseyn *et al.* Human mucin gene MUC5B, the 10.7 -kb large central exon encodes various alternate subdomains resulting in a super-repeat. Structural evidence for a 11p15.5 gene family. *J. Biol. Chem* 272 (1997), 3168-3178.
5. J. L. Desseyn *et al.* Evolutionary history of the 11p15 human mucin gene family. *J. Mol Evol.* 46 (1998), 102-106.
6. J. L. Desseyn *et al.* Evolution of the large secreted gel-forming mucins, *Mol. Biol. Evol.* 17 (2000), 1175-1184.
7. J. Felsentein, PHYLIP, version 3.57c, Dept. of Genetics, Univ. of Washington, Seattle.

8. S. Ho and Y. Kim. Carbohydrate antigens on cancer-associated mucin-like molecules. *Semin. Cancer Biol.* 2(1991), 389-400.

9. X. Huang, Multiple Sequence Alignment, *http://genome.cs.mtu.edu/map/map.html.*

10. D. Jaitly *et al.*. Methods for reconstructing the history of tandem repeats and their application to the human genome. To appear on *J. Computer and Systems Sciences.*

11. J. Kim *et al.* Homology-Driven Assembly of a Sequence-Ready Mouse BAC Contig Map Spanning Regions Related to the 46-Mb Gene-Rich Euchromatic Segments of Human Chromosome 19. *Genomics* 74(2001), 129-141.

12. M. Li, J Tromp, and L. Zhang. On the nearest neighbor interchange distance between evolutionary trees. *Journal of Theoret. Biol.* 182(1996), 463-467.

13. A. Lim and L. Zhang. WebPHYLIP: A web interface to PHYLIP. *Bioinformatics* 12(1999), 1068-1069.

14. P. Pigny *et al. Genomics* 8(1996), 340-352

15. D. L. Swofford *et al.* Phylogeny inference. In *Molecular Systematic* (edited by D.M. Hillis *et al.* pages 407-514. Sinauer Associate Inc., MA, 1996.

16. M. Tang, M. Waterman, and S. Yooseph. Zinc finger gene clusters and tandem gene duplication. *RECOMB01*, 297-304, 2001.

17. The *Arabidopsis* Genome Initiative. *Nature* 408(2000), 796-815.

18. The *C. elegans* Sequencing Consortium. *Science* 282 (1998), 2012-2018.

19. L. Wang, and D. Gusfield. Improved approximation algorithms for tree alignment. *J. of Algorithms* 25(1997), 255-273, 1997.

20. L. Wang, T. Jiang, and E. Lawler. Approximation algorithms for tree alignment with a given phylogeny. *Algorithmica* 16(3), 302-315, 1996.

Genome Rearrangement Phylogeny Using Weighbor

Li-San Wang

Department of Computer Sciences, University of Texas
Austin, TX 78712 USA
lisan@cs.utexas.edu

Abstract. Evolution operates on whole genomes by operations that change the order and strandedness of genes within the genomes. This type of data presents new opportunities for discoveries about deep evolutionary rearrangement events. Several distance-based phylogenetic reconstruction methods have been proposed [12,21,19] that use neighbor joining (NJ) [16] with the expected breakpoint or inversion distances after k rearrangement events. In this paper we study the variance of the breakpoint and inversion distances. The result is combined with Weighbor [5], an improved version of NJ using the variance of true evolutionary distance estimators, to yield two new methods, Weighbor-IEBP and Weighbor-EDE. Experiments show the new methods have better accuracy than all previous distance-based methods, and are robust against model parameter misspecifications.

1 Background

Distance-Based Phylogenetic Reconstruction. A (phylogenetic) tree T on a set of taxa S is a tree representation of the evolutionary history of S: T is a tree leaf-labeled by S, such that the internal nodes reflect past speciation events. Given a tree T on a set S of genomes and given any two leaves i, j in T, we denote by $P(i, j)$ the path in T between i and j. We let λ_e denote the number of evolutionary events on the edge e during the evolution of the genomes in S within the tree T; we call it the *true evolutionary distance* between the two endpoints. We can then define the matrix of true evolutionary distances, $d_{ij} = \sum_{e \in P(i,j)} \lambda_e$, which is additive (a distance D is additive if it satisfies the *four-point condition*: for every distinct four points $\{i, j, l, m\}$, $D_{ij} + D_{lm} \leq \max\{D_{il} + D_{jm}, D_{im} + D_{jl}\}$). Given an additive matrix, many distance-based methods are guaranteed to reconstruct the tree T and the edge weights.

Neighbor Joining and Its Variants BioNJ and Weighbor. Neighbor joining (NJ) is the most popular distance-based tree inference method. The input to the method is a matrix of estimated leaf-to-leaf distances $\{D_{ij} | 1 \leq i, j \leq n\}$ on n leaves. Starting with these leaves as one-node subtrees, the algorithm creates new subtrees iteratively by picking two subtrees using a least-squares criterion of the distances to other roots of the subtrees (the pairing step), and updates

R. Guigó and D. Gusfield (Eds.): WABI 2002, LNCS 2452, pp. 112–125, 2002.
© Springer-Verlag Berlin Heidelberg 2002

the distances of the root of the new subtree to other roots of the subtrees (the distance update step) according to some least-squares criterion [7].

In reality, we do not get exact distance estimates between leaves due to the random nature of evolution. Atteson showed NJ is guaranteed to reconstruct the true tree topology if the input distance matrix is sufficiently close to a distance matrix defining the same tree topology [1]. Consequently, techniques that yield a good estimate of the matrix $\{d_{ij}\}$ are of significant interest.

In this paper we will use two modified versions of neighbor joining called BioNJ [7] and Weighbor [5]. Both methods use the variance of the true distance estimators in the pairing (Weighbor only) and distance update steps to improve the accuracy of the tree reconstruction.

Genome Rearrangement Evolution. Modern laboratory techniques yield the ordering and strandedness of genes on a chromosome, allowing us to represent each chromosome by an ordering of signed genes (where the sign indicates the strand). Evolutionary events can alter these orderings through rearrangements such as inversions and transpositions, collectively called genome rearrangements. Because these events are rare, they give us information about ancient events in the evolutionary history of a group of organisms. In consequence, many biologists have embraced this new source of data in their phylogenetic work [6,14,15]. Appropriate tools for analyzing such data remain primitive when compared to those developed for DNA sequence data; thus developing such tools is becoming an important area of research.

The genomes of some organisms have a single chromosome or contain single chromosome organelles (such as mitochondria [4] or chloroplasts [14,15]) whose evolution is largely independent of the evolution of the nuclear genome for these organisms. When each genome has the same set of genes and each gene appears exactly once, a genome can be described by an ordering (circular or linear) of these genes, each gene given with an orientation that is either positive (g_i) or negative ($-g_i$). If the genome is circular, we always represent the genome "linearly" so g_1 is positive and at the first position. A close inspection shows a circular genome with n genes, when represented linearly, is identical to a linear genome with $n - 1$ genomes.

Let G be the genome with signed ordering $g_1, g_2, \ldots, g_k$. An *inversion* between indices a and b, for $a \leq b$, produces the genome with linear ordering $(g_1, g_2, \ldots, g_{a-1}, -g_b, -g_{b-1}, \ldots, -g_a, g_{b+1}, \ldots, g_k)$. A *transposition* on the (linear or circular) ordering G acts on three indices, a, b, c, with $a \leq b$ and $c \notin [a, b]$, picking up the interval $g_a, g_{a+1}, \ldots, g_b$ and inserting it immediately after g_c. Thus the genome G above (with the assumption of $c > b$) is replaced by $(g_1, \ldots, g_{a-1}, g_{b+1}, \ldots, g_c, g_a, g_{a+1}, \ldots, g_b, g_{c+1}, \ldots, g_k)$. An *inverted transposition* is a transposition followed by an inversion of the transposed subsequence.

The Generalized Nadeau-Taylor Model. The Generalized *Nadeau-Taylor* (GNT) model [21] of genome evolution uses only genome rearrangement events which do not change the gene content. The model assumes that the number of each of the three types of events obeys a Poisson distribution on each edge, that the relative probabilities of each type of event are fixed across the tree, and that events of a given type are equiprobable. Thus we can represent a GNT model tree as a triplet $(T, \{\lambda_e\}, (\alpha, \beta))$, where the (α, β) defines the relative probabilities of transpositions and inverted transpositions (hence an event is an inversion with probability $1 - \alpha - \beta$). For instance, $(\frac{1}{3}, \frac{1}{3})$ indicates that the three event classes are equiprobable, while the pair $(0, 0)$ indicates that only inversions happen.

Estimating True Evolutionary Distances Using Genome Rearrangements. Let us be given a set of genome rearrangement events with a particular weighting scheme; the weight of a sequence of rearrangement events from this set is the sum of the weights of these events. The *edit distance* between two gene orders is the minimum of the weights of all sequences of events from the given set that transform one gene order into the other. For example, the *inversion distance* is the edit distance when only inversions are permitted and all inversions have weight 1. The inversion distance can be computed in linear time [2], and the transposition distance is of unknown computational complexity [3].

Another type of genomic distance that received much attention in the genomics community is the *breakpoint distance*. Given two genomes G and G' on the same set of genes, a *breakpoint* in G is an ordered pair of genes (g_a, g_b) such that g_a and g_b appear consecutively in that order in G, but neither (g_a, g_b) nor $(-g_b, -g_a)$ appear consecutively in that order in G'. The number of breakpoints in G relative to G' is the *breakpoint distance* between G and G'. The breakpoint distance is easily calculated by inspection in linear time.

Estimating the true evolutionary distance requires assumption about the model; in the case of gene-order evolution, the assumption is that the genomes have evolved from a common ancestor under the Nadeau-Taylor model of evolution. The technique in [4], applicable only to inversions, calculates this value exactly, while Approx-IEBP [21] and EDE [12], applicable to very general models of evolution, obtain approximations of these values, and Exact-IEBP [19] calculates the value exactly for any combination of inversions, transpositions, and inverted transpositions. These estimates can all be computed in low polynomial time.

Variance of Genomic Distances. The following problem has been studied in [17,19]: given any genome G with n genes, what is the expected breakpoint distance between G and G' when G' is the genome obtained from G by applying k rearrangements according to the GNT model? Both papers approach the problem by computing the probability of having a breakpoint between every pair of genes; by linearity of the expectation the expected breakpoint distance can be obtained by n times the aforementioned probability. Each breakpoint can be characterized as a Markov process with $2(n-1)$ states. But the probability of a breakpoint is a sum of $O(n)$ terms that we do not know yet how to further simplify.

However the variance cannot be obtained this way since breakpoints are not independent (under any evolutionary model) by the following simple observation: the probability of having a breakpoint for each breakpoint position is nonzero, but the probability of the breakpoint distance being 1 is zero (the breakpoint distance is always 0 or at least 2). Thus, to compute the variance (or the second moment) of the breakpoint distance we need to look at two breakpoints at the same time. This implies we have to study a Markov process of $O(n^2)$ states and a sum of $O(n^2)$ terms that is hard to simplify. As for the inversion distance, even the expectation is still an open problem.

Estimating the variance of breakpoint and inversion distances is important for several reasons. Based on these estimates we can compute the variances of the `Approx-IEBP` and `Exact-IEBP` estimators (based on the breakpoint distance), and the `EDE` estimator (based on the inversion distance). It is also informative when we compare estimators based on breakpoint distances to other estimators, e.g. the inversion distance and the `EDE` distance. Finally, variance estimation can be used in distance-based methods to improve the topological accuracy of tree reconstruction.

Outline of this Paper. We start in Section 2 by presenting a stochastic model approximating the breakpoint distance, and derive the analytical form of the variance of the approximation, as well as the variance of the `IEBP` estimators. In Section 3 the variance of the inversion and the `EDE` distances are obtained through simulation. Based on these variance estimates we propose four new methods, called `BioNJ-IEBP`, `Weighbor-IEBP`, `BioNJ-EDE`, and `Weighbor-EDE`. These methods are based on `BioNJ` and `Weighbor`, but the variances in these algorithms have been replaced with the variances of `IEBP` and `EDE`. In Section 4 we present our simulation study to verify the accuracy of these new methods.

2   Variance of the Breakpoint Distance

The Approximating Model. We first define the following notation in this paper: $\binom{a}{b}$ is the number of choosing b objects from a (the binomial coefficient) when $a \geq b \geq 0$; $\binom{a}{b}$ is set to 0 otherwise.

We motivate the approximating model by the case of inversion-only evolution on signed circular genomes. Let n be the number of genes, and b be the number of breakpoints of the current genome G. When we apply a random inversion (out of $\binom{n}{2}$ possible choices) to G, we have the following cases according to the two endpoints of the inversion [10]:

1. None of the two endpoints of the inversion is a breakpoint. The number of breakpoints is increased by 2. There are $\binom{n-b}{2}$ such inversions.
2. Exactly one of the two endpoints of the inversion is a breakpoint. The number of breakpoints is increased by 1. There are $b(n-b)$ such inversions.
3. The two endpoints of the inversion are two breakpoints. There are $\binom{b}{2}$ such inversions. Let g_i and g_{i+1} be the left and right genes at the left breakpoint, and let g_j and g_{j+1} be the left and right genes at the right breakpoint. There are three subcases:
   (a) None of $(g_i, -g_j)$ and $(-g_{i+1}, g_{j+1})$ is an adjacency in G_0. The number of breakpoints is unchanged.
   (b) Exactly one of $(g_i, -g_j)$ and $(-g_{i+1}, g_{j+1})$ is an adjacency in G_0. The number of breakpoints is decreased by 1.

(c) $(g_i, -g_j)$ and $(-g_{i+1}, g_{j+1})$ are adjacencies in G_0. The number of break-points is decreased by 2.

When $b \geq 3$, out of the $\binom{b}{2}$ inversions from case 3, case 3(b) and 3(c) count for at most b inversions; this means given that an inversion belongs to case 3, with probability at least $1 - b/\binom{b}{2} = \frac{b-3}{b-1}$ it does not change the breakpoint distance; this probability is close to 1 when b is large. Furthermore, when $b << n$ almost all the inversions belong to case 1 and 2. Therefore, when n is large, we can drop cases 3(b) and 3(c) without affecting the distribution of breakpoint distance drastically.

The approximating model we use is as follows. Assume first the evolutionary model is such that each rearrangement creates r breakpoints on an unrearranged genome (for example, $r = 2$ for inversions and $r = 3$ for transpositions and in-verted transpositions). Let us be given n boxes, initially empty. At each iteration r boxes will be chosen randomly (without replacement); we then place a ball into each of these r boxes if it is empty. The number of nonempty boxes after k iterations, b_k, can be used to estimate the number of breakpoints after k rear-rangement events are applied to an unrearranged genome. This model can also be extended to approximate the GNT model: at each iteration, with probability $1 - \alpha - \beta$ we choose 2 boxes, and with probability $\alpha + \beta$ we choose 3 boxes.

Mean and Variance of the Approximating Model. Fix n (the number of boxes) and k (the number of times we choose r boxes). Consider the expansion of the following expression

$$S = ((x_1 x_2 + x_1 x_3 + \cdots + x_{n-1} x_n)/\binom{n}{2})^k.$$

Each term corresponds to the number of ways of choosing $r = 2$ boxes for k times where the total number of times box i is chosen is the power of x_i, and the coefficient of that term is the total probability of these ways. For example, the coefficient of $x_1^3 x_2 x_3^2$ in S (when $k = 6$) is the probability of choosing box 1 three times, box 2 once, and box 3 twice. Let u_i be the coefficient of the terms with i distinct symbols; $\binom{n}{i} u_i$ is the probability i boxes are nonempty after k iterations. The identity of u_i for all terms of the same set of power indices holds as long as the probability of each box being chosen is identical; in other words, S is not changed by permuting $\{x_1, x_2, \ldots, x_n\}$ arbitrarily.

To solve for u_i exactly for all k is difficult and unnecessary. Instead we can find the expectation and variance of b_k directly. Actually the following results give all moments of b_k. Let $S(a_1, a_2, \ldots, a_n)$ be the value of S when we substitute x_i by a_i, $1 \leq i \leq n$, and let S_j be the value of S when $a_1 = a_2 = \cdots = a_j = 1$ and $a_{j+1} = a_{j+2} = \cdots = a_n = 0$. For integers j, $0 \leq j \leq n$, we have

$$\sum_{i=0}^{j} \binom{j}{i} u_i = S(\underbrace{1, 1, 1, \ldots, 1}_{j\ 1's}, 0, \ldots, 0) = S_j.$$

Let

$$Z_a = \sum_{i=0}^{n} i(i-1) \cdots (i-a+1) \binom{n}{i} u_i = \sum_{i=a}^{n} n(n-1) \cdots (n-a+1) \binom{n-a}{i-a} u_i$$

for all a, $1 \leq a \leq n$. We want to express Z_a by some linear combination of S_i, $0 \leq i \leq n$. The following lemma, which is a special case of equation (5.24) in [9], finds the coefficients of the linear combination.

Lemma 1. *Let a be some given integer such that $1 \leq a \leq n$. Let us be given $\{u_i : 0 \leq i \leq n\}$ that satisfy $\sum_{j=0}^{i} \binom{i}{j} u_j = \sum_{j=0}^{n} \binom{i}{j} u_j = S_i$, where $0 \leq i \leq n$. We have $\sum_{i=n-a}^{n} (-1)^{n-i} \binom{a}{n-i} S_i = \sum_{j=0}^{n} \binom{n-a}{j-a} u_j$.*

The following results follow from Lemma 1; we state them without proof due to space limitations.

Theorem 1. *For all a, $1 \leq a \leq n$,*

$$Z_a = n(n-1)\cdots(n-a+1) \sum_{i=n-a}^{n} (-1)^{n-i} \binom{a}{n-i} S_i.$$

Corollary 1. *(a) $Eb_k = Z_1 = n(1 - S_{n-1})$.*
(b) $Var\, b_k = nS_{n-1} - n^2 S_{n-1}^2 + n(n-1)S_{n-2}$.

These results work for all integers r, $1 \leq r \leq n$. When there are more than one type of rearrangement events with different r's we can change S accordingly. For example, let $\gamma = \alpha + \beta$; for the GNT model we can set

$$S = \left(\frac{1-\gamma}{\binom{n}{2}} \Big(\sum_{1 \leq i_1 < i_2 \leq n} x_{i_1} x_{i_2} \Big) + \frac{\gamma}{\binom{n}{3}} \Big(\sum_{1 \leq i_1 < i_2 < i_3 \leq n} x_{i_1} x_{i_2} x_{i_3} \Big) \right)^k. \tag{1}$$

Mean and Variance of the Breakpoint Distance under the GNT Model. We begin this section by finding the mean and variance for b_k with respect to the GNT model. By substituting into equation (1):

$$S_{n-1} = (1 - \frac{2+\gamma}{n})^k, \quad S_{n-2} = \left(\frac{(n-3)(n-2-2\gamma)}{n(n-1)} \right)^k.$$

For the GNT model, we have the following results:

$$\frac{d}{dk} Eb_k = -nS_{n-1}(\frac{1}{k} \ln S_{n-1}) = -nS_{n-1} \ln(1 - \frac{2+\gamma}{n}) \tag{2}$$

$$Var\, b_k = nS_{n-1} - n^2 S_{n-1}^2 + n(n-1)S_{n-2}$$
$$= (nS_{n-1} - n^2 S_{n-1}^2 + n(n-1)S_{n-2}) \tag{3}$$

Using BioNJ and Weighbor. Both BioNJ and Weighbor are designed for DNA sequence phylogeny using the variance of the true evolutionary distance estimator. In BioNJ, the distance update step of NJ is modified so the variances of the new distances are minimized. In Weighbor, the pairing step is also modified to utilize the variance information. We use the variance for the GNT model in this section and the expected breakpoint distance in [19] in the two methods. The new methods are called BioNJ-IEBP and Weighbor-IEBP. To estimate the

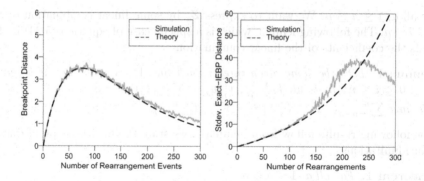

Fig. 1. Accuracy of the estimator for the variance. Each figure consists of two sets of curves, corresponding to the values of simulation and theoretical estimation. The number of genes is 120. The number of rearrangement events, k, range from 1 to 220. The evolutionary model is inversion-only GNT; see Section 1 for details. For each k we generate 500 runs. We then compute the standard deviation of b_k for each k, and those of $\widehat{k}(b_k)$ for each k, and compare them with the values of the theoretical estimation.

true evolutionary distance, we use `Exact-IEBP`, though we can also use Equation (2) which is less accurate. Let $\widehat{k}(b)$ denote the `Exact-IEBP` distance given the breakpoint distance is b; $\widehat{k}(b)$ behaves as the inverse of Eb_k, the expected breakpoint distance after k rearrangement events. The variance of $\widehat{k}(b)$ can be approximated using a common statistical technique, the delta method [13], as follows:

$$\text{Var } \widehat{k}(b) \simeq (\frac{d}{dk}Eb_k)^{-2}\text{Var } b_k = \frac{\left(1 - nS_{n-1} + (n-1)(\frac{S_{n-2}}{S_{n-1}})\right)}{nS_{n-1}(\ln(1 - \frac{2+\gamma}{n}))^2}.$$

When the number of rearrangements are below the number of genes (120 in the simulation), these results are accurate approximations to the mean and variance of the breakpoint distance under the GNT model, as the simulation in Figure 1 shows. As the number of rearrangements k is so high the breakpoint distance is close to the maximum (the resulting genome is random with respect to the genome before evolution), the simulation shows the variance is much lower than the theoretical formula. This is due to the application of the delta method: while the method assumes the `Exact-IEBP` distance is continuous, in reality it is a discrete function. The effect gets more obvious as k is large: different values of k all give breakpoint distances close to the maximum, yet the `Exact-IEBP` can only return one estimate for k, hence the very low variance. This problem is less serious as n increases.

3   Variance of the Inversion and EDE Distances

The EDE Distance. Given two genomes having the same set of n genes and the inversion distance between them is d, we define the EDE distance as $nf^{-1}(\frac{d}{n})$:

here n is the number of genes, and f, an approximation to the expected inversion distance normalized by the number of genes, is defined as (see [12]):

$$f(x) = \min\{x, \frac{ax^2 + bx}{x^2 + cx + b}\}$$

We simulate the inversion-only GNT model to evaluate the relationship between the inversion distance and the actual number of inversion applied. Regression on simulation results suggests $a = 1$, $b = 0.5956$, and $c = 0.4577$. As the rational function is inverted, we take the larger (and only positive) root:

$$x = \frac{-(b - cy) \pm \sqrt{(b - cy)^2 + 4(a - y)by}}{2(a - y)}.$$

Let $y = \frac{d}{n}$. Thus

$$f^{-1}(y) = \max\{y, \frac{-(b - cy) \pm \sqrt{(b - cy)^2 + 4(a - y)by}}{2(a - y)}\}.$$

Here the coefficients do not depend on n, since for different values of n the curves of the normalized expected inversion distance are similar.

Regression for the Variance. Due to the success of nonlinear regression in the derivation of EDE, we use the same technique again for the variance of the inversion distance (and that of EDE). However for different numbers of genes, the curves of the variance are very different (see Figure 2). From the simulation it is obvious the magnitudes of the curves are inversely proportional to the number of genes (or some kind of function of it).

We use the following regression formula for the standard deviation of the inversion distance normalized by the number of genes after nx inversions are applied:

$$g_n(x) = n^q \frac{ux^2 + vx}{x^2 + wx + t}.$$

The constant term in the numerator is zero because we know $g(0) = 0$. Let r be the value such that rn is the largest number of inversions applied; we use $r = 2.5$. Note that

$$\ln(\frac{1}{rn} \sum_0^{rn} g_n(x)) \simeq \ln(\frac{1}{r} \int_0^r g_n(x) dx) = q \ln n + \ln(\frac{1}{r} \int_0^r \frac{ux^2 + vx}{x^2 + wx + t} dx)$$

is a linear function of $\ln n$. Thus we can obtain q as the slope in the linear regression using n as the independent variable and $\ln(\frac{1}{rn} \sum_0^{rn} g_n(x))$ as the independent variable (see Figure 2(b); simulation results suggest the average of the curve indeed is inversely proportional to $\ln n$). When q is obtained we apply nonlinear regression to obtain u, v, w, and t using the simulation data for 40, 80, 120, and 160 genes. The resultant functions are shown as the solid curves in Figure 2, with coefficients $q = -0.6998$, $u = 0.1684$, $v = 0.1573$, $w = -1.3893$, and $t = 0.8224$.

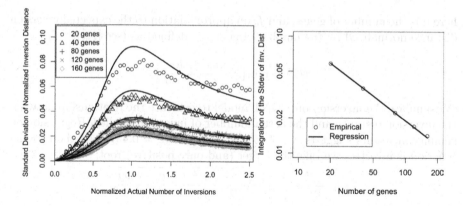

Fig. 2. *Left: simulation (points) and regression (solid lines) of the standard deviation of the inversion distance.* Right: regression of coefficient q (see Section 3); for every point corresponding to n genes, the y coordinate is the average of all data points in the simulation.

Variance of the EDE Distance. Let X_k and Y_k be the inversion and EDE distances after k inversions are applied to a genome of n genes, respectively. We again use the delta method. Let $x = \frac{k}{n}$. Since $X_k = nf(\frac{Y_k}{n})$, we have

$$\left|\frac{dY_k}{dX_k}\right|^{-1} = \left|\frac{dX_k}{dY_k}\right| = \frac{1}{n}\left|\frac{dX_k}{d(Y_k/n)}\right| = f'(x) = \frac{d}{dx}\left(\min\{x, \frac{x^2 + bx}{x^2 + cx + b}\}\right).$$

The point where $x = \dfrac{x^2 + bx}{x^2 + cx + b}$ is when $x = 0.5423$. Therefore

$$f'(x) = \begin{cases} 1 & \text{if } 0 \le x < 0.5423 \\ \frac{d}{dx}(\frac{x^2 + bx}{x^2 + cx + b}) = \frac{x^2(c - b) + 2bx + b^2}{(x^2 + cx + b)^2} & \text{if } x \ge 0.5423 \end{cases}$$

and $Var(Y_k) \simeq \left|\frac{dY_k}{dX_k}\right|^2 Var(X_k) = (f'(x))^{-2}(ng_n(x))^2 = (ng_n(\frac{k}{n})/f'(\frac{k}{n}))^2$.

4   Simulation Study

We present the results of two simulation studies in this section. The first experiment compares the accuracy of our new methods with other distance-based methods for genome rearrangement phylogeny. In the second experiment we show the robustness of `Weighbor-IEBP` against errors in the parameters of the GNT model, (α, β).

4.1   Experiment 1: Accuracy of the New Methods

In this experiment we compare the trees reconstructed using `BioNJ-IEBP`, `Weighbor-IEBP`, `BioNJ-EDE`, and `Weighbor-EDE` to neighbor joining trees using different distance estimators. The following four distance estimators are used

Table 1. Settings for Experiments 1 and 2.

Parameter	Value
1. Number of genes	120 (plant chloroplast genome)
2. Model tree generation	Uniformly Random Topology
(See the Model Tree paragraph in Section 4.1 for details.)	
4. GNT Model parameters $(\alpha, \beta)^{\dagger}$	$(0,0), (\frac{\cdot}{\cdot}, \frac{\cdot}{\cdot})$
5. Datasets for each setting	30

† The probabilities that a rearrangement is an inversion, a transposition, or an inverted transposition are $1 - \alpha - \beta$, α, and β, respectively.

with neighbor joining: (1) BP, the breakpoint distance, (2) INV [2], the inversion distance, and (3) Exact-IEBP [19] and (4) EDE [12], true evolutionary distance estimators based on BP and INV, respectively. The procedure of neighbor joining combined with distance X will be denoted by NJ(X). According to past simulation studies [21,12,19], NJ(EDE) has the best accuracy, followed closely by NJ(Exact-IEBP). See Table 1 for the settings for the experiment.

Quantifying Error. Given an inferred tree, we compare its "topological accuracy" by computing "false negatives" with respect to the "true tree" [11,8]. During the evolutionary process, some edges of the model tree may have no changes (*i.e. evolutionary events) on them. Since reconstructing such edges is at best guesswork, we are not interested in these edges. Hence, we define the true tree to be the result of* contracting *those edges in the model tree on which there are no changes.*

We now define how we score an inferred tree, by comparison to the true tree. For every tree there is a natural association between every edge and the bipartition on the leaf set induced by deleting the edge from the tree. Let T be the true tree and let T' be the inferred tree. An edge e in T is "missing" in T' if T' does not contain an edge defining the same bipartition; such an edge is called a false negative. *Note that the external edges (*i.e. edges incident to a leaf*) are trivial in the sense that they are present in every tree with the same set of leaves. The* false negative rate *is the number of false negative edges in T' with respect to T divided by the number of internal edges in T.*

Software. We use PAUP* 4.0 [18] to compute the neighbor joining method and the false negative rates between two trees. We have implemented a simulator [12,21] for the GNT model. The input is a rooted leaf-labeled model tree $(T, \{\lambda_e\})$, and parameters (α, β). On each edge, the simulator applies random rearrangement events to the circular genome at the ancestral node according to the model with given parameters α and β. We use the original Weighbor and BioNJ implementations [5,7] (downloadable from the internet) and make modifications so they use the new variance formulas.

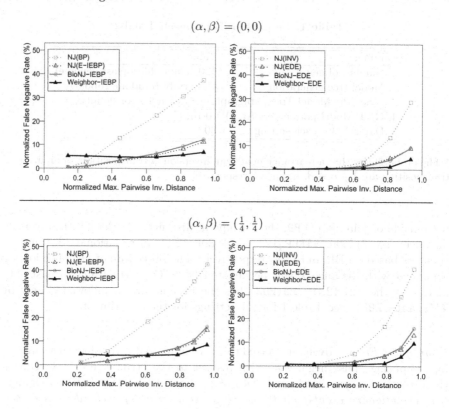

Fig. 3. The topological accuracy of various distance-based tree reconstruction methods. The number of genomes is 160. See Table 1 for the settings in the experiment.

Model Trees. The model trees have topologies drawn from the uniform distribution[1], and edge lengths drawn from the discrete uniform distribution on intervals $[1, b]$, where b is one of the following: 3, 6, 12, 18 (higher values of b makes the variance of the edge lengths higher). Then the length of each edge is scaled by the same factor so the diameter of the tree (the maximum pairwise leaf-to-leaf distance on the tree) is 36, 72, 120, 180, 360 (so low- to high-evolutionary rates are covered).

Discussion. The results of the simulation are in Figure 3. Due to space limitations, and because the relative order of accuracy remains the same, we describe our results only for a subset of our experiments. For each setting for simulation, we group methods based on the genomic distances they are based on: breakpoint or inversion distance. In either group, the relative order of accuracy is roughly the same. Let Y be the true distance estimator based on a genomic distance X;

---

[*] This is easily done and well known by the community by add one leaf at a time to produce the whole tree. At each iteration, we choose an edge from the current tree (each edge has the same probability to be chosen) and attach the new leaf to it.

e.g. Y=IEBP if X=BP, and Y=EDE if X=INV. The order of the methods, starting from the worst, is (1) NJ(X), (2) BioNJ-Y, (3) NJ(Y), and (4) Weighbor-Y (except for very low evolutionary rates when Weighbor-IEBP is worst, but only by a few percents). The differences between NJ(Y) and BioNJ-Y are extremely small. Weighbor-EDE has the best accuracy over all methods.

When we compare methods based on breakpoint distance and methods based on inversion distance, the latter are always better (or no worse) than the former if we compare methods of the same complexity: NJ(INV) is better than NJ(BP), NJ(EDE) is better than NJ(Exact-IEBP), BioNJ-EDE is better than BioNJ-IEBP, and Weighbor-EDE is better than Weighbor-IEBP. This suggests INV is a better statistic than BP for the true evolutionary distance under the GNT model, even when transpositions and inverted transpositions are present. This is not surprising as INV, just like BP, increases by a small constant when a rearrangement event from the GNT model is applied. Also, though their maximum allowed values are the same (the number of genes for circular signed genomes), the fact the average increase in INV is smaller[2] than the average increase in BP gives INV a wider effective range.

Note Weighbor-IEBP outperforms NJ(Exact-IEBP) when the normalized maximum pairwise inversion distance, or the diameter of the dataset, exceeds 0.6; Weighbor-IEBP (based on BP) is even better than NJ(EDE) (based on the better INV) when the diameter of the dataset exceeds 0.9. This suggests the Weighbor approach really shines under high amounts of evolution.

Running Time. NJ, BioNJ-IEBP, and BioNJ-EDE all finish within 1 second for all settings on our Pentium workstations running Linux. However, Weighbor-IEBP and Weighbor-EDE take considerably more time; both Weighbor-IEBP and Weighbor-EDE take about 10 minutes to finish for 160 genomes. As a side note, fast maximum parsimony methods for genome rearrangement phylogeny, MPME/MPBE, almost always exceed the four-hour running time limit in the experiment in [20]; Weighbor-EDE is almost as good as these two methods except for datasets with very high amount of evolution.

4.2 Experiment 2: Robustness of Weighbor-IEBP

In this section we demonstrate the robustness of the Weighbor-IEBP method when the model parameters are unknown. The settings are the same in Table 1. The experiment is similar to the previous experiment, except here we use

---

[.] An inversion creates two breakpoints; a transposition and an inverted transposition can be realized by three and two inversions, respectively, and they all create three breakpoints each. Thus under the GNT model with model parameters (α, β) and assumption that genome G has only a small breakpoint $(BP(G, G.))$ and inversion $(INV(G, G.))$ distance from the reference (ancestral) genome $G.$, the average increase in $BP(G, G.)$ after a random rearrangement is applied to G is $2(1 - \alpha - \beta) + 3\alpha + 3\beta = 2 + \alpha + \beta$ and the average increase in $INV(G, G.)$ is $(1 - \alpha - \beta) + 3\alpha + 2\beta = 1 + 2\alpha + \beta$. The latter is always smaller, and the two quantities are equal only when $\alpha = 1$, i.e. only transpositions occur.

both the correct and the incorrect values of (α, β) for the Exact-IEBP distance and the variance estimation. Due to space limitations the results are not shown here. The false negative curves are similar even when different values of α and β are used. These results suggest that Weighbor-IEBP is robust against errors in (α, β).

5   Conclusion and Future Work

In this paper we study the variance of the breakpoint and inversion distances under the Generalized Nadeau-Taylor model. We then used these results to obtain four new methods: BioNJ-IEBP, Weighbor-IEBP, BioNJ-EDE, and Weighbor-EDE. Of these Weighbor-IEBP and Weighbor-EDE yield very accurate phylogenetic trees, and are robust against errors in the model parameters. Future research includes analytical estimates of the expectation and variance of the inversion distance, and the difference between the approximating model and the breakpoint distance under the true GNT model.

Acknowledgements

The author thanks Tandy Warnow for suggesting the direction of research which leads to this paper, and the two anonymous referees for their helpful comments.

References

1. K. Atteson. The performance of the neighbor-joining methods of phylogenetic reconstruction. *Algorithmica*, 25(2/3):251–278, 1999.
2. D. A. Bader, B. M. E. Moret, and M. Yan. A linear-time algorithm for computing inversion distances between signed permutations with an experimental study. *J. Comput. Biol.*, 8(5):483–491, 2001.
3. V. Bafna and P. Pevzner. Sorting permutations by transpositions. In *Proc. 6th Annual ACM-SIAM Symp. on Disc. Alg. SODA95*, pages 614–623. ACM Press, 1995.
4. M. Blanchette, M. Kunisawa, and D. Sankoff. Gene order breakpoint evidence in animal mitochondrial phylogeny. *J. Mol. Evol.*, 49:193–203, 1999.
5. W. J. Bruno, N. D. Socci, and A. L. Halpern. Weighted neighbor joining: A likelihood-based approach to distance-based phylogeny reconstruction. *Mol. Biol. Evol.*, 17:189–197, 2000. http://www.t10.lanl.gov/billb/weighbor/.
6. S. R. Downie and J. D. Palmer. Use of chloroplast DNA rearrangements in reconstructing plant phylogeny. In P. Soltis, D. Soltis, and J.J. Doyle, editors, *Molecular Systematics of Plants*, volume 49, pages 14–35. Chapman & Hall, 1992.
7. O. Gascuel. BIONJ: an improved version of the nj algorithm based on a smple model of sequence data. *Mol. Biol. Evol.*, 14:685–695, 1997. http://www.crt.umontreal.ca/~olivierg/bionj.html.
8. O. Gascuel. Personal communication, April 2001.
9. R. L. Graham, D. E. Knuth, and O. Patashnik. *Concrete Mathematics*. Addison-Wesley, 1994. 2nd ed.

10. S. Hannenhalli and P. Pevzner. Transforming cabbage into turnip (polynomial algorithm for genomic distance problems). In *Proc. 27th Annual ACM Symp. on Theory of Comp. STOC95*, pages 178–189. ACM Press, NY, 1995.

11. S. Kumar. Minimum evolution trees. *Mol. Biol. Evol.*, 15:584–593, 1996.

12. B. M. E. Moret, L.-S. Wang, T. Warnow, and S. Wyman. New approaches for reconstructing phylogenies based on gene order. In *Proc. 9th Intl. Conf. on Intel. Sys. for Mol. Bio. (ISMB 2001)*, pages 165–173. AAAI Press, 2001.

13. G. W. Oehlert. A note on the delta method. *Amer. Statist.*, 46:27–29, 1992.

14. R. G. Olmstead and J. D. Palmer. Chloroplast DNA systematics: a review of methods and data analysis. *Amer. J. Bot.*, 81:1205–1224, 1994.

15. L. A. Raubeson and R. K. Jansen. Chloroplast DNA evidence on the ancient evolutionary split in vascular land plants. *Science*, 255:1697–1699, 1992.

16. N. Saitou and M. Nei. The neighbor-joining method: A new method for reconstructing phylogenetic trees. *Mol. Biol. & Evol.*, 4:406–425, 1987.

17. D. Sankoff and M. Blanchette. Probability models for genome rearrangements and linear invariants for phylogenetic inference. *Proc. 3rd Int'l Conf. on Comput. Mol. Bio. (RECOMB99)*, pages 302–309, 1999.

18. D. Swofford. *PAUP* 4.0*. Sinauer Associates Inc, 2001.

19. L.-S. Wang. Improving the accuracy of evolutionary distances between genomes. In *Lec. Notes in Comp. Sci. No. 2149: Proc. 1st Workshop for Alg. & Bio. Inform. WABI 2001*, pages 175–188. Springer Verlag, 2001.

20. L.-S. Wang, R. K. Jansen, B. M. E. Moret, L. A. Raubeson, and T. Warnow. Fast phylogenetic methods for the analysis of genome rearrangement data: An empirical study. In *Proc. 7th Pacific Symp. Biocomputing (PSB 2002)*, pages 524–535, 2002.

21. L.-S. Wang and T. Warnow. Estimating true evolutionary distances between genomes. In *Proc. 33th Annual ACM Symp. on Theory of Comp. (STOC 2001)*, pages 637–646. ACM Press, 2001.

Segment Match Refinement and Applications

Aaron L. Halpern, Daniel H. Huson*, and Knut Reinert

Informatics Research, Celera Genomics Corp.
45 W Gude Drive, Rockville, MD 20850, USA
Phone: +240-453-3000, Fax: +1-240-453-3324
{Aaron.Halpern,Knut.Reinert}@celera.com

Abstract. Comparison of large, unfinished genomic sequences requires fast methods that are robust to misordering, misorientation, and duplications. A number of fast methods exist that can compute local similarities between such sequences, from which an optimal one-to-one correspondence might be desired. However, existing methods for computing such a correspondence are either too costly to run or are inappropriate for unfinished sequence. We propose an efficient method for refining a set of segment matches such that the resulting segments are of maximal size without non-identity overlaps. This resolved set of segments can be used in various ways to compute a similarity measure between any two large sequences, and hence can be used in alignment, matching, or tree construction algorithms for two or more sequences.

1 Introduction

Given two (possibly incomplete) genomic sequences of two related organisms, or two separately derived sequences from the same organism, one may wish to know how the sequences relate to one another. In this paper, we introduce techniques that facilitate analyses for two particular questions. The first question is how much of each sequence is not represented in the other, under the restriction that positions in each sequence can be involved in at most one selected match. This may be of interest, e.g., in determining the relative degrees of completion of two sequencing projects. The second question is how the sequences may be most economically derived from one another by various rearrangements, duplications, insertions or deletions. This gives both a "transformation distance" [15] and an edit script implying specific differences between the sequences, whether these result from true evolutionary events or are artifacts of a sequence assembly project.

We assume the existence of a pre-computed set of local *segment matches* between two sequences. Such sets can be obtained efficiently by various techniques, e.g. [2,4,5,9]. However, a set of local segment matches is not disjoint: more than one match may cover a given interval, whereas the two analyses described above require either a one-to-one matching or a many-to-one matching.

---

* New address: WSI-AB, Tübingen University, Sand 14, 72076 Tübingen, Germany. Phone: +49-7071-29-70450, email: huson@informatik.uni-tuebingen.de

R. Guigó and D. Gusfield (Eds.): WABI 2002, LNCS 2452, pp. 126–139, 2002.

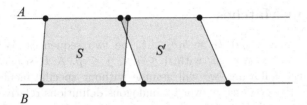

Fig. 1. Two partly overlapping matches.

Methods for rapidly constructing a nearly optimal global sequence alignment from such a set of matches have been presented [4,3], but such a global alignment ignores the treatment of transpositions, inversions, and duplications.

Another approach to selecting near-optimal many-to-one matches has been to find the set of matches that gives maximum coverage of the primary axis of interest without overlapping on that axis [14,15]. For our purposes, this treatment is incomplete in two respects. Firstly, such an approach may block the use of one match that only partially overlaps a larger match, where a solution making use of parts of two matches would be better than a solution making use of either one, as in Figure 1; although the treatment of matches which overlap on the primary axis is briefly mentioned in [15], no details are given. Secondly, such an approach does not readily extend to the problem of determining a one-to-one match, i.e. to choosing an optimal set of matches that don't overlap on either axis.

Finally, we note that the problem of determining an optimal one-to-one set of matches is not directly a classical matching problem, nor is it a form of "generalized sequence alignment" [13].

In the following, we introduce an efficient and provably optimal approach to the problem of constructing one-to-one or many-to-one match sets by minimally "refining" (subdividing) matches until all overlaps between the refined matches are "resolved": The projections of any two refined matches onto each sequence are either disjoint or are identical. Given such a resolved set of matches, several analyses that are otherwise difficult at best become quite straightforward, including not only the one-to-one matching problem and the many-to-one matching problems introduced above but also sequence alignment based on conserved segments [8]. For instance, we note that the one-to-one correspondence on the refined set is reduced to the classical matching problem. In addition, it is straightforward to compute the transformation distance or an alignment between the sequences if desired.

The paper is organized as follows. We first introduce segment matches in Section 2 and then discuss refinements and resolved refinements in Section 3. In Section 4 we then present our main result, an efficient algorithm for computing a resolved refinement. Applications of the algorithm to the "matched sequence problem", in the context of transformation distances and to the "maximal weight trace problem" are discussed in Section 5. Finally, in Section 6 we illustrate the feasibility of the discussed methods on real data.

2 Segment Matches

Let $A = a_1a_2\ldots a_p$ and $B = b_1b_2\ldots b_q$ be two sequences. We call $A_{ij} = a_{i+1}a_{i+2}\ldots a_j$ an A-*segment*, with $0 \leq i < j \leq p$. A B-*segment* is defined similarly. In the following, we will assume without specific mention that any definitions relative to A carry over to analogous definitions relative to B.

A *segment match* $S = (A_{ij}, B_{kl})$ consists of an A-segment A_{ij} and a B-segment B_{kl}. We call i, j, k and l the *left-A*, *right-A*, *left-B* and *right-B* positions of S, respectively. Given a set of segment matches Σ, let $\text{supp}_A(\Sigma)$ denote the A-support of Σ, i.e. the set of left-A and right-A positions of all matches $S \in \Sigma$.

We assume that every segment match $S = (A_{ij}, B_{kl})$ comes with two *projection maps* $\alpha_S : [i,j] \to [k,l]$ and $\beta_S : [k,l] \to [i,j]$ that reflect e.g. the alignment associated with S, as depicted in Figure 2.

In the case of a direct (i.e. orientation preserving) match, we require that α_S and β_S are *non-crossing*, i.e. that the two maps are monotonically increasing and that we have $\alpha_S(h) \leq h' \Leftrightarrow h \leq \beta(h')$ for all $h \in [i,j]$ and $h' \in [k,l]$.

In the case of a transposition, we require that the projections are *(completely) crossing*, i.e. that they are monotonically decreasing and that we have $\alpha_S(h) \leq h' \Leftrightarrow h \geq \beta(h')$ for all $h \in [i,j]$ and $h' \in [k,l]$.

When the alignments in matches are gapless, one can use linear interpolation in place of the α and β projections, i.e. by setting

$$\alpha_S(h) := \begin{cases} k + \frac{l-k}{j-i}(h - i), & \text{if } S \text{ is a direct match, and} \\ l - \frac{l-k}{j-i}(h - i), & \text{else,} \end{cases}$$

for every $S = (A_{ij}, B_{kl}) \in \Sigma$ and $i \leq h \leq j$.

A match $S' = (A_{i'j'}, B_{k'l'})$ is called a *submatch of* S, if $i \leq i' \leq j' \leq j$, and we have $\alpha_{S'}(i') = k'$ or $\beta_{S'}(k') = i'$, and $\alpha_{S'}(j') = l'$ or $\beta_{S'}(l') = k'$. Here, we assume that α'_S, and β'_S, equals the restriction of α_S to $[i',j']$, and of β_S to $[k',l']$, respectively.

3 Resolved Refinement

Consider a match $S = (A_{ij}, B_{kl}) \in \Sigma$. A set of matches $\Sigma' = \{S_1, S_2, \ldots, S_r\}$ is called a *refinement of* S, if each $S' \in \Sigma'$ is a submatch of S and Σ' *tiles* S, i.e. we the following two disjoint unions:

$$(i,j] = \bigcup_{(A_{i'j'}, B_{k'l'}) \in \Sigma'} (i',j'] \quad \text{and} \quad (k,l] = \bigcup_{(A_{i'j'}, B_{k'l'}) \in \Sigma'} (k',l'].$$

Definition 1. *A set Σ' of matches is called a* refinement *of $\Sigma = \{S_1, S_2, \ldots, S_n\}$, if there exists a partitioning $\Sigma' = \Sigma'_1 \dot\cup \Sigma'_2 \dot\cup \ldots \dot\cup \Sigma'_n$ such that Σ'_i is a refinement of S_i for every $i = 1, 2, \ldots, n$.*

We are particularly interested in refinements of Σ that have the property that the segments of any two matches are either disjoint or are identical, and capture this property as follows:

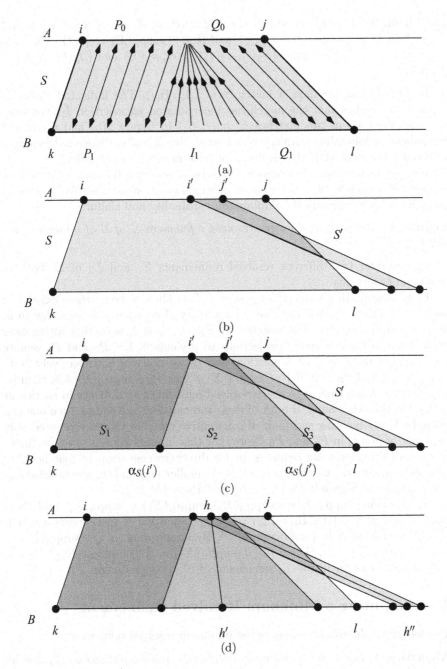

Fig. 2. Refining and resolving matches. (a) A valid match $S = (A_{ij}, B_{kl})$, together with arrows depicting α_S (pointing down) and β_S (pointing up). (b) The set $\Sigma = \{S, S'\}$ is unresolved. (c) The set $\{S., S., S., S'\}$ is a minimum resolved refinement of Σ. (d) This set is a non-minimum resolved refinement of Σ.

Definition 2. *A set of matches Σ is called* resolved, *if for any $S = (A_{ij}, B_{kl}) \in \Sigma$ we have $[i, j] \cap supp_A(\Sigma) = \{i, j\}$ and $[k, l] \cap supp_B(\Sigma) = \{k, l\}$. For technical reasons, we also require $\{\alpha_S(i), \alpha_S(j)\} \subseteq supp_B(\Sigma)$ and $\{\beta_S(k), \beta_S(l)\} \subseteq supp_A(\Sigma)$.*

In Figure 2(a) we show a match $S = (A_{ij}, B_{kl})$. The indicated choice of α_S and β_S implies that the match has incorporated an insertion I_1. We show an unresolved set of two matches $\Sigma = \{S, S'\}$ in Figure 2(b) and a resolved refinement of minimum cardinality in Figure 2(c). Finally, Figure 2(d) shows a resolved refinement of Σ that is does not have minimum cardinality.

Any set of matches Σ possesses a (trivial) resolved refinement obtained by simply refining every $S \in \Sigma$ into a set of single position matches. We are interested in resolved refinements of minimum cardinality and claim:

Lemma 1. *There exists a unique resolved refinement $\bar{\Sigma}$ of Σ of minimum cardinality.*

Proof. Consider two different resolved refinements Σ_1 and Σ_2 of Σ, both of minimum cardinality.

First, assume that $supp_A(\Sigma_1) \neq supp_A(\Sigma_2)$. Then, w.l.o.g., there exists $h \in supp_A \Sigma_1 \setminus supp_A \Sigma_2$. We say that an arbitrary A-position p *is attracted to* h, if there exists a sequence of matches $S_1, S_2, \ldots, S_r \in \Sigma$ such that alternating application of their α and β projections to p yields h. Let P_A and P_B denote the "basin of attraction" of A-positions and B-positions for h, respectively. Note that $P_A \neq \emptyset$ and $P_B \neq \emptyset$. If $P_A \cap supp_A(\Sigma) \neq \emptyset$ or $P_B \cap supp_B(\Sigma) \neq \emptyset$, then by Definition 2, h must occur in *every* resolved refinement of Σ, hence in particular in Σ_2. On the other hand, if both of these intersections are empty, then one can simplify Σ_1 by merging each pair of consecutive matches that is separated only by a pair of points in P_A and P_B. Note that this simplification is possible due to the non-crossing/crossing requirement for direct/reverse segment matches. We thus obtain a resolved refinement of Σ of smaller cardinality, a contradiction. For example, in Figure 2(d), $P_A = \{h\}$ and $P_B = \{h', h''\}$.

Now, assume $supp_A(\Sigma_1) = supp_A(\Sigma_2)$, $supp_B(\Sigma_1) = supp_B(\Sigma_2)$ and there exists a match $S = (A_{ij}, B_{kl}) \in \Sigma_1 \setminus \Sigma_2$. By Definition 1, there exists a match $S' \in \Sigma$ such that S is a submatch of S'. By assumption, $[i, j] \cap supp_A(\Sigma_2) = [i, j] \cap supp_A(\Sigma_1) = \{i, j\}$ and $[k, l] \cap supp_B(\Sigma_2) = [k, l] \cap supp_B(\Sigma_1) = \{k, l\}$, and so S must occur in the Σ_2 refinement of S', a contradiction.

4   Computing a Minimum Resolved Refinement

The following algorithm computes the minimum resolved refinement:

Algorithm 1 *Input: set of segment matches Σ and projections α_S, β_S for all $S \in \Sigma$. Output: resolved refinement $\bar{\Sigma}$ of Σ of minimum cardinality. We maintain a bipartite graph $G = (V_A \dot{\cup} V_B, E \subseteq \Sigma \times V_A \times V_B)$ with node sets $V_A \subseteq \{a_0, \ldots, a_p\}$ and $V_B \subseteq \{b_0, \ldots, b_q\}$ representing A- and B-positions, respectively, and a set E of (labeled) edges that keep track of pairs of positions projected onto each other by a given sequence match.*

First, for a set W_A of A-positions, we define the A-refinement step as follows:
   If this is not the first execution of this step:
      Set $W_B = \emptyset$.
   For each $h \in W_A \setminus V_A$:
      For each match $S = (A_{ij}, B_{kl}) \in \Sigma$ such that $i \leq h \leq j$:
         With $h' := \alpha_S(h)$, add the edge (S, h, h') to G (creating nodes as necessary) and insert h' into W_B.
Second, for a set W_B of B-positions, we define the B-refinement step as follows:
   Set $W_A = \emptyset$.
   For each $h' \in W_B \setminus V_B$:
      For each match $S = (A_{ij}, B_{kl}) \in \Sigma$ such that $k \leq h' \leq l$:
         With $h := \beta_S(h')$, add the edge (S, h, h') to G (creating nodes as necessary) and insert h into W_A.
Third:
   Initialize $V_A = \emptyset$, $W_A = supp_A(\Sigma)$, $V_B = \emptyset$ and $W_B = supp_B(\Sigma)$.
   Repeat until no further refinement is necessary (i.e. until either $W_A = \emptyset$ or $W_B = \emptyset$):
      Apply the A-refinement step to W_A and then the B-refinement step to W_B.
Finally:
   Lexicographically order the set of edges E, using (S, i, k) or $(S, i, -k)$, depending on whether α_S and β_S are non-crossing or crossing, respectively. Then, set $\bar{\Sigma} := \{(A_{ij}, B_{kl}) \mid (S, i, j)$ and (S', k, l) are consecutive and $S = S'\}$.

Lemma 2. *Algorithm 1 computes the minimum resolved refinement $\bar{\Sigma}$ of Σ.*

Proof. In the first iteration of the A-refinement step, all original A-positions are projected onto B using all segment matches. Then, all original and new B-positions are projected onto A. We then repeatedly project the newly generated positions back and forth until no new positions in A or B are reached. At this stage, V_A and V_B are the A- and B-supports, respectively, of a refinement of Σ. Because every position was produced by a chain of projections starting from a position coming from an original match in Σ, there are no superfluous matches and the resulting set has minimum cardinality.

In the final extraction step, for each match $S \in \Sigma$, the refinement of S based on the set M of all position pairs in S is computed.

Lemma 3. *Given a set of n segment matches Σ between two sequences $A = a_1 a_2 \ldots a_p$ and $B = b_1 b_2 \ldots b_q$, with $p \geq q$. The graph G in Algorithm 1 has at most $O(kp)$ edges and can be computed in at most $O(p \log^2 n + kp \log p)$ steps, where k is the maximal number of segment matches containing any given position.*

Proof. The maximum number of different A-positions is p. To insert an A-position h into V_A, we first determine which segment matches contain it in

$O(\log^2 n + k)$ steps, using a range tree [10,17]. For each of the $O(k)$ matches $S \in \Sigma$ that contains h, determine whether $\alpha_S(h)$ is contained in V_B in $O(\log p)$ steps. Putting this together, we obtain a bound of $O(p(\log^2 n + k + k \log p)) = O(p \log^2 n + kp \log p)$. Note that each of $O(p)$ positions gives rise to at most k edges in G.

Lemma 4. *Given a set of n segment matches Σ between two sequences $A = a_1 a_2 \ldots a_p$ and $B = b_1 b_2 \ldots b_q$, with $p \geq q$. The minimum resolved refinement $\bar{\Sigma}$ of Σ can be computed in at most $O(p \log^2 n + kp \log kp)$ steps, where k is the maximal number of segment matches containing any given position.*

Proof. The preceding result implies that the ordering step in Algorithm 1 will take at most $O(kp \log kp)$ computations.

Note that the size of a minimum resolved refinement depends on the length of the input sequences. For example, a set of only four segment matches between segments of length d, say, can give resize to $4d$ matches, as demonstrated in Figure 3.

5   Applications

In this section we describe several comparison algorithms for large genomic sequences that operate on sets of resolved matches and discuss their worst case time bounds.

5.1   The Matched Sequence Problem

Given two sequences A and B and a set of segment matches between them. The *Maximal Matched Sequence Problem* (MMSP) is to compute a set of non-intersecting matches Σ' that are all submatches of Σ, such that the amount of sequence covered by matched segments is maximized.

Let Σ be a resolved set of segment matches between sequences A and B. The associated *match* graph is an edge-weighted bipartite graph $G = (V, E, \omega : E \to \mathbb{N})$. For the node set $V = V_A \cup V_B$, we define V_A, or V_B, as the set of all intervals $(i, j]$, or $(k, l]$ respectively, for which there exists a segment match $(A_{ij}, B_{kl}) \in \Sigma$. We connect any two such nodes $(i, j] \in V_A$ and $(k, l] \in V_B$ by an edge $e \in E$, if $S = (A_{ij}, B_{kl}) \in \Sigma$, and set $\omega(e)$ to the "length" of S.

Lemma 5. *Given a resolved set of n segment matches Σ. The Maximal Matched Sequence Problem can be solved in at most $O(n^2 \log n)$ steps.*

Proof. Any maximal weight matching of the match graph $G = (V, E)$ gives rise to a solution. Such a matching can be computed in $O(v(e + v \log(v)))$ time, where $v = |V|$ and $e = |E|$. Since $v \leq 2n$ and $e = n$, the claim follows.

The preceding results imply:

Theorem 2. *Let Σ be a set of n matches between sequences $A = a_1 a_2 \ldots a_p$ and $B = b_1 b_2 \ldots b_q$, with $p \geq q$. The Maximal Matched Sequence Problem can be solved in at most $O((kp)^2 \log(kp))$ steps, where k is the maximal number of matches containing any given position.*

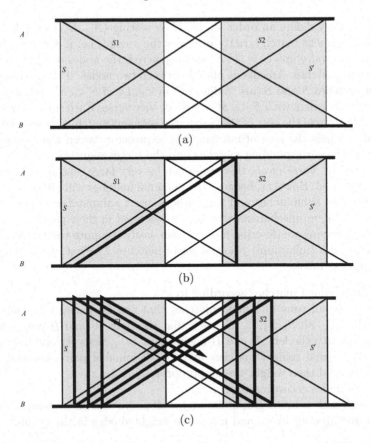

Fig. 3. (a) Matches S and S' (shown in grey) map two A-segments of length d onto two B-segments of length d. Diagonal match $S.$ maps the B-segment of S onto the A-segment of $S.$, whereas $S.$ maps the A-segment of S onto the B-segment of $S.$ with a shift of h base pairs. (b) The left-B position of $S.$ forces a refinement of match S' and then, subsequently, of match $S..$ (c) This leads to a chain of refinements, producing $O(\frac{d}{h})$ matches.

Proof. The minimum resolved refinement of Σ has size $O(kp)$ and can be computed in $O(p \log^2(n) + kp \log(kp))$ steps. On the set of refined matches, the MMSP problem can be solved in at most $O((kp)^2 \log(kp)))$ steps. Hence, the total running time is at most $O((kp)^2 \log(kp) + p \log^2(n) + kp \log(kp)) = O((kp)^2 \log(kp))$.

5.2   The Transformation Distance Problem

The *transformation distance* introduced in [15] provides a measure for comparing two large sequences A and B based on computing an optimal script that uses copy, reverse-copy and insertion operations to build B from segments of A. The input considered by Varré *et al.* consists of a set of segment matches Σ that are so-called *factors*, i.e. matches between segments of A and B allowing at most one mismatch. They define an order $<_0$ on Σ by setting $(A_{ij}, B_{kl}) <_0 (A_{i'j'}, B_{k'l'})$ if the first segment match strictly precedes the second, i.e. if $j < i'$ and $l < k'$. A weighted *script graph* $G = (V, E)$ is constructed, the nodes of which represent the segment matches. An edge is placed between two nodes, if the corresponding segment matches S and S' are "adjacent" in $<_0$, i.e. if $S <_0 S'$ and no segment match $S'' \in \Sigma$ exists with $S <_0 S'' <_0 S'$, or vice versa. Each node is assigned a weight that reflects the cost of the associated segment match, whereas the weight of an edge equals the cost of inserting the sequence between the two segments covered by the corresponding matches into the target sequence.

As noted by Varré *et al.*, this graph can be very large, and so only maximal matches are used. However, because the maximal matches will often overlap, one needs access to submatches and the authors insert submatches in the graph as necessary, using an mechanism that is not described in their paper.

In the following we describe how one can easily compute the transformation distance given a (minimum) resolved set of matches Σ. First, we construct the *script graph* as follows:

1. Sort the refined matches according to $<_0$.
2. Determine all runs of refined matches that can be joined to create longer matches, i.e. all $S_1, S_2, \ldots, S_k$ such that the right A- and B-positions of S_i coincide with the left A- and B-positions of S_{i+1}, respectively. Insert nodes for all original refined matches and for all joinable pairs, triplets, etc. of matches and then weight them appropriately [15].
3. Sort all nodes according to $<_0$.
4. For each node n in the graph, determine all nodes that correspond to matches that are adjacent in $<_0$ and insert the weighted edge in the graph.

Obviously this graph is minimum in the sense that our refinement is minimum and we only add new nodes that correspond to two refined matches that are joinable. In other words, we insert all original matches and all those submatches that are *necessary* to compute the transformation distance. (We note that in the case of an affine weight function for joinable matches one can omit the insertion of new nodes in step 2 and instead adjust the edge weights between joinable nodes, which gives rise to a problem that is computationally much easier to solve.)

Theorem 3. *Let Σ be a set of n matches between sequences $A = a_1 a_2 \ldots a_p$ and $B = b_1 b_2 \ldots b_q$, with $p \geq q$. Let m be the length of the longest run of 'joinable' matches. Then the Transformation Distance Problem can be solved in at most $O(kpm^2 \log(kpm))$ steps.*

Proof. As mentioned above we can compute the minimum resolved refinement of Σ in $O(p(\log^2(n) + kp \log(kp))$ steps. In step 1 we have to sort the $O(kp)$ matches. Determining the runs of joinable matches can be done in linear time. There can be at most $O(kp/m)$ such runs where each run gives rise to $O(m^2)$ new nodes for a total of $O(kpm)$ new nodes. In step 3 we have to sort the nodes in time $O(kpm \log(kpm))$. Finally we insert edges between adjacent nodes in step 4. Each node in a joinable run of length m is adjacent to $O(m)$ nodes. So in the worst case we have to insert $O(kpm^2)$ edges. Hence the edge insertion has a dominating run time of $O(kpm^2 \log(kpm))$.

5.3   The Trace Problem

If the sequences under consideration are small enough to assume a minor impact of translocations and duplications, then the refined set of segments can be used to compute a *maximal weight trace*.

The original formulation of the trace problem was introduced by Kececioglu [7]. Two characters of distinct strings in S are said to be *aligned* if they are placed into the same column of the alignment array. One can view the character positions of the k input strings in S as the vertex set V of a k-partite graph $G = (V, E)$ called the input *alignment graph*. The edge set E represents pairs of characters that one would like to have aligned in an alignment of the input strings. We say that an edge is *realized* by an alignment if the endpoints of the edge are placed into the same column of the alignment array. The subset of E realized by an alignment is called the *trace* of $\hat{S}$, denoted by $\text{trace}(\hat{S})$.

The original formulation of [7] assumes single character matches, where Reinert et al. [8] generalized it to allow for block matches. However, they used a refinement into single character matches. This immediately affects the time complexity of their separation method in the branch-and-cut algorithm described in [8].

Similar to the transformation distance, this formulation can also immediately benefit from a minimum fully refined set of matches by reducing the size of the input alignment graph.

If we apply a suitable scoring scheme for the refined matches (e.g. [1,11,16]), we can construct the alignment graph analogous to the match graph in Section 5.1 (note that it can easily be extended to the comparison of multiple sequences). Then the following holds:

Theorem 4. *Let Σ be a set of n matches between sequences $A = a_1 a_2 \ldots a_p$ and $B = b_1 b_2 \ldots b_q$, with $p \geq q$. The Trace Problem can be solved in at most $O(p(\log^2(n) + kp \log(kp))$ steps.*

Proof. As mentioned above we can compute the minimum resolved refinement of Σ in $O(p(\log^2(n) + kp \log(kp))$ steps resulting in a minimum resolved refinement of size $O(kp)$. As noted in the proof of Lemma 5 the number of distinct nodes in the alignment graph is $O(kp)$. The trace problem for two sequences can be reduced to computing the heaviest common subsequence which is possible in time $O((kp + kp) \log kp)$ [6]. This time is dominated by the construction of the refinement.

6   Illustration

To illustrate the strengths and weaknesses of the match refinement method described above, we use it to evaluate an alternative, greedy approach to finding a one-to-one matching of two large sequence sets.

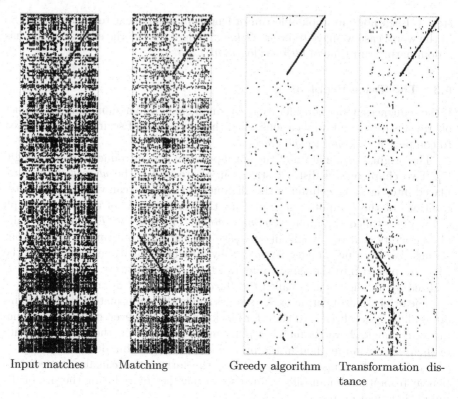

| Input matches | Matching | Greedy algorithm | Transformation distance |

Fig. 4. Illustration of method on two Drosophila assemblies.

The greedy technique was developed earlier to evaluate progress in assembling the human genome (Halpern, Huson, Delcher, Sutton, and Myers, unpublished), for which we needed a pipeline that is extremely fast, reasonable in memory usage, and is capable of finding matches to most repeated elements but also chooses the intuitively "correct" match between repeats. Such a pipeline might be useful for assessing a modification in an assembly algorithm, in determining the relative degrees of completion of two separate genome sequencing projects [12], or in identifying corresponding pieces between incremental assemblies for the purpose of annotation transfer.

In a nutshell, the method conducts a suffix-tree-based all-against-all search of two assemblies [5], and then greedily determines a set of one-to-one matches. The greedy phase first constructs a priority queue of all resulting segment matches ordered by decreasing length and initializes the set of selected (one-to-one) matches to null. Then, the match at the head of the priority queue is popped and compared against the set of selected matches. If it does not overlap any selected match, then it is added to the selected set, otherwise, the match is trimmed to eliminate all such overlaps and the shorted match is reinserted into the queue. This is repeated until the match at the head of the priority queue is shorter than some cutoff.

Although the results of the greedy method appeared quite good on informal inspection, the question arose just how close to maximal the obtained results were. The method described in this paper permitted us to answer this problem definitively for a given dataset.

For illustration here, we computed an initial match set of gapless alignments between two (partial) versions of the drosophila genome. The first was obtained from the BDGP (http://www.fruitfly.org/sequence/dlMfasta.shtml), corresponding to 50,107,720 bp of two finished chromosome arms (2L and 3R). The second was a unpublished assembly of whole genome shotgun sequencing data conducted at Celera in late 2001. The suffix-tree method yielded 292,769 alignments seeded with a perfect match of at least 50 bp. These matches covered 49,953,627 bp of the finished sequence (i.e., they covered all but 154,093 bp). Precisely 43,740,070 bp were covered by a single match, but the remainder was covered two or more times; since the total length of all matches was 125,863,932, a relatively small fraction of the sequence is clearly involved in a large number of matches.

The refinement method described in this paper transformed the match set into 17,167,891 distinct matches. This took a total of 880 seconds on a Unix workstation (667 Mhz Alpha processor) and shows that even for a large data set the refinement is computed quickly for a realistic set of input matches. Thus, the initial match set, with mean length 429 bp, was cut up into much smaller pieces (mean length 7 bp) by the refinement. However, since most of the original sequence was covered by a single match, there should still be matches of considerable size; indeed, the maximum length refined match was 466,877 bp, and of the finished sequence covered by matches, half is covered by a refined match of length at least 97,008 bp (this is known as the "N50" value). Again, the complexity in the match set is restricted to the repetitive regions of the sequence.

An optimal matching was determined on the refined match set, as described above. The resulting set of matches involved 262,885 refined matches and covered 49,875,599 bp, leaving 232,121 bp of the finished sequence uncovered. Due to the refinement process, groups of selected matches may correspond to (parts of) original matches, so it may be of greater interest to consider runs of adjacent, consistent refined matches as a single matching block; 84211 such blocks were present in the optimal matching, with a mean length of 592 bp, and an N50 value of 97,164 bp.

Given the availability of the optimal matching, why would one ever choose the greedy method? The optimal matching makes no effort to favor long runs of consistent matches (i.e. large blocks), and may choose more or less at random among a set of refined matches covering a given copy of a repeated element. In contrast, the greedy method attempts to maximize the length of the individual matches, rather than the sum of their lengths, as can be seen by comparing the number of selected matches. For the dataset at hand, the greedy method selected 1042 matches (881 trimmed) totaling 49,754,205 bp, leaving 353,515 bp of the finished sequence uncovered. The mean length is 47,749 bp and the N50 is 136,332. The difference between the optimal matching and the greedy method with respect to the contiguity of the selected matches can be seen in Figure 4.

Whether the difference in total coverage between the greedy and optimal matchings is large enough to be of concern is perhaps in the eye of the beholder. Both methods find matches for > 99% of the total finished sequence, but if one is interested in exactly how much is not covered, it may be of concern that the greedy method leaves nearly twice as many bases uncovered (that were part of an initial match) as the optimal matching.

In the end, for most purposes, the greedy match set is probably sufficient and preferable. However, the availability of the optimal method means that we can determine a (relatively tight) upper-bound on the maximum one-to-one coverage against which to compare the greedy set. This points to a general role for the optimal method in evaluating the consequences of various heuristics for generating initial matches or resolving them.

We can also easily determine a transformation distance trace. As noted in Section 5.2, by using an affine cost for insertions, we obtain explanations of one sequence by the other that favor consistent runs of adjacent matches, and thus at least partially avoid the fracturing of matches that is exhibited by the optimal matching. Accordingly, we computed the affine-cost transformation distance from the WGS assembly to the finished sequence. The resulting segment set, which is one-to-many (one segment from the WGS assembly may be matched to more than one region of the finished assembly), covers all but 154,093 bp of the finished sequence, i.e. *all* of the sequence covered by the initial match set. This coverage is accomplished by 3,374 matching blocks (mean 14811, N50 123,513), shown in Figure 4.

In the experiment above, we restricted our inputs to ungapped alignments. This permits the use of linear interpolation rather than the α and β projections to determine exact base-to-base matches in the alignments. It should be carefully noted that application of such linear interpolation to gapped alignments leads to a spurious refinement in which the refined matches are incorrect in detail and also much smaller than necessary (frequently a single basepair), and so greatly increases the computational burden. The projection maps are used precisely to avoid this problem.

7   Conclusions

We have presented a method for transforming a set of overlapping segment matches between two sequences to a refined set of matches reflecting the same alignments that have a special property: any two matches are either non-overlapping or have identical projections on one of the sequences. This transformation simplifies the statement of various analyses such that they can be solved to optimality. We implemented the proposed algorithm and illustrated its practicability on genome-sized sequences.

References

1. S. F. Altschul and B. W. Erickson. Locally optimal subalignments using nonlinear similarity functions. *Bull. Math. Biol.*, 48:633–660, 1986.
2. S. F. Altschul, W. Gish, W. Miller, E. W. Myers, and D. J. Lipman. Basic local alignment search tool. *Journal of Molecular Biology*, 215:403–410, 1990.
3. S. Batzoglou, L. Pachter, J. P. Mesirov, B. Berger, and E. S. Lander. Human and mouse gene structure: Comparative analysis and application to exon prediction. *Genome Research*, 10:950–958, 2000.
4. A. L. Delcher, S. Kasif, R. D. Fleischmann, J. Peterson, O. White, and S. L. Salzberg. Alignment of whole genomes. *Nucleic Acids Research*, 27(11):2369–2376, 1999.
5. Delcher, A. and others. unpublished.
6. G. Jacobson and K.-P. Vo. Heaviest increasing/common subsequence problems. In *Proceedings 3rd Annual Symposium on Combinatorial pattern matching (CPM)*, pages 52–66, 1992.
7. J. D. Kececioglu. The maximum weight trace problem in multiple sequence alignment. In *Proc. 4-th Symp. Combinatorial Pattern Matching*, number 684 in Lecture Notes in Computer Science, pages 106–119. Springer-Verlag, 1993.
8. J. D. Kececioglu, H.-P. Lenhof, K. Mehlhorn, P. Mutzel, K. Reinert, and M. Vingron. A polyhedral approach to sequence alignment problems. *Discrete Applied Mathematics*, 104:143–186, 2000.
9. S. Kurtz and C. Schleiermacher. REPuter: fast computation of maximal repeats in complete genomes. *Bioinformatics*, 15(5):426–427, 1999.
10. G. S. Luecker. A data structure for orthogonal range queries. *Proc. 19th IEEE Symposium on Foundations of Computer Science*, pages 28–34, 1978.
11. B. Morgenstern, W. R. Atchley, K. Hahn, and A. Dress. Segment-based scores for pairwise and multiple sequence alignments. In *Proceedings of the Sixth International Conference on Intelligent Systems for Molecular Biology (ISMB-98)*, 1998.
12. E. W. Myers, G. G. Sutton, H. O. Smith, M. D. Adams, and J. C. Venter. On the sequencing and assembly of the human genome. *Proc Natl Acad Sci U S A*, 99(7):4145–4146, 2002.
13. P. A. Pevzner and M. S. Waterman. Generalized sequence alignment and duality. *Advances in Applied Mathematics*, 14:139–171, 1993.
14. S. Schwartz, Z. Zhang, K. A. Frazer, A. Smit, C. Riemer, J. Bouck, R. Gibbs, R. Hardison, and W. Miller. PipMaker–a web server for aligning two genomic dna sequences. *Genome Research*, 10:577–586, 2000.
15. J.-S. Varré, J.-P. Delahaye, and E. Rivals. Transformation distances: a family of dissimilarity measures based on movements of segments. *Bioinformatics*, 15(3):194–202, 1999.
16. W. J. Wilbur and D. J. Lipman. The context dependent comparison of biological sequences. *SIAM J. Applied Mathematics*, 44(3):557–567, 1984.
17. D. E. Willard. New data structures for orthogonal queries. *SIAM Journal of Computing*, pages 232–253, 1985.

Extracting Common Motifs
under the Levenshtein Measure:
Theory and Experimentation

Ezekiel F. Adebiyi and Michael Kaufmann

Wilhelm-Schickard-Institut für Informatik, Universität Tübingen
72076 Tübingen, Germany
{adebiyi,mk}@informatik.uni-tuebingen.de

Abstract. Using our techniques for extracting approximate non-tandem repeats[1] on well constructed maximal models, we derive an algorithm to find common motifs of length P that occur in N sequences with at most D differences under the Edit distance metric. We compare the effectiveness of our algorithm with the more involved algorithm of Sagot[17] for Edit distance on some real sequences. Her method has not been implemented before for Edit distance but only for Hamming distance[12,20]. Our resulting method turns out to be simpler and more efficient theoretically and also in practice for moderately large P and D.

1   Introduction

The problem we consider here, called common motifs extraction, is stated as follows: Given a set of $N \geq 2$ subject strings $\{S_i, | \ i = 1...N\}$, each of average length n over a finite alphabet Σ and a threshold D, find common motifs of length P that occur with at most D differences (or simply D-occurrence) in $1 \leq q \leq N$ distinct sequences of the set. We consider this problem with respect to the Levenshtein metric (or simply Edit distance). We consider here the simple unit measure, whereby, the Edit distance between two strings $S_1[1..i]$ and $S_2[1..j]$, $\delta(i,j)$ is defined as follows:

$$\delta(i,j) = \min[\delta(i-1,j)+1, \delta(i,j-1)+1, \delta(i-1,j-1)+t(i,j)],$$

where $t(i,j)$ is 0, if $S_1(i) = S_2(j)$ else 1. As far as we know, this work will be the first that discusses this problem with respect to Edit distance not only theoretically (as been done before) but also experimentally.

The above problem has found applications in finding the DNA binding sites. There are various methods to extract common motifs. See Adebiyi[2] for an overview. Most of the approach made use of the Hamming distance metric. It is only the work of Rocke and Tompa[15] and Sagot[17] that considered extracting common motifs under Edit distance but did so theoretically only. Various pointers[13,9] have indicated that extracting common motifs under Edit distance is also an important biological problem. It is this open problem that we consider in this paper.

R. Guigó and D. Gusfield (Eds.): WABI 2002, LNCS 2452, pp. 140–156, 2002.

Our contributions are in two folds: First, we have implemented the interesting but involved algorithm of Sagot[17] for motifs finding for Edit distance. The theoretical running time of this algorithm is $O(N^2 n P^D |\Sigma|^D)$. It has only been implemented before for Hamming distance[20,12]. Second, we present an alternative algorithmic approach also for Edit distance based on the concept described by Adebiyi et al.[1]. The resulting algorithm is simpler and more efficient for moderately large P and D in theory as well as in practice. It runs in $O(N \cdot n + D(N \cdot n)^{1+pow(\varepsilon)})$ expected time for $D \leq 2\log_{|\Sigma|}(N \cdot n)$, where $\varepsilon = D/P$ and $pow(\varepsilon)$ is an increasing, concave function that is 0 when $\varepsilon = 0$ and about 0.9 for DNA and protein sequences. We implemented this algorithm in C and compare its effectiveness using our implementation of Sagot's algorithm on sequences that include B. Subtilis[7], yeast S. cerevisiae[10] and H. Pylori[18].

This paper is structured as follows. In sections 2 and 3, we present the algorithm of Sagot and our algorithm for Edit distance, and in section 4, we discuss our experimental experience with the algorithms from sections 2 and 3 and also, with our implementation of Sagot's algorithm for Hamming distance.

2 Basic Concepts and Sagot's Algorithm for Edit Distance

Here, we will describe in detail the algorithm of Sagot[17] for Edit distance, since we use some of the concepts later on for the new algorithm, and start with some definitions. A *model* of length P is a pattern over Σ^P for some positive $P \geq 1$ and a *valid model* is a model that present a single representation for all reoccurrences of a motif. A valid model is elsewhere also known as a consensus. The distance between a valid model and the reoccurrences of the motif that it represents is not more than D. In particular, we refer to a motif and its reoccurrences in the sequences using its valid model.

Sagot's method is basically based on the concept of validating all words/models (they are elsewhere known as *external objects*, and are the candidates for motifs) over Σ^P that occur in $1 \leq q \leq N$ distinct sequences of the set $\{S_i,| i = 1...N\}$. The generation or *SPELLING* after Sagot of each character of the models is done simultaneously as we search them in the preprocessed generalized suffix tree (henceforth known as GST, see below for the definition) of the sequences. GST is the suffix tree that represent the suffixes of multiple sequences. The SPELLING of a model stops, if we have reached the length required for models or the models may not be further extended while remaining valid. If we introduce errors (Hamming or Edit distance), for any model, we may also stop the descent down a path in GST as soon as too many insertions, deletions, or substitutions have been accumulated along it. The models of length P, which have a chance to be valid, are those in the D-neighborhood of the words present in the sequences of the set. Let u be a word present in the sequences. Model X is in the D-neighborhood of u, if the distance between X and u is not more than D. For Hamming distance, Sagot[17] proved that the D-neighborhood of a word of a length P contains

$$V(D, P) = \sum_{i=1}^{D} \binom{P}{i}(|\Sigma| - 1|)^i \le P^D |\Sigma|^D$$

elements. Recently, Crochemore and Sagot[5] indicated that this also bounds the number of D-neighborhood for Edit distance from above.

To introduce the concept of generalized suffix tree, for brevity, we adopt the combinatorial definitions given by Blumer[3].

Definition 1. *A substring, x, of a string S, occurs in m left (right) contexts if x occurs at least m times, with these occurrences preceded (followed) by at least m distinct characters.*

Definition 2. *Given a finite set s of strings and x, y in s, y is a minimal extension of x (in s) if x is a prefix of y and there is no z in s shorter than y such that x is a prefix of z and z a prefix of y.*

We can now define a suffix tree for a single string S of size n as follows,

Definition 3. *Let s be the set of all substrings of S. The suffix tree (henceforth known as ST) for S is the directed graph (V, E), where V is the set of substrings of S that are either suffixes of S or that occur in S in at least two right contexts, and E is the set of all directed edges (x, y), x, y in s, such that y is a minimal extension of x.*

In particular, ST of a string S of size n is a tree with $O(n)$ edges and internal vertices and n external vertices (the leaves). Each leaf in the suffix tree is associated with an index i ($1 \le i \le n$). The edges are labeled with characters so that the concatenation of the labeled edges on the path from the root to the leaf with index i is the ith suffix of S. Here, ST is used to represent a single string. This can be extended to represent multiple strings $\{S_i, | i = 1...N\}$. The resulting tree is called GST. When two or more input strings have the same suffix, GST has several leaves corresponding to that suffix, but each corresponds to a different input string.

Before giving a detailed description of Sagot algorithm for Edit distance, note that SPELLING all occurrences of a model if errors are not allowed leads to a node x in the suffix tree, while it leads to more than one node $\{x_1, x_2, ...\}$, once errors are allowed. Her algorithm is based on the following recurrence formula.

Lemma 1. *[5] Let x be a node in the suffix tree and err $\le D$. (x, err) is a node-occurrence of $m' = m\alpha$ with $m \in \Sigma^k$ and $\alpha \in \Sigma$ (i.e node x is a substring approximate to m' within distance err) if and only if, one of the following conditions is verified:*

(match) *(father(x),err) is a node-occur. of m and the label of father(x) to x is α;*

(substitution) *(father(x), err-1) is a node-occurrence of m and the label of the arc from father(x) to x is $\beta \ne \alpha$;*

(deletion) *(x, err-1) is a node-occurrence of m;*

(insertion) *(father(x), err-1) is a node-occurrence of $m\alpha$.*

To perform the SPELLING step, some additional information has to be added to the nodes of the GST. The first preprocessing task, is to calculate for each node x, the number $CSS_x[8]$ of different sequences of the set $\{S_i, | \ i = 1...N\}$, that the leaves in the subtree rooted at x refer to. This has been shown to be solvable in $O(N \cdot n)$ time by Hui. Furthermore, to each x, a boolean array $Colors_x$ of dimension N is associated, whose entries are used to determine which sequences are common to two or more nodes of the tree. The i-th entry of $Colors_x$ is 1 if at least one leaf in the subtree rooted at x represents a suffix of string S_i and 0 otherwise. $Colors_x$ for all nodes x may be obtained by a simple traversal of the tree.

The pseudo-code of the Edit distance algorithm, as outlined by Sagot[17] but implemented by us, are presented below. The following auxiliary structures are used:

1. a set Ext_m of symbols by which a model may be extended at the next step;
2. a set Occ_m of node-occurrences of a model m. Recall that these correspond to classes of occurrences. Each node-occurrence x is represented by a pair (x, x_{err}) where x_{err} is the number of insertions, deletions or mismatches between m and the label of the path leading from the root to x in the tree;
3. a variable CSS_x as defined above;
4. a boolean array $Colors_x$ as also defined above;
5. $minseq$ that indicates the minimum of CSS_x for all node-occurrences x of the extended model;
6. a variable $maxseq$ that indicates the sum of CSS_x for all node-occurrences x of the extended model;
7. a $Color_m$, whose i-th entry is 1 if m occurs in string S_i and 0 otherwise;
8. a function KeepModel(m), that either stores all information concerning a valid model of required length for printing later, or immediately prints this information.

In procedure $SPELLMODELS$ (fig. 1), the models m are spelled as they are searched against the words in the sequences via the paths that encoded them in GST. Note that the Edit operations can be rewritten in any of the following forms: $a \longrightarrow \varepsilon$ (a deletion), $\varepsilon \longrightarrow a$ (an insertion), or $a \longrightarrow b$ (a change or substitution), where $a, b \in \Sigma$, $a \neq b$, and ε is the empty string. Let the characters at the left-hand side of the binary operator $\longrightarrow$ come from the substrings of the node occurrences, while that on the right-hand side come from the models. To spell the models, while allowing its re-occurrences to contain gaps, sub-procedure $TREAT$ in $SPELLMODELS$ of fig. 1 simulates the adding of the last row of a dynamic programming matrix of the model m against the nodes of GST[19].

$TREAT$ is called recursively as TREAT(Occ_m, $Occ_{m\alpha}$, $Ext_{m\alpha}$, $Colors_{m\alpha}$, $minseq$, $maxseq$, x, x_{err}, α, $level + 1$) and works as follows. At $level = 0$, the edit script of deleting a character from the model $m\alpha$ is carried out if $x_{err} < D$ and parameters $maxseq$, $minseq$, $Colors_{m\alpha}$, $Ext_{m\alpha}$ and $Occ_{m\alpha}$ are updated accordingly using the node x in concern. At $level > 0$, the nodes x' of the subtree rooted at x are pruned accordingly using their min_{err}, i.e, the minimum error that exist between them and the model considering insertions, deletions and substitutions. And if $min_{err} \leq D$, parameters $maxseq$, $minseq$, $Colors_{m\alpha}$, $Ext_{m\alpha}$ and $Occ_{m\alpha}$ are again updated accordingly using the nodes x and x' in concern.

Note that in all cases, the number of distinct sequences, the model occurs in, is in the range between $minseq$ and $maxseq$. For simplicity, we assumed that motifs of a fixed length P are been sought for. The function $SPELLMODELS$ is called with the arguments $(0, \lambda, Occ_\lambda = \{(root, 0)\}, Ext_\lambda)$, where $Ext_\lambda =$

$$\begin{cases} \Sigma & : & D > 0 \\ label_b \; for \; branches \; b \; from \; root & : & otherwise \end{cases}$$

Note that the number of all possible models in line 4 that does not necessarily represent a motif is $|\Sigma|^P$ and the loop of line 8 is executed $O(N \cdot n)$[3] times. In fact, the number of nodes x, considered in loop of line 8, are further reduced in sub-procedure $TREAT$. The models which might be valid (those representing motifs) are those in the D-neighborhood of the words in the sequences of the set $\{S_i,| \; i = 1...N\}$. Therefore, the nested loops starting in line 4 and line 8 are executed $O(N \cdot n \cdot V(D, P))$ times, where $V(D, P) \leq P^D |\Sigma|^D$. Within these loops, the dominating operation is the operation of line 14 and this can be carried out in $O(N)$ time. Therefore, it follows that for all possible models in Σ^P, the running time of the algorithm $SPELLMODELS(l, P, Occ_m, Ext_m)$ is $O(N^2 n P^D |\Sigma|^D)$.

1.   Procedure SPELLMODELS(l, P, Occ_m, Ext_m)
2.   if $(l = P)$ then KeepModel(m)
3.   else if $(l < P)$
4.   for each symbol α in Ext_m
5. $maxseq = 0$, $minseq = \infty$
6. $Colors_{m\alpha}$ is initialized with no colors
7. $Ext_{m\alpha} = 0$, $Occ_{m\alpha} = 0$
8.   for each pair (x, x_{err}) in Occ_m
9.   remove(x, x_{err}) from Occ_m
10.   TREAT($Occ_m, Occ_{m\alpha}, Ext_{m\alpha}, Colors_{m\alpha}, minseq, maxseq, x, x_{err}, \alpha$, $level = 0$)
11.   if $(maxseq < q)$ return (no hope)
12.   else if $(minseq \geq q)$
13.   SPELLMODELS($l + 1, m\alpha, Occ_{m\alpha}, Ext_{m\alpha}$)
14.   else if the number of bits at 1 in $Colors_{m\alpha}$ is not less than q
15.   SPELLMODELS($l + 1, m\alpha, Occ_{m\alpha}, Ext_{m\alpha}$)

Fig. 1. Pseudocode of the procedure for SPELLING models corresponding to common motifs when gaps as well as mismatches are allowed.

3   An Alternative Algorithmic Approach

We develop another algorithmic approach by a careful generation of potential models using the adapted techniques of Adebiyi et al.[1]. Any string S of size n is a combination of exact repeated and non-repeated substrings respectively. Two types of exact repeats exist: maximal and non-maximal repeats. An exact repeat β is maximal if $a\beta c$ and $b\beta d$ occur in S for some $a \neq b$ and $c \neq d$, $a, b, c, d \in \Sigma$. Otherwise it is non-maximal. Note that a maximal repeat might be a proper substring of another maximal repeat. Therefore, a supermaximal repeat is a maximal repeat that does not occur as a substring of any other maximal repeat. Models are constructed by partitioning them into repeats-oriented equivalence classes and avoiding those models that are contained in another maximal models or that will extend to the same motif. Here, we use the ideas developed in Adebiyi et al.[1]. The number of such models is linear in the total length of the input strings.

Using the foregone details, the algorithm is based on a clever construction of models that are extended into common motifs.

3.1   The Algorithm

The algorithm is divided into two steps. First, we extract the models, second, we extend these models to derive common motifs. We now discuss how the models are constructed. We begin with formal definitions[6] for identifying *maximal and supermaximal repeats respectively* in a suffix tree, ST.

Definition 4. *For each position i in string S, character $S(i-1)$ is called the* left character *of i. The left character of a leaf of ST is the left character of the suffix position represented by that leaf.*

Definition 5. *A node v of ST is called* left diverse, *if at least two leafs in v's subtree have different left characters. By definition, a leaf cannot be left diverse.*

Lemma 2. *The string α, labeling the path to a node v of ST is a* maximal repeat *if and only if v is left diverse. A left diverse internal node v represents a supermaximal repeat α, if and only if all of v's children are leaves and each has a distinct left character.*

Lemma 3. *[1] Each maximal or non-maximal repeat is contained in some supermaximal repeats.*

We have the following observations, the first one is straight forward.

Observation 1. A model substring must be extendable in all strings in $\{S_i, | \ i = 1...N\}$. That is, its left index in each string shifted by $P - 1$ positions to the right, must be less than or equal to the end index of each string.

Observation 2. A substring that is completely non-maximal in all strings in $\{S_i, | \ i = 1...N\}$ cannot be a model, since it is contained in some supermaximal repeats.

Our models now should satisfy the two observations above and are defined using the following definitions. Note that the keywords *maximal* used in the naming of our models, have different meaning from their use with *repeat* as given in Lemma 2 above.

Definition 6. A repeat model *is a repeat (*i.e., *non-maximal, maximal or supermaximal) in at least two of the subject strings.*

Definition 7. A maximal model *is a non repeated pattern occurring in at least q of the subject strings and a maximal repeat when considering its occurrences in at least q of the subject strings.*

Let R and M be the sets of *repeat and maximal models* respectively. In general, we will call R and M models, the set of maximal models. Blumer[3] showed that the minimum expected length of a repeated pattern is $f \log_{|\Sigma|} n$ for $f \leq 1$. In Adebiyi et al.[1], we derived that the expected length of the longest repeat is $2 \log_{|\Sigma|} n + o(\log n)$.

This implies that, the length of each model l defined above satisfied $l = \Theta(\log_{|\Sigma|} n)$. This information will be useful for the efficient extension step of each model.

3.2   Step 1: Extracting the Maximal Models

We will use the above definitions to extract models from a given set of $N \geq 2$ subject strings $\{S_i, | \ i = 1...N\}$, each of average length n using a generalized suffix tree. To do this, we extend the algorithm for finding exact maximal and supermaximal repeats as described by Gusfield[6], using the ideas of Hui[8] for solving the CSS_x problem described in section 2. This is encapsulated in procedure FINDMODELS(N, n).

The main auxiliary structures and functions used by Hui include LeafList and lastleaf(x), where x is a leaf in GST. LeafList is the list of leaves ordered according to a post-order traversal of GST. For each leaf that belongs to a sequence in $\{S_i, | \ i = 1...N\}$, lastleaf(x) is the last leaf which precedes x in LeafList and belongs to the same sequence. If no leaf precedes x in LeafList, Nil=lastleaf(x). We add the following preprocessing to extract our maximal models.

Let $lca(x, y)$ be the least common ancestor of x and y, where x and y are vertices in GST. An internal vertex v is a <u>r</u>epeat <u>r</u>epresentative <u>p</u>air-lca (*of a sequence in* $\{S_i, | \ i = 1...N\}$, if there exists a leaf x of a sequence S_i , such that $v = lca(y = lastleaf(x), x)$ and lastleaf(y)=Nil. Let $RRPcount(v)$ denotes the number of such situations for each node v. Therefore, to verify the conditional part of Definition 6 for a node u, the problem to solve is equivalent to the calculation of its number of Non-Nil lastleaf that refers to distinct sequences in $\{S_i, | \ i = 1...N\}$. This value for a node u, is called $distinct(u)$ and is recursively calculated as $distinct(u) = \sum_{v \in subtree(u)} RRPcount(v) = RRPcount(u) + \sum_{w \in W} distinct(w)$, where W is the set of children of u.

1. **procedure** FINDMODELS(N, n)
2. Build GST
3. Create the LeafList
4. Compute the lastleaf() values
5. Compute the RRPcount() values
6. Compute the distinct() values
7. **For** each node x (*in bottom-up format*)
8.   if(!non-maximal$_x$)
9.   if($distinct \geq 2$ and $CSS_x \geq 2$)
10.   Append x to R
11.   else if($distinct == 0|1$ and $true == TM$)
12.   Append x to M

Fig. 2. Extracting maximal models.

The property expressed in Observation 2 above, can be verified by adding a boolean variable non-maximal$_x$ to each node x. The boolean variable non-maximal$_x$ is 0, if the node is at least once maximal in a sequence and 1 if non-maximal in all strings in $\{S_i,| \ i = 1...N\}$. This can be done by a simple combination of the original algorithm of Gusfield and the auxiliary structure and function LeafList and lastleaf. To verify the properties from Definition 7 for a node u, distinct(u) must be zero or one and what remains is to test its maximality (henceforth TM) with respect to its occurrences in at least q strings in $\{S_i,| \ i = 1...N\}$. This involves collecting and scanning of the left characters of the suffixes represented by its children. Using the above facts, we can now compactly present FINDMODELS(N, n).

Lemma 4. All the models defined in Definitions 6 and 7 can be extracted from GST in $O(N \cdot n)$ expected time.

Proof. The preprocessing tasks before extracting models include Creating LeafList, computing lastleaf(), RRPcount() and distinct() values respectively. lastleaf() values are calculated via a single scan of LeafList, therefore, since the set of leaves is not more than $N \cdot n$, creating LeafList and computing lastleaf values can be done in $O(N \cdot n)$ time. Given an arbitrary pair of leaves, their *lca* can be computed in $O(1)$ time[16]. After initializing all $RRPcount()$ values by zero, for every leaf x, we compute $u = lca(x, y = lastleaf(x))$ and if $lastleaf(y) = Nil$, which means, it has not been counted, we increment $RRPcount(u)$ by 1. Therefore $RRPcount()$ values can be computed in $O(N \cdot n)$ time for all leaves. What then follows is the computation of all distinct() values by its recursive definition given above via a post-order traversal. This is also achieved in $O(N \cdot n)$ time.

In lines 8-12, we now extract our maximal models. This is also achievable in $O(N \cdot n)$ time because the total number of nodes and leaves in a GST is bounded by $O(N \cdot n)$[3]. □

3.3 Step 2: Extending the Maximal Models

The algorithm we use to extend the models is similar to the algorithm used in Adebiyi et al.[1] to extend seeds to find approximate repeats in one long subject string of length n. To extend a seed W, given P, the length of short substrings sought for, that repeat with at most D-differences, $i.e.$ insertions, deletions, and mismatches, we split W into a left and right part W^l and W^r, where $|W^l| = \log_{|\Sigma|} n$. Next, using SAM (Sublinear Algorithm of Myers[11,1]), we find all locations of all approximate repeats $(W^l)'$ of W^l. Their right indices are stored in set H. Next, we check whether we can extend $(W^l)'$ to the right to get approximate repeats for $W = W^l W^r$. Therefore, for every index H_i in H corresponding to a word W_i^l with differences D_i, we consider the word W_i^r starting from H_i and having length $P - \log_{|\Sigma|} n$. Let E_W denote the set of possible rightward extensions of W^r to a word of length $P - \log_{|\Sigma|} n$. Note that $|E_W|$ is the number of occurrences of the seed W in string S, which is, fortunately, at most $|\Sigma|$ for supermaximal repeats. We compute the pairwise alignments of W_i^r with all elements E_j of E_W and, if the difference between any of the two substrings is at most $D_i' = D - D_i$, then the extension works and we can add $W_i = W_i^l W_i^r$ to the output list SA for the short approximate repeats. The procedure is called ADDWR(D_i', H_i, P).

To extract common motifs, the situation is different, the subject string is of size $N \cdot n$. We assume in the following exposition that $P = 2\log_{|\Sigma|}(N \cdot n)$. Note that if $P \leq \log_{|\Sigma|}(N \cdot n)$, we just directly apply SAM and we do not need any extension by subprocedure ADDWR described below. For each model, after we ran SAM on their left part, we sort the set H, so that we can partition them into N subclasses, depending on where they occur in each subject string S_j, $j = 1...N$. That is, $H_{ij} \in H$, $i = 1...|H|$ and $j = 1...N$. Then in subprocedure ADDWR, for every index H_{ij} in H corresponding to a word W_{ij}^l with differences D_i, we consider the words W_{ij}^r starting from H_{ij} of length $P - \log_{|\Sigma|}(N \cdot n)$. Next, we can no longer assume that $|E_W|$ is at most $|\Sigma|$, since supermaximal, maximal and non-maximal repeats are all used to construct our models. Note that the concept of maximality is still paramount in our models construction, so that $|E_W|$ in expectation, is the number of maximal substring occurrences divided by the number of internal nodes in a generalized suffix tree. Therefore $|E_W|$ is constant in expectation. The only exceptional case is when a model appears as maximal in one sequence but non-maximal in the others. In this case, its logarithmic length can be used to show that in expectation, its number of occurrences is still bounded by the number of maximal substring occurrences. Finally, we compute the pairwise alignment of W_{ij}^r with all elements E_o of E_W and, if the difference between substrings of W_{ij}^r s.t. $j = 1...t$, $t \geq q$ and one element of E_o is at most $D_i' = D - D_i$, then $W_{ij} = W_{ij}^l W_{ij}^r$ are the locations of a common motif in the subject strings $\{S_j \mid j = 1...t\}$. This one element of E_o is stored in the set of Common Motifs, CM.

The operation described above is encapsulated in figure 3 below.

1. **procedure** FINDSMOTIF(D, P, N, n)
2. $R, M \longleftarrow$ FINDMODELS(N, n)
3.   **for** W_i, $i = 1...|R \cup M|$ **do**
4. $W^l, W^r \longleftarrow$ SPLITW(W_i)
5. $H \longleftarrow$ SAM(D^l, W^l)
6.   **Sort H**
7.   **for each** H_{ij}, $i = 1...|H|$, $j = 1...N$ **do**
8. $CM \longleftarrow$ ADDWR(D^r, H_{ijr}, P)

Fig. 3. Extracting common motifs.

Theorem 1. *Given a set of $N \geq 2$ subject sequences $\{S_i, | \ i = 1...N\}$, each of average length n over a finite alphabet Σ and threshold D, all common motifs, each of size $P = O(\log(N \cdot n))$ can be found in $O(N \cdot n + D(N \cdot n)^{1+pow(\varepsilon)} \log(N \cdot n))$ expected time for $D \leq 2\log_{|\Sigma|} N \cdot n$, where $\varepsilon = D/P$, $pow(\varepsilon) = \log_{|\Sigma|}(c + 1)/(c - 1) + \varepsilon \log_{|\Sigma|} c + \varepsilon$ and $c = \varepsilon^{-1} + \sqrt{1 + \varepsilon^{-2}}$.*

Proof. The subprocedure FINDSMODELS includes the building of the generalized suffix tree and finding the models. GST can be built in $O(N \cdot n)$ time. Lemma 4 shows that we can also extract the maximal models in $O(N \cdot n)$ time. Therefore the expected running time of FINDMODELS(N, n) is $O(N \cdot n)$.

Note that the number of maximal models is bounded by $O(N \cdot n)$ and the operation perform in SPLITW can be done for each model in constant time. SAM running time for each model is $O(D(N \cdot n)^{pow(\varepsilon)} \log(N \cdot n))$. It is known[11] that the size of the output H of SAM for each query model is $O(N \cdot n^{pow(\varepsilon)})$. Therefore the time to sort set H is at most $O(N \cdot n^{pow(\varepsilon)} \log(N \cdot n))$, while ADDWR requires for all indices in H at most $O(N \cdot n^{pow(\varepsilon)} \log_{|\Sigma|}^2(N \cdot n))$ time. Therefore, the total running time for calling SAM and ADDWR for all models is bounded by $O(D(N \cdot n)^{1+pow(\varepsilon)} \log(N \cdot n))$, for all values of N, n, and Σ. The bound stated in the lemma then follows.   □

We show the correctness of our method using the following proposition.

Proposition 1. *It is true that the above algorithm in Fig. 3 with the above complexity solves the problem of extracting common motifs under the Levenshtein measure.*

Proof. In our construction of maximal models, we optimally ensure that all possible valid node-occurrences of common motifs are covered. To specifically establish it, we make use again of the algorithm of Sagot[17]. For simplicity, we assume that motifs of fixed length P are sought for. The basic ingredients in the validating algorithm include CSS_x, $Colors_x$ (see section 2 for their definitions) and the simultaneous SPELLING of each characters of the 'external objects', the models, over Σ^P and the recursively *grouping and validating* of its reoccurrences in each sequence using the given basic parameters P, D and q.

In the following, when we refer to repeats, we mean the reoccurrences of a pattern in the same sequence. The nodes in a generalized suffix tree can be categorized into two: 1) The ones that represent patterns that are repeated but also occur (or not) in other sequences. 2) The ones that represent patterns that are non-repeated but also occur (or not) in other sequences. The set of *Repeat* and *maximal models* (see Definition 6 and 7) covers the first and the second group respectively. This is done in line 2.

The calculation of CSS_x is performed in line 6, while $Colors_x$ is technically performed in lines 3-5. Finally, the grouping and validating processes are performed in line 7-8. These lines also check to make sure that too many insertions, deletions and substitutions have not been accumulated. The basic reason of using 'external objects' as models in Sagot's algorithm, is because the real motif (in some literature, consensus) itself may not exist in the set of sequences, but it may be represented by one of its occurrences. This implies that the set of occurrences of a motif may share approximately a common pattern. In our algorithm above, the concept of D-neighborhood also used in SAM but in a different way, allows us to find approximate motif reoccurrences that are related via an approximately common pattern instead of an exact pattern. □

Remark: Comparison of the running times $O(N^2 n P^D |\Sigma|^D)$ of Sagot algorithm and $O(N \cdot n + D(N \cdot n)^{1+pow(\varepsilon)} \log(N \cdot n))$ of our algorithm shows that our algorithm is a major improvement for moderately large P and D that normally arise in practice. Note further that for small values of D (for example, $D = 1$) as in some cases in practice, the complexity of Sagot's algorithm will therefore scale linearly and thus its complexity is better than our algorithm in this case.

4   Experimental Experience

We implemented Sagot's algorithm[17] for Hamming and also for Edit distance. The three real sequences we used here consist of sequences from B. Subtilis[7], collection of promoter regions upstream the genes SWI4, CLN3, CDC6, CDC46 and CDC47 of yeast S. cerevisiae that is known to contain a shared cell-cycle dependent promoter[10], obtained from NCBI database, and those extracted from H. pylori[18] obtained from $ftp : //ftp.tigr.org/pub/data/h\_pylori$. We found minor inconsistencies in the extraction of sequences of B. Subtilis[20]. We found out that the required set of B. Subtilis contains 131 sequences of average length 100bp (instead of 99bp), for a total of 13073 (instead of 13099) nucleotides. The set of extracted sequences of yeast we used, contains 5 sequences and the largest of them is 2565bp. The total nucleotides for this set is 8639. The set (called Set C in Vanet et al.[20]) of the extracted sequences of H. pylori consists of non-coding regions upstream from genes coding for either a ribosomal RNA or a ribosomal protein, or from operons including ribosomal protein genes. This set consists of 26 sequences, the largest sequence has 2997bp and the total size of the set is 29324 nucleotides.

4.1   Efficiency Discussion

Let A and B denote the algorithms of Sagot[17] for Hamming and Edit distance respectively and C be our new algorithm for Edit distance. We compare here the running times and the number of common motifs returned by each algorithm[1].

Figures 4 to 7 of the appendix show that the algorithms of Sagot are clearly inefficient for moderately large values of P and D as theoretically analyzed before.

Note that algorithms A and B report much more redundant motifs than C. This is due to the fact that many of the motifs found by A and B correspond approximately to the same words in the sequences[20]. Algorithm C does not have this effect. In table 1 of the appendix, for $D \geq 4$, the numbers of resulting motifs for Hamming and Edit distances do not differ because of the lack of sufficient node-recurrences. This behavior depends on the set of sequences under consideration as we observed different behaviors for the sequences sets of H. pylori and Yeast. Finally for $P \geq 10$, $D = 5$ and $q = 65$ in table 2 of the appendix, the sensitivity of Edit distance compared to Hamming distance becomes obvious as no motif was found with Hamming distance (Algorithm A). We will analyze in the next section the common motifs extracted for the Edit distance. Here also for algorithm B, the number of node-reoccurrences decreases dramatically, so that the number of models returned decreased from 1126894 to 50680. These effects lead to the flatness observed in Figures 7 for the run-times of algorithms A and B. We predict also that this behavior depends on the set of sequences under consideration.

4.2   Effectiveness Discussion

We begin this section by presenting the common motifs found by algorithm A in comparison to the results of Vanet et al.[20]. We found no significant error in the common motifs reported, because all common motifs reported by them were also found with differences only in the number of times they occur. All the other common motifs found by our algorithm A correspond to variants of those reported. Vanet et al.[20] also observed this in their report. This result is given in table 3 below.

Algorithm C compares favorably with the validating algorithm B. The common motifs that played the pivoting role here, are those returned from B. Subtilis and set C of H. pylori in Vanet et al.[20]. Table 3 shows the results on B. Subtilis. An elaborate experimental results on the sequences of H. pylori can be found in table 5.6 of [2]. There, we ran algs. B and C to find the pivoter motifs using $D = 1$, $q = 13$, and $P = 6, 7, 8, 9$. Note that, if the common motifs (more than 3) returned do not exactly match the pivoting common motifs, we represent them with one of their variant. Note that G,C-GTAAAA means GGTAAAA, CG-TAAAA, TATA-TT,AT means also TATATT, TATAAT and blank entry means

---

[1] The programs were run on a SUN-sparc computer under Solaris 2.5.1 with a 366 MHz-Processor and 360Mbytes of main memory

the pivoter pattern or its variants are not found. We observe that in table 3 (for Algs. B and C), a slightly smaller q provides the pivoter motifs that both algorithms cannot find at $q = 65$. This is not the case in table 5.6[2] as both algorithms find either exact or variants of the pivoter motifs. Therefore we concluded that the differences (although mostly small) in the number of the pivoter motifs returned by the two algorithms, when both returned them exactly, is simply due to various optimizations performed in SAM with the ones, we did in deriving our models in algorithm C. This does not affect the effectiveness of algorithm C[2].

Algorithm A failed to predict the corresponding motif in yeast as we can see in table 4. Vanet et al.[20] recognized this problem as they stated that suggestions from biologists concerning possible extension of their algorithm application to eukaryotes are being sought. Algorithm B ran for more than 16 hours. We stopped its execution due to lack of storage to store all the motifs found. Algorithm C got two corresponding motifs within a reasonable time. We observed that its predicted motifs distance themselves away from the published motif by a Hamming distance of 4.

Lastly, in table 5 below, we analyze and present some common motifs extracted in the case, where the Hamming distance failed. Our example here is for $D = 3$, $P = 10$ and $q = 65$ and they can be found around the TATA-box and known patterns like (TATAAT and TTGACT)[7] and (TAAAAT and GAAAAA)[12].

We use a combination of the data shuffling methods[9] and the χ^2-test with one degree of freedom to evaluate the statistical significance of the common motifs found. The statistical significance of each motif is evaluated using a contingency table[14]. The first table listed the number of occurrences of each motif per sequence without shuffling the sequences. This is the observed hypothesis. The second table corresponds to what is expected when the sequences are shuffled. This is the expected hypothesis otherwise known as the null hypothesis. For each motif, we are now to find the probability of getting this motif, given the null hypothesis. To obtain the χ^2 values in table 5, we performed a thousand random shufflings preserving both the mono- and dinucleotide frequency distributions of the original sequences.

Using the Hamming distance, Vanet et al.[20] obtained statistical significance common motifs whose length are greater than six, with at most one substitution and a quorum of $q = 65$ from B. Subtilis sequences. The length of the motifs they obtained is actually either six or seven. This confirm our results in table 2 that algorithm A failed to predict common motifs for $P \geq 10$. The χ^2 values of the motifs that Vanet et al.[20] obtained for B. Subtilis is in the range of 11-31. Note that the χ^2 values above 10.8 means that the corresponding motifs that have such χ^2 values, have the probability of happening by chance below 10^{-3}.

The χ^2 values of the common motifs for B. Subtilis under the Edit distance that we show in table 5 is in the range of 50-400. These are higher than the χ^2 values of Vanet et al.[20] on B. Subtilis sequences for Hamming distance. These higher values are expected, since the number of occurrences of statistical significant motifs per sequence is higher, under the Edit distance than under the Hamming distance.

5   Conclusion

We have implemented and tested the more involved algorithm of Sagot[17] for Edit distance. Furthermore, we have also presented an algorithm that is efficient both in theory and in practice for finding moderately long motifs that distance itself away from its reoccurrences under the Edit distance metric. Its effectiveness compares favorably with the algorithm of Sagot[17]. Our algorithm reports also less false positive motifs. Additionally, we have demonstrated the relevance (in practice) of the Edit distance metric in the extraction of motifs. The sensitivity of Edit distance and the significance of its common motifs, presented in Table 5, that eluded the Hamming distance gives evidence that some important motifs may have eluded existing motifs finder algorithms that are mostly based on Hamming distance.

Our results are important and significant but left some important problems open. The results of table 5 below are preliminary. More random shufflings in thousands need to be performed on sequences of organisms like the Human and the ground Squirrel, e.t.c, that may have common motifs caused by substitutions, insertions, and deletions[9]. Another statistic, like the Z-score should be used to certified the results obtained. The motifs, we consider here are "consensus sequences" that must occur at specific positions. The main limitation of consensus sequences is that they do not allow for different positions to be conserved to different degrees, or for some differences to the consensus to be penalized more than others. An interesting work will be to introduce a better model that have some contribution from each base at each position, and the sum of all the contributions to be above some threshold. This model is nicely handled by a weight matrix representation, called Position Weight Matrices (PWM)[4]. Finally, it will be interesting to apply our ideas here to the extraction of structured motifs[12,20].

Acknowledgement

We thank J. D. Helmann for providing pointer to the sequences of B. subtilis. We also thank Tao Jiang and Kay Nieselt-Struwe for some useful discussions.

References

1. E. F. Adebiyi, T. Jiang, and M. Kaufmann. *An efficient algorithm for finding short approximate non-tandem repeats (Extended Abstract).* Bioinformatics, 17(1):S5-S13, 2001.
2. E. F. Adebiyi. *Pattern Discovery in Biology and Strings Sorting: Theory and Experimentation.* Ph. D Thesis, 2002.

3. A. Blumer and A. Ehrenfeucht and others. *Average size of suffix trees and DAWGS.* Discrete Applied Mathematics, 24, 37-45, 1989.

4. J.-M. Claverie and S. Audic. *The Statistical significance of nucleotide position-weight matrix matches.* Computer Applications in Biosciences 12(5), 431-439, 1996.

5. M. Crochemore and M.-F. Sagot. *Motifs in sequences: localization and extraction.* In Handbook of Computational Chemistry, Crabbe, Drew, Konopka, eds., Marcel Dekker, Inc., 2001. To appear.

6. D. Gusfield. *Algorithms on strings, trees and sequences.* Cambridge University Press, New York, 1997.

7. J. D. Helmann. *Compilation and analysis of Bacillus Subtilis σ^A-dependent promoter sequences: evidence for extended contact between RNA polymerase and upstream promoter DNA.*, Nucleic Acids Research, 23(13): 2351-2360, 1995.

8. L. C. K. Hui. *Color set size problem with applications to string matching.* In CPM Proceeding, vol. 644 of LNCS, 230-243, 1992.

9. S. Karlin, F. Ost, and B. E. Blaisdell. *Patterns in DNA and amino acid sequences and their statistical significance.* In M. S. Waterman, editor, Mathematical Methods for DNA Sequences, 133-158, 1989.

10. C. J. McInerny, J. F. Patridge, G. E. Mikesell, D. P. Creemer, and L. L. Breeden. *A novel Mcm1-dependent element in the SWI4, CLN3, CDC6, CDC46, and CDC47 promoters activates M/G_1-specific transcription.* Genes and Development, 11: 1277-1288, 1997.

11. E. Myers. *A sub-linear algorithm for approximate keyword matching.* Algorithmica 12, 4-5, 345-374, 1994.

12. L. Marsan and M. F. Sagot. *Extracting structured motifs using a suffix tree-algorithms and application to promoter consensus identification.* RECOMB 2000.

13. P. Pevzner and S.-H. Sze. *Combinatorial approaches to finding subtle signals in DNA sequences.* ISMB, 269-278, 2000.

14. W. H. Press, S. A. Teukolsky, W. T. Vetterling, and B. P. Flannery. *Numerical Recipes.* In The Art of Scientific Computing, Cambridge University Press, Cambridge.

15. E. Rocke and M. Tompa. *An algorithm for finding novel gaped motifs in DNA sequences.* RECOMB, 228-233, 1998.

16. B. Schieber and U. Vishkin. *On Finding Lowest Common Ancestors: Simplification and Parallelization.* SIAM Journal on Computing, 17:1253-1262, 1988.

17. M.-F. Sagot. *Spelling approximate repeated or common motifs using a suffix tree.* LNCS 1380: 111-127, 1998.

18. J. F. Tomb et al. *The complete genome sequence of the gastric pathogen Helicobacter pylori.* Nature, 388, 539-547, 1997.

19. E. Ukkonen. *Approximate string matching over suffix trees.* LNCS 684: 228-242, 1993.

20. A. Vanet, L. Marsan, A. Labigne and M.-F. Sagot. *Inferring regulatory elements from a whole genome. an analysis of Helicobacter pylori σ^{80} family of promoter signals.* J. Mol. Biol., 297, 335-353, 2000.

Appendix: Figures and Tables

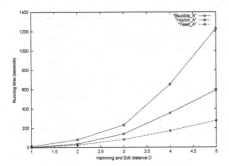

Fig. 4. The running times of algorithm A for $P = 7, 8, 8$, $q = 65, 13, 3$ and various values of D on B. Sublitis, H. Pylori and Yeast sequences.

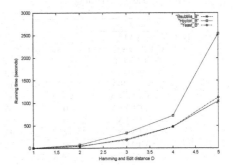

Fig. 5. The running times of algorithm B for $P = 7, 8, 8$, $q = 65, 13, 3$ and various values of D on B. Sublitis, H. Pylori and Yeast sequences.

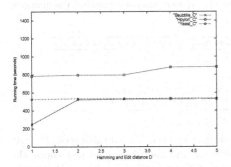

Fig. 6. The running times of algorithm C for $P = 7, 8, 8$, $q = 65, 13, 3$ and various values of D on B. Sublitis, H. Pylori and Yeast sequences.

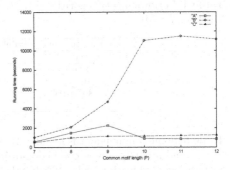

Fig. 7. The running times of algorithms A, B, C for $D = 5$, $q = 65$ and various values of P on B. Sublitis sequences.

Table 1. The number of common motifs returned by each algorithm for $P=7$, $q=65$ and various values of D on B. Subtilis sequences.

D/Algorithms	A	B	C
1	129	41	34
2	6895	12225	2520
3	15493	16384	2572
4	16384	16384	2572
5	16384	16384	2572

Table 2. The number of common motifs returned by each algorithm for $D=5$, $q=65$ and various values of P on B. Subtilis sequences.

P/Algorithms	A	B	C
7	16384	16384	2572
8	65283	65536	4945
9	152941	262144	6618
10	0	1008661	7425
11	0	1126894	7687
12	0	50680	7626

Table 3. The common motifs returned for $D = 1$, $q = 65$ and $P = 6, 7$ from B. Subtilis. Columns 3 and 4 give the results for Hamming distance, while columns 5 and 6 give the results for Edit distance.

	B. Subtilis	Alg. A	Alg.[20]	Alg. B	Alg. C
Family 1	TATAATA	75	94	(TATAAAA)67	102
	GTATAAT	(GTAAAAA) 68	74	(G,C-GTAAAA) 69,72	82
	TGTTATA	67	66		86
	ATAATAT	73	82		95
	ACAATA	81	108	77	
	TAAAATA	79	95	72	(TAAAATG) 74
	TATTATA	70	76	(TATAAAA) 67	85
	GATATA	79	98	81	(GAAATA) 73
	TATAGT	77	95	(TATA-TT,AT,AA)80,81,81	(TATACA) 76
Family 2	TTTTACA	70	76		84
	GTGACA	(GTGAAA) 72	68		
	GTTGAC	72	66	(CGTTGA) 78	
	TTTACAA	73	75		(TTTACAT) 67
Classical Motifs[7]	TATAAT			81	76
	TTGACA				

Table 4. Performance of algorithms A, B, and C on an eukaryotic promoter sequences, Yeast for $P = 16$.

		Running time	Published motif is TTtCCcnntnaGGAAA[10]
Algorithm	D	for last D	Correspond Motif(s) predicted
A	2-5	111.06 sec	-
B	2-8	>16.7 hrs	-
C	4	1440.48 sec	TTTCACCACTAGAAGA,
			TTCAAAAATGAGGAAA

Table 5. Some common motifs found in B. Subtilis sequences for Edit distance ($P = 10$, $D = 3$, and $q = 65$). The second column listed the number of distinct sequences, the patterns occur. The third column listed the same for the shuffled versions of B. Subtilis sequences, average over 1000 simulations. The last column gives each pattern χ^2 value. All motifs have the probability of happening by chance below 10^{-3}.

	# of distinct sequences each Common Motifs occurs:		
Common Motifs	Observed	Expected	χ^2
AAGTATAATG	72	27	218.33
ACGGTATATA	65	10	54.42
CCTATATTTA	65	13	54.05
GATTATAATG	65	15	170.41
GGGTGCTATA	65	15	44.75
CCGTGAAAAA	78	23	332.21
AGTTATAATG	72	27	218.33
ACGTTGAATA	66	18	65.97
CCGTAAAATG	66	18	123.11
CCGTGAAAAA	78	23	332.21
TTATATAATG	66	12	47.45

Fast Algorithms for Finding Maximum-Density Segments of a Sequence with Applications to Bioinformatics

Michael H. Goldwasser[1], Ming-Yang Kao[2,*], and Hsueh-I Lu[3,**]

[1] Department of Computer Science, Loyola University Chicago
6525 N. Sheridan Rd., Chicago, IL 60626
mhg@cs.luc.edu
www.cs.luc.edu/~mhg
[2] Department of Computer Science, Northwestern University
Evanston, IL 60201
kao@cs.northwestern.edu
www.cs.northwestern.edu/~kao
[3] Institute of Information Science, Academia Sinica
128 Academia Road, Section 2, Taipei 115, Taiwan
hil@iis.sinica.edu.tw
www.iis.sinica.edu.tw/~hil

Abstract. We study an abstract optimization problem arising from biomolecular sequence analysis. For a sequence $A = \langle a_1, a_2, \ldots, a_n \rangle$ of real numbers, a *segment* S is a consecutive subsequence $\langle a_i, a_{i+1}, \ldots, a_j \rangle$. The *width* of S is $j - i + 1$, while the *density* is $(\sum_{i \leq k \leq j} a_k)/(j - i + 1)$. The *maximum-density segment* problem takes A and two integers L and U as input and asks for a segment of A with the largest possible density among those of width at least L and at most U. If $U = n$ (or equivalently, $U = 2L - 1$), we can solve the problem in $O(n)$ time, improving upon the $O(n \log L)$-time algorithm by Lin, Jiang and Chao for a general sequence A. Furthermore, if U and L are arbitrary, we solve the problem in $O(n + n \log(U - L + 1))$ time. There has been no nontrivial result for this case previously. Both results also hold for a weighted variant of the maximum-density segment problem.

1 Introduction

Non-uniformity of nucleotide composition within genomic sequences was first revealed through thermal melting and gradient centrifugation experiments [19,22]. The GC of the DNA sequences in all organisms varies from 25% to 75%. GC-ratios have the greatest variations among bacteria's DNA sequences, while the typical GC-ratios of mammalian genomes stay in 45-50%. The GC content of human DNA varies widely throughout the genome, ranging between 30% and 60%. Despite intensive research effort in the past two decades, the underlying

* Supported in part by NSF grant EIA-0112934.
** Supported in part by NSC grant NSC-90-2218-E-001-005.

R. Guigó and D. Gusfield (Eds.): WABI 2002, LNCS 2452, pp. 157–171, 2002.
© Springer-Verlag Berlin Heidelberg 2002

causes of the observed heterogeneity remain contested [4,2,7,10,32,34,11,16,8,5]. Researchers [25,31] observed that the compositional heterogeneity is highly correlated to the GC content of the genomic sequences. Other investigations showed that gene length [6], gene density [36], patterns of codon usage [29], distribution of different classes of repetitive elements [30,6], number of isochores [2], lengths of isochores [25], and recombination rate within chromosomes [12] are all correlated with GC content. More research related to GC-rich segments can be found in [23,24,18,33,27,15,35,13,20] and the references therein.

Although GC-rich segments of DNA sequences are important in gene recognition and comparative genomics, only a couple of algorithms for identifying GC-rich segments appeared in the literature. A widely used window-based approach is based upon the GC-content statistics of a fixed-length window [14,25,26,9]. Due to the fixed length of windows, these practically fast approaches are likely to miss GC-rich segments that span more than one window. Huang [17] proposed an algorithm to accommodate windows with variable lengths. Specifically, by assigning $-p$ points to each AT-pair and $1 - p$ points to each GC-pair, where p is a number with $0 \leq p \leq 1$, Huang gave a linear-time algorithm for computing a segment of length no less than L whose score is maximized. As observed by Huang, however, this approach tends to output segments that are significantly longer than the given L.

In this paper, we study a previous abstraction of the problem. Let $A = \langle a_1, a_2, \ldots, a_n \rangle$ be a sequence of real numbers. A *segment* S of A is a consecutive subsequence $\langle a_i, a_{i+1}, \ldots, a_j \rangle$. The *width* of S is $j - i + 1$, i.e., the number of entries in S. The *density* of S is simply the average $(\sum_{i \leq k \leq j} a_k)/(j - i + 1)$. Let L and U be positive integers with $L \leq U$. The *maximum-density segment* problem takes A, L, and U as input and asks for a segment of A with the largest possible density among those of width at least L and at most U.

In its most basic form, the sequence A corresponds to the given DNA sequence, where $a_i = 1$ if the corresponding nucleotide in the DNA sequence is G or C; and $a_i = 0$ otherwise. In the work of Huang, sequence entires took on values of p and $1 - p$ for some real number $0 \leq p \leq 1$. More generally, we can look for regions where a given set of patterns occur very often. In such applications, a_i could be the relative frequency that the corresponding DNA character appears in the given patterns. Further natural applications of this problem can be designed for sophisticated sequence analyses such as mismatch density [28], ungapped local alignments [1], and annotated multiple sequence alignments [31].

Nekrutendo and Li [25], and Rice, Longden and Bleasby [26] employed algorithms for the case where $L = U$. This case is trivially solvable in $O(n)$ time. More generally, when $L \neq U$, the problem is trivially solvable in $O(n(U - L + 1))$ time. Huang [17] studied the case where $U = n$, i.e., there is effectively no upper bound on the width of the desired maximum-density segments. He observed that an optimal segment exists with width at most $2L - 1$. Therefore, this case is equivalent to the case with $U = 2L - 1$ and can be solved in $O(nL)$ time in a straightforward manner. Recently, Lin, Jiang, and Chao [21] gave an $O(n \log L)$-time algorithm for this case based on right-skew partitions of a sequence.

Other related work include algorithms for the problem of computing a segment $\langle a_i, \ldots a_j \rangle$ with a maximum sum $a_i + \cdots + a_j$ instead of a maximum density. Bentley [3] gave an $O(n)$-time algorithm for the case where $L = 0$ and $U = n$. Within the same linear time complexity, Huang [17] solved the case with arbitrary L and $U = n$. More recently, Lin, Jiang, and Chao [21] solved the case with arbitrary L and U.

In this paper, we report the following new results:

- Case 1: $U = n$ (or equivalently, $U = 2L - 1$). The maximum-density segment problem can be solved in $O(n)$ time. This result exploits a non-nesting structure of locally optimal segments to improve upon the algorithm of Lin, Jiang, and Chao [21].
- Case 2: U and L are arbitrary. The maximum-density segment problem can be solved in $O(n + n \log(U - L + 1))$ time.

We can generalize these two new results in a weighted variant of the problem on an input sequence A of n objects. Each object is represented by a pair of two real numbers (a_i, w_i) for $i = 1, \ldots, n$, with $w_i > 0$ denoting the *width*. With these definitions, the *density* of a segment $\langle a_i, \ldots a_j \rangle$ becomes $(\sum_{i \leq k \leq j} a_k)/(\sum_{i \leq k \leq j} w_k)$. The width constraint now requires that $L \leq w_i + \cdots + w_j \leq U$. We can solve this weighted variant with the same running times as in Case 1 with $w_i > 0$ and in Case 2 with $w_i \geq 1$, respectively. Note that computationally, this weighted variant can be used to compress a real number sequence A to reduce its density analysis time in practice or theory.

The remainder of this paper is organized as follows. Section 2 introduces some notation and definitions. In Section 3, we carefully review the previous work of Lin, Jiang and Chao [21], in which they introduce the concept of right-skew partitions. Our main results are presented in Section 4, and we discuss the generalization of these results to the weighted variant in Section 5. We conclude the paper with some open problems in Section 6.

2 Notation and Preliminaries

Recall that $A = \langle a_1, a_2, \ldots, a_n \rangle$ denotes the input sequence of real numbers. We let $A(i, j)$ denote the segment $\langle a_i, \ldots, a_j \rangle$, $w(i, j) = j - i + 1$ the *width* of $A(i, j)$, and $\mu(i, j)$ the density of $A(i, j)$. We note that the prefix sums of the input sequence can be precomputed in $O(n)$ time. With these, a value $\mu(i, j)$ can be computed in $O(1)$ time, as

$$\mu(i, j) = \left(\sum_{1 \leq k \leq j} a_k - \sum_{1 \leq k \leq i-1} a_k \right) / w(i, j).$$

3 Right-Skew Segments

Lin, Jiang and Chao [21] define segment $A(i, k)$ to be *right-skew* if and only if for all $i \leq j < k$, $\mu(i, j) \leq \mu(j + 1, k)$. They prove that every A can be uniquely partitioned into right-skew segments $A_1 A_2 \ldots A_m$ such that $\mu(A_x) > \mu(A_y)$ for any $x < y$. They call this decomposition the *decreasingly right-skew partition of A*.

If $A(i,k)$ is not right-skew, its decomposition must consist of the right-skew segment $A(i,j)$ for some $i \leq j < k$, followed by the unique partition of $A(j+1,k)$. Because of this common structure, the decreasing right-skew partitions for all $A(i,n)$ can be simultaneously represented by keeping a *right-skew pointer*, $p[i]$ for each $1 \leq i \leq n$, where $A(i,p[i])$ is the first right-skew segment for the decomposition of $A(i,n)$. They implicitly use dynamic programming to construct all such right-skew pointers in $O(n)$ time.

In order to find a maximum-density segment of width at least L, they proceed by independently searching for the "good partner" of each index i. The good partner of i is the index j that maximizes $\mu(i,j)$ while satisfying $w(i,j) \geq L$. In order to find each good partner, they make use of the following three lemmas.

Lemma 1 (Atomic). *Let B, C and D be three real sequences with $\mu(B) < \mu(C) < \mu(D)$. Then $\mu(BC) < \mu(BCD)$.*

Lemma 2 (Bitonic). *Let B be a real number sequence and $C_1 C_2 \cdots C_m$ the decreasingly right-skew partition of a sequence C which immediately follows B. Let $\mu(BC_1 C_2 \cdots C_k) = \max\{\mu(BC_1 C_2 \cdots C_i) \mid 0 \leq i \leq m\}$. Then $\mu(BC_1 C_2 \cdots C_i) > \mu(BC_1 C_2 \cdots C_{i+1})$ if and only if $i \geq k$.*

Lemma 3. *Given a real number sequence B, let C denote the shortest segment of B with width at least L realizing the maximum density. Then the width of C is at most $2L - 1$.*

Without any upper bound on the desired segment length, the consequence of these lemmas is an $O(\log L)$-time algorithm for finding a good partner for arbitrary index i. Since only segments of width L or greater are of interest, the segment $A(i, i + L - 1)$ must be included. Lin, Jiang and Chao [21] make use of the decreasingly right-skew partition of $A(i + L, n)$ when searching for the good partner of i. If part of a right-skew segment improves the density of a candidate segment, then Lemma 1 assures that the entire right-skew segment should be included (C represents part of a complete right-skew segment CD in the statement of the lemma). Therefore, the good partner for i must be $i + L - 1$ or else the rightmost index of one of the right-skew segments from the partition of $A(i + L, n)$. Lemma 2 shows that the inclusion of each successive right-skew segment leads to a bitonic sequence of densities, thus binary search can be used to locate the good partner. Finally, Lemma 3 assures that at most L right-skew segments need be considered for inclusion, and thus the binary search for a given i runs in $O(\log L)$ time. The result is an $O(n \log L)$ algorithm for arbitrary L and $U = n$.

4   Improved Algorithms

Our results make use of decreasingly right-skew partitions, as reviewed in Section 3. Our improvements are based upon the following observation. An exact good partner for an index i need not be found if it can be determined that such a partner would result in density no greater than that of a segment already considered. This observation allows us to replace the binary searches used by Lin, Jiang and Chao [21], at cost of $O(\log L)$ time each, with sequential searches that result in an *amortized* cost of $O(1)$ time each. Before developing our complete algorithms, we introduce the sweep-line data structure which manages the search for good partners.

4.1   A Sweep-Line Data Structure

Our data structure augments the right-skew pointers for a given interval with additional information used to speed up searches for good partners. In order to achieve improved efficiency, the structure supports searches limited in two ways:

1. The structure can be used to find matches for an arbitrary collection of values for i, however the queries must be made in decreasing order.
2. When asked to find the match for a left index, the structure will only find the true good partner in the case that the good partner has index less than or equal to all previously returned indices.

The data structure contains the following information, relative to fixed parameters $1 \leq x \leq y \leq n$:

1. A (static) array, $p[k]$ for $x \leq k \leq y$, where $A(k, p[k])$ is the right-skew segment starting the decreasingly right-skew partition of $A(k, y)$.
2. A (static) sorted list, $S[k]$, for each $x \leq k \leq y$, containing all indices j for which $p[j] = k$.
3. Two indices ℓ and r (for "left" and "right"), whose values are non-increasing as the algorithm progresses.
4. A variable, b (for "bridge"), which is the minimum index such that $\ell \leq b \leq r \leq p[b]$.

The data structure is initialized with procedure `Initialize`(x, y) for $1 \leq x \leq y \leq n$, as defined in Figure 1. An example of an initialized structure is given in Figure 2. Two other supported operations are given in Figures 3 and 4.

Lemma 4 (Nesting). *There cannot exist indices j and k such that $j < k \leq p[j] < p[k]$.*

Proof. For contradiction, assume that j is the maximum such value achieving a violation. Consider the point when $p[j]$ is determined in `Initialize`. The value $p[j]$ is initially set equal to j at line 3, and is increased by each execution of line 5. We consider the first time that $j < k \leq p[j] < p[k]$. Initially, $p[j] = j$, so the violation must be created at line 5. Immediately before that line, it must be that $p[j] < k$ yet $k \leq p[p[j] + 1] < p[k]$. If $p[j] + 1 < k \leq p[p[j] + 1] < p[k]$, this would be a contradiction to $j < p[j] + 1$ being the maximum violation of the lemma. Otherwise, it must be that $p[j] + 1 = k$, but then $p[p[j] + 1] = p[k] \not< p[k]$, leading to another contradiction. Therefore, no such violation can occur.   □

```
         procedure Initialize(x, y)              assumes 1 ≤ x ≤ y ≤ n
1            for i ← y downto x do
2                S[i] ← ∅
3                p[i] ← i
4                while (p[i] < y) and μ(i, p[i]) ≤ μ(i, p[p[i] + 1]) do
5                    p[i] ← p[p[i] + 1]
6                end while
7                Insert i at beginning of S[p[i]]
8            end for
9            ℓ ← y; r ← y; b ← y
```

Fig. 1. Data structure's Initialize(x, y) operation.

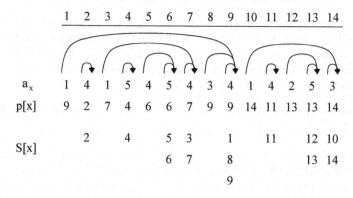

Fig. 2. Example of data structure after Initialize(1, 14).

Lemma 5. *If b is the minimum value satisfying ℓ ≤ b ≤ r ≤ p[b], then A(b, p[b]) is the segment of the decreasingly right-skew decomposition of A(ℓ, y) which contains a_r.*

Proof. Assume b achieves such a minimum. The decomposition of A(ℓ, y) will equal A(ℓ, p[ℓ]), A(p[ℓ] + 1, p[p[ℓ] + 1]), and so on, until reaching right endpoint y. We claim that A(b, p[b]) must be part of that decomposition. If not, there must be some other A(m, p[m]) with m < b ≤ p[m]. By Lemma 4, it must be that p[m] ≥ p[b], but such an m would violate the minimality of b.   □

Lemma 6. *Whenever line 1 of FindMatch is evaluated, b is the minimum value satisfying ℓ ≤ b ≤ r ≤ p[b].*

Proof. We show this by induction over time. When initialized, ℓ = b = r = p[b] = y, and thus b is the only satisfying value. The only time this invariant can be broken is when the value of ℓ or r changes. ℓ is changed only when decremented at line 2 of DecreaseI. Since it still remains that ℓ ≤ b ≤ r ≤ p[b], the only possible violation of the invariant is if new index ℓ satisfies ℓ ≤ r ≤ p[ℓ]. This is exactly the condition handled by lines 3–4.

```
     procedure DecreaseI(i)
1        while (w(i, ℓ − 2) ≥ L) and (ℓ − 2 ≥ x) do
2            ℓ ← ℓ − 1
3            if p[ℓ] ≥ r then
4                b = ℓ
5            end if
6        end while
```

Fig. 3. Data structure's DecreaseI(i) operation.

```
     procedure FindMatch(i)
1        while (r ≥ ℓ) and (μ(i, b − 1) > μ(i, p[b])) do
2            r ← b − 1
3            if (r ≥ ℓ) then
4                find minimum k ∈ S[r] such that k ≥ ℓ
5                b ← k
6            end if
7        end while
8        return r
```

Fig. 4. Data structure's FindMatch(i) operation.

Secondly, we note that r is modified only at line 2 of FindMatch. Immediately before this line is executed, we consider all k such that $\ell \leq k < b$, and ask where $p[k]$ might lie. By Lemma 4, it cannot be the case that $b \leq p[k] < p[b]$. If $p[k] \geq p[b] \geq r$, then k would violate the minimality of b assumed at line 1. Therefore, it must be that any k with $\ell \leq k < b$ has $p[k] \leq b - 1$. After r is set to $b - 1$, we ask what values k satisfy $\ell \leq k \leq r \leq p[k]$. The only possibility will be those k for which $p[k] = b - 1$. $S[b - 1]$ contains exactly those values of k, and so lines 4–5 restore the invariant regarding b. □

Lemma 7. *For a given call to* FindMatch(i), *let* m_0 *be the most recently returned value from* FindMatch, *or* y *if this is the first such call. Let* $A(i, m)$ *be a maximum-density segment that starts with* i, *has width at least* L, *and with* $m \in [x, y]$. *Then* FindMatch(i) *returns the value* $\min(m, m_0)$ *so long as* DecreaseI(i) *was executed immediately prior.*

Proof. Since we are interested in segments of width at least L ending in $[x, y]$, any candidate segment must include interval $A(i, k)$ where k is the minimum value satisfying both $k \geq x$ and $w(i, k) \geq L$. Line 2 of the procedure DecreaseI ensures that variable $\ell = k + 1$ upon exit. As discussed in Section 3, the optimal such m must either be k or else among the right endpoints of the right-skew segments in the decomposition of $A(\ell, y)$.

Upon entering FindMatch, $r = m_0$, and by Lemmas 5–6, $A(b, p[b])$ is the right-skew segment containing a_r in the decomposition of $A(\ell, y)$. If $\mu(i, b-1) \leq \mu(i, p[b])$, the good partner must have index at least $p[b] \geq r$, by Lemma 2. In this case, the while loop is never entered, and the procedure returns $m_0 = \min(m, m_0)$.

In any other case, the while loop of line 1, combined with Lemmas 2 and 6, assures that the returned value of r is precisely $m = \min(m, m_0)$.   □

Lemma 8. *The data structure supports its operations with amortized running time of $O(y - x + 1)$ for* Initialize(x, y), *and $O(1)$ for each of* DecreaseI(i) *and* FindMatch(i).

Proof. With the exception of lines 2, 7 and 9, the initialization procedure is simply a restatement of the algorithm given by Lin, Jiang and Chao [21] for constructing the right-skew pointers. An $O(y - x + 1)$-time worst case bound was proven by those authors.

To account for DecreaseI we note that ℓ is initialized to value y at line 9 of Initialize, is modified only when decremented at line 2 of DecreaseI, and remains at least $x + 1$ due to line 1 of DecreaseI. Therefore the loop executes $O(y - x + 1)$ times and this cost can be amortized against the initialization cost. An $O(1)$ amortized cost can account for checking the initial test condition before entering the loop.

In analyzing the cost of FindMatch we note that variable r is initialized to value y at line 9 of Initialize. By Lemma 6, $b \leq r \leq p[b]$, and so we note that the execution of line 2 of FindMatch always results in a strict decrease in the value of r. Therefore, the while loop of that routine executes $O(y - x + 1)$ times. The only step within that loop which cannot be bounded by $O(1)$ in the worst case is that of line 4. However, since each k appears in list $S[r]$ for a distinct value of r, the overall cost associated with line 4 is bounded by $O(y - x + 1)$. This, and the remaining cost of the while loop, can be amortized against the initialization cost.   □

4.2   Maximum-Density Segment with Width at Least L

In this section, we consider the problem of finding a segment with the maximum possible density among those of width at least L. We present a linear time algorithm which makes use of the data structure developed in Section 4.1. The key to the correctness of the algorithm is the following lemma.

Lemma 9. *For a given j, assume $A(j, k)$ is a maximum-density segment, of those starting with index j and having width at least L. For a given $i < j$, assume $A(i, m)$ is a maximum-density segment, of those starting with index i and having width at least L. If $m > k$, then $\mu(j, k) \geq \mu(i, m)$.*

Proof. A typical such configuration is shown in Figure 5. Since both $A(j, k)$ and $A(j, m)$ have width at least L, the optimality of $A(j, k)$ guarantees that $\mu(j, k) \geq \mu(j, m)$. This implies that $\mu(j, k) \geq \mu(j, m) \geq \mu(k + 1, m)$. Since both

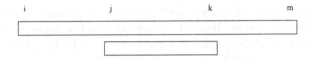

Fig. 5. Segments in proof of Lemma 9.

---

1   Call `Initialize(1, n)` to create data structure
2   $i_0 \leftarrow$ maximum index such that $w(i_0, n) \geq L$
3   **for** $i \leftarrow i_0$ **downto** 1 **do**
4   `DecreaseI(i)`
5      $g[i] \leftarrow$ `FindMatch(i)`
6   **end for**
7   **return** $(k, g[k])$ which maximizes $\mu(k, g[k])$ for $1 \leq k \leq i_0$

Fig. 6. Algorithm for finding maximum-density segment with width at least L.

$A(i, k)$ and $A(i, m)$ have width at least L, the optimality of $A(i, m)$ guarantees that $\mu(i, m) \geq \mu(i, k)$, which in turn implies $\mu(k + 1, m) \geq \mu(i, m) \geq \mu(i, k)$. Combining these inequalities, $\mu(j, k) \geq \mu(j, m) \geq \mu(k + 1, m) \geq \mu(i, m)$, thus proving the claim that $\mu(j, k) \geq \mu(i, m)$. □

Theorem 1. *A maximum-density segment of $A = \langle a_1, a_2, \ldots, a_n \rangle$ of those with width at least L can be found in $O(n)$ time.*

Proof. The algorithm is given in Figure 6. The correctness is a direct result of Lemmas 7 and 9. The running time bound is a direct consequence of Lemma 8. □

4.3   Maximum-Density Segment with Width at Least L, at Most U

In this section, we consider the problem of finding an interval with the maximum possible density among those of width at least L and at most U. When $U \geq 2L - 1$, the upper bound is inconsequential by Lemma 3, and the linear-time algorithm from the previous section suffices. For $L < U < 2L - 1$, however, there is a subtle problem if trying to apply the previous algorithm directly.

Without any upper bound on the width, a good partner for index i was chosen in the range $A(p_i, n)$ where p_i is the minimum index such that $w(i, p_i) \geq L$. The decreasingly right-skew decomposition of $A(p_i + 1, n)$ was used in the search for the good partner. In similar spirit, when later searching for a good partner for $i - 1$, the decreasingly right-skew decomposition of $A(p_{i-1} + 1, n)$ was used. The beauty of the framework was the fact that the right-skew pointers simultaneously represent the decomposition for all segments of the form $A(p + 1, n)$, for fixed endpoint n. Furthermore, Lemma 1 ensured that the right-skew segments could be treated atomically.

With an upper bound on the width, however, an atomic right-skew segment from the decomposition of $A(p_i + 1, n)$ may violate the upper limit. It appears that the more relevant decomposition, for a given i, is that of $A(p_i + 1, q_i)$ where q_i the maximum index with $w(i, q_i) \leq U$. Sadly, it is not apparent whether the structure of that decomposition can be reused as the right end of the interval changes for other values of i.

In the remainder of this section, we present an $O(n + n \log(U - L + 1))$-time algorithm in the setting with both a lower bound and upper bound on the segment width. For a given index i, we know that the good partner lies in a range $A(p_i, q_i)$, as defined above. Rather than directly compute the decreasingly right-skew partition of $A(p_i + 1, q_i)$ for each i, we will compute decompositions for a collection of smaller blocks which exactly cover the range $A(p_i, q_i)$ yet can be reused for other values of i. We build a sweep-line data structure independently for each block and make use of a lemma which guarantees that we find the global solution to the problem, even if we do not find the true good partner for each i.

For ease of notation, we assume, without loss of generality, that the overall sequence A is padded with values so that it has length n which is a power of two. We consider n blocks of size 1, $n/2$ blocks of size 2, $n/4$ blocks of size 4, and so on until $n/2^\beta$ blocks of size 2^β, where $\beta = \lfloor \log_2(U - L + 1) \rfloor$. Specifically, we define block $B_{j,k} = A(1 + j * 2^k, (j + 1) * 2^k)$ for all $0 \leq k \leq \beta$ and $0 \leq j < n/2^k$. We begin with the following lemma.

Lemma 10. *For any interval $A(p, q)$ with cardinality at most $U - L + 1$, we can compute, in $O(1 + \beta)$ time, a collection of $O(1 + \beta)$ disjoint blocks such that $A(p, q)$ equals the union of the blocks.*

Proof. The algorithm `CollectBlocks` is given in Figure 7, and a sample result is shown in Figure 8. We leave it as an exercise to verify the claim.   □

Before presenting the main result for this section, we need the following variant of Lemma 9.

Lemma 11. *For a given j, assume $A(j, k)$ is a maximum-density segment, of those starting with index j, having width at least L, and ending with index in range $[x, y]$. For a given $i < j$, assume $A(i, m)$ is a maximum-density segment, of those starting with index i, having width at least L, and ending with index in range $[x, y]$. If $m > k$ then $\mu(j, k) \geq \mu(i, m)$.*

Proof. Identical to Lemma 9, as both k and m lie in range $[x, y]$.   □

Theorem 2. *A maximum-density segment of $A = \langle a_1, a_2, \ldots, a_n \rangle$ of those with width at least L and at most U can be found in $O(n + n \log(U - L + 1))$ time.*

Proof. The algorithm is given in Figure 9. First, we discuss the correctness. Assume that the global maximum is achieved by $A(i, m)$. We must simply show that this pair, or one with equal density, was considered at line 14. Based on the width constraints, it must be that $p \leq m \leq q$, for p and q as defined in lines 9–10. By Lemma 10, m must lie in some $B_{j,k}$ returned by `CollectBlocks`(p, q).

procedure CollectBlocks(p, q)

```
1     s ← p; k ← 0
2     while (2^k + s − 1 ≤ q) do
3         while (2^{k+1} + s − 1 ≤ q) do
4             k ← k + 1
5         end while
6         j ← ⌈s/2^k⌉ − 1
7         Add block B_{j,k} to the collection
8         k ← k + 1
9     end while
10    while (s ≤ q) do
11        while (2^k + s − 1 > q) do
12            k ← k − 1
13        end while
14        j ← ⌈s/2^k⌉ − 1
15        Add block B_{j,k} to the collection
16        k ← k − 1
17    end while
```

Fig. 7. Algorithm for finding collection of blocks to cover an interval.

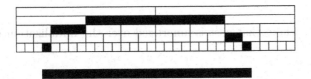

Fig. 8. A collection of blocks for a given interval.

Because $\mu(i, m)$ is a global maximum, the combination of Lemmas 7 and 11 assures us that FindMatch at line 13 will return m (or an index achieving the same density).

We conclude by showing that the running time is $O(n + n\beta)$. Notice that for a fixed k, blocks $B_{j,k}$ for $0 \leq j < n/2^k - 1$ comprise a partition of the original input A, and thus the sum of their cardinalities is $O(n)$. Therefore, for a fixed k lines 3–5 run in $O(n)$ time by Lemma 8, and overall lines 2–6 run in $O(n\beta)$ time. Lines 9–10 can be accomplished in $O(1)$ time as $p = i + L - 1$ and $q = i + U - 1$ by definition. Each call to CollectBlocks from line 11 runs in $O(1 + \beta)$ time by Lemma 10, and produces $O(1 + \beta)$ blocks. Therefore, the body of that loop, lines 12–17, executes $O(n + n\beta)$ times over the course of the entire algorithm.

Finally, we must account for the time spent in all calls to DecreaseI and FindMatch from lines 12–13. Rather than analyze these costs chronologically, we account for these calls by considering each block $B_{j,k}$ over the course of the algorithm. For a given block, the parameter values sent to DecreaseI are indeed strictly decreasing. Therefore by Lemma 8, each of these calls has an amortized

```
1   β ← ⌊log₂(U − L + 1)⌋; μ_max ← −∞
2   for k ← 0 to  β do
3       for j ← 0 to  n/2^k − 1 do
4           B_{j,k} ← Initialize(1 + j * 2^k, (j + 1) * 2^k)
5       end for
6   end for
7   i₀ ← maximum index such that w(i₀, n) ≥ L
8   for i ← i₀ downto 1 do
9       p ← minimum index such that w(i, p) ≥ L
10      q ← maximum index such that w(i, q) ≤ U
11      foreach B_{j,k} in CollectBlocks(p, q) do
12          DecreaseI(i)
13          temp ← FindMatch(i)
14          if μ(i, temp) > μ_max then
15              μ_max ← μ(i, temp)
16              record achieving endpoints, (i, temp)
17          end if
18      end foreach
19  end for
```

Fig. 9. Algorithm for maximum-density segment with width at least L and at most U.

cost of $O(1)$, where that cost is amortized over the initialization costs for the blocks.   □

5   Extensions to the Weighted Case

We can generalize the maximum-density segment problem to an input A of n objects, where each object is represented by a pair of two real numbers (a_i, w_i) for $i = 1, \ldots, n$, with $w_i > 0$. We use the term "width" for w_i rather than "weight" to avoid possible confusion in context, as this parameter appears as the denominator in the following "density" formula:

$$\mu(i, j) = \left(\sum_{i \leq k \leq j} a_k \right) \Big/ \left(\sum_{i \leq k \leq j} w_k \right).$$

We note that the prefix sums can separately be computed for a_k and w_k in $O(n)$ time, after which we can compute $w(i, j)$ in $O(1)$ time as

$$\left(\sum_{1 \leq k \leq j} w_k - \sum_{1 \leq k \leq i-1} w_k \right),$$

and $\mu(i, j)$ as described in Section 2. In this setting, we now ask whether we can find the segment with maximum possible density of those having $L \leq w(i, j) \leq U$. We offer the following two theorems.

Theorem 3. *A maximum-density segment of* $A = \langle(a_1, w_1), (a_2, w_2), \ldots,$ $(a_n, w_n)\rangle$ *of those with width at least L can be found in $O(n)$ time.*

Theorem 4. *If $w_i \geq 1$ for all i, a maximum-density segment of $A =$* $\langle(a_1, w_i), (a_2, w_2), \ldots, (a_n, w_n)\rangle$ *of those with width at least L and at most U can be found in $O(n + n \log(U - L + 1))$ time.*

The algorithms as presented in Section 4 can be used essentially unchanged with the new computations for $w(i, j)$ and $\mu(i, j)$. The only additional algorithmic change required is that for a given left index i, we must be able to calculate the effective range for the good partner search. Namely, we must find p which is the minimum index satisfying $w(i, p) \geq L$ and q which is the maximum index satisfying $w(i, q) \leq U$. These values can be precomputed for all i in $O(n)$ overall time by sweeping a pair of indices across the full sequence.

As far as the correctness, we note that the facts about right-skew segments provided by Lin, Jiang and Chao [21] do not in any way involve the width of segments, and so those results are unaffected by the width parameters. Those parameters only have effect once lower and upper bounds on the width are applied to the problem. All of our proofs from Section 4 were written so as to be valid with this weighted variant.

The running time bounds for the algorithms remain intact in this weighted version as well. The caveat that $w_i \geq 1$ from Theorem 4 is needed to guarantee that the cardinality of the interval $[p, q]$ is at most $(U - L + 1)$, so as to apply Lemma 10.

6   Conclusions

We have shown that if $U = n$ (or equivalently, $U = 2L - 1$), we can solve the maximum-density segment problem in $O(n)$ time. If U and L are arbitrary, we can solve the problem in $O(n + n \log(U - L + 1))$ time. These results hold for both the unweighted and weighted versions of the problem. Recently, the techniques of this paper have been expanded by the authors to achieve linear time when U and L are arbitrary; details will appear in a subsequent paper.

A further research direction is to generalize the input sequence A to an array of dimension two or higher. This generalization conceivably may have applications to image analysis in bioinformatics and other fields.

Acknowledgements

We wish to thank Yaw-Ling Lin for helpful discussions.

References

1. N. N. Alexandrov and V. V. Solovyev. Statistical significance of ungapped sequence alignments. In *Proceedings of Pacific Symposium on Biocomputing*, volume 3, pages 461–470, 1998.

2. G. Barhardi. Isochores and the evolutionary genomics of vertebrates. *Gene*, 241:3–17, 2000.

3. J. L. Bentley. *Programming Pearls*. Addison-Wesley, Reading, MA, 1986.

4. G. Bernardi and G. Bernardi. Compositional constraints and genome evolution. *Journal of Molecular Evolution*, 24:1–11, 1986.

5. B. Charlesworth. Genetic recombination: patterns in the genome. *Current Biology*, 4:182–184, 1994.

6. L. Duret, D. Mouchiroud, and C. Gautier. Statistical analysis of vertebrate sequences reveals that long genes are scarce in GC-rich isochores. *Journal of Molecular Evolution*, 40:308–371, 1995.

7. A. Eyre-Walker. Evidence that both G+C rich and G+C poor isochores are replicated early and late in the cell cycle. *Nucleic Acids Research*, 20:1497–1501, 1992.

8. A. Eyre-Walker. Recombination and mammalian genome evolution. *Proceedings of the Royal Society of London Series B, Biological Science*, 252:237–243, 1993.

9. C. A. Fields and C. A. Soderlund. gm: a practical tool for automating DNA sequence analysis. *Computer Applications in the Biosciences*, 6:263–270, 1990.

10. J. Filipski. Correlation between molecular clock ticking, codon usage fidelity of DNA repair, chromosome banding and chromatin compactness in germline cells. *FEBS Letters*, 217:184–186, 1987.

11. M. P. Francino and H. Ochman. Isochores result from mutation not selection. *Nature*, 400:30–31, 1999.

12. S. M. Fullerton, A. B. Carvalho, and A. G. Clark. Local rates of recombination are positively correlated with GC content in the human genome. *Molecular Biology and Evolution*, 18(6):1139–1142, 2001.

13. P. Guldberg, K. Gronbak, A. Aggerholm, A. Platz, P. thor Straten, V. Ahrenkiel, P. Hokland, and J. Zeuthen. Detection of mutations in GC-rich DNA by bisulphite denaturing gradient gel electrophoresis. *Nucleic Acids Research*, 26(6):1548–1549, 1998.

14. R. C. Hardison, D. Drane, C. Vandenbergh, J.-F. F. Cheng, J. Mansverger, J. Taddie, S. Schwartz, X. Huang, and W. Miller. Sequence and comparative analysis of the rabbit alpha-like globin gene cluster reveals a rapid mode of evolution in a G+C rich region of mammalian genomes. *Journal of Molecular Biology*, 222:233–249, 1991.

15. W. Henke, K. Herdel, K. Jung, D. Schnorr, and S. A. Loening. Betaine improves the PCR amplification of GC-rich DNA sequences. *Nucleic Acids Research*, 25(19):3957–3958, 1997.

16. G. P. Holmquist. Chromosome bands, their chromatin flavors, and their functional features. *American Journal of Human Genetics*, 51:17–37, 1992.

17. X. Huang. An algorithm for identifying regions of a DNA sequence that satisfy a content requirement. *Computer Applications in the Biosciences*, 10(3):219–225, 1994.

18. K. Ikehara, F. Amada, S. Yoshida, Y. Mikata, and A. Tanaka. A possible origin of newly-born bacterial genes: significance of GC-rich nonstop frame on antisense strand. *Nucleic Acids Research*, 24(21):4249–4255, 1996.

19. R. B. Inman. A denaturation map of the 1 phage DNA molecule determined by electron microscopy. *Journal of Molecular Biology*, 18:464–476, 1966.

20. R. Jin, M.-E. Fernandez-Beros, and R. P. Novick. Why is the initiation nick site of an AT-rich rolling circle plasmid at the tip of a GC-rich cruciform? *The EMBO Journal*, 16(14):4456–4466, 1997.

21. Y. L. Lin, T. Jiang, and K. M. Chao. Algorithms for locating the length-constrained heaviest segments, with applications to biomolecular sequence analysis. *Journal of Computer and System Sciences*, 2002. To appear.

22. G. Macaya, J.-P. Thiery, and G. Bernardi. An approach to the organization of eukaryotic genomes at a macromolecular level. *Journal of Molecular Biology*, 108:237–254, 1976.

23. C. S. Madsen, C. P. Regan, and G. K. Owens. Interaction of CArG elements and a GC-rich repressor element in transcriptional regulation of the smooth muscle myosin heavy chain gene in vascular smooth muscle cells. *Journal of Biological Chemistry*, 272(47):29842–29851, 1997.

24. S.-i. Murata, P. Herman, and J. R. Lakowicz. Texture analysis of fluorescence lifetime images of AT- and GC-rich regions in nuclei. *Journal of Hystochemistry and Cytochemistry*, 49:1443–1452, 2001.

25. A. Nekrutenko and W.-H. Li. Assessment of compositional heterogeneity within and between eukaryotic genomes. *Genome Research*, 10:1986–1995, 2000.

26. P. Rice, I. Longden, and A. Bleasby. EMBOSS: The European molecular biology open software suite. *Trends in Genetics*, 16(6):276–277, June 2000.

27. L. Scotto and R. K. Assoian. A GC-rich domain with bifunctional effects on mRNA and protein levels: implications for control of transforming growth factor beta 1 expression. *Molecular and Cellular Biology*, 13(6):3588–3597, 1993.

28. P. H. Sellers. Pattern recognition in genetic sequences by mismatch density. *Bulletin of Mathematical Biology*, 46(4):501–514, 1984.

29. P. M. Sharp, M. Averof, A. T. Lloyd, G. Matassi, and J. F. Peden. DNA sequence evolution: the sounds of silence. *Philosophical Transactions of the Royal Society of London Series B, Biological Sciences*, 349:241–247, 1995.

30. P. Soriano, M. Meunier-Rotival, and G. Bernardi. The distribution of interspersed repeats is nonuniform and conserved in the mouse and human genomes. *Proceedings of the National Academy of Sciences of the United States of America*, 80:1816–1820, 1983.

31. N. Stojanovic, L. Florea, C. Riemer, D. Gumucio, J. Slightom, M. Goodman, W. Miller, and R. Hardison. Comparison of five methods for finding conserved sequences in multiple alignments of gene regulatory regions. *Nucleic Acids Research*, 27:3899–3910, 1999.

32. N. Sueoka. Directional mutation pressure and neutral molecular evolution. *Proceedings of the National Academy of Sciences of the United States of America*, 80:1816–1820, 1988.

33. Z. Wang, E. Lazarov, M. O'Donnel, and M. F. Goodman. Resolving a fidelity paradox: Why Escherichia coli DNA polymerase II makes more base substitution errors in at- compared to GC-rich DNA. *Journal of Biological Chemistry*, 2002. To appear.

34. K. H. Wolfe, P. M. Sharp, and W.-H. Li. Mutation rates differ among regions of the mammalian genome. *Nature*, 337:283–285, 1989.

35. Y. Wu, R. P. Stulp, P. Elfferich, J. Osinga, C. H. Buys, and R. M. Hofstra. Improved mutation detection in GC-rich DNA fragments by combined DGGE and CDGE. *Nucleic Acids Research*, 27(15):e9, 1999.

36. S. Zoubak, O. Clay, and G. Bernardi. The gene distribution of the human genome. *Gene*, 174:95–102, 1996.

FAUST: An Algorithm for Extracting Functionally Relevant Templates from Protein Structures

Mariusz Milik, Sandor Szalma, and Krzysztof A. Olszewski*

accelrys
9685 Scranton Road
San Diego, CA 92121, USA
{mlk,ssz,kato}@accelrys.com

Abstract. FAUST (Functional Annotations Using Structural Templates) is an algorithm for: extraction of functionally relevant templates from protein structures and using such templates to annotate novel structures. Proteins and structural templates are represented as colored, undirected graphs with atoms as nodes and interatomic distances as edge weights. Node colors are based on chemical identities of atoms. Edge labels are equivalent if interatomic distances for corresponding nodes (atoms) differ less than a threshold value. We define FAUST structural template as a common subgraph of a set of graphs corresponding to two or more functionally related proteins. Pairs of functionally related protein structures are searched for sets of chemically equivalent atoms whose interatomic distances are conserved in both structures. Structural templates resulting from such pair wise searches are then combined to maximize classification performance on a training set of irredundant protein structures. The resulting structural template provides new language for description of structure—function relationship in proteins. These templates are used for active and binding site identification in protein structures. We are demonstrating here structural template extraction results for the highly divergent family of serine proteases. We compare FAUST templates to the standard description of the serine proteases active site pattern conservation and demonstrate depth of information captured in such description. Also, we present preliminary results of the high-throughput protein structure database annotations with a comprehensive library of FAUST templates.

1 Introduction

Proteins are basic elements of every living organism. They control how the cell is build, how information and material are transported in the cell, which biological reactions are catalyzed and which are inhibited, etc. In recent years protein sequences and protein structures have been deposited in public databases at an enormous pace. This unprecedented growth of data puts structural biology

---

* to whom correspondence should be addressed

R. Guigó and D. Gusfield (Eds.): WABI 2002, LNCS 2452, pp. 172–184, 2002.
© Springer-Verlag Berlin Heidelberg 2002

on a verge of paradigm change, because the speed at which data are produced excludes the possibility of manual screening and analysis. The majority of functional annotations of sequence and structural data are based on the evolutionary distance between two protein sequences. When the given protein sequence is similar to that of some protein with already known biological function, the function of the former may be assumed by similarity. It has been estimated that for proteins that are at least 30–40% identical, such annotation transfer is highly accurate [12]. For more distant evolutionary relationships, structural information becomes very important for precise functional annotation.

Protein structures are better conserved in the evolutionary process than protein sequences, i.e., many changes in protein sequences may be accepted in the process of evolution, as long as these changes do not drastically alter the overall structure of the protein and its function. Many examples of proteins exist with similar overall shapes (protein topology) and very similar fragments of structures (domains), but with very limited similarity in sequences. Additionally, different species could evolve proteins performing identical or similar biological function independently, from different ancestor proteins. Such proteins will share similarity in atom configuration in the neighborhood of the functionally important fragment of structure (active site, binding site, etc.) but often have completely different topology and no sequential similarity.

Unfortunately, structural data are not keeping pace with the speed of genome sequencing. As of now, we know the structures of about 10,000 proteins [4]. Only about 500 among them are dissimilar enough to be considered different folds. On the other hand, between 1,000 and 10,000 different folds are estimated to exist in nature [1,5]. Structural genomics (using high-throughput protein structure determination methods) is well poised to fill the gap between sequence and structural data. However, since the same structural scaffold can be used to support multiple functional sites, knowing protein structure does not guarantee correct functional annotation. Hence there is a significant need to facilitate characterization, extraction and scanning of protein function using functionally relevant parts of protein (structural motifs/templates).

We define the structural template as a part of the protein, which performs the specific biological function, e.g., catalyzing peptide bond hydrolysis. We distinguish this template from protein "scaffold" used to fix the active atoms in their active configuration. Functionally relevant structural templates are often conserved for proteins that exhibit the same function. Identification of such templates is equivalent to the maximal common subgraph problem for protein structure graphs. Unfortunately, strict solution to this problem leads to multiple applications of graph isomorphism algorithm that would make the problem computationally intractable. In the algorithm described in this paper, we have used supervised learning to extract common structural templates from protein structures by iterative changes in a proposed seed template. By appropriate choice of training sets of protein and application of pattern extraction method, we were able to extract basic structural features related to the specific biological function of the analyzed family of proteins. In example presented here we concentrated on one type of proteins: enzymes, whose biological function is to catalyze specific chemical reactions. Obviously, this method may be used to analyze proteins participating in cell signaling, transport, storage, etc.

2   Representation

2.1   Similar Approaches

Three-dimensional structural patterns have proved useful for protein function annotation. Different algorithms for structural pattern detection and scanning have already been proposed based on different active site representations. Artymiuk *et al.* [2] used subgraph isomorphism algorithm and reduced representation for side chains, and distances and angles between side chains. Fischer *et al.* [8] proposed geometric hashing and C-α only representation. Wallace *et al.* [13] also used geometric hashing but expanded side chain representation. The representation of Russel [11] is closest to the one used here, although he used side chain conservation to reduce search space. Fetrow *et al.* [6,7] used a combination of C-α only representation and local properties; in this case patterns are created based on an expert knowledge and analysis of scientific literature. Unfortunately, all of these methods rely on the RMS distance between substructures to assess the quality of the template, while for more distantly related evolutionary cousins often no rigid body transformation exists that can superimpose their essential active site elements. We have introduced an active site representation and a descriptive language that takes into account the malleability of the active site template. Pattern detection for all of the above methods is manual and often requires prior knowledge of active site whereabouts. In contrast, our goal was to design an automated pattern detection approach; hence we are not using any prior knowledge about the active site.

2.2   Protein Structure Representation

Most of the interesting protein structures contain thousands of atoms, making protein structure comparison very expensive. Our approach is to select a subset of atoms that will still preserve most of the information relevant for the protein function. The nature of this selection depends on the specific biological function of the chosen protein family. For example, one may select polar atoms from amino-acid side chains, when one is interested in catalytic function or ion-binding sites. Surface-exposed atoms are appropriate for signal proteins, receptors, and membrane proteins, while atoms involved in protein-protein interfaces may be chosen for protein complexes. Entity selection does not have to be restricted to atoms and may for example consists of groups of atoms or whole residues or abstract geometrical entities, like electrostatic fields around the protein. We present an example application to templates extracted from the positions of polar atoms from the side chains of globular protein structures. This feature selection appears to be favorable to the extraction of structural templates related to the catalytic function of enzymes. It may be viewed as a reduction of the electrostatic field around the protein to abstract point charges centered on polar side-chain atoms.

Protein structures were represented as undirected, colored graphs P_i. In this representation, nodes a_i^k of the graph are formed by protein atoms, and are colored by atom properties (for example type of atom or atom PDB name or corresponding residue type, etc.). Edges of the graph between a_i^k and a_i^l are labeled with Euclidean distances between the atoms k and l. Our template extraction method may be formulated as: having a set of graphs defined as above, find all nontrivial common subgraphs, preserving coloring and edge labels. By nontrivial templates we define subgraphs that are created by more than five atoms belonging to at least three different amino acid residues. These constrains are used here to filter out clusters of atoms created by pairs of residues connected by salt bridges, disulfide bonds etc. which are abundant in protein structures. We hypothesize that clusters containing such pairs are in most cases related to protein structural scaffold or to overall solubility of the protein molecule, rather than to the specific biological function of the protein.

We have focused initially on finding structural templates within a single protein chain, even though, there are known examples of active sites build by multiple chains.

3 The Algorithm

3.1 Clique Prescreening

Given our representation of protein structures as graphs P_i our objective is to identify nontrivial (as defined above), common subgraphs of two or more graphs. We start with searching for nontrivial cliques in a correspondence graph of two protein graphs. Clique is defined as a set of nodes in an undirected graph, in which every node has an edge to every other node. The protein correspondence graph $C(P_i, P_j)$ for P_i and P_j is defined as a graph that consists of nodes labeled by all pairs of P_i^k and P_j^l nodes. Nodes in the $C(P_i, P_j)$ are connected iff corresponding nodes in P_i and P_j are connected by edges with the same label (here, distance).

The primitive entity in our definition of structural template is an atom clique, which introduces more constraints designed to help prune the search tree. Atom clique $C_n(P_1, P_2)$ is defined as a clique in the corresponding graph $C(P_1, P_2)$ for which the following is true:

1. atoms a_1^i is extracted from structural graph P_1 and atoms a_2^j from P_2
2. atoms belonging to the same pair: a_1^i and a_2^j have the same chemical identity
3. distances between atoms in clique are compatible; i.e., for every two atoms from graph P_1 and their complementary atoms from graph P_2, the difference in the distances is less that a given threshold (1Å)
4. distance between any pair of atoms in a clique is less than a threshold value (8Å)

5. clique atoms from structure corresponding to graph P_1 and clique atoms from structure corresponding to graph P_2 form configurations with the same chirality, i.e., for every pair of 4 corresponding atoms both quadruplets have the same chirality

In other words, atom clique establishes correspondence between two compact (Rule 4) subsets of atoms, selected from two protein structures (Rule 1), in the way that interatomic distances (Rule 3) and overall chirality (Rule 5) are conserved. Our experience indicate that Rules 3, 4, 5 are quite effective at pruning search tree, leading to manageable computing time, e.g., for two medium size proteins (300 residues) clique search process takes time in range from couple of minutes to half hour. This time depends mostly on overall structural similarity of the compared proteins. Using this, highly restrictive, definition of primitive atom clique, we are able to compare two protein structures and extract prescreened atom cliques, containing 4 or more atom pairs, in algorithm sketched below:

1. Start from two protein structures (P_1 and P_2) sharing a common, structure-related feature (for example: active site, binding site, metal coordination site...)
2. For every pair of atoms from structure P_1 search for all analogous pairs of atoms from structure P_2 that conform to the atom clique definition,
3. continue extending the new clique by adding new atom pairs to it, always checking atom clique definition, until no new extension is possible.
4. If the final clique candidate contains four or more atom pairs—add it to the list of extracted cliques
5. Extracted cliques are checked for isomorphism with already extracted cliques and the bigger clique is retained
6. Atom cliques containing less than 5 nodes that do not share any atom pairs with any other cliques are excluded from the list.

The above procedure may be summarized as a greedy, deep-first tree traversing procedure and is used to search exhaustively sets of atoms extracted from protein structures containing up to one thousand atoms each.

3.2   Clique Merging

After the prescreening procedure, we obtain a set of atom cliques, containing both functionally or structurally relevant atom pairs, as well as random cliques, created by pure coincidence. Our analysis of atom cliques shows that 4 atom cliques are to small to differentiate between correlation and coincidence. In order to extract more relevant sets of atom pairs, we join overlapping extracted cliques into larger graphs. However, because some distance constraints are not always fulfilled for these graphs—we no longer require conservation of all distances and they can be no more represented by atom cliques. Thus the presented approach provides an important malleability feature to the active site description language, and allows us to capture similarities between active sites that are very difficult to compare otherwise, since they cannot be superimposed by the rigid body transformation.

In general, a collection of atom cliques from the protein correspondence graph identified by the prescreening procedure may not be consistent, i.e., one atom from structure P_1 may create pairs with many different atoms from structure P_2 and vice versa. We are interested in one-to-one mapping between atoms of both compared structures; conflicting cliques are therefore removed from our list to create such mapping. It is done in an iterative way, where one of the conflicting is removed from the list and the cliques are *de novo* joined into graphs. The clique to be removed is chosen so that remaining cliques have maximal possible coverage in assigning atoms from structure P_1 to atoms from structure P_2. This procedure is repeated until maximal one-to-one mapping between chosen atoms from structures P_1 and P_2 is obtained.

Sometimes such a procedure may create two or more mutually exclusive versions of graphs of comparative cardinality. It may be an indication that at least one of compared structures contains multiple active sites of similar structural features. In this case, multiple versions of the graphs are kept for further analysis.

We repeat the above extraction procedure for another pair of protein structures from the same family (P_1, P_3) in order to focus even more our search for functionally relevant structural templates. The first protein structure from this pair, P_1, is the same in both searches; the second one, P_3, is chosen to be as distant as possible (in terms of sequence similarity) to proteins P_1 and P_2 while keeping the same functional assignment.

As a result of the above procedure, we have two sets of atoms extracted from protein structure P_1 by comparison with structures of proteins P_2 and P_3, where structures P_1, P_2 and P_3 share the same functional assignment. These sets are projected to form template graphs in $P_{1.}$. Nodes in template graphs are connected by an edge if they belong to the same clique extracted from proteins P_1 and P_2 (or proteins P_1 and P_3 in the case of the second set). An exact subgraph search algorithm was then used for these graphs, to find a subset of atoms from protein P_1 with configuration shared with both proteins P_2 and P_3. The template extraction algorithm works best for comparison of structures of unrelated proteins expressing some common feature or function. We hypothesize, since configuration of the template atoms is conserved in divergent protein structures sharing this function, the selected set have a large probability to contain functionally relevant atoms of protein associated with graph $P_{1.}$ This common subgraph is then used as a seed for the FAUST structural template for the given protein.

3.3 Structural Template Relevance

Functional relevance of the structural template may be supported by testing its classification power against a testing set of protein structures with assigned biological functions. We used a subset of protein structures chosen from Protein Structure Data Bank [8]. To decrease redundancy in the testing set, the selected subset contains only structures of proteins with sequence identity less than 95%

[13]. Using a structural template candidate, we can scan all protein structures from our testing set for existence of a matching group of atoms.

If the structural template containing protein was previously classified to have the template related function—we call it a "true positive" (TP) hit. If this protein was not classified to have this function—we call it a "false positive" (FP) hit. "True negative" (TN) and "false negative" (FN) hits are defined analogously. The quality of a structural template may be assessed by its sensitivity, defined as TP/(TP+FN), and its specificity, defined as TN/(TN+FP).

When the specificity value for the tested template is low—then it means that the template is too generic and template extraction process should be repeated for the given protein family using another pair of proteins—provided that the family is big enough. When the sensitivity value is low—it means that the template does not recognize many of function-performing proteins—the template is too specific.

3.4   FAUST Template Refinement

In the case when the template is too specific, we perform cliques extraction and graph development procedure for protein pair containing the previously used protein P_1 and a new protein classified as false negative by the current template. The newly developed graph is then compared with the modified template and again maximal common subgraph was extracted. As the result, the elements of the original template graphs (nodes and edges) that are responsible for the misclassification of the false negative protein structure are removed from the modified template definition.

In general one can manually remove nodes or edges from the graph. The choice of the removed elements may be based on the existing knowledge of the expected active site or, when it is not accessible, by analysis of protein structures classified as false negative by the template.

Generally, this procedure may give three outcomes:

1. The resulting template has acceptable sensitivity and selectivity values on the testing set—in this case, the procedure is finished.
2. The resulting template is too general (low selectivity) or even empty graph (no common subgraph)—in this case, the active/binding site for the searched family of proteins may be structurally diverse and we need to subdivide this family into smaller subfamilies, extracting independent templates for them.
3. The extracted template is still too selective—in this case the nodes and edges deletion procedure must be repeated with another protein structure classified as false negative.

4   Applications

4.1   Serine Proteinase Family

We applied our algorithm to extract FAUST template from structures of proteins from serine endopeptidases family, classified in Enzyme Classification Database

Table 1. List of protein structures used for pattern extraction. All selected proteins belong to serine endopeptidase family.

PDB name	E.C. number	SwissProt entry	Protein name
8gch	3.4.21.1	CTRA_BOVIN	chymotrypsin.
1aq7	3.4.21.4	TRY1_BOVIN	trypsin.
1tom	3.4.21.5	THRB_HUMAN	thrombin.
1boq	3.4.21.12	PRLA_LYSEN	alpha-lytic endoptidase
1ton	3.4.21.35	KLK2_RAT	tonin
1b0e	3.4.21.36	EL1_PIG	pancreatic elastase.
1hne	3.4.21.37	ELNE_HUMAN	leukocyte elastase.
1pjp	3.4.21.39	MCT1_HUMAN	chymase.
1arb	3.4.21.50	API_ACHLY	lysyl endopeptidase.
1st3	3.4.21.62	SUBB_BACLE	subtilisin.
1tec	3.4.21.66	THET_THEVU	thermitase.
1sgc	3.4.21.80	PRTA_STRGR	streptogrisin A.
1sgp	3.4.21.81	PRTB_STRGR	streptogrisin B.
1hgp	3.4.21.82	GLUP_STRGR	glutamyl endoptidase II

[3] as E.C. 3.4.21.*. This class of proteins called proteinases or endopeptidases is a group of enzymes that catalyze hydrolysis of covalent peptide bonds, i.e., bonds that form the backbones (main chains) of protein molecules. In the case of serine endopeptidases, the mechanism is based on nucleophilic attack of the targeted peptide bond by the oxygen atom from a serine side chain. The nucleophilic property of the attacking group is improved by the presence of a histidine held in proton acceptor state by closely coordinated acidic aspartate side chain. These spatially aligned side chains of serine, histidine and aspartate build the charge relay system, very often called "catalytic triad" in literature. This system is common to all serine endopeptidase enzymes and is strictly related to their functional definition. Existence of this well explored and defined active site in the serine endopeptidase enzymes makes them a very good training set for robustness testing of our structural template extraction algorithm. Additionally, proteins belonging to this family are very important for cell function. Some of them control blood clotting, other are involved in signaling pathways, enzyme activation and protein degradation in different cellular or extracellular compartments.

4.2 Serine Proteinase Template Extraction

Table 1 lists the 14 structures that we have chosen; together, they span all variety of proteins belonging to the serine endopeptidase family.

The FAUST algorithm was used on the set of representative endopeptidase structures. We extracted common atom templates containing at least 5 atoms. Figure 1 presents all such cliques extracted by comparison between chymotrypsin structure (8gch) and the remaining representative structures. Template 1 is derived from of 8gch and 1st3 structures (denoted 8gch/1st3); Template 2 from 8gch/1arb, 8gch/1sgp, and 8gch/1hne; Template 3 from 8gch/1sgp; Template 4 from 8gch/1hneE; Template 5 from 8gch/1arb and 8gch/1sgp; Template 6 from 8gch/1sgp and 8gch/1hne; and Template 7from 8gch/1hne. Table 2 show the mapping from template node to chymotrypsin 8gch.

All templates extracted from these structures are constructed around the charge relay system of serine endopeptidases, containing Ser, His, and Asp residues, i.e., all of the atoms selected in the template extraction procedure lie in the neighborhood of the experimentally characterized active site of these proteins. For most of the analyzed pairs of structures, some additional residues from the neighborhood of the charge relay system are included in templates. Some of them (like Ser 214 in 8gch and 1hne) are connected with selectivity of substrate binding to serine endopeptidases; others still await proper analysis of their functional importance.

4.3   Template Validation and Manual Refinement

In order to validate functional relevance of FAUST templates, we prepared a diverse testing set containing structures of 314 enzymes with well-defined biological function. In the template testing process, we searched all these structures for groups of atoms analogous to the extracted template, again using greedy tree search. Table 3 presents results of this database search for the smallest templates extracted by analysis of structures proteins from serine endopeptidases family.

This template contains five atoms, which are conserved in all compared structures from this family (labeled "Template 1" in Fig. 1). All atoms of Template 1 belong to residues that form the experimentally characterized charge relay system—the active site for enzymes from serine endopeptidase family.

Initially, nine template occurrences in our database were classified as false positive hits and the template was detected in all but one proteins belonging to serine endopeptidase family (false negative). The structure from this family, where our template was not found (false negative hit) was 1ton structure of tonin. However, this structure is known to be inactive [9], since zinc ion binding perturbs the structure of the active site. Fujinaga and James used zinc to aid

Table 2. Template graphs node to atom mappings, based on the 8gch chymotrypsin mapping.

Template node id	Atom name	Residue name
1	ND1	His 57
2	NE2	HIS 57
3	OG	Ser 195
4	OD1	Asp 102
5	OD2	Asp 102
6	SG	Cys 58
7	OG1	Thr 54
8	SG	Cys 42
9	OH	Tyr 94
0	OG	Ser 214

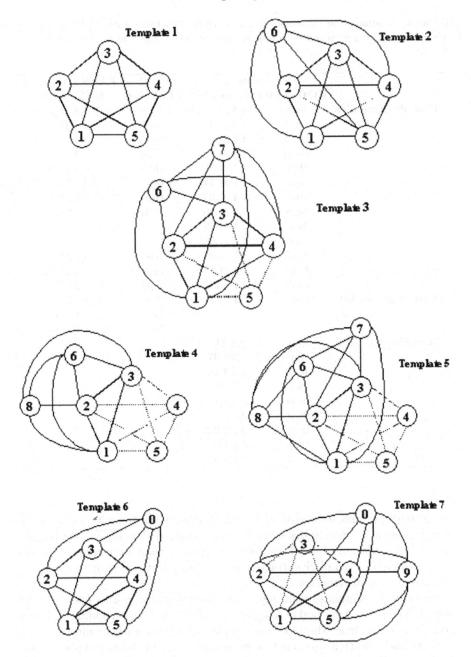

Fig. 1. All atom templates containing 5 or more atoms, extracted from chymotrypsin structure 8gch by common graph comparisons with other representative structures from serine endopeptidase family.

Table 3. Scanning the database of test protein structures with FAUST templates created by merging prescreened common atom cliques. (1ton is not a real false positive, cf. text.)

	Template 1		Optimized FAUST template	
	PDB code	E.C. number	PDB code	E.C. number
True Positive hits	8gch	3.4.21.1	8gch	3.4.21.1
	1aq7	3.4.21.4	1aq7	3.4.21.4
	1tom	3.4.21.5	1tom	3.4.21.5
	1boq	3.4.21.12	1boq	3.4.21.12
	1b0e	3.4.21.36	1b0e	3.4.21.36
	1hne	3.4.21.37	1hne	3.4.21.37
	1pjp	3.4.21.39	1pjp	3.4.21.39
	1arb	3.4.21.50	1arb	3.4.21.50
	1st3	3.4.21.62		
	1tec	3.4.21.66		
	1sgc	3.4.21.80	1sgc	3.4.21.80
	1sgp	3.4.21.81	1sgp	3.4.21.81
	1hgb	3.4.21.82	1hpg	3.4.21.82
False Negative hits	1ton	3.4.21.35	1ton	3.4.21.35
			1st3	3.4.21.62
			1tec	3.4.21.66
False Positive hits	1fmt	2.1.2.9		
	1bs0	2.3.1.47		
	1cgt	2.4.1.19		
	1dek	2.7.4.13		
	1auo	3.1.1.1		
	1byd	3.2.1.2		
	1mrj	3.2.2.22		
	1ivy	3.4.16.5		
	1qj4	4.1.2.39		

the protein crystallization process [9]. Interaction of zinc ion with aspartate was strong enough to change distances between atoms belonging to our template and classify it as inactive. This result is again a good indication of possible application of our template extraction method to selectively assign function and activity to protein structures.

Since other templates were more selective we attempted to optimize Template 1. The manual method of template modification relies on adding or deleting nodes and edges from the modified template, together with monitoring its selectivity and sensitivity parameters by scanning the modified template against testing set of protein structures. To achieve optimal template, one additional atom had to be added to the Template 1, we presented above. This atom is sulfur atom from CYS 58 residue from 8gch chymotrypsin structure. Results of test database scanning with this modified template are presented in Table 2. The modified template becomes much more selective. There are no false positive hits for this template. Unfortunately, two additional serine endopeptidase structures:

1st3 (subtilisin) and 1tec (thermitase) were classified as not containing this template (false negative hits). As it often happens in machine learning algorithms, template optimization is related to a delicate interplay between its selectivity and sensitivity. Using this version of the template we have performed search of serine endopeptidases function in the AtlasStore protein structure database that contain over 220 thousands models of proteins from human genome [10]. We have identified over 350 additional novel putative serine proteases. More extensive analysis of the extracted templates and their functional and structural relevance will be presented in following publications.

Verification of statistical relevancy of candidate templates is an open issue. Unfortunately, in the case of most enzyme families there are not enough divergent protein structures to create statistically meaningful training, testing and verification sets. Therefore, we decided to use values of false positives and false negatives on the training set of structures as template quality indicators (c.f. Table 3). Additionally, templates extracted from different protein pairs belonging to the same family may be used to partial verification of the pattern. For example in the case of the presented serine proteases family, we extracted additional pattern by comparison of structure 1a0j_A (trypsin, 3.4.21.4) with 1d6w_A (thrombin, 3.4.21.5). The extracted, selective pattern contains atoms from residues: His 57, Cys58, Asp102, Ser195 and Ser214. The extracted template is not identical, but compatible with templates presented above.

4.4   Automated Template Refinement

Since manual modification of templates is not practical for large-scale template derivation projects, we have used template projection approach to refine structural templates. Using this technique we were able to prepare a comprehensive library of structural templates based on over a hundred families or subfamilies of enzymes. The detailed list of all derived templates is available from authors upon request.

5   Summary

Supervised machine learning method combined with approximated graph isomorphism search problem was used here to extract common structural patterns from protein structures. By appropriate choice of training sets of protein structures and application of FAUST template extraction method, we are able to extract basic structural features related to the specific biological function of the analyzed family of proteins. FAUST method is based on heuristic solution of the graph isomorphism problem to identify sets of corresponding atoms from the compared protein structures. The algorithm focuses on groups of atoms that conserve subset of their pairwise distances. The method can be used for detection and analysis functionally important fragments of proteins—like active sites, binding sites, etc. The extracted structural templates are used to assign biological function to proteins in protein sequence and structure databases.

References

1. N.N. Alexandrov and N. Go, "Biological meaning, statistical significance, and classification of local spatial similarities in nonhomologous proteins," *Protein Science* 3:866, 1994.

2. P.J. Artymiuk, A.R. Poirette, H.M. Grindley, D.W. Rice, and P. Willet, "A graph-theoretic approach to identification of the three-dimensional patterns of amino-acid side chains in protein structures," *J. Mol. Biol.* 243:327, 1994.

3. A. Bairoch, "The ENZYME database in 2000." *Nucleic Acids Res.* 28:304, 2000.

4. H.M. Berman, J. Westbrook, Z. Feng, G. Gilliland, T.N. Bhat, H. Weissig, I.N. Shindyalov, and P.E. Bourne, "The protein data bank." *Nucleic Acids Research,* 28:235, 2000.

5. C. Chotia, "One thousand families for the molecular biologist," *Nature* 357:543, 1992.

6. J.S. Fetrow, A. Godzik, and J. Skolnick, "Functional analysis of the E. Coli genome using the sequence-to-structure-to-function paradigm: Identification of proteins exhibiting the glutaredoxin/thioredoxin disulfide oxidoreductase activity," *J. Mol. Biol.* 1998.

7. J.S. Fetrow and J. Skolnick, "Method for prediction of protein function from sequence using sequence-to-structure-to-function paradigm with application to glutaredoxins/thioredoxins and T_1 ribonucleases," *J. Mol. Biol.* 1998.

8. D. Fischer, H. Wolfson, S.L. Lin, and R. Nussinov, "Three-dimensional, sequence-order independent structural comparison of a serine protease against the crystallographic database reveals active site similarities," *Protein Sci.* 3:769, 1994.

9. M. Fujinaga and M.N.G. James, "Rat submaxillary gland serine protease, tonin, structure solution and refinement at 1.18Å resolution," *J. Mol. Biol.* 195:373, 1987.

10. D.H. Kitson, A. Badredtinov, Z.-Y. Zhu, M. Velikanov, D.J. Edwards, K. Olszewski, S. Szalma, and L. Yan, "Functional annotation of proteomic sequences based on consensus of sequence and structural analysis," *Briefings in Bioinformatics,* 3:32, 2002.

11. R.B. Russel, "Detection of protein three-dimensional side chain pattern." *J. Mol. Biol.* 279:1211, 1998.

12. A.E. Todd, C.A. Orengo, and J.M. Thornton, "Evolution of function in protein superfamilies from a structural perspective, " *J. Mol. Biol.* 307:1113, 2001.

13. A.C. Wallace, N. Borkakoti, and J.M. Thornton, "TESS, a geometric hashing algorithm for deriving 3D coordinate templates for searching structural databases: Application to enzyme active sites. " *Protein Sci.* 6:2308, 1997.

Efficient Unbound Docking of Rigid Molecules

Dina Duhovny[1,*], Ruth Nussinov[2,3,**], and Haim J. Wolfson[1]

[1] School of Computer Science, Tel Aviv University, Tel Aviv 69978, Israel
{duhovka,wolfson}@post.tau.ac.il
[2] Sackler Inst. of Molecular Medicine, Sackler Faculty of Medicine
Tel Aviv University
[3] IRSP - SAIC, Lab. of Experimental and Computational Biology, NCI - FCRDC
Bldg 469, Rm 151, Frederick, MD 21702, USA
ruthn@fconyx.ncifcrf.gov

Abstract. We present a new algorithm for unbound (real life) docking of molecules, whether protein–protein or protein–drug. The algorithm carries out rigid docking, with surface variability/flexibility implicitly addressed through liberal intermolecular penetration. The high efficiency of the algorithm is the outcome of several factors: (*i*) focusing initial molecular surface fitting on localized, curvature based surface patches; (*ii*) use of Geometric Hashing and Pose Clustering for initial transformation detection; (*iii*) accurate computation of shape complementarity utilizing the *Distance Transform*; (*iv*) efficient steric clash detection and geometric fit scoring based on a multi-resolution shape representation; and (*v*) utilization of biological information by focusing on *hot spot* rich surface patches. The algorithm has been implemented and applied to a large number of cases.

1 Introduction

Receptor-ligand interactions play a major role in all biological processes. Knowledge of the molecular associations aids in understanding a variety of pathways taking place in the living cell. Docking is also an important tool in computer assisted drug design. A new drug should fit the active site of a specific receptor. Although electrostatic, hydrophobic and van der Waals interactions affect greatly the binding affinity of the molecules, shape complementarity is a necessary condition. The docking problem is considered difficult and interesting for a number of reasons. The combinatorial complexity of the possible ways to fit the surfaces is extremely high. The structures of the molecules are not exact, containing experimental errors. In addition, molecules usually undergo conformational changes upon association, known as induced fit. Docking algorithms must be tolerant to those difficulties, making the docking task one of the most challenging problems in structural bioinformatics.

---

* To whom correspondence should be addressed
** The publisher or recipient acknowledges right of the U.S. Government to retain a nonexclusive, royalty-free license in and to any copyright covering the article.

R. Guigó and D. Gusfield (Eds.): WABI 2002, LNCS 2452, pp. 185–200, 2002.
© Springer-Verlag Berlin Heidelberg 2002

There are two instances in the docking task - 'bound' and 'unbound'. In the 'bound' case we are given the co-crystallized complex of two molecules. We separate them artificially and the goal is to reconstruct the original complex. No conformational changes are involved. Successful application of an algorithm to 'bound' docking cases is necessary to test its validity, yet it does not ensure success in the real-life 'unbound' docking prediction, where we are given two molecules in their native conformations. In this case the docking algorithm should consider possible conformational changes upon association. Most of the docking algorithms encounter difficulties with this case, since shape complementarity is affected [14].

The goal of docking algorithms is to detect a *transformation* of one of the molecules which brings it to optimal fit with the other molecule without causing steric clash. Naturally, optimality here depends not only on geometric fit, but also on biological criteria representing the resulting complex stability. Molecular docking algorithms may be classified into two broad categories: (*i*) brute force enumeration of the transformation space; (*ii*) local shape feature matching.

Brute force algorithms search the entire 6-dimensional transformation space of the ligand. Most of these methods [33,30,31,32,11,2,3] use brute force search for the 3 rotational parameters and the FFT (Fast Fourier Transform, [19]) for fast enumeration of the translations. The running times of those algorithms may reach days of CPU time. Another brute force algorithm is the 'soft docking' method [17] that matches surface cubes. There are also non-deterministic methods that use genetic algorithms [18,12].

Local shape feature matching algorithms have been pioneered by Kuntz [20]. In 1986 Connolly [6] described a method to match local curvature maxima and minima points. This technique was improved further in [23,24] and was also applied to unbound docking [25]. Additional algorithms that employ shape complementarity constraints, when searching for the correct association of molecules were also developed [21,13,10]. Some algorithms are designed to handle flexible molecules [28,27,9].

Our method is based on local shape feature matching. To reduce complexity we, first, try to detect those molecular surface areas which have a high probability to belong to the binding site. This reduces the number of potential docking solutions, while still retaining the correct conformation. The algorithm can treat receptors and ligands of variable sizes. It succeeds in docking of large proteins (antibody with antigen) and small drug molecules. The running times of the algorithm are of the order of seconds for small drug molecules and several minutes for large proteins. In addition, we improve the shape complementarity measure, making the function more precise and reducing the complexity of its computation.

2   Methods

Our docking algorithm is inspired by object recognition and image segmentation techniques used in computer vision. We can compare docking to assembling a jigsaw puzzle. When solving the puzzle we try to match two pieces by picking one piece and searching for the complementary one. We concentrate on the patterns that are unique for the puzzle element and look for the matching patterns in the rest of the pieces. Our algorithm employs a similar technique. Given two molecules, we divide their surfaces into patches according to the surface shape. These patches correspond to patterns that visually distinguish between puzzle pieces. Once the patches are identified, they can be superimposed using shape matching algorithms. The algorithm has three major stages:

1. **Molecular Shape Representation** - in this step we compute the molecular surface of the molecule. Next, we apply a segmentation algorithm for detection of geometric patches (concave, convex and flat surface pieces). The patches are filtered, so that only patches with 'hot spot' residues are retained [15].
2. **Surface Patch Matching** - we apply a hybrid of the Geometric Hashing [34] and Pose-Clustering [29] matching techniques to match the patches detected in the previous step. Concave patches are matched with convex and flat patches with any type of patches.
3. **Filtering and Scoring** - the candidate complexes from the previous step are examined. We discard all complexes with unacceptable penetrations of the atoms of the receptor to the atoms of the ligand. Finally, the remaining candidates are ranked according to a geometric shape complementarity score.

2.1   Molecular Shape Representation

Molecular Surface Calculation. The first stage of the algorithm computes two types of surfaces for each molecule. A high density Connolly surface is generated by the MS program [5,4]. The calculated surface is preprocessed into two data structures: distance transform grid and multi-resolution surface (see Appendix). Those data structures are used in the scoring routines. In addition, the distance transform grid is used further in the shape representation stage of the algorithm. Next, a sparse surface representation [22] is computed. It is used in the segmentation of the surface into geometric patches.

The sparse surface consists of points nicknamed 'caps', 'pits' and 'belts', where each cap point belongs to one atom, a belt to two atoms and a pit to three atoms. These correspond to the face centers of the convex, concave and saddle areas of the molecular surface [22]. The gravity center of each face is computed as a centroid and projected to the surface in the normal direction.

Detection of Geometric Patches. The input to this step is the sparse set of critical points. The goal is to divide the surface to patches of almost equal area of three types according to their shape geometry: concavities, convexities and flats. We construct a graph induced by the points of the sparse surface. Each node is labeled as a 'knob', 'flat' or a 'hole' according to their curvature (see below). We compute connected components of knobs, flats and holes. Then we apply the *split* and *merge* routines to improve the component partitioning of the surface to patches of almost equal area.

Surface Topology Graph. Based on the set of sparse critical points, the graph $G_{top} = (V_{top}, E_{top})$ representing the surface topology is constructed in the following way:

$$V_{top} = \{Sparse\ Critical\ Points\}$$
$$E_{top} = \{(u, v) \mid if\ u\ and\ v\ belong\ to\ the\ same\ atom\}$$

The number of edges in the graph is linear, since each pit point can be connected by an edge to at most three caps and three belts. Each belt point is connected to two corresponding caps (see figure 2 (a)).

Shape Function Calculation. In order to group the points into local curvature based patches we use the shape function defined in [6]. A sphere of radius R is placed at a surface point. The fraction of the sphere inside the solvent-excluded volume of the protein is the *shape function* at this point. The shape function of every node in the graph is calculated. The radius of the shape function sphere is selected according to the molecule size. We use 6Å for proteins and 3Å for small ligands. As a result every node is assigned a value between 0 and 1. In previous techniques [23] points with shape function value less than $\frac{1}{3}$, named 'knobs' and points with value greater than $\frac{2}{3}$, named 'holes', were selected as critical points. All 'flat' points with function value between those two were ignored. As can be seen from the histogram of the shape function values of the trypsin molecule (PDB code 2ptn) in figure 1, a large number of points are 'flats'. In fact about 70% of the points are flats and about 30% are knobs or holes. (These statistics are typical for other molecules as well.) Consequently, the problem was, that the number of matching sets of quadraples/triples/pairs of knobs versus holes was very low [6]. Here we sort the shape function values and find two cut-off values that split the nodes to three equal sized sets of knobs, flats and holes.

Simultaneously with the shape function calculation, the *volume normal* orientation is computed using the same probe sphere. The solvent-accessible part is defined as the complement of the solvent-excluded part of the probe sphere. Given a probe sphere we define the *volume normal* to be the unit vector at the surface point in the direction of the gravity center of solvent-accessible part.

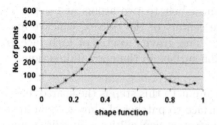

Fig. 1. Histogram of shape function values for the trypsin molecule (PDB code 2ptn).

Patch Detection. The idea is to divide the surface of the molecule into non-intersecting patches of critical points. The geometric patch is defined as a connected set of critical points of the same type (knobs, holes or flats). By 'connected' we mean that points of the patch should correspond to a connected subgraph of G_{top}. To assure better matching of patches, they must be of almost equal sizes. For each type of points (knobs, flats, holes) the graphs G_{knob}, G_{flat} and G_{hole} are constructed as subgraphs of G_{top} induced on the corresponding set of nodes.

The algorithm for finding connected components in a graph is applied to each of the three graphs. The output components are of different sizes, so *split* and *merge* routines are applied to achieve a better partition to patches. We present some definitions that are used below:

1. The **geodesic distance** between two nodes of the connected component is a weighted shortest path between them, where the weight of every edge is the Euclidean distance between the corresponding surface points.
2. The **diameter of the component** is defined as the largest geodesic distance between the nodes of the component. Nodes s and t that give the diameter are called **diameter nodes**. In case the diameter nodes are not unique we arbitrarily choose one of the pairs as the single diameter.

For each connected component we compute the diameter and it's nodes by running the APSP (*All Pairs Shortest Paths* [7]) algorithm on the component. We use two thresholds for the diameter of the components as follows:

- If the diameter of the connected component is more than *low_patch_thr* and less than *high_patch_thr*, the component represents a valid geometric patch.
- If the diameter of the connected component is larger than *high_patch_thr* the *split* routine is applied to the component.
- If the diameter of the connected component is less than *low_patch_thr* the points of this component are merged with closest components.

Note that a connected component is not always a patch, since the component may be split to a number of patches or merged with other patches.

Split routine. Given a component C, the distance matrix from the APSP algorithm and it's diameter nodes s, t we split it into two new components S and T that correspond to the Voronoi cells [8] of the points s, t. If the diameter of each new component is within the defined thresholds, the component is added to the list of valid patches, otherwise it is split again.

Merge routine. The goal is to merge points of small components with 'closest' big patches. Those points correspond to the components that are usually located between hole and knob patches, where surface topology changes quickly from concave to convex. For each point of small components we compute the geodesic distance to every valid patch using the Dijkstra [7] algorithm. The point is added to the closest patch.

At this stage most of the surface is represented by three types of patches of almost equal areas (see figure 2(b)). Now, we attempt to detect those patches, which are most likely to appear in the binding sites of the molecules.

Detection of Active Sites. The success of the docking can be significantly improved by determining the active site of the molecules. Knowledge of the binding site of at least one molecule greatly reduces the space of possible docking interactions. There are major differences in the interactions of different types of molecules (e.g. enzyme–inhibitor, antibody–antigen). We have developed filters for every type of interaction and focus only on the patches that were selected by the appropriate filter.

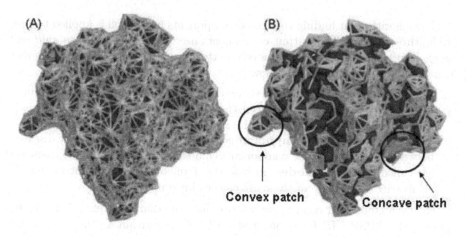

Fig. 2. (a) Surface topology graphs for trypsin inhibitor (PDB code 1ba7). The caps, belts and pits are connected with edges. (b) Geometric patches: the patches are in light colors and the protein is dark.

Hot Spot Filtering. A number of studies have shown that protein-protein interfaces have conserved polar and aromatic 'hot-spot' residues. The goal is to use the statistics collected [15,16] in order to design a patch filter. The filter is supposed to select patches that have high probability of being inside the active site. Residue hot spots have experimentally been shown (via alanine scanning mutagenesis) to contribute more than 2 Kcal/mol to the binding energetics. In order to measure it, we compute the *propensity of a residue in a patch* $Pr(res_i, patch)$ as the fraction of the residue frequency in the patch compared to the residue frequency on the molecular surface. Subsequently, we choose patches with the high propensities of hot spot residues which are: (*i*) Tyr, Asp, Asn, Glu, Ser and Trp for antibody; (*ii*) Arg, Lys, Asn and Asp for antigen; (*iii*) Ser, Gly, Asp and His for protease; and (*iv*) Arg, Lys, Leu, Cys and Pro for protease inhibitor.

Antibody-Antigen interactions: Detection of CDRs. It is well known that antibodies bind to antigens through their hypervariable (HV) regions, also called complementarity-determining regions (CDRs) [1]. The three heavy-chain and three light-chain CDR regions are located on the loops that connect the β strands of the variable domains. We detect the CDRs by aligning the sequence of the given antibody to a consensus sequence of a library of antibodies. The docking algorithm is then restricted to patches which intersect the CDR regions.

2.2   Surface Patch Matching

Given the patches of a receptor and a ligand, we would like to compute hypothetical docking transformations based on local geometric complementarity. The idea is that knob patches should match hole patches and flat patches can match any patch. We use two techniques for matching:

1. *Single Patch Matching:* one patch from the receptor is matched with one patch from the ligand. This type of matching is used for docking of small ligands, like drugs or peptides.
2. *Patch-Pair Matching:* two patches from the receptor are matched with two patches from the ligand. We use this type of matching for protein-protein docking. The motivation in patch-pair matching is that molecules interacting with big enough contact area must have more than one patch in their interface. Therefore matching two patches simultaneously will result in numerically more stable transformations. For this purpose we develop a concept of neighboring patches: two patches are considered as neighbors if there is at least one edge in G_{top} that connects the patches.

Both utilize the Computer Vision motivated Geometric Hashing [34] and Pose Clustering [29] techniques for matching of critical points within the patches. At first local features of the objects are matched to compute the candidate transformations that superimpose them. After matching of the local features a clustering step is applied to select poses (hypothetical transformations) with strong evidence, i.e. transformations consistent with a large enough number of matching local features.

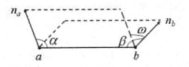

Fig. 3. Base formed by two points and their normals.

Generation of Poses. We implement the matching step using Geometric Hashing [34]. There are two stages in the matching algorithm: preprocessing and recognition. We denote by a **base** a minimal set of critical features which uniquely defines a rigid transformation. A transformation invariant shape signature is calculated for each base. In the preprocessing stage, each base is stored in a hash-table according to its signature. In the recognition stage, the bases of the receptor are computed. The base signature is used as a key to the hash-table. For each ligand base found in the hash-table a candidate transformation that superimposes local features of the receptor and the ligand is computed.

We use two points (a, b) and their volume normals (n_a, n_b) as the base for transformation calculation [23]. The base signature $(dE, dG, \alpha, \beta, \omega)$ is defined as follows (see figure 3):

- The Euclidean and geodesic distance between the two points: dE and dG.
- The angles α, β formed between the line segment ab and each of the normals n_a, n_b.
- The torsion angle ω formed between the plane of a, b, n_a and the plane of a, b, n_b.

Two signatures, one from the receptor and the other from the ligand, are compatible if their signatures are close enough. Note that we do not demand here matching of a knob to a hole, since it is ensured by matching knob patches with hole patches.

Clustering Poses. Matching of local features may lead to multiple instances of 'almost' the same transformation (similar pose). Therefore, clustering is necessary to reduce the number of potential solutions. We apply two clustering techniques: clustering by transformation parameters and RMSD clustering. Clustering by transformation parameters is coarse but very fast, and is applied first. After the number of transformations is significantly reduced, we run RMSD clustering, which is more exact, but also much slower.

2.3   Filtering and Scoring

Since the transformation is computed by local matching of critical points within patches, it may result in unacceptable steric clashes between the receptor and ligand atoms. We should filter out all those transformations. In addition, we need to rank the rest of the solutions.

Steric Clashes Test. In this stage the distance transform grid is extensively used. For each candidate transformation we perform the steric clash test as follows. The transformation is applied on the surface points of the ligand. Next we access the distance transform grid of the receptor with the coordinates of every surface point. If the distance is less than *penetration threshold* for each surface point, the transformation is retained for the next step, otherwise the transformation is disqualified.

Geometric Scoring. The general idea is to divide the receptor into shells according to the distance from the molecular surface. For example, in [23,10] a receptor was divided into 3 shells and a grid representing them was constructed as follows: interior(I) grid voxels corresponding to interior atoms (no surface point generated for them), exterior(E) voxels for exterior atoms and surface(S) voxels for surface MS dots. The score of the transformation was a weighted function of ligand surface points in each range: $S\text{-}4E\text{-}10I$. We have generalized this approach and made it more accurate using a distance transform grid (see Appendix). Each shell is defined by a range of distances in the distance transform grid. Instead of using 3 shells we can use any number of shells and the score is a weighted function of the number of ligand surface dots in each shell. In the current algorithm implementation 5 shells are used: $[-5.0, -3.6)$, $[-3.6, -2.2)$, $[-2.2, -1.0)$, $[-1.0, 1.0)$, $[1.0-)$. In the scoring stage for each candidate transformation we count the number of surface points in each shell. The geometric score is a weighted average of all the shells, when we prefer candidate complexes with large number of points in the $[-1.0, 1.0)$ shell, and as little as possible points in the 'penetrating' shells: $[-5.0, -3.6)$, $[-3.6, -2.2)$.

Those algorithms provide very accurate filtering and scoring, but are also very slow, especially when high-density surface is used for the ligand. In order to speed-up this part of the program we use a multi-resolution molecular surface data structure (see Appendix). We construct two trees: one of high density using Connolly's MS Surface and the other of lower density, using the sparse surface [22] as the lowest level. The tree based on the sparse surface is used for the primary scoring of the transformations. The tree based on the denser MS surface is used for penetration (steric clash) check and for fine geometric scoring.

Given a set of candidate transformations from the matching step, we first check the steric clashes. Only transformations with 'acceptable' penetrations (less than 5Å) of the atoms are retained. These transformations are scored with the low density based tree. We select 500 high scoring transformations for every patch and re-score them using the high density surface.

The remaining transformations can be further re-ranked according to biological criteria.

3   Results and Discussion

It is not trivial to detect data to test unbound docking algorithms. Note that we need to know three structures. The structures of the two molecules to be docked as well as the structure of their complex, so that we could reliably test our prediction results. Thus, the maximal currently available benchmark contains only few tens of molecule pairs.

Table 1 lists the 35 protein-protein cases from our test set. Our dataset includes 22 enzyme/inhibitor cases and 13 antibody/antigen cases. In 21 cases the unbound structures of both proteins were used in docking, in the other 14 cases the unbound structure for only one molecule was available. Our method is completely automated. The same set of parameters is used for each type of interactions. In the enzyme/inhibitor docking we match 'hole' patches of the enzyme with 'knob' patches of the inhibitor, since the inhibitor usually blocks the active site cavity of the enzyme. In the antibody/antigen docking we match all patch types, since the interaction surfaces are usually flat compared to enzyme/inhibitor. In addition we restrict our search to the CDRs of the antibody.

The final results are summarized in table 2. All the runs were made on a PC workstation (Pentium©II 500 MHz processor with 512MB internal memory). First, we present the results for the bound cases. As can be seen the correct solution was found for all the cases. In 31 out of the 35 examples, the lowest RMSD achieved is below 2Å. The first ranked result is the correct one in 26 cases. In other 9 cases the correct solution is ranked among the first 30 results. Columns 4 and 5 describe the steric clashes that occur when the unbound structures are superimposed on the bound complex. It shows the extent of shape complementarity in every example. Column 4 lists the maximal penetration of the ligand and receptor surfaces into each other. In some cases it is more than 5Å. Column 5 lists the number of residues of the receptor and the ligand respectively that cause deep steric clashes (more than 2.2Å surface penetration). The number of those residues is more then 10 in 4 cases. In the unbound-bound cases the numbers are significantly lower, allowing us to reach better results. We do not list the penetration level for the bound complexes since the numbers are very small: the maximal surface penetration is below 2.5Å. In the results obtained for the unbound cases we succeeded in finding solutions with RMSD under 5Å in all but 3 cases. In those cases (PDB codes 1DFJ, 2SNI, 1DQJ) shape complementarity

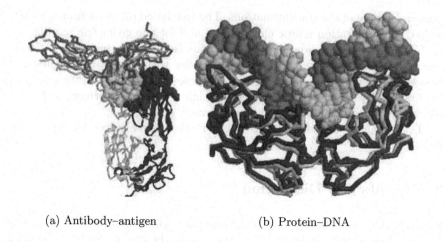

(a) Antibody–antigen   (b) Protein–DNA

Fig. 4. (a) Unbound docking of the Antibody Fab 5G9 with Tissue factor (PDB codes 1FGN,1BOY). The antibody is depicted in ribbons representation and the CDRs are in spacefill. The antigen and the solution obtained by our program are depicted in backbone representation (RMSD 2.27Å, total rank 8). (b) Protein-DNA docking: unbound-bound case (PDB codes 1A73,1EVX). Best RMSD obtained 0.87, rank 2. The DNA is shown in spacefill. Our solution superimposed on the native complex is shown in backbone representation.

is affected by many side-chain movements. The rank is under 350 in 17 out of 22 enzyme/inhibitor cases and is under 1000 in 11 out of 13 antibody/antigen cases. The best ranked result for antibody Fab 5G9/Tissue factor is shown in figure 4(a).

The program was tested on additional examples (Protein-DNA, Protein-Drug) which are not shown here. In figure 4(b) we show one of these results.

The rank of the correct solution depends on a number of factors:

1. **Shape complementarity** - steric clashes introduce noise to the scoring functions, reducing the score and the rank of the correct solution.
2. **Interface shape** - it is much easier to find correct association in molecules with concave/convex interface (enzyme/inhibitor cases) rather than with flat interfaces (antibody/antigen cases). The shape complementarity is much more prominent in the concave/convex interfaces and therefore easier to detect.
3. **Sizes of the molecules** - the larger the molecules the higher the number of results.

In our algorithm the division to patches and selection of energetic hot spots reduce the area of the matched surfaces, therefore the number of possible docking configurations is decreased and the ranking is improved.

Table 1. The dataset of Enzyme/Inhibitor and Antibody/Antigen test cases. We list the PDB codes of the complex and the unbound structures, protein names and the number of amino acids in every protein. The unbound-bound cases are marked with *.

Complex	Receptor	Ligand	Description	Rec. size	Lig. size
1ACB	5CHA	1CSE(I)	α-chymotrypsin/Eglin C	236	63
1AVW	2PTN	1BA7	Trypsin/Sotbean Trypsin inhibitor	223	165
1BRC	1BRA	1AAP	Trypsin/APPI	223	56
1BRS	1A2P	1A19	Barnase/Barstar	108	89
1CGI	1CHG	1HPT	α-chymotrypsinogen/pancreatic secretory trypsin inhibitor	226	56
1CHO	5CHA	2OVO	α-chymotrypsin/ovomucoid 3rd Domain	236	56
1CSE	1SCD	1ACB(I)	Subtilisin Carlsberg/Eglin C	274	63
1DFJ	2BNH	7RSA	Ribonuclease inhibitor/Ribonuclease A	456	124
1FSS	2ACE	1FSC	Acetylcholinesterase/Fasciculin II	527	61
1MAH	1MAA	1FSC	Mouse Acetylcholinesterase/inhibitor	536	61
1PPE*	2PTN	1PPE	Trypsin/CMT-1	223	29
1STF*	1PPN	1STF	Papain/Stefin B	212	98
1TAB*	2PTN	1TAB	Trypsin/BBI	223	36
1TGS	2PTN	1HPT	Trypsinogen/Pancreatic secretory trypsin inhibitor	223	56
1UDI*	1UDH	1UDI	Virus Uracil-DNA glycosylase/inhibitor	228	83
1UGH	1AKZ	1UGI	Human Uracil-DNA glycosylase/inhibitor	223	83
2KAI	2PKA	6PTI	Kallikrein A/Trypsin inhibitor	231	56
2PTC	2PTN	6PTI	β-trypsin/ Pancreatic trypsin inhibitor	223	56
2SIC	1SUP	3SSI	Subtilisin BPN/Subtilisin inhibitor	275	108
2SNI	1SUP	2CI2	Subtilisin Novo/Chymotrypsin inhibitor 2	275	65
2TEC*	1THM	2TEC	Thermitase/Eglin C	279	63
4HTC*	2HNT	4HTC	α-Thrombin/Hirudin	259	61
1AHW	1FGN	1BOY	Antibody Fab 5G9/Tissue factor	221	211
1BQL*	1BQL	1DKJ	Hyhel - 5 Fab/Lysozyme	217	129
1BVK	1BVL	3LZT	Antibody Hulys11 Fv/Lysozyme	224	129
1DQJ	1DQQ	3LZT	Hyhel - 63 Fab/Lysozyme	215	129
1EO8*	1EO8	2VIR	Bh151 Fab/Hemagglutinin	219	267
1FBI*	1FBI	1HHL	IgG1 Fab fragment/Lysozyme	226	129
1JHL*	1JHL	1GHL	IgG1 Fv Fragment/Lysozyme	224	129
1MEL*	1MEL	1LZA	Vh Single-Domain Antibody/Lysozyme	132	127
1MLC	1MLB	1LZA	IgG1 D44.1 Fab fragment/Lysozyme	220	129
1NCA*	1NCA	7NN9	Fab NC41/Neuraminidase	221	388
1NMB*	1NMB	7NN9	Fab NC10/Neuraminidase	229	388
1WEJ	1QBL	1HRC	IgG1 E8 Fab fragment/Cytochrome C	220	104
2JEL*	2JEL	1POH	Jel42 Fab Fragment/A06 Phosphotransferase	228	85

Table 2. Results obtained for the test set. We list the PDB code of each complex, results for bound cases, level of penetration in the unbound complexes superimposed on bound and results for the unbound cases. [a] Best RMSD(Å) result. In the brackets is its rank in the list of the results for the patch it was found in. [b] Best ranking of the result with RMSD under 5Å among all the results. [c] Maximal penetration between the surfaces in Å. [d] Number of residues that penetrate the surface with distance greater than 2.2Å for receptor and ligand respectively. [e] The running time of matching and scoring step for the unbound cases.

	Bound cases		Penetrations in unbound		Unbound cases		
PDB code	RMSD(Å) (patch rank)[a]	Best rank[b]	penetration distance[c]	# of residues[d]	RMSD(Å) (patch rank)[a]	Best rank[b]	CPU (min)[e]
1ACB	1.05(23)	1	2.43	2,0	1.12(34)	10	59:48
1AVW	0.82(1)	1	2.79	7,3	1.92(223)	330	55:11
1BRC	1.07(96)	1	4.76	7,0	4.76(168)	179	15:10
1BRS	0.66(1)	1	2.79	3,1	3.19(85)	143	24:09
1CGI	1.21(2)	1	3.30	4,4	1.74(338)	135	21:06
1CHO	1.77(198)	2	3.28	7,0	1.25(4)	5	14:25
1CSE	1.43(1)	1	2.99	3,0	1.76(478)	603	28:34
1DFJ	1.84(15)	15	4.81	16,9	7.04(10)	-	35:16
1FSS	2.35(115)	4	3.14	5,7	1.64(360)	142	32:38
1MAH	0.71(2)	1	3.25	6,2	2.47(24)	736	21:07
1PPE*	0.98(1)	1	2.18	0,0	0.96(1)	1	07:31
1STF*	0.56(1)	1	1.94	0,0	1.32(1)	1	30:39
1TAB*	1.63(21)	1	2.62	1,0	1.72(35)	81	07:01
1TGS	0.71(2)	1	4.30	8,5	2.82(445)	573	17:17
1UDI*	1.35(2)	1	3.02	2,4	1.83(6)	73	20:33
1UGH	0.84(1)	1	3.31	6,8	2.48(151)	44	17:06
2KAI	0.89(2)	1	4.67	14,10	3.21(235)	224	20:59
2PTC	0.74(3)	1	4.07	9,2	1.86(222)	10	13:45
2SIC	1.49(3)	3	2.91	7,6	1.30(6)	122	29:59
2SNI	2.22(1)	1	5.64	12,5	6.95(392)	-	14:09
2TEC*	0.93(27)	1	2.61	1,0	0.77(29)	241	31:29
4HTC*	1.76(3)	1	2.72	2,2	1.54(1)	1	09:04
1AHW	0.83(1)	1	3.02	4,3	1.73(98)	8	52:06
1BQL*	0.96(3)	1	2.30	0,0	0.66(1)	1	22:51
1BVK	1.32(10)	1	2.01	0,0	2.91(185)	577	08:37
1DQJ	1.32(1)	1	4.08	12,13	5.24(244)	-	20:56
1EO8*	1.19(1)	1	2.75	3,4	2.27(168)	1071	33:22
1FBI*	1.46(2)	1	4.01	7,1	1.05(228)	282	20:41
1JHL*	1.30(167)	3	1.94	0,0	2.91(134)	274	27:01
1MEL*	1.45(2)	1	2.39	1,0	1.20(3)	2	07:30
1MLC	1.17(112)	3	4.23	7,9	3.10(397)	689	15:55
1NCA*	1.69(1)	1	1.74	0,0	1.92(16)	43	20:30
1NMB*	2.79(17)	17	1.40	0,0	2.07(44)	218	15:09
1WEJ	1.47(11)	11	2.16	0,0	2.09(260)	417	16:06
2JEL*	3.67(4)	4	2.38	3,0	3.37(45)	112	08:39

4   Conclusions

We have presented a novel and highly efficient algorithm for docking of two molecules. While here we have shown results obtained by applying the algorithm to the docking of two protein molecules, the algorithm can be applied to receptor–drug cases as well (not shown here). The attractive running times of the algorithm and the high quality of the results compared to other state of the art method [26,12,3] are the outcome of several components. First, the algorithm divides the molecular surface into shape-based patches. This division addresses both the efficiency and at the same time, distinguishes between residue types (polar/non-polar) in the patches. Further, we make use of residue hot spots in the patches. Second, the method utilizes distance transform to improve the shape complementarity function. Third, it implements faster scoring, based on multi-resolution surface data structure. Our improved shape complementarity function further contributes to the quality of the results. While here the docking is rigid, the utilization of the last three components enables us to permit more liberal intermolecular penetration (up to 5 Å here).

References

1. C. Branden and J. Tooze. *Introduction to Protein Structure*. Garland Publishing, Inc., New York and London, 1991.
2. J.C. Camacho, D.W. Gatchell, S.R. Kimura, and S. Vajda. Scoring docked conformations generated by rigid body protein–protein docking. *PROTEINS: Structure, Function and Genetics*, 40:525–537, 2000.
3. R. Chen and Z Weng. Docking unbound proteins using shape complementarity, desolvation, and electrostatics. *PROTEINS: Structure, Function and Genetics*, 47:281–294, 2002.
4. M.L. Connolly. Analytical molecular surface calculation. *J. Appl. Cryst.*, 16:548–558, 1983.
5. M.L. Connolly. Solvent-accessible surfaces of proteins and nucleic acids. *Science*, 221:709–713, 1983.
6. M.L. Connolly. Shape complementarity at the hemoglobin $\alpha_1\beta_1$ subunit interface. *Biopolymers*, 25:1229–1247, 1986.
7. T. H. Cormen, C. E. Leiserson, and R. L. Rivest. *Introduction to Algorithms*, chapter 26. The MIT Press, 1990.
8. M. de Berg, M. van Kreveld, M. Overmars, and O. Schwarzkopf. *Computational Geometry: Algorithms and Applications*. Springer-Verlag, 2000.
9. T.J.A. Ewing, Makino S., Skillman A.G., and I.D. Kuntz. Dock 4.0: Search strategies for automated molecular docking of flexible molecule databases. *J. Computer-Aided Molecular Design*, 15:411–428, 2001.
10. D. Fischer, S. L. Lin, H.J. Wolfson, and R. Nussinov. A geometry-based suite of molecular docking processes. *J. Mol. Biol.*, 248:459–477, 1995.
11. H.A. Gabb, R.M. Jackson, and J.E. Sternberg. Modelling protein docking using shape complementarity, electrostatics, and biochemical information. *J. Mol. Biol.*, 272:106–120, 1997.
12. E.J. Gardiner, P. Willett, and P.J. Artymiuk. Protein docking using a genetic algorithm. *PROTEINS: Structure, Function and Genetics*, 44:44–56, 2001.

13. B.B. Goldman and W.T. Wipke. Molecular docking using quadratic shape descriptors (qsdock). *PROTEINS: Structure, Function and Genetics*, 38:79–94, 2000.

14. I. Halperin, B. Ma, H. Wolfson, and R. Nussinov. Principles of docking: An overview of search algorithms and a guide to scoring functions. *PROTEINS: Structure, Function and Genetics*, 47:409–443, 2002.

15. Z. Hu, B. Ma, H.J Wolfson, and R. Nussinov. Conservation of polar residues as hot spots at protein–protein interfaces. *PROTEINS: Structure, Function and Genetics*, 39:331–342, 2000.

16. R.M. Jackson. Comparison of protein-protein interactions in serine protease-inhibitor and antibody-antigen complexes: Implications for the protein docking problem. *Protein Science*, 8:603–613, 1999.

17. F. Jiang and S.H. Kim. Soft docking : Matching of molecular surface cubes. *J. Mol. Biol.*, 219:79–102, 1991.

18. G. Jones, P. Willet, R. Glen, and Leach. A.R. Development and validation of a genetic algorithm for flexible docking. *J. Mol. Biol.*, 267:727–748, 1997.

19. E. Katchalski-Katzir, I. Shariv, M. Eisenstein, A.A. Friesem, C. Aflalo, and I.A. Vakser. Molecular Surface Recognition: Determination of Geometric Fit between Protein and their Ligands by Correlation Techniques. *Proc. Natl. Acad. Sci. USA*, 89:2195–2199, 1992.

20. I.D. Kuntz, J.M. Blaney, S.J. Oatley, R. Langridge, and T.E. Ferrin. A geometric approach to macromolecule-ligand interactions. *J. Mol. Biol.*, 161:269–288, 1982.

21. H.P. Lenhof. Parallel protein puzzle: A new suite of protein docking tools. In *Proc. of the First Annual International Conference on Computational Molecular Biology RECOMB 97*, pages 182–191, 1997.

22. S. L. Lin, R. Nussinov, D. Fischer, and H.J. Wolfson. Molecular surface representation by sparse critical points. *PROTEINS: Structure, Function and Genetics*, 18:94–101, 1994.

23. R. Norel, S. L. Lin, H.J. Wolfson, and R. Nussinov. Shape complementarity at protein-protein interfaces. *Biopolymers*, 34:933–940, 1994.

24. R. Norel, S. L. Lin, H.J. Wolfson, and R. Nussinov. Molecular surface complementarity at protein-protein interfaces: The critical role played by surface normals at well placed, sparse points in docking. *J. Mol. Biol.*, 252:263–273, 1995.

25. R. Norel, D. Petrey, H.J. Wolfson, and R. Nussinov. Examination of shape complementarity in docking of unbound proteins. *PROTEINS: Structure, Function and Genetics*, 35:403–419, 1999.

26. P.N. Palma, L. Krippahl, J.E. Wampler, and J.G Moura. Bigger: A new (soft)docking algorithm for predicting protein interactions. *PROTEINS: Structure, Function and Genetics*, 39:372–384, 2000.

27. M. Rarey, B. Kramer, and Lengauer T. Time-efficient docking of flexible ligands into active sites of proteins. In *3'rd Int. Conf. on Intelligent Systems for Molecular Biology (ISMB'95)*, pages 300–308, Cambridge, UK, 1995. AAAI Press.

28. B. Sandak, H.J. Wolfson, and R. Nussinov. Flexible docking allowing induced fit in proteins. *PROTEINS: Structure, Function and Genetics*, 32:159–174, 1998.

29. G. Stockman. Object recognition and localization via pose clustering. *J. of Computer Vision, Graphics, and Image Processing*, 40(3):361–387, 1987.

30. I.A. Vakser. Protein docking for low resolution structures. *Protein Engineering*, 8:371–377, 1995.

31. I.A. Vakser. Main chain complementarity in protein recognition. *Protein Engineering*, 9:741–744, 1996.

32. I.A. Vakser, O.G. Matar, and C.F. Lam. A systematic study of low resolution recognition in protein–protein complexes. *Proc. Natl. Acad. Sci. USA*, 96:8477–8482, 1999.

33. P.H. Walls and J.E. Sternberg. New algorithms to model protein-protein recognition based on surface complementarity; applications to antibody-antigen docking. *J. Mol. Biol.*, 228:227–297, 1992.

34. H.J. Wolfson and I. Rigoutsos. Geometric hashing: An overview. *IEEE Computational Science and Eng.*, 11:263–278, 1997.

Appendix

Distance Transform Grid

Distance transform grid is a data structure for efficient queries of type *distance from surface*. The molecule is represented by a 3D grid, where each voxel (i, j, k) holds a value corresponding to the distance transform $DT(i, j, k)$. There are three types of voxels: surface (MS surface point maps to the voxel), interior (inside the molecule) and exterior (outside the molecule). The distances are zero at the surface voxels and change as the distance from the surface increases/decreases. The distance transform is negative for inside molecule voxels and positive for outside voxels.

Supported Queries.
- *distance from surface:* Given a point p, we access the grid with its coordinates and return a value of the voxel corresponding to the point. Clearly this query consumes $O(1)$ time.
- *shape function and volume Normal:* A sphere of radius R is placed at a given point p. We count the ratio of negative grid voxels inside the sphere and the total number of sphere voxels. The volume normal is computed in the same manner. We compute the gravity center of the positive grid voxels inside the sphere. This sets the direction of the normal.

Multi-resolution Surface

In order to support fast geometric scoring and filtering of the transformations, we construct multi-resolution data structure for the ligand's surface dots. We build a surface tree, where each layer represents the ligand surface at a different resolution. This data structure allows us to work with very high MS surface density and to reduce greatly the number of queries in the receptor distance transform grid. The main idea is that if the point falls outside the interface, there is no need to check where the adjacent points fall.

Supported Queries.
- *isPenetrating:* Given a transformation and a *penetration threshold* we check whether the surface points of the ligand penetrate the surface of the receptor with more then the given threshold.

- *maxPenetration:* Given a transformation find the maximum surface penetration.
- *score:* Given a transformation and a list of ranges, the goal is to count the number of lowest level surface points in each range.
- *interface:* Selects the interface surface points for a given transformation. The output is all the nodes with the distance transform value less then *interface threshold.*

All the queries employ the DFS search with iterative implementation using a stack. The complexity of the queries is proportional to high level size + interface size.

A Method of Consolidating and Combining EST and mRNA Alignments to a Genome to Enumerate Supported Splice Variants

Raymond Wheeler

Affymetrix, 6550 Vallejo St., Suite 100, Emeryville, CA 94608, USA

Abstract. Many researchers have focused their attention on alternative splice forms of genes as a means to explain the apparent complexity of human and other higher eukaryotic organisms. The exon-intron structure implied by aligning ESTs and mRNAs to genomic sequence provides a useful basis for comparing different splice forms. There are software tools that enumerate all possible exon-intron combinations implied by these alignments, but this approach has potential drawbacks including a combinatorial explosion in the number of splice forms and the generation of many unsupported splice forms. *altMerge* retains information about exon combinations present in the input alignments. It uses this information to combine compatible subsets of the alignments into a single representative form. Substructures of these representative forms are then exchanged to enumerate only those alternative splice forms that are consistent with the input alignments.

1 Introduction

Since EST and mRNA sequences are transcripts, we expect that the areas where these sequences align well to the genome represent exons, and that the long genomic inserts in the alignment represent introns (See Figure 1). Given this interpretation, we can characterize two types of exon boundaries. There are exon boundaries that border on inferred introns (exon boundaries internal to the alignment), and there are two terminal exon boundaries that are simply the extent of the transcript alignment. We assume that the inferred exon/intron boundaries correspond to exon 5' and 3' splice sites. Since we believe these sites correspond to actual gene features, we consider them fixed boundaries, and we give them the name *hard edges*. Since many of the aligned sequences are fragments of an original transcript, the endpoints of the alignment can lay virtually anywhere within the gene. Therefore, the two terminal alignment endpoints do not necessarily correspond to any gene feature, and are therefore named *soft ends*, since they are merely suggestions of exon boundaries.

There has been some work done in using cDNA alignments to infer individual introns associated with alternative splice forms [2]. However there seems to be little, if any, published work on the use of combined, overlapping alignments to reconstruct full alternative transcripts. Jim Kent, of University of California at

R. Guigó and D. Gusfield (Eds.): WABI 2002, LNCS 2452, pp. 201–209, 2002.

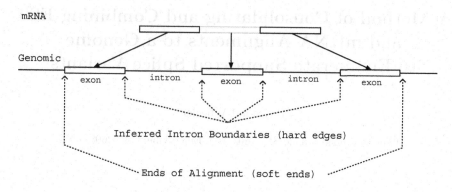

Fig. 1. Alignment anatomy

Example 1

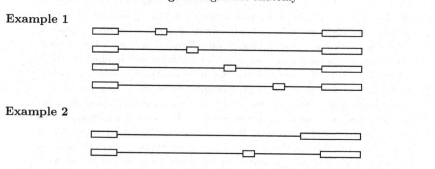

Example 2

Fig. 2. Examples of possible mutually exclusive exon combinations. This shows a cartoon of 6 different cDNA alignments to a genome. The rectangles represent matching blocks, i.e. exons, while the connecting lines represent gaps in the alignment, i.e. introns

Santa Cruz, has an unpublished algorithm that catalogues all logically possible exon combinations based on cDNA alignments to a genome [3][4]. Simply enumerating all possible exon combinations however, can produce large numbers of undesirable splicing possibilities. For example, some genes exhibit a mutually exclusive splicing behavior, as seen in Drosophila dscam [5]. A cartoon representing this mutually exclusive exon behavior is seen in Figure 2.

In Example 1 in Figure 2, assume that the four different middle exons are mutually exclusive, i.e. no more than one of them is ever transcribed. Likewise for Example 2, assume that the presence of the middle exon is always accompanied by a shifted acceptor site for the rightmost exon. In each of these examples, it would be inexpedient to derive all possible exon combinations from their alignments (16 possible for Example 1, and 4 possible for Example 2). If a gene exhibits many examples of mutually exclusive splicing combinations in different loci, then the impact of an "all exon combinations approach" increases exponentially (See Figure 3).

The program *altMerge* attempts to address the issue of mutually exclusive exon combinations. It stores information about exon combinations in the input

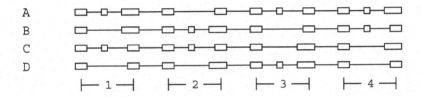

Fig. 3. Cartoon of 4 cDNAs aligned to a genome. Assume the gene has 4 loci containing 4 splice variants each. Permitting all possible exon combinations would yield 256 possibilities

alignments and preserves unique gene substructures. How *altMerge* achieves this will be discussed in the next section.

The following definitions will be used for the rest of the paper. A transcript that has been aligned to the genome has an *alignment region* defined by the section of the genomic strand between the 5' and 3'-most ends of the alignment. Two transcript alignments are said to *overlap* if their alignment regions overlap. We say that two alignments *agree* when they overlap and there is no conflict with respect to their hard edge locations. Conversely, we say that two alignments *disagree* when they overlap and there is some disparity in the alignment location of any hard edge. In the *Input* section of Figure 4, note that alignment A agrees with alignments E, F, and G, but disagrees with alignments B, C, or D.

2 Algorithm

The *altMerge* algorithm works in 2 phases. In the first phase, all sets of agreeing alignments are merged into one Representative Splice Form (RSF). A single alignment may contribute to many different representatives. The output of this phase is a set of Representative Splice Forms, one for each possible splice variant consistent with the input alignments. This set of Representative Splice Forms has two key properties. 1) Every distinct pair of overlapping Representative Splice Forms disagrees at some intron boundary. 2) There is an RSF for each input alignment that agrees with all of its hard edges. These properties suggest a sense of minimality and completeness, respectively. The set is minimal in that there are no completely redundant RSFs, while it is complete in that every exon combination suggested by the input is represented in at least one RSF.

In the second phase of the algorithm, some RSFs are "extended" by other compatible, overlapping representatives. Specifically, one RSF X can be *extended* by another RSF Y only if X and Y overlap, and their 3' and 5'-most overlapping exons agree. RSF X is extended by Y by catenating the portions of Y that do not overlap X onto ends of X. The result is an enhanced set of Representative Splice Forms, wherein exon-intron substructures of one RSF are presented in the context of genomically adjacent substructures in another RSF. An example of a sample run of *altMerge* and the results of its two phases can be seen in Figure 4.

Input:

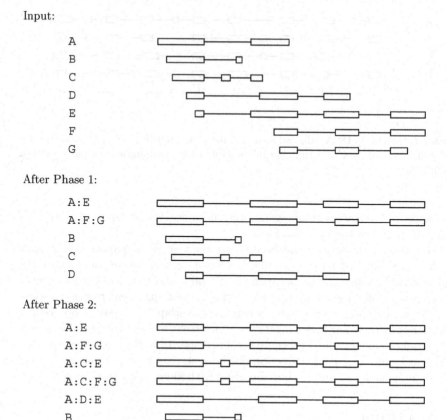

After Phase 1:

After Phase 2:

Fig. 4. Sample *altMerge* run. Phase 1 shows results after merging agreeing alignments. Phase 2 shows results after extending

In Figure 4, phase 1, notice that alignments B, C, and D do not merge with any other alignments. This is because each of these alignments infers an intron boundary that is unique, and therefore their associated RSFs are identical to the input sequences. The other alignments A, E, F, and G each agree with at least one other alignment, so they are merged into RSFs with multiple members, namely $A{:}E$ and $A{:}F{:}G$. Notice that Step 1 serves the purpose of filtering out redundant alignments. Both alignments F and G suggest the same splice form, but after merging there is only 1 RSF representing that splice form.

In phase 2 of Figure 4, the Representative Spice Forms defined by alignments C and D are extended by other RSFs. Notice that alignment B is not extended because its 3'-most overlapping exon does not agree with any other RSF. Representative Splice Form D is extended by RSF $A{:}E$ because the first exon of D agrees with the first exon of $A{:}E$, and the last exon of D agrees with the third exon of $A{:}E$. Accordingly, the parts of $A{:}E$ that do not overlap D are catenated

Input:

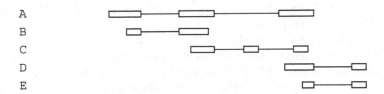

Results if alignment B is discarded early:

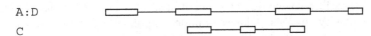

Results if alignment B is discarded later:

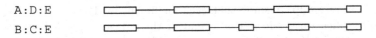

Fig. 5. Different *altMerge* outputs, depending on when redundant alignment B is discarded

onto the respective ends of D. The result is shown in Figure 4 Phase 2 as output $A{:}D{:}E$. Similarly, RSF C is extended by both $A{:}E$ and $A{:}F{:}G$ because the 5' and 3'-most overlapping exons of C meet the extension criteria with respect to each of these RSFs.

2.1   Phase 1 – Merging

The purpose of this step is to create Representative Splice Forms, as well as prune away any redundant splice form information. To this end, the merge step performs a pairwise comparison of each overlapping cDNA aligned to the same strand. If the intron boundaries agree, then either a new, longer alignment (an incomplete RSF) is created out of the two (see Figure 4, sequences A and B), or nothing happens because one alignment is *redundant*. An alignment X is *redundant* to an alignment Y if they agree, and the alignment boundaries of X are entirely within the alignment boundaries of Y, inclusive. Even though one alignment is found to be redundant to another alignment, both of them are still compared to the other overlapping alignments. This might seem like unnecessary extra work, but consider Figure 5. Input alignments B and E are redundant with alignments A and D, respectively. If alignments B and E are removed from the merging process, the splice variant implied by alignment C will not be represented as fully as it could. If alignments B and E are allowed to merge with C, however, a more complete picture of splice variation emerges.

Alignments that have been merged (or, incomplete Representative Splice Forms: iRSF) are consequently compared to the other input alignments. For example, say that there were 10 input alignments, and alignment 1 is compared with alignments 2 through 10, merging with alignments 4 and 7. The incomplete RSF (iRSF) 1:4 is then compared with alignments 5 through 10, and iRSF 1:7 is compared with alignments 8 through 10. If iRSF 1:4 merged with alignment 6, then iRSF 1:4:6 would be compared to alignment 7 through 10 and so on. There is no need to compare each iRSF to *all* of the input alignments. For example, there is no need to compare RSF 1:4 with alignment 2 because this merging would be taken care of by another iRSF 1:2 merging with alignment 4.

Each comparison may yield a new item that must then be used in other comparisons, possibly spawning yet other items. This fact makes for a non-trivial time-complexity analysis. A lower-bounded exponential running time for this algorithm cannot be avoided however, since the space complexity is exponential. A particular set of input may yield an an exponentially greater number of RSFs. Suppose there are n input alignments broken into two groups, with $n/2$ alignments in each group. Assume that each alignment in the first group represents a unique splice variation at the 5' end of the transcript, and likewise for the second group at the 3' end of the transcript. If each 5' alignment can be merged with each 3' alignment, then the result is $(n/2)^2$ RSFs. Similarly, if the input is broken up into 3 groups of $n/3$, and each alignment from group 1 can merge with each alignment in group 2, and so on, then there could be $(n/3)^3$ possible RSFs. In general if the input were divided up into x groups, the number of RSF produced is determined by the following formula.

$$\#\text{RSFs} = (n/x)^x \tag{1}$$

Consider the situation in which there are y groups with n/y alignments in each group. Solving Equation (1) for $x = n/y$ yields $\#\text{RSFs} = y^{n/y}$. Since it is possible to produce an output that is exponentially larger than the input, we know that the time complexity of the algorithm is exponentially lower bounded in the worst case.

The pseudo-code for the merging portion of the algorithm is shown in Figure 6. As input alignments and iRSFs are merged with other alignments, the result of the merging is pushed on a "todo_list" along with an integer index, indicating where within the list of input alignments the subsequent merging attempts should begin. The todo_list is processed until it is empty, and then another input sequence is used as the basis for comparison.

2.2   Phase 2 – Extending

This step involves a pairwise comparison of all the RSFs output from the merging phase. For each pair, the left-most and right-most overlapping exons are considered. If there are no disagreements with the first and last overlapping exons, then the shorter ended alignments are extended in such a way that they match the longer alignments. An example of this can be seen in Figure 4.

The processing of input in the extension procedure is essentially identical to the merge procedure. The input RSFs are compared pairwise for suitability

```
bool merged;
for i = 0 to sizeOf(cdna_array) {
  for j = (i + 1) to sizeOf(cdna_array) {
    merged = 0;
    if (overlap(cdna_array[i], cdna_array[j])) {
      merged |= merge(cdna_array[i], cdna_array[j], todo_list);
    }
  }
  if (!merged) {
    push(done_list, cdna_array[i]);
  }
  else { /* there must be a merged alignment on the todo_list */
    while (todo_list != NULL) {
      todo = pop(todo_list);
      merged = 0;
      for k = todo.start_idx to sizeOf(cdna_array) {
        if (overlap(todo, cdna_array[k])) {
          merged |= merge(todo, cdna_array[k], todo_list);
        }
      }
      if (!merged) {
        push(done_list, todo);
      }
    }
  }
}
```

Fig. 6. Pseudo-code for merging algorithm. **cdna_array** is an array of cDNA alignments sorted by increasing start points. **todo_list** is a list of merged alignments (incomplete RSFs) that are scheduled to be compared to the other alignments in **cdna_array**. Associated with each incomplete RSF is an array index (start_idx) indicating where in **cdna_array** the comparisons should begin. **merged** is a function that attempts to merge two alignments to produce an incomplete RSF. If it succeeds, it pushes the merged result onto the supplied **todo_list** to be processed later. **merged** returns true if a merge occurred, false otherwise

for extension. If, for example, RSF X can be extended by RSF Y, a new RSF $(X:Y)$ is created and placed on a "**todo_list**". The items on the **todo_list** are then used as candidates for further extensions with subsequent input RSFs. The pseudo-code for this step is essentially identical to the pseudo-code for the merging step, as is therefore omitted.

3   Results

altMerge was run on Human chromosome 1 using the April 2001 release of the human genome. First, we attempted to align all of dbEST (April 28 , 2001), Genbank (April 25, 2001; release 123.0), and RefSeq (April 30, 2001). We found that

2,435,289 sequences had a run of 30 consecutive matching bases. The alignments were filtered using the following criteria.

- At least 90% of the transcript must align perfectly.
- For each transcript, only the alignment with the greatest number of perfectly aligning bases was used.
- The alignment must infer the presence of an intron (a genomic insert of > 20 bases).
- The dinucleotide splice pairs at the intron boundaries must be one of the following: gt—ag, at—ac, or gc—ag.

The last criteria is due to the fact that both the genome and the aligning transcripts can be very noisy. This results in some alignments suggesting very improbable intron boundaries. That is why the alignments are restricted to only those that contain the most commonly recognized dinucleotide splice pairs.

After filtering, there were 75,303 alignments that mapped to 2264 regions of chromosome 1. *altMerge* took 688 seconds (wall clock) to process these alignments, running on a 500MHz PentiumIII with 384MB RAM. Here are the results for *altMerge* compared to an enumeration of all possible exon combinations.

	altMerge	all exon combinations
Number of transcripts produced	12714	164397
Average transcripts per gene	5.6	72.6
Maximum transcripts per gene	622	79858

altMerge produced approximately ten times fewer transcripts than the all-exon-combination approach. It did this without dismissing any information, since every intron implied by the input alignments was represented in at least one of the output *altMerge* transcripts.

4   Discussion

Another variation to this algorithm that should be considered is to allow an RSF to be extended "in one direction only". As described, an RSF is extended in both directions by another RSF if and only if the first and last overlapping exons agree. It is conceivable that one may want to extend an RSF "to the left" if only the first overlapping exon agrees, or likewise to the right if only the last overlapping exon agrees. The drawback for using this approach is that the ends of the cDNA alignments can be error prone, due to low quality bases in the cDNA, errors in the genomic assembly, or the use of incompletely spliced cDNA. For example, if a relatively good EST alignment had errors at its 3' end, it might be undesirable to let those errors be extended by possibly many other 5' splice variants. So, this one-way extension approach probably would be best used only if the user had high confidence in the input sequences and used a fairly strict alignment method.

Another variation could be to include polyA site annotations, which would effectively cause the soft ends of some alignments to be treated as hard edges, since they correspond to an annotated end of a transcript.

There are some alternative splice variations that are missed by *altMerge* that might be of some interest. For example, it is possible that a splice variant might be represented by only a single alignment A. This alignment might meet the criteria for extension with respect to other alignments. However, alignment A may be first merged with another alignment B, and the resulting RSP $A : B$ does not fit the criteria for extension. In this way, the splice form represented by alignment A might lack representation in some valid splice forms. One possible solution to this would be to use an approach of extending once, then merging once in an iterative fashion, rather than performing all merges and then all extensions.

5   Conclusion

altMerge is a conservative, yet effective algorithm that addresses the issue of consolidating and combining aligned cDNAs to create a concise set of unique examples of alternative splice transcripts of a gene. *altMerge* is capable of extending small, unique cDNA alignments that represent an alternative splicing into much larger alignments, consistent with the other well-supported splice variants. The result is a minimal set of splice variant alignments, extended as far as possible given the input alignment data.

Because *altMerge* yeilds a relatively conservative set of hypothetical splice variants, it makes the job of validating these variants in a lab more productive. Indeed, splice variants produced by *altMerge* have been validated using laboratory methods. For example, 3 novel variants of the *bcl-x* gene were confirmed in a lab using RT-PCR [1].

References

1. Helt, G., Harvey, D. (Affymetrix): Personal communication. (2000)
2. Kan, Z., Rouchka, E., Gish, W., States, D.: Gene structure prediction and alternative splicing analysis using genomically aligned ESTs. Genome Res. **11** (2001) 889–900.
3. Kent, J. (University of California at Santa Cruz): *altGraph*. Personal communication. (2000).
4. Mironov, A., Fickett, J., Gelfand, M.: Frequent alternative splicing of human genes. Genome Research **9** (1999) 1288–1293.
5. Schmucker et al.: Drosophila dscam is an axon guidance receptor exhibiting extraordinary molecular diversity. Cell **101** (2000) 671–684.

A Method to Improve the Performance of Translation Start Site Detection and Its Application for Gene Finding

Mihaela Pertea and Steven L. Salzberg

The Institute for Genomic Research, Rockvile MD 20850, USA
{mpertea,salzberg}@tigr.org

Abstract. Correct signal identification is an important step for all ab initio gene finders. The aim of this work is to introduce a new method for detecting translation start sites in genomic DNA sequences. By using interpolated context Markov model to capture the coding potential of the region immediately following a putative start codon, the novel method described achieves an 84% accuracy of the start codon recognition. The implementation of this technique into GlimmerM succeeds in improving the sensitivity of the gene prediction by 5%.

1 Introduction

An important step in a genome project is the identification of the protein coding regions. Most approaches focus on determining the coding function of nucleotide sequences by statistical analysis, but determining the regulatory regions that signal the different processes in the gene to the protein pathway can prove a much more difficult task than identifying exons [1]. Such regions include the translation start site that marks the start of the coding region and whose accurate identification is essential in determining the correct protein product of a gene. While the translation start site in DNA genomic sequences is most often determined by the codon ATG, the identification signal is very weak in the context sequences surrounding the start codon. Shine-Dalgarno [2] and Kozak sequences [3] describe consensus motifs in the context of the translation initiation both in bacteria and in eukaryotes, and several methods use them to improve the translation start site prediction [4,5,6,7,8]. However these consensus patterns have very limited differentiation power of the true start codons from other occurrences of ATGs in the DNA sequence. The lack of strong identification factors produces an overwhelming number of false positives that given the small number of true start codons makes the detection of the true translation start sites a very challenging problem.

2 Related Work

Prediction of protein translation start sites from the genomic DNA sequences has a relatively moderate success. Most previous methods used for the identification of the start sites are designed to work for almost full-length mRNAs.

R. Guigó and D. Gusfield (Eds.): WABI 2002, LNCS 2452, pp. 210–219, 2002.

This implies that with the exception of the first coding exons, the location of all the other coding regions in the DNA sequence is already known. For instance NetStart [9], ATGpr [10] or the more recent method of Zien *et al.* based on support vector machines [11] are more tuned towards EST data, and there is little evidence about their performance on genomic sequences where intron and intergenic regions are introduced. On the cDNA data, their reported performance does not exceed 85% on the test set. A new method that combines conserved motifs sensitivity and non-coding/coding sensitivity was introduced recently by Hatzigeorgiou [12]. According to his test data, the accuracy of the prediction was 82.5%, where 94% of the translation initiation sites are predicted correctly, which is above the previous rate of 85% reported in Argawal and Bafna [13]. Here, the accuracy of the prediction is taken to be the average of the prediction on positive and negative data. Even so, his algorithm is designed for analysis of full-length mRNAs sequences, and no sensitivity and specificity are reported in the case of genomic DNA sequences.

The methods presented above attempt to help the gene identification problem by achieving proper annotation of the cDNA that are then used to provide reliable information for the structural annotation of genomic sequences. However cDNA data is never complete. The most comprehensive EST projects will miss transcripts expressed only transiently or under unusual conditions. Other methods that would use similarity to homologue proteins might also prove successful in determining the correct gene structure, but even closely related proteins may have different start sites [14,15]. Nevertheless, only about 50-70% of the genes in a new genome are similar to an existing functionally characterized gene, so in most cases, a gene finder must be trained specifically for each organism.

The accurate prediction of the translation start site is a very important step of *de novo* recognition of protein coding genes in genomic DNA. The most frequently used techniques in the sequence analysis, like for instance hidden Markov models [16] may not be very successful in the detection of the start sites, given the small number of positive examples that won't be enough to accurately estimate the complex parameters of these techniques. Therefore, many gene finders use positional weight matrices for start site detection [17,18,19]. Salzberg [20] extends these positional probabilities to first-order Markov dependencies. GlimmerM, our gene finder system [21], uses the scoring function introduced by Salzberg. According to this scoring function, if T is a true site, then $P(T|S)$, the probability of T given a sequence $S = (s_1, s_2, \ldots, s_n)$, could be computed using Bayes' Law:

$$P(T|S) = P(S|T)P(T)/P(S) \tag{1}$$

In this equation, the underlying prior probability $P(T)$ could be treated as a constant, while $P(S/T)$ is in fact a 1-state Markov chain model, computed by multiplying the conditional probability of each successive base, given the previous base in the sequence. The probability $P(S)$ is estimated as

$$P(s_1) \prod_{i=2}^{n} P(s_i|s_{i-1}) \tag{2}$$

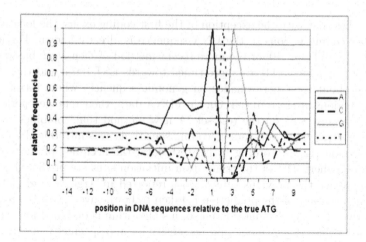

Fig. 1. Profile of base composition around the translation start sites in the Arabidopsis database. Position 1 in the sequence is the position of nucleotide A in the start codon ATG.

where the 16 prior conditional probabilities are computed based on the entire data set. Below, when not ambiguous, we refer to this method that uses the scoring function of Salzberg by the first-order Markov model.

The method we present here to detect translation start sites improves upon our previous technique [20] by including a module based on the interpolated context Markov models introduced in [22] that is sensitive to the coding region following a potential start codon. Preliminary results on *Arabidopsis* and human data sets show that the translation start site detection as well as the gene finding accuracy is improved after the implementation of this method into GlimmerM.

3   Start Site Detection Module

An initial screening of all ATGs is done with the first-order Markov method presented above. A simple statistic on our collected *Arabidopsis* data set presented in Figures 1 and 2 show a different base composition around the true start sites compared to other non-start ATGs present in the data. The difference is especially apparent in the immediate region previous to the start codon, and diminishes rapidly at the beginning of the coding region. The method presented in [20] should be sensitive to this specific profile of the true ATGs and we applied it on a window of 19 bp around the start site, where 12 bp were in the non-coding region before the ATG. Since no pattern is apparent after the start codon we used interpolated context Markov models (ICMs) as introduce in [22] to determine the coding potential of the region following a potential translation initiation site.

The choice of ICMs was motivated by their similarity to probabilistic decision trees [22] that have been previously used for modeling splice site junctions [17]

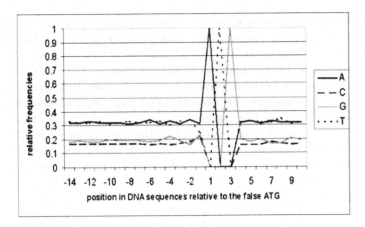

Fig. 2. Profile of base composition around the non-start ATGs in the Arabidopsis database. Position 1 in the sequence is the position of nucleotide A in the non-start ATG.

whose identification is very similar to identifying start sites. Briefly, the way an ICM works can be summarized as follows. Given a context $C = b_1 b_2 \ldots b_k$ of length k, the ICM computes the probability distribution of b_{k+1} using the bases in C that have the maximum mutual information in the training data set. More specific the construction of the ICM is done by considering all the oligomers of length $k + 1$ that occur in the training data set. If X_i is a random variable which model the distribution of the base in the ith position in the oligomer, then the mutual information values $I(X_1; X_{k+1})$, $I(X_2; X_{k+1})$, ..., $I(X_k; X_{k+1})$ are computed where

$$I(X;Y) = \sum_i \sum_j P(x_i, y_j) log \frac{P(x_i)P(y_j)}{P(x_i, y_j)} \qquad (3)$$

The variable that has the highest mutual information with the base at position $k + 1$ is selected, and the set of oligomers is partitioned based on the four nucleotide values at that position. This way the ICM can be viewed as a tree in which each node consists of a set of oligomers that provide a probability distribution for the base at position $k + 1$. The process is repeated on each subset of oligomers until the number of oligomers in a subset is too small to estimate the probability of the last base position. The probability of predicting the base b_{k+1} is then computed by weighting the distribution probabilities at all nodes in the ICM tree that correspond to the observed oligomer. The weights for different distribution probabilities are chosen according to the statistical significance of the distributions in the training data, as described in [23].

The method used for training of our start site detection module was as follows: given complete gene models with precisely known gene structure, we extracted only the coding portions between a start and a stop codon. We then compute the frequencies of all possible oligomers of length 1 to $k + 1$ in each of the three reading frames where the reading frame is given by the last base. Using these frequencies one can build three separate ICMs, one for each codon position. Let's denote by ICM^0, ICM^1, and ICM^2 these three ICMs, where ICM^i is computed for all oligomers in reading frame i. For any potential start site resulted after the initial screening of all ATGs using first order Markov dependencies, we extract from the genomic DNA sequence a window of length w, starting at this putative start codon. Let $S = s_0 s_1 \ldots s_{w-1}$ be this window, i.e. $s_0 s_1 s_2$ are the potential start codon ATG. Because the presence of the ATG, already defines the reading frame of the putative coding region starting as this ATG, we can use the three ICMs to compute the score of the sequence S:

$$Score_{ICM}(S) = \sum_{x=0}^{w-1} log\left(ICM^{x \bmod 3}(S_x)\right) \qquad (4)$$

where S_x is the oligomer ending at position x.

We compute a second score of S to capture the probability that the sequence was generated by a model of independent probabilities for each base, which to some extent represents the probability that the sequence is random:

$$Score_{rand}(S) = \sum_{x=0}^{w-1} log\left(P(s_x)\right) \qquad (5)$$

where $P(s_x)$ represents the probability of apparition of nucleotide s_x in the genomic DNA sequence. We will say that the ATG at the start of the sequence S is followed by a coding sequence only if $Score_{ICM}(S) - Score_{rand}(S) > 0$.

The choice of the parameter w will naturally influence the number of true translational start sites that will be determined by our algorithm as having a coding potential, since shorter coding regions then w in the DNA sequence following the start site will result in a lower score of the ICM module. In order not to miss any true translational start site, our algorithm[1] works as follows: for any encountered ATG in the genomic DNA sequence, we check if it has a coding potential. If it does then it is accepted as a true start site. If it does not, then we predict it as being a translation start site only if the score computed by the first order Markov method returns a score above some fixed threshold.

4   Results

To evaluate the performance of our translation start site predictor we needed databases of confirmed genes that were accurately annotated. We chose two two organisms of reference for training and testing our new system: the model plant *Arabidopsis thaliana* and human. A description of the two data sets used, and results of the start site detection module are presented below.

---

[1] The program can be obtained by request from the authors.

Data Sets

The *Arabidopsis* data set was extracted from a set of very high-quality gene models obtained from 5000 full-length transcripts sequenced by Ceres, Inc and released to the public in March 2001 (`ftp://ftp.tigr.org/pub/data/a_thaliana/ceres`), and at GenBank as accession numbers AY084215-AY089214) [24]. Since homology in the data set could influence the results, we further refined this reference set of gene models for *A.thaliana* by performing pairwise alignments between all genes to remove sequences with more than 80% homology. This resulted in a non-redundant data set of 4046 genes that we used for training and testing of our algorithm. 9000 bp were included in average at both 5' and 3' borders for each gene.

For the human database, we used the reference set provided by the Homo sapiens RefSeq records available from NCBI at `http://www.ncbi.nlm.nih.gov/genome/guide/human/`. The RefSeq database is a non-redundant set of reference sequences, including constructed genomic contigs, curated genomic regions, mRNAs, and proteins. To be even more accurate, we only selected reviewed records which have been manually processed by NCBI staff or collaborating groups. As of September, 2001, when we collected this data, the reviewed database contained 3537 accessions for completely annotated gene models. 90 sequences for which the start codon of the gene model was within the first 12 bp were eliminated.

Accuracy of Translation Start Site Prediction

A total of 3237 genes which represents about 80% of the genes in the *Arabidopsis* database were randomly extracted to form the *A.thaliana* training data. The remaining 809 genes were used for testing. We also randomly selected about 80% of the genes from the human database for the training data and 20% for the test data set. The total numbers of true and translation start sites from the two databases on both training and test data are shown in Table 1.

Table 1. Number of true and false translation start sites in the data sets for *A.thaliana* and human.

Database	Training data		Test data	
	True start sites	Non-start ATGs	True start sites	Non-start ATGs
Arabidopsis	3237	105009	809	24583
Human	2824	99194	617	21353

To test the system we had to choose a value for the context length k, used by the ICM model. The ICM constructed by Glimmer [22] uses a context length of 12. Based on the bacterial gene finder's success, we also chose $k = 12$ for our system. By choosing different threshold for the score obtained with the first

Table 2. False negative and false positive rates for the translation start site detection on the *A.thaliana* data set. MM and ICM refer to the first-order Markov method and to our improved technique based on ICM respectively, while the window length to determine the coding potential is referred by *w*; *thr* denotes different threshold values computed with the MM method.

w	thr	Training data FN(%)	FP(%) MM	ICM	Test data FN(%)	FP(%) MM	ICM	Acc MM	ICM
	-2.95	0.34	88.6	73.7	0.12	86.7	69.9	56.6	65
50	-1.99	0.65	83.5	62.2	0.62	70.8	63	64.3	68.2
	-0.99	1.36	73.9	50.4	2.10	54.1	51	71.9	73.5
	0.01	2.47	63.8	41.5	3.96	47.6	41.5	74.2	77.3
	-3.26	0.25	90.8	75.1	0.12	86.7	73.9	56.6	63
	-1.75	1.11	77.8	53.9	0.99	62.9	55.4	68.1	71.8
100	-0.75	2.32	65.5	40.4	2.60	52.4	40.9	72.5	78.3
	0.05	4.11	53.4	32.1	4.70	42.7	32.3	76.3	81.5
	1	8.31	38.2	25.6	9.89	30.2	25.6	80	82.3
	-2.75	0.71	83	65.9	0.37	80.6	65.7	59.5	67
200	-0.5	5.68	45.8	30.6	4.57	43.5	31.4	76	82
	0.5	11.5	31.1	20.1	11.9	25	20.2	81.6	**84**
	1.49	20.8	17.4	14	19.9	16.4	13.7	81.9	83.2

order Markov method we can obtain different rates of true translation start site that are missed, and of false positives. A complete ROC (receiver operating characteristics) was computed for the start site detection module. Different false negative and false positive rates are shown in Tables 2 and 3 for the *Arabidopsis* and human databases, respectively. The translation start site model's parameters were estimated using only the training data. We tried several values for the parameter *w* that specifies the length of the window starting at a given ATG necessary for determining the coding potential of the putative start codon. Bigger values of *w* resulted in general in a slightly better accuracy of the predictor. In fact the accuracy was improved as the window length approached the average length of the first coding exon in the database (defined as the coding portion starting at the translational start codon in the first exon containing a coding part in the gene structure), then tended to decline slowly as *w* was further increased.

As Tables 2 and 3 show, the recognition performance is much improved over the previous technique by using the new start site detection module. On both human and Arabidopsis data, the accuracy of the prediction is above 83.95%, which is above the reported accuracy on test data of any previous method know to us for the identification of start sites in genomic DNA sequences.

Improvements in the Gene Finding Prediction

To study the impact of our new translation start site recognition module on the performance of gene recognition, we implemented it in our GlimmerM system. The gene finder was trained on the *A.thaliana* training data set and tested on

Table 3. False negative and false positive rates for the translation start site detection on the human data set. MM and ICM refer to the first-order Markov method and to our improved technique based on ICM respectively, while the window length to determine the coding potential is referred by w; thr denotes different threshold values computed with the MM method.

w	thr	Training data FN(%)	FP(%) MM	ICM	Test data FN(%)	FP(%) MM	ICM	Acc MM	ICM
	-2.95	0.36	90.8	66.8	1.62	82.1	74.6	58.1	61.9
	-0.99	0.87	83	48.5	2.11	78.3	52.3	59.8	72.8
150	0.01	1.42	78.3	40.4	2.92	73.1	41.2	62	78
	1	2.14	75	36.9	3.24	72.7	35.6	62	80.6
	1.49	2.41	74.2	35.6	4.05	72.2	33.9	61.9	81
	2	2.89	70.1	34.9	4.21	69.6	32.7	63	81.5
	-2.75	0.24	93.8	64	0.49	92.7	73.5	53.4	63
	-0.74	1.07	81.8	43	0.97	86.7	47.1	56.2	76
300	1.05	2.06	75	32.2	3.08	72.7	31.4	62.1	82.8
	2.5	3.64	67.3	29.2	3.89	72.6	27.4	61.8	**84.4**
	4	4.35	64.9	28.7	4.86	66.4	26.5	64.4	84.3
	7.22	5.10	62.1	28.6	6.16	58.2	26.3	67.8	83.8

Table 4. Gene finding accuracy for GlimmerM using the first-order Markov model and the improved ICM model, on 809 test genes from *A.thaliana*

	GlimmerM with MM	GlimmerM with ICM
Total length of sequences	1140669bp	1140669bp
Number of correct predicted start codons	704	743
Number of correct predicted genes	294	309
Number of true exons	3224	3224
Number of false predicted exons	1521	1480
Number of correct predicted exons	2148	2169
Sensitivity and specificity (%)	93.14+93.08	93.25+93.44

the 809 *Arabidopsis* gene pool of test data, where the upstream and downstream sequence for each gene was chopped to 100bp to compute accurate specificities values. It should be noted though that gene prediction can be influenced by upstream or downstream sequences. The results shown in Table 4 are only preliminary, and more conclusive tests should be done on different data sets and using different values for the start site detection parameters. Our results are computed for $w = 50$, with a threshold for the first-order Markov model that will only miss 1% of the total number of true start sites. Nevertheless, by introducing the new start site detection module the number of start sites that were predicted correctly increased by 6%, and the sensitivity of the gene prediction increased by 5%.

5 Conclusions

A key component of the most successful gene finding algorithms is the ability to recognize the DNA and RNA sequence patterns that are critical for transcription, splicing, and translation. If all of these signals could be detected perfectly, then the protein coding region could be identified simply by removing the introns, concatenating all the exons, and reading off the protein sequence from start to stop. Unfortunately there is no completely accurate method to identify any of these signals. In this study we addressed the problem of translational start detection and succeeded to improve the accuracy of both start site and gene recognition in genomic DNA sequences. One of the most important limitations in *ab initio* gene finding is the scarcity of known data available for training. The first-order Markov method that was previously used by our gene finder, was limited to be trained on small windows around the start codon. In our new method the interpolated context Markov models will estimate their probabilities from all the coding regions in the genes from the training data, which represents an advantage when the number of genes available for training is reduced.

Acknowledgements

This work was supported in part by grant R01-LM06845 from the National Institutes of Health and grant MCB-0114792 from the National Science Foundation.

References

1. Black, D.L.: Protein Diversity from Alternative Splicing: A challange fro Bioinformatics and Post-Genome Biology. Cell **103** (2000) 367-370.
2. Shine, J., Dalgarno, L.: The 3'-terminal sequence of Escherichia coli 16S ribosomal RNA: complementarity to nonsense triplets and ribosome binding sites. Proc Natl Acad Sci U S A **71** (1974) 1342-1346.
3. Kozak, M.: An analysis of 5'-noncoding sequences from 699 vertebrate messenger RNAs. Nucleic Acids Res. **15** (1987) 8125-8148.
4. Hayes, W.S., Borodovsky, M.: Deriving ribosomal binding site (RBS) statistical models from unannotated DNA sequences and the use of the RBS model for N-terminal prediction. Pac Symp Biocomput. (1998) 279-290.
5. Frishman, D., Mironov, A., Mewes, H.W., Gelfand, M.: Combining diverse evidence for gene recognition in completely sequenced bacterial genomes. Nucleic Acids Res. **26** (1998) 2941-2947.
6. Hannenhalli, S.S., Hayes, W.S., Hatzigeorgiou, A.G., Fickett, J.W.: Bacterial start site prediction. Nucleic Acids Res. **27** (1999) 3577-3582.
7. Suzek, B.E., Ermolaeva, M.D., Schreiber, M., Salzberg, S.L.: A probabilistic method for identifying start codons in bacterial genomes. Bioinformatics **17** (2001) 1123-1130.
8. Joshi, C.P., Zhou, H., Huang, X., Chiang, V.L.: Context sequences of translation initiation codon in plants. Plant Mol Biol. **35** (1997) 993-1001.

9. Pedersen, A.G., Nielsen, H.: Neural network prediction of translation initiation sites in eukaryotes: perspectives for EST and genome analysis. Proc Int Conf Intell Syst Mol Biol. **5** (1997) 226-233.
10. Salamov, A.A., Nishikawa, T., Swindells, M.B.: Assessing protein coding region integrity in cDNA sequencing projects. B ioinformatics **14** (1998) 384-390.
11. Zien, A., Ratsch, G., Mika, S., Scholkopf, B., Lengauer, T., Muller, K.R.: Engineering support vector machine kernels that recognize translation initiation sites. Bioinformatics **16** (2000) 799-807.
12. Hatzigeorgiou, A.G.: Translation initiation start prediction in human cDNAs with high accuracy. Bioinformatics **18** (2002) 343-350.
13. Agarwal, P., Bafna, V.: The ribosome scanning model for translation initiation: implications for gene prediction and full-length cDNA detection. Proc Int Conf Intell Syst Mol Biol. **6** (1998) 2-7.
14. Stromstedt, M., Rozman, D., Waterman, M.R.: The ubiquitously expressed human CYP51 encodes lanosterol 14 alpha-demethylase, a cytochrome P450 whose expression is regulated by oxysterols. Arch Biochem Biophys.**329** (1996) 73-81.
15. Croop, J.M., Tiller, G.E., Fletcher, J.A., Lux, M.L., Raab, E., Goldenson, D., Son, D., Arciniegas, S., Wu, R.L.: Isolation and characterization of a mammalian homolog of the Drosophila white gene. Gene **185** (1997) 77-85.
16. Durbin, R., Eddy, S.R., Krogh, A., Mitchison, G.: Biological Sequence Analysis: Probabilistics Models of Proteins and Nucleic Acids. Cambridge University Press (1998).
17. Burge, C. and Karlin, S.: Prediction of complete gene structures in human genomic DNA. Journal of Molecular Biology **268** (1997) 78-94.
18. Henderson, J., Salzberg, S., Fasman, K.H. (1997) Finding genes in DNA with a Hidden Markov Model. J Comput Biol. **4** (1997) 127-141.
19. Salzberg, S., Delcher, A., Fasman, K., and Henderson, J.: A Decision Tree System for Finding Genes in DNA. Journal of Computational Biology **5** (1998) 667-680.
20. Salzberg, S.L.: (1997) A method for identifying splice sites and translational start sites in eukaryotic mRNA. Computational Applied Bioscience **13** (1997) 365-376.
21. Salzberg, S.L., Pertea, M. , Delcher, A.L., Gardner, M.J., and Tettelin, H.: Interpolated Markov models for eukaryotic gene finding. Genomics **59** (1999) 24-31.
22. Delcher, A.L., Harmon, D., Kasif, S., White, O., Salzberg, S.L.: Improved microbial gene identification with GLIMMER. Nucleic Acids Res. **27** (1999) 4636-4641.
23. Salzberg, S., Delcher, A., Kasif, S., and White, O.: Microbial gene identification using interpolated Markov models. Nucl. Acids Res. **26** (1998) 544-548.
24. Haas, B.J., Volfovsky, N., Town, C.D., Troukhan, M., Alexandrov, N., Feldmann, K.A., Flavell, R.B., White, O., and Salzberg, S.L.: Full-Length Messenger RNA Sequences Greatly Improve Genome Annotation (to be published).

Comparative Methods for Gene Structure Prediction in Homologous Sequences

Christian N.S. Pedersen[1,*] and Tejs Scharling[1]

BiRC[**], Department of Computer Science, University of Aarhus
Ny Munkegade, Building 540, DK-8000 Århus C, Denmark
{cstorm,tejs}@birc.dk

Abstract. The increasing number of sequenced genomes motivates the use of evolutionary patterns to detect genes. We present a series of comparative methods for gene finding in homologous prokaryotic or eukaryotic sequences. Based on a model of legal genes and a similarity measure between genes, we find the pair of legal genes of maximum similarity. We develop methods based on genes models and alignment based similarity measures of increasing complexity, which take into account many details of real gene structures, e.g. the similarity of the proteins encoded by the exons. When using a similarity measure based on an exiting alignment, the methods run in linear time. When integrating the alignment and prediction process which allows for more fine grained similarity measures, the methods run in quadratic time. We evaluate the methods in a series of experiments on synthetic and real sequence data, which show that all methods are competitive but that taking the similarity of the encoded proteins into account really boost the performance.

1 Introduction

At the molecular level a gene consists of a length of DNA which encodes a protein, or in some cases, a tRNA or rRNA molecule. A gene on a prokaryotic genomic sequence is a stretch of nucleotides flanked by start and stop codons. A gene on a eukaryotic genomic sequence is also a stretch of nucleotides flanked by start and stop codons, but it is further divided into an alternating sequence of blocks of coding nucleotides (exons) and non-coding nucleotides (introns). Each intron starts at a donor site and ends at an acceptor site.

Identifying the location and structure of genes in the genome of an organism is an important task, and a lot of work have been devoted to develop efficient and good computational gene finders. Many methods focus on constructing signal sensors for identifying e.g. possible acceptor and donor site positions in eukaryotic sequences, and context sensors for classifying smaller sequences as being e.g. coding or non-coding. Constructing such signal and context sensors

---

[*] Partially supported by the Future and Emerging Technologies programme of the EU under contract number IST-1999-14186 (ALCOM-FT).

[**] Bioinformatics Research Center (BiRC), www.birc.dk, funded by Aarhus University Research Fundation.

R. Guigó and D. Gusfield (Eds.): WABI 2002, LNCS 2452, pp. 220–234, 2002.
© Springer-Verlag Berlin Heidelberg 2002

based on statistics of known genes, e.g. nucleotides frequencies or shared motifs, have lead to various methods, see [6] for an overview. Neural networks have been successfully used in signal sensors, e.g. NetGene [4], and hidden Markov models have been used in context sensors, and to make sure that the predicted gene structures obey a proper gene syntax, e.g. HMMgene [14] and Genscan [5]. The increasing number of sequenced genomes has lead to methods which use sequence homology and evolutionary patterns as context sensors and guidelines for identifying genes. These comparative gene finders search for genes by comparing homologous DNA sequences looking for similar segments which can be assembled into genes, or by using proteins from related organisms to search for gene structures which encode similar proteins, see e.g. [1, 2, 3, 8, 13, 16].

In this paper we consider a simple but flexible approach to comparative gene finding: Given two homologous DNA sequences, we identify the most likely pair of orthologous genes based on a model of legal gene structures and a similarity measure between genes. For eukaryotic gene pairs, we furthermore identify the most likely exon structures. We develop methods based on gene models and alignment based similarity measures of increasing complexity which allow us to take into account many details of real gene structures that are often neglected in comparative methods. E.g. that insertions and deletions in exons do not only happen at codon boundaries, that stop codons do not occur in exons, and that improper modeling of insertion and deletions in exons may produce stop codons. We develop methods where the gene similarity measure is based on an existing alignment of the sequences, as well as methods where the gene similarity measure is the optimal score of an alignment of a pair of legal genes, which is computed as an integral part of the prediction process.

Using an existing alignment is simple but only allows us to assign similarity to legal gene pairs which are aligned by the given alignment. Computing the alignments of legal gene pairs as an integral part of the prediction process is more complicated but allows us to assign similarity of all legal gene pairs. In both cases it is possible to employ ideas from existing alignment score functions, which we use this to construct a gene similarity measure that explicitly models that the concatenated exons encode a protein by using a DNA- and protein-level score function as proposed by Hein in [10]. All our methods can be implemented using dynamic programming such that gene finding using a similarity measure based on an existing alignment takes time proportional to the number of columns in the given alignment, and gene finding using a similarity measures based on integrating the alignment and prediction process takes time $O(nm)$, where n and m are the lengths of the two sequences.

To evaluated the quality of the develop methods, we have performed a series of experiments on both simulated and real pairs of eukaryotic sequences which show that all methods are competitive. The best results are obtained when integrating the alignment and prediction process using a similarity measure which takes the encoded proteins into account. In all the experiments we used very simple pattern based signal sensors to indicate the possible start and stop positions of exons and introns. An increased performance might be achieved by using better signal sensors, e.g. splice site detectors such as NetGene [4], as a preprocessing step. However, since our focus is to examine the quality of our pure comparative approach to gene finding, we do not consider this extension in this paper. The comparative approach can also be extended by comparison of multiple sequences which is beyond the scope of this paper.

The rest of the paper is organized as follows. In Sect. 2 we introduce legal gene structures and similarity measures. In Sect. 3 we present our basic methods for gene finding in prokaryotic and eukaryotic sequences. In Sect. 4 we consider a number of extensions of the basic methods. In Sect. 5 we evaluate our methods through a series of experiments on synthetic and real sequence data.

2   Preliminaries

Let $a = a_1 a_2 \cdots a_n$ and $b = b_1 b_2 \cdots b_m$ be two homologous genomic sequences which contain a pair of orthologous genes of unknown structure. We want to identify the most likely pair of genes based on a model of legal gene structures and a gene similarity measure between pairs of legal genes.

Model of Legal Gene Structures. An important feature in biological and computational gene identification is functional signals on the genomic sequence. Typically a signal is made up of a certain pattern of nucleotides, e.g. the start codon *atg*, and these signal flanks the structural elements of the gene. Unfortunately most signals are not completely determined by a simple two or three letter code (not even the start-codon). Various (unknown) configurations of nucleotides in the neighborhood determine if e.g. *atg* is a start codon. We leave the discussion of how to identify potential signals for Sect. 5, and settle here with a more abstract definition of the relevant signals. We consider four types of signals: start codons, stop codons, acceptor sites, and donor sites. Donor and acceptor sites are only relevant for eukaryotic sequences. Potential start and stop codons on sequence a are given by the indicator variables $G_{a,i}^{start}$ and $G_{a,i}^{stop}$, where $G_{a,i}^{start}$ is true when a_i is the last nucleotide of a potential start codon and $G_{a,i}^{stop}$ is true when a_i is the first nucleotide of a potential stop codon. Similarly, potential donor and acceptor sites are given by the indicator variables D_i^a and A_i^a, where D_i^a is true when a_i is the first nucleotide of a potential donor site, and A_i^a is true when a_i is the last nucleotide of a potential acceptor site.

A *legal prokaryotic gene* on a genomic sequence a is a substring $\alpha = a_i \cdots a_j$, where $G_{a,i-1}^{start}$ and $G_{a,j+1}^{stop}$ are true. See Fig. 1. Similarly, a *legal eukaryotic gene* on a genomic sequence a is a substring $\alpha = a_i \cdots a_j$, where $G_{a,i-1}^{start}$ and $G_{a,j+1}^{stop}$ are true, that is divided into substrings which are alternating labeled intron and exon, such that each intron is a substring $a_h \cdots a_k$, for some $i \leq h < k \leq j$, where D_h^a and A_k^a are true. The *exon structure* of a eukaryotic gene α is the sequence of substrings $\alpha_1, \alpha_2, \ldots, \alpha_k$ that are labeled exons. See Fig. 2. In Sect. 4 we refine our definition of legal gene structures to exclude gene structures which contain in-frame stop codons, and gene structures where the coding nucleotides do not constitute an integer number of codons, i.e. a number of coding nucleotides which is not a multi-plum of three.

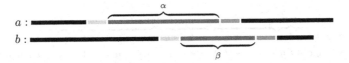

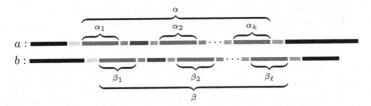

Fig. 1. We want to find a pair of legal prokaryotic genes, α and β, of highest similarity. A legal prokaryotic gene is a substring flanked by certain signals. In the simplest case these signals can be modeled as just the start and stop codons.

Fig. 2. We want to find a pair of legal eukaryotic genes, α and β, of highest similarity, and their exon structures $\alpha_1, \ldots, \alpha_k$ and $\beta_1, \ldots, \beta_\ell$. A legal eukaryotic gene is a substring flanked by certain signals. In the simplest case these signals can be modeled as just the start and stop codons. A legal exon structure is a sequence of substrings of the gene fulfilling that the remaining substrings of the gene, the introns, all start at a donor site and end at an acceptor site.

Gene Similarity Measures. Let $\mathrm{Sim}(\alpha, \beta)$ be a similarity measure which quantifies the similarity between two genes α and β. We use two alignment based approaches to define $\mathrm{Sim}(\alpha, \beta)$. The first approach is to use an existing alignment. Let $(\hat{a}, \hat{b})$ be an alignment of a and b, where $\hat{a}$ and $\hat{b}$ is the first and second row of the alignment matrix respectively. We say that two substrings α and β of a and b respectively are aligned by $(\hat{a}, \hat{b})$ if there is a sub-alignment $(\hat{\alpha}, \hat{\beta})$ of $(\hat{a}, \hat{b})$, i.e. a consecutive block of columns, which is an alignment of α and β. We define the similarity of α and β, $\mathrm{Sim}(\alpha, \beta)$, as the score of the alignment $(\hat{\alpha}, \hat{\beta})$ induced by $(\hat{a}, \hat{b})$ cf. a given alignment score function. Note that this only defines the similarity of substrings which are aligned by $(\hat{a}, \hat{b})$. The advantage of defining the similarity measure based on an existing alignment is of course that we can use any existing and well proven alignment method to construct the alignment. A drawback is that it only defines the similarity of substrings aligned by $(\hat{a}, \hat{b})$, which implies that we cannot consider the similarity of all pairs of possible legal gene structures. To circumvent this problem, we as the second approach simply define the similarity of α and β as the score of an optimal alignment of α and β.

3   Methods

In this section we present methods for solving the following problem. Given two sequences $a = a_1 a_2 \cdots a_n$ and $b = b_1 b_2 \cdots b_m$, find a pair of legal genes α and β in a and b respectively such that $\mathrm{Sim}(\alpha, \beta)$ is maximized. For eukaryotic genes we also want to find their exon structures. For ease of presentation, the methods

$$
\begin{array}{c}
\overset{\text{start}}{\overbrace{}} \quad \overset{\text{exon}}{\overbrace{}} \qquad\quad \overset{\text{intron}}{\overbrace{}} \qquad \overset{\text{exon}}{\overbrace{}} \overset{\text{stop}}{\overbrace{}}
\end{array}
$$

$$
\hat{a} = a\ g\ t\ c\ atg\ c\ a\ a\ t\ a\ c\ g\ t\ t\ a\ g\ -\ -\ -\ -\ c\ t\ c\ taa\ a\ g\ g\ -\ -
$$

$$
\hat{b} = -\ -\ -\ g\ c\ atg\ c\ -\ -\ -\ t\ c\ -\ -\ -\ -\ -\ g\ t\ a\ g\ g\ t\ c\ t\ g\ a\ t\ c\ a\ g\ a
$$

$$
\text{log-odds ratio} = \delta_{c,c} + 3\cdot\lambda + \delta_{a,t} + \delta_{c,c} + \gamma + 9\cdot 0 + \gamma + \delta_{c,g} + \delta_{t,t} + \delta_{c,c}
$$

Fig. 3. Calculating the log-odds ratio of an alignment of two eukaryotic sequence with known gene structures. Only coding columns and switching between coding and non-coding models contribute to the score. The score of non-coding columns are zero.

in this section do not take into account that in frame stop codons must not occur in exons, that the combined length of exons must be a multiple of three, that a codon can be split by an intron, or that the exons encode a protein. These extensions are considered in Sect. 4.

To define $\mathrm{Sim}(\alpha, \beta)$ we use a column based alignment score which allows us to assign scores to alignments of α and β by summing the score of each column in the alignment. The assumption of coding regions being well-conserved and non-coding regions being divergent can be expressed in two probabilistic models, one for coding alignment columns $P\binom{x}{y} = p_{xy}$, and one for non-coding alignment columns $P\binom{x}{y} = q_x q_y$, where q_x is the overall probability of observing x in an alignment. Note that an alignment column can represent an insertion or deletion, i.e. either x or y can be a gap. Since the chance of inserting or deleting a nucleotide depends on the background frequency of the nucleotide but not its kind, we set $p_{x-} = p_{-x} = \sigma q_x$, for some constant σ. We formulate these models in terms of their log-odds ratio, and use this as a scoring scheme.

We observe that the log-odds ratio of a non-coding column is $-\log\frac{q_x q_y}{q_x q_y} = 0$. The log-odds ration for a coding column consisting of two nucleotides is described by $\delta_{x,y} = -\log\frac{p_{xy}}{q_x q_y}$, and a column including a gap by $\lambda = -\log\frac{\sigma}{q_-}$. Hence, the cost of a coding column is given by a substitution cost $\delta_{x,y}$ and a gap cost λ, and the cost of a non-coding column is zero. Switching between the coding and non-coding models is penalized by γ which denotes the probability (in log-odds ratio) of observing a splice-site. In Sect. 4 we consider extensions of this simple score function. Given an alignment $(\hat{a}, \hat{b})$ and a known gene structure, its log-odds ratio can be calculated by summing over columns aligning the genes, see Fig. 3.

Finding Prokaryotic Genes

When working with prokaryotic sequences a and b, we know that legal genes are substrings of coding nucleotides. Hence, we only have to use the coding model when computing the similarity between possible pairs of genes.

Prokaryotic Method 1 – Using an Existing Alignment. Let $\mathcal{A} = (\hat{a}, \hat{b})$ be an alignment of a and b. A sub-alignment $\mathcal{A}_{i,j}$ from column i to j is an alignment of the substrings $\hat{a}_i \cdots \hat{a}_j$ and $\hat{b}_i \cdots \hat{b}_j$ with gaps removed. We denote these aligned substrings $\alpha = a_{\mathrm{a}(i)} \cdots a_{\mathrm{a}(j)}$ and $\beta = b_{\mathrm{b}(i)} \cdots b_{\mathrm{b}(j)}$ respectively, where a($\cdot$) and b($\cdot$) are functions which translate column indices in $\mathcal{A}$ to string indices in a and b respectively. We say that $\mathcal{A}_{i,j}$ is a legal sub-alignment if the aligned substrings α and β are legal genes, i.e. if $G^{start}_{\mathrm{a}(i)-1,\mathrm{b}(i)-1}$ and $G^{stop}_{\mathrm{a}(j)+1,\mathrm{b}(j)+1}$ are true. The score $\mathrm{Sim}(\hat{a}_i \cdots \hat{a}_j, \hat{b}_i \cdots \hat{b}_j)$ of $\mathcal{A}_{i,j}$ is the sum of the score of each of the $j - i + 1$ columns. We want to find a legal sub-alignment of maximum score.

We define $S(i) = \max_{h < i} \mathrm{Sim}(\hat{a}_h \cdots \hat{a}_i, \hat{b}_h \cdots \hat{b}_i)$ where $G^{start}_{\mathrm{a}(h)-1,\mathrm{b}(h)-1}$ is true, i.e. $S(i)$ is the maximum score of a subalignment ending in column i of substrings starting at legal gene start positions. Since our alignment score is column based, we can compute $S(i)$ using dynamic programming based on the recursion:

$$S(i) = \max \begin{cases} S(i-1) + \delta_{\hat{a}_i, \hat{b}_i} \\ 0 & \text{if } G^{start}_{\mathrm{a}(i),\mathrm{b}(i)} \end{cases} \tag{1}$$

$$i'' = \operatorname*{argmax}_{i} \left\{ S(i) \mid G^{stop}_{\mathrm{a}(i)+1,\mathrm{b}(i)+1} \right\}$$

where i'' is the index of the last column in a legal sub-alignment of maximal score. By finding the maximum $i' < i''$ where $S_{i'} = 0$ and $G^{start}_{\mathrm{a}(i'),\mathrm{b}(i')}$ is true, we have that $\mathcal{A}_{i'+1,i''}$ is a legal sub-alignment of maximal score, and that the corresponding pair of substrings is a pair of legal genes of maximum similarity. Computing $S(i)$ and finding i' and i'' takes time proportional to the number of columns in $\mathcal{A}$, i.e. the total running time is $O(n + m)$.

Prokaryotic Method 2 – Maximizing over All Possible Gene Pairs. We define the similarity $\mathrm{Sim}(\alpha, \beta)$ between legal genes α and β as the score of an optimal alignment of α and β using the column based alignment score function introduced above. Finding a pair of legal genes of maximum similarity is similar to the local alignment problem [19] with the additional condition that we only maximize the similarity over pairs of substrings which are legal genes.

We define $S(i, j) = \max_{h < i, k < j} \mathrm{Sim}(a_h \cdots a_i, b_k \cdots b_j)$ where $G^{start}_{h-1,k-1}$ is true, i.e. $S(i, j)$ is the maximum similarity of a pair of substrings ending in positions i and j and starting at legal gene start positions. Since our similarity measure is a column based alignment score, we can compute $S(i, j)$ using dynamic programming based on the recursion:

$$S(i, j) = \max \begin{cases} S(i-1, j-1) + \delta_{a_i, b_j} \\ S(i-1, j) + \lambda \\ S(i, j-1) + \lambda \\ 0 & \text{if } G^{start}_{i,j} \end{cases} \tag{2}$$

$$(i'', j'') = \operatorname*{argmax}_{i,j} \left\{ S(i, j) \mid G^{stop}_{i+1,j+1} \right\}$$

where (i'', j'') are the last positions in a pair of legal genes of maximum similarity. We can find the start positions of these genes by back-tracking through the recursive computation of $S(i'', j'')$ searching for the maximum indices $i' < i''$

and $j' < j''$ where $S(i', j') = 0$ and $G^{start}_{i',j'}$ is true. The substrings $a_{i'+1} \cdots a_{i''}$ and $b_{j'+1} \cdots b_{j''}$ is by construction legal genes of maximum similarity. Computing $S(i, j)$ and finding (i'', j'') takes time $O(nm)$ using dynamic programming, backtracking through the recursive computation to find (i', j') takes time $O(n + m)$, i.e. the total running time is $O(nm)$. The space consumption is $O(nm)$ but can be reduced to $O(n + m)$ using the technique in [12].

Finding Eukaryotic Genes

When working with eukaryotic sequences a and b, we not only want to find a pair of legal genes α and β of highest similarity, but also their most likely exon structures. We solve this problem by making it possible to switch between the coding and non-coding score model at splice-sites when computing the similarity between possible pairs of genes.

Eukaryotic Method 1 – Using an Existing Alignment. Let $\mathcal{A} = (\hat{a}, \hat{b})$ be an alignment of a and b. A legal sub-alignment $\mathcal{A}_{i,j}$ is defined as above. The score of a column in $\mathcal{A}_{i,j}$ is computed by the coding or non-coding score model cf. these rules: (1) the score of the first column, i.e. column i, can be computed using the coding model. (2) the score of column k can be computed using the score model used to compute the score of column $k-1$. (3) the score of column k can be computed in the non-coding model if $D^a_{a(k)}$ and $D^b_{b(k)}$ are true, i.e. if we are at a donor site. (4) the score of column k can be computed in the coding model if $A^a_{a(k)-1}$ and $A^b_{b(k)-1}$ are true, i.e. if we have just passed an acceptor site.

The score $\text{Sim}(\hat{a}_i \cdots \hat{a}_j, \hat{b}_i \cdots \hat{b}_j)$ of $\mathcal{A}_{i,j}$ is defined as the sum of the score of each of the $j + i - 1$ columns maximized over all possible divisions into coding and non-coding columns. The division of $\mathcal{A}_{i,j}$ into coding or non-coding columns which yields the maximum score also yields the exon structures of the aligned substrings. We want to find a legal sub-alignment of maximum score and the corresponding division into coding and non-coding columns. We define $S(i) = \max_{h<i} \text{Sim}(\hat{a}_h \cdots \hat{a}_i, \hat{b}_h \cdots \hat{b}_i)$ where $G^{start}_{a(h)-1,b(h)-1}$ is true, and column i is either a coding column or the last column in an intron, i.e. $S(i)$ is the maximum score of a sub-alignment ending in column i of substrings starting at legal gene start positions and ending in an exon or at an acceptor site. We can compute $S(i)$ using dynamic programming based on the recursions:

$$I(i) = \max \begin{cases} \mathcal{I} : I(i-1) \\ \mathcal{I} : S(i-1) + \gamma \text{ if } D^a_{a(i)} \wedge D^b_{b(i)} \end{cases}$$

$$S(i) = \max \begin{cases} \mathcal{E} : S(i-1) + \delta_{\hat{a}_i, \hat{b}_i} \\ \mathcal{I} : I(i-1) + \gamma & \text{if } A^a_{a(i)} \wedge A^b_{b(i)} \\ 0 & \text{if } G^{start}_{a(i),b(i)} \end{cases} \tag{3}$$

$$i'' = \underset{i}{\text{argmax}} \left\{ S(i) \mid G^{stop}_{a(i)+1,b(i)+1} \right\}$$

where i'' is the index of the last column in a legal sub-alignment of maximal score. Recall that the constant γ is the probability (in log-odds ratio) of observing a splice-site. We can find the start positions of these genes by back-tracking through the recursive computation of $S(i'')$ searching for the maximum index $i' < i''$ where $S(i') = 0$ and $G^{start}_{a(i'),b(j')}$ is true. By construction the sub-alignment $\mathcal{A}_{i'+1,i''}$ is a legal sub-alignment of maximal score, and the corresponding pair of substrings is a pair of legal genes of maximum similarity. The exon structures of these genes is determined while back-tracking, where the annotation, $\mathcal{E}$ or $\mathcal{I}$, of each applied recursion indicates if the corresponding column in $\mathcal{A}_{i'+1,i''}$ is coding or non-coding. Computing $S(i)$ and finding i' and i'' takes time proportional to the number of columns in $\mathcal{A}$, i.e. the total running time is $O(n+m)$.

Eukaryotic Method 2 – Maximizing over All Possible Gene Pairs. We define the similarity $\text{Sim}(\alpha, \beta)$ between legal genes α and β as the score of an optimal alignment of α and β, where we use the coding and non-coding score models, and make it possible to switch between them at donor and acceptor sites. Since the non-coding score model sets the score of a non-coding column to zero, we can ignore the possibility of non-coding substitution columns when computing an optimal alignment. Hence, we only have to consider alignments where non-coding nucleotides are in gap columns cf. the alignment in Fig. 3. Intuitively, this approach makes it possible to skip introns by allowing free gaps starting at a donor site and ending at an acceptor site. We define $S(i,j) = \max_{h<i,k<j} \text{Sim}(a_h \cdots a_i, b_k \cdots b_j)$ where $G^{start}_{h-1,k-1}$ is true and a_i and b_j are either coding or the last symbols in introns. Hence, $S(i,j)$ is the maximum similarity of a pair of substrings ending in positions i and j and starting at legal gene start positions, that would be a pair of legal genes if $G^{stop}_{i+1,j+1}$ was true. We can compute $S(i,j)$ using dynamic programming based on the recursions:

$$I^a(i,j) = \max \begin{cases} \mathcal{I} : I^a(i-1,j) \\ \mathcal{I} : S(i-1,j) + \gamma \text{ if } D^a_i \end{cases}$$

$$I^b(i,j) = \max \begin{cases} \mathcal{I} : I^a(i,j-1) \\ \mathcal{I} : S(i,j-1) + \gamma \text{ if } D^b_j \end{cases}$$

$$S(i,j) = \max \begin{cases} \mathcal{E} : S(i-1,j-1) + \delta_{a_i,b_j} \\ \mathcal{E} : S(i-1,j) + \lambda \\ \mathcal{E} : S(i,j-1) + \lambda \\ \mathcal{I} : I^a(i-1,j) + \gamma & \text{if } A^a_i \\ \mathcal{I} : I^b(i,j-1) + \gamma & \text{if } A^b_j \\ 0 & \text{if } G^{start}_{i,j} \end{cases} \tag{4}$$

$$(i'',j'') = \underset{i,j}{\text{argmax}} \left\{ S(i,j) \mid G^{stop}_{i+1,j+1} \right\}$$

where $I^a(i,j)$ and $I^b(i,j)$ is the maximum similarity of a pair of substrings ending in positions i and j inside an intron in sequence a and b respectively, and starting at legal gene start positions. The pair (i'',j'') are the last positions in a pair of legal genes of maximum similarity. We can find the start positions of these genes by back-tracking through the recursive computation of $S(i'', j'')$ searching for the maximum indices $i' < i''$ and $j' < j''$ where $S(i', j') = 0$ and $G_{i',j'}^{start}$ is true. The substrings $a_{i'+1} \cdots a_{i''}$ and $b_{j'+1} \cdots b_{j''}$ is by construction legal genes of maximum similarity. The exon structures of these genes is determined when back-tracking where the annotation, $\mathcal{E}$ or $\mathcal{I}$, of each applied recursion indicates if the symbols in the corresponding column are coding or non-coding. The back-tracking also yields an optimal alignment of the identified gene pair. Similar to *Prokaryotic Method 2*, the time and space complexity of the method is $O(nm)$, where the space consumption can be improved to $O(n + m)$.

4   Extensions

The model of legal gene structures and the coding and non-coding score functions presented in the previous sections are deliberately made simple. However, there are some obvious extensions to both of them. The codon structure is an important feature of gene structures. The coding parts of a gene can be completely divided into subsequent triplets of nucleotides denoted codons. Each codon encodes an amino acid which evolves slower than its underlying nucleotides. If a codon is a stop codon the translation, and thus the gene, must end. Moreover, the codon structure implies that deleting a number of nucleotides which is not a multi-plum of three changes all encoded amino acids downstream of the deletion (called a frame shift). This is believed to be a very rare event. The coding score function should describe a more evolved treatment of evolutionary events, e.g. take changes on the encoded amino acids into account when scoring an event, and allow insertion/deletion of consecutive nucleotides.

Taking the Encoded Amino Acids into Account. The alignment score function by Hein in [11] is a reasonable and fairly simple way of taking the protein level into account. The idea is to think of an amino acid encoded as being "attached" to the middle nucleotide of the codon encoding it. When matching two middle nucleotides a_2 and b_2, i.e. nucleotides in position two in predicted codons $a_1a_2a_3$ and $b_1b_2b_3$, the similarity between the amino acids A and B encoded by these codons is taken into account. For example, if $\delta_{a,b}$ is the similarity between two nucleotides and $\Delta_{A,B}$ is the similarity between two amino acids, then the cost of matching two middle nucleotides a_2 and b_2 in codons encoding amino acids A and B is $\delta_{a_2,b_2} + \Delta_{A,B}$, while the cost of matching any two non-middle nucleotides a_i and b_j is δ_{a_i,b_j}. By keeping track of the codon-position of the current nucleotides, we can distinguish the middle nucleotide from the other two and thus incorporate Δ.

Avoiding Stop Codons and Obeying Length Requirements. Keeping track of the codon-position as explained above also enables us to identify and exclude possible stop codons. Furthermore by only allowing genes to end in nucleotides that are in codon position three, we ensure that the total length of the coding regions is a multi-plum of three.

Avoiding Frame Shifts. Frame shifts can be avoided if we only allow gaps to come in triplets. That is to consider three columns in an alignment at the time when producing gaps in coding regions. When working on a given alignment there is however the problem that frame shifts may be introduces by errors in the preceding alignment step. If we forbid these frame shifts, we might discard the true genes because of an alignment error. Instead we can penalize "frame shifts" (alignment errors) with an additional cost ϕ.

Affine Gap-Cost. In the probabilistic model underlying our coding score model it was assumed that gap-symbols, reflecting insertions or deletions of nucleotides through evolution, were independent. However, it is a well-known fact that insertions and deletions often involve blocks of nucleotides, thus imposing a dependence between neighboring gaps. In the alignment literature an affine gap-cost function is often used to model this aspect of evolution. Affine gap cost means that k consecutive gap-columns has similarity $\mu + \lambda \cdot k$ instead of $\lambda \cdot k$, i.e. probability $p_{gap} \cdot (p_{x,-})^k$ instead of $(p_{x,-})^k$ cf. [7]. By using the technique by Gotoh in [9] we can incorporate affine gap cost in all our methods without increasing the asymptotic time complexity.

Keeping Track of Codons Across Exon Boundaries. In a eukaryotic gene, a codon can be split between two exons. This implies that nucleotides in a codon not necessarily are consecutive in the sequence as there between any two nucleotides in a codon can be a number of nucleotides forming an intron. The problem of keeping track of the codon-position of the current nucleotide becomes more difficult when this is possible. It can be solved by doing some additional bookkeeping on the different ways to end an exon (with respect to nucleotides in an possibly unfinished codon). For example, the recursion for $I_1^a(i,j)$ can be divided into sixteen recursions $I_{1,x,y}^a(i,j)$, for $x,y \in \{a,g,t,c\}$, where each represents a different way for two exons to end with one "hanging" nucleotide each. Similarly, $I_{2,x_1x_2,y_1y_2}^a(i,j)$ represent the cases where each exon has two hanging nucleotides. Handling hanging nucleotides properly is a technically hard extension, but it only increases the running time of our methods by a constant factor.

The implementations of our gene finding method used for experiments in Sect. 5 incorporate the above five extensions. The details of adding these extensions to the recursions presented in Sect. 3 are shown in [17].

5   Experiments

To evaluate the quality of our comparative gene finding method, we examine the performance of eukaryotic gene finding based on *Eukaryotic Method 1* and *Eukaryotic Method 2* including the extensions of Sect. 4, on two different data sets; a set of simulated sequence pairs, and a set of human-mouse sequence pairs. The performance of the two methods is evaluated using two different alignment score functions; one score function which only considers the nucleotide level, and one score function which also incorporates the protein level. In total we thus examine four different implementations of eukaryotic gene finding. The implementations are available as GenePair via `http://www.birc.dk/Software/`.

Data Set 1 – Simulated Sequences. The first data set consists of a time series of simulated sequence pairs. For steps of 0.1 in the time interval $[0\,;2]$ we simulate 20 sequence pairs. Each simulation starts with a random generated DNA-sequence (1000 nucleotides in length) with uniform background frequencies. Four regions (total length of about 250 nucleotides) of the sequence are exons and thus bounded with appropriate signals, i.e. start/stop-codons and splice sites. Given a time parameter two copies of the sequence undergo a Jukes-Cantor substitution process and a insertion/deletion Poisson-process with geometrical length distribution. Non-synonymous mutations changing amino acid σ to σ' are kept with probability $\exp(-d(\sigma, \sigma')/c)$, where d is the Grantham distance matrix from PAML [20]. The parameter c lets us control the dN/dS ratio, i.e. the difference between the evolutionary patterns in coding and non-coding regions. Mutations that change signals or introduce frame shifts are not allowed. Simulation parameters are chosen such that evolution time 1 yields coding regions with a mutation rate of 0.25 per nucleotide, a dN/dS ratio of 0.20, and 20 per cent gaps.

Data Set 2 – Human and Mouse Sequences. The second data set is the ROSETTA gene set from [2], which consists of 117 orthologous sequence pairs from human and mouse. The average length of the sequences is about 4750 nucleotides of which 900 are coding, and on average there are four exons per gene. Estimates in [18] give a dN/dS ratio of 0.12 and a mutation rate of 0.22 per nucleotide.

Parameters. Parameter estimation is a crucial aspect of most gene finding and alignment methods. Since our gene finding methods are based on probabilistic models of sequence evolution, we can do maximum likelihood estimation of the parameters if the true relationship between data are known, i.e. if the alignments and gene structures of the sequence pairs are known. This is the case for the simulated sequence pairs in Data Set 1. For each time step we simulate an additional 100 sequence pairs for the purpose of parameter estimation. The situation is more subtle when working with the human and mouse sequences in Data Set 2 since the alignments and gene structures are unknown. Usually the evolutionary distance between two genes is estimated and a corresponding set of predefined parameters are used, e.g. PAM similarity matrices. For our experiments, we use the parameters estimated in [2].

Signals. Much work is concerned about identifying functional sites of a gene, and several customized programs exist for identifying particular sites, see e.g. [15]. In our gene finding methods, we need to identify the start/stop codons and the donor/acceptor sites of a gene. We could perform this identification by using a customized program on the genes in question to predict the positions of potential signals, i.e. determine the sets D_i^x, A_i^x, $G_{x,i}^{start}$ and $G_{x,i}^{stop}$. However, this approach would make it difficult to compare the accuracy of our gene finding method with

Table 1. Signals are defined as short patterns of nucleotides.

Signal	Pattern	Definition
Start codon	atg	$G_{x,i}^{start} = \{i \mid x_{i-3}x_{i-2}x_{i-1} = \text{atg}\}$
Stop codon	taa, tag, tga	$G_{x,i}^{stop} = \{i \mid x_{i+1}x_{i+2}x_{i+3} \in \{\text{taa,tag,tga}\}\}$
Donor site	gt	$D_i^x = \{i \mid x_i x_{i+1} = \text{gt}\}$
Acceptor site	ag	$A_i^x = \{i \mid x_{i-1}x_i = \text{ag}\}$

the accuracy of other methods, since a good prediction might be the result of a good signal detection by the customized program. To avoid this problem, we use a very general and simple definition of functional sites, where each site is just a pattern of two or three nucleotides as summarized in Table 1.

Performance Evaluation. To compare prediction performance, we measure the nucleotide level sensitivity Sn = TP/(TP + FN), i.e. the fraction of true coding nucleotide predicted to be coding, and the nucleotide level specificity Sp = TP/(TP + FP), i.e. the fraction of predicted nucleotides actually true, for each method. TP (true positives) is the number of coding nucleotides predicted to be coding, FP (false positives) is the number of non-coding nucleotides predicted to be coding, and FN (false negatives) is the number of coding nucleotides predicted to be non-coding. Neither Sn or Sp alone constitute a good measure of global accuracy. As a global measure we use the correlation coefficient, $CC = (\text{TP} \cdot \text{TN} - \text{FP} \cdot \text{FN})/\sqrt{(\text{TP} + \text{FP}) \cdot (\text{TN} + \text{FN}) \cdot (\text{TP} + \text{FN}) \cdot (\text{TN} + \text{FP})}$, a well-known measure in the gene finding literature [6]. All three measures take their values between 0 and 1, where values close to 1 indicate good predictions.

Experiments on Data Set 1. This data set consists of a set of simulated sequence pairs with known alignments gene structures. For each sequence pair we use the four implementations of our gene finding approach to predict the gene structures of the sequences. The two implementations which use a given alignment (*Eukaryotic Method 1*) are tested with the *true* alignment known from the simulations, and a *computed* alignment obtained by a standard alignment method [9]. The true alignment is presumable the best possible alignment, and the computed alignment is presumable the best possible inferred alignment because optimal alignment parameters is estimated and used. The other two implementations *integrate* the alignment and prediction steps (*Eukaryotic Method 2*).

Figure 4 and 5 show the plots of the accuracy statistics for the predictions. For all four implementations, we observe over-prediction for small evolutionary distances. This is a consequence of our founding assumption of non-coding regions being fare divergent, which does not hold when the overall evolutionary distance is small. In general, very closely related sequences will always cause problems in a pure comparative gene finder because different evolutionary patterns cannot be observed. The experiments also indicate that better predictions are made by combining the alignment and prediction steps instead of using a pre-inferred alignment (unless when the true alignment is used). This is confirmed

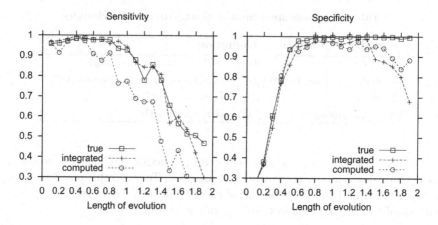

Fig. 4. DNA level score: Nucleotide sensitivity and specificity of the predictions made on simulated sequences. The alignment score function only considers the nucleotide level. Each data point is the average over 20 predictions.

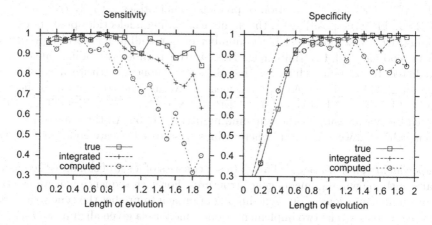

Fig. 5. DNA and protein level score: Nucleotide sensitivity and specificity of the predictions made on simulated sequences. The alignment score function also considers the protein level. Each data point is the average over 20 predictions.

by Fig. 6 which shows the global accuracy of the four implementations by plotting the correlation coefficient of the predicted structures. The experiments also indicate that including the protein level in the alignment score function results in improved predictions.

Experiments on Data Set 2. This data set consists of a set of 117 sequence pairs from human and mouse with known gene structures but unknown alignments. Again we use the four implementations of our gene finding approach in a total of six different settings to predict the gene structures of the 117 sequence pairs. The two implementations which use a given alignment are tested with a *com-*

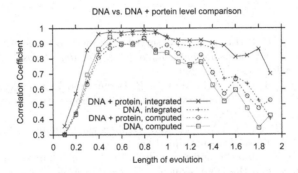

Fig. 6. Correlation coefficient of predictions made on the time series of simulated sequences. Each data point is the average over 20 predictions.

Table 2. Accuracy statistics for gene predictions in the human/mouse data set. The statistics for GLASS/ROSETTA and Genscan are from [2].

Gene finding method	Specificity	Sensitivity	Correlation
DNA level, computed alignment	0.89	0.92	0.88
DNA level, glass alignment	0.92	0.93	0.90
DNA level, integrated alignment	0.88	0.97	0.91
DNA and protein level, computed alignment	0.89	0.93	0.88
DNA and protein level, glass alignment	0.92	0.93	0.90
DNA and protein level, integrated alignment	0.92	0.98	0.94
GLASS/ROSETTA	0.97	0.95	–
Genscan	0.89	0.98	–

puted alignment obtained by a standard alignment method [9], and an alignment found using the GLASS alignment program from [2]. Table 2 shows the prediction accuracy of our gene finders measured in sensitivity, specificity and correlation coefficient. The table also shows the prediction accuracy of two existing gene prediction programs, GLASS/ROSETTA [2] and Genscan [5]. GLASS/ROSETTA is a comparative approach which first aligns two genomic sequences using GLASS and then predicts genes on the basis of the aligned sequences using ROSETTA. The idea behind ROSETTA is similar to our similarity measure that is based on an given alignment, but with a more elaborated signal prediction scheme. Genscan is a HMM based gene prediction program. It analyzes each sequence separately and does not utilize any comparative methods.

The prediction results on the 117 pairs of human and mouse sequences lead to the same conclusions as the prediction results on the simulated sequences. Predictions made by combining the alignment and prediction steps instead of using a pre-inferred alignment yield better results. Including the protein level in the alignment score function further improves the prediction accuracy. The experiments show that our gene finding methods can compete with existing gene finding programs, and in some cases outperform them.

References

1. V. Bafna and D. H. Huson. The conserved exon method for gene finding. In *Proceedings of the 8th International Conference on Intelligent Systems for Molecular Biology (ISMB)*, pages 3–12, 2000.

2. S. Batzolou, L. Pachter, J. P. Mesirov, B. Berger, and E. S. Lander. Human and mouse gene structure: Comparative analysis and application to exon prediction. *Genome Research*, 10:950–958, 2000.

3. P. Blayo, P. Rouzé, and M.-F. Sagot. Orphan gene finding - an exon assembly approach. Unpublished manuscript, 1999.

4. S. Brunak, J. Engelbrecht, and S. Knudsen. Prediction of human mRNA donor and acceptor sites from the DNA sequence. *Journal of Molecular Biology*, 220:49–65, 1991.

5. C. Burge and S. Karlin. Prediction of complete gene structures in human genomic DNA. *Journal of Molecular Biology*, (268):78–94, 1997.

6. M. Burset and R. Guigó. Evaluation of gene structure prediction programs. *Genomics*, 34:353–367, 1996.

7. R. Durbin, S. Eddy, A. Krogh, and G. Mitchison. *Biological Sequence Analysis: Probabilistic Models of Proteins and Nucleic Acids*, chapter 1-6. Cambridge University Press, 1998.

8. M. S. Gelfand, A. A. Mironov, and P. A. Pevzner. Gene recognition via spliced sequence alignment. *Proceedings of the National Academy of Science of the USA*, 93:9061–9066, 1996.

9. O. Gotoh. An improved algorithm for matching biological sequences. *Journal of Molecular Biology*, 162:705–708, 1982.

10. J. Hein. An algorithm combining DNA and protein alignment. *Journal of Theoretical Biology*, 167:169–174, 1994.

11. J. Hein and J. Støvlbæk. Combined DNA and protein alignment. In *Methods in Enzymology*, volume 266, pages 402–418. Academic Press, 1996.

12. D. S. Hirschberg. A linear space algorithm for computing maximal common subsequences. *Communication of the ACM*, 18(6):341–343, 1975.

13. I. Korf, P. Flicek, D. Duan, and M. R. Brent. Integrating genomic homology into gene structure prediction. *Bioinformatics*, 17:140–148, 2001.

14. A. Krogh. A hidden Markov model that finds genes in e. coli DNA. *Nucleic Acids Research*, 22:4768–4778, 1994.

15. L. Milanesi and I. Rogozin. Prediction of human gene structure. In *Guide to Human Genome Computing*, chapter 10. Academic Press Limited, 2nd edition, 1998.

16. L. Pachter, M. Alexandersson, and S. Cawley. Applications of generalized pair hidden Markov models to alignment and gene finding problems. In *Proceedings of the 5th Annual International Conference on Computational Molecular Biology (RECOMB)*, pages 241–248, 2001.

17. C. N. S. Pedersen and T. Scharling. Comparative methods for gene structure prediction in homologous sequences. Technical Report RS-02-29, BRICS, June 2002.

18. J. S. Pedersen and J. Hein. Gene finding with hidden Markov model of genome structure and evolution. Unpublished manuscript, submitted to *Bioinformatics*.

19. T. F. Smith and M. S. Waterman. Identification of common molecular subsequences. *Journal of Molecular Biology*, 147:195–197, 1981.

20. Z. Yang. *Phylogenetic Analysis by Maximum Likelihood (PAML)*. University College London, 3.0 edition, may 2000.

MultiProt – A Multiple Protein Structural Alignment Algorithm

Maxim Shatsky[1,*], Ruth Nussinov[2,3,**], and Haim J. Wolfson[1]

[1] School of Computer Science, Tel Aviv University, Tel Aviv 69978, Israel
{maxshats,wolfson}@post.tau.ac.il
[2] Sackler Inst. of Molecular Medicine, Sackler Faculty of Medicine
Tel Aviv University
[3] IRSP - SAIC, Lab. of Experimental and Computational Biology, NCI - FCRDC
Bldg 469, Rm 151, Frederick, MD 21702, USA
ruthn@fconyx.ncifcrf.gov

Abstract. We present a fully automated highly efficient technique which detects the multiple structural alignments of protein structures. Our method, *MultiProt*, finds the common geometrical cores between the input molecules. To date, only few methods were developed to tackle the structural multiple alignment problem. Most of them require that all the input molecules be aligned, while our method does not require that all the input molecules participate in the alignment. Actually, it efficiently detects high scoring partial multiple alignments for all possible number of molecules from the input. To demonstrate the power of the presented method we provide a number of experimental results performed by the implemented program. Along with the known multiple alignments of protein structures, we present new multiple structural alignment results of protein families from the *All beta proteins* class in the SCOP classification.

1   Introduction

The increasing number of determined protein structures rises the need to develop appropriate tools for molecule structure analysis. This need will become especially acute with the advance of the Structural Genomics Project.

Many methods have been proposed to solve the pairwise structural alignment of protein molecules [26,17,27,19,25]. For comprehensive reviews see [14,5]. Obviously, multiple structural alignment is a significantly more powerful tool than pairwise structure alignment, almost for the same reasons that multiple sequence alignment contains significantly more information than pairwise sequence alignment [8]. In particular, such a tool enables us to detect structural cores of protein families for homology modeling and threading applications, structurally

---

* To whom correspondence should be addressed
** The publisher or recipient acknowledges right of the U.S. Government to retain a nonexclusive, royalty-free license in and to any copyright covering the article.

conserved active sites and more. Nevertheless, there are only few methods that address the multiple structure alignment problem.

The *SSAPm* method of Orengo and Taylor [20] performs all pairwise structural alignments applying the double dynamic programming SSAP method. The highest scoring pair is selected as a seed for the multiple alignment. The consensus molecule is defined as the average of the selected pair. Then, the method starts iteratively to add the best-fitting structure to the consensus molecule. The consensus molecule is recomputed after each iteration. The solution obtained by this method depends on the order of joining molecules to the multiple alignment.

The *MUSTA* algorithm [13,12] tries to solve the multiple structural alignment problem without using pairwise alignments and dealing with multiple molecules at once. The method is based on the *Geometric Hashing* technique. This technique was successfully applied for docking algorithms [7,22], and a number of pairwise structure alignment methods [19,23]. One of the strengths of this technique is its efficiency and independence on the order of the structural features representing the molecules. However, there are several disadvantages of the *MUSTA* algorithm. The first one - the algorithm requires that all input molecules participate in the multiple structural alignment, i.e. the algorithm is incapable to detect the structural cores between a non-predefined subset of the input molecules. The second one - its runtime is practical only for sets of about 10-15 molecules.

In our method the goal is to solve the multiple structural alignment problem with detection of partial solutions and making the algorithm very efficient and practical for tens (and may be up to hundreds) of protein molecules.

2   Problem Definition

Let us discus the MSTA (Multiple STructural Alignment) problem in more detail. The definition of the MSTA problem is not a straightforward issue. Given m input molecules, should all m molecules participate in the alignment or only a subset of the m molecules? Certainly, we wish to detect the best subsets which give good multiple structure alignments. Consider an example where among 100 input molecules there are 40 structurally similar molecules from family "A", 50 structurally similar molecules from family "B" and additional 10 molecules, which are structurally dissimilar to any other molecule in the input. Naturally, we require from a MSTA algorithm to detect simultaneously the similarity between the 40 molecules from family "A" and between the 50 molecules from family "B". Also, there might be only a sub-structure (motif, domain) that is similar between some molecules. Thus, partial similarities between the input molecules should also be reported by the MSTA algorithm.

Obviously, the number of all possible solutions could be exponential in the number of input molecules. For example, consider proteins which contain α-helices. Each pair of α-helices could be structurally aligned (at least partially, if they are different in their lengths). Any combination of α-helices from different molecules gives us some solution to the MSTA problem. The number of such combinations is exponential. Thus, it is not practical to output (even if the algorithm is capable to detect) all possible solutions.

Let us start from the pairwise structure alignment. Given two sets of points in 3-D space (e.g. two proteins represented by the coordinates of their C_α atomic centers), a geometrical pattern detection problem is to find two subsets of points, one from each input set (molecule), such that these two subsets are congruent. In computational geometry this problem is called the *largest common point set* detection or LCP [21]. Obviously, it is unrealistic to detect an exact congruence, thus a more practical problem is finding LCP which is ε-congruent, i.e. the distance between the matching subsets is less than some threshold ε. Distance measuring is done using some metric. For example, the *bottleneck* metric [4], where ε is the maximal distance between the corresponding points. Another commonly used metric is the *Root Mean Square Deviation (RMSD)* [10,11]. There are two major aspects to the ε-congruent LCP problem with the *bottleneck* or *RMSD* metric. The first is the *correspondence* task, i.e. detection of corresponding point pairs. The second is *computation of an optimal 3-D rigid transformation* that superimposes one set onto the other. Of course, these two aspects are interrelated. For the *RMSD* metric, given a correspondence between the subsets, we can calculate the transformation that minimizes the *RMSD* in linear time [10].

Therefore, the multiple-LCP task is to detect the subsets of points, one from each of the input sets, which are ε-congruent (the ε-congruence for multiple sets is redefined in the sequel). Not surprisingly, the multiple-LCP problem is a hard one. It is NP-hard even in the one dimensional space for the case of an exact congruence, i.e. $\varepsilon = 0$ [1].

In our method we try to give an efficient heuristic solution to the MSTA problem, while answering more practical requirements, than the multiple-LCP problem. In our approach we define the MSTA problem as:

(*) *Given m molecules, a parameter κ and a threshold value ε, for each r $(2 \leq r \leq m)$, find the κ largest ε-congruent multiple alignments containing exactly r molecules.*

The above definition (*) is general and can be applied for comparison of any 3-D objects represented as unconnected point sets in 3-D space. However, we wish to utilize the fact that a protein structure can be represented as an ordered set of points, e.g. by the sequence of the centers of the C_α atoms. Thus, we exploit the natural assumption that any solution for MSTA of proteins should align, at least short, contiguous fragments (minimum 3 points) of input atoms. For example, these fragments could be secondary structure elements (but not necessarily), which could be aligned between the input molecules. First we detect all possibly aligned fragments of maximal length between the input molecules. Then, we select solutions that give high scoring global structural similarity based on the (*) definition. Aligning protein fragments is not a new idea. It has been previously applied in several methods for pairwise protein structural alignment [27,25]. In our method we use the algorithm which detects structurally similar fragments of maximal length, i.e. fragment pairs which cannot be extended while preserving ε-congruence [23,24].

The distance functions used are the *RMSD* and the *bottleneck*. At the first stage of the algorithm we establish an initial, local, correspondence between point subsets, and calculate the 3-D transformation that minimizes the *RMSD* between these sub-sets [10]. Then, when the transformation is established, we calculate the global similarity based on the *bottleneck* distance.

3   Methods

Our method is based on the pivoting technique, i.e. there is a pivot molecule that has to participate in all the alignments. In other words, the rest of the molecules are aligned with respect to the pivot molecule. In order not to be dependent on the choice of the pivot, we iteratively choose every molecule to be the pivot one.

> Input: m molecules $S = \{M_1...M_m\}$
> $for\ i = 1\ to\ m - 1$
> $\quad M_{pivot} = M_i$
> $\quad S' = S \setminus M_{pivot}$
> $\quad Alignments = MSTA\_With\_Pivot(M_{pivot}, S')$
> $\quad Select\_High\_Scoring\_Alignments(Alignments)$
> end

The goal of the *MSTA_With_Pivot* algorithm is to detect the largest structural cores between the input molecules with respect to the pivot molecule. The algorithm requires that the pivot molecule participates in the multiple alignments, but it does not require that all the input molecules from the set S' are included in the multiple alignment. Thus, the method detects the best *partial*[1] multiple alignments.

In practice, to speed up the algorithm we do not include previous pivot molecules in the set S', i.e. $S' = \{M_{p+1}...M_m\}$. Suppose that at iteration i molecule M_k, $k < i$, is included into the set $S' = \{M_k, M_{i+1}...M_m\}$. If one of the multiple alignment solutions includes molecule M_k, then this alignment has already been detected in the k'th iteration (would we have an optimal algorithm implemented in function *MSTA_With_Pivot*), thus, we don't include previous pivots into the set S'. In our experimental results the difference in the solution size, including previous pivot molecules into S' and not including (this is a program parameter), differed in about 10%. So, a user can choose between the trade-off of speed and accuracy. In the sequel we refer to the more efficient version, where previous pivots are excluded, i.e. $S' = \{M_{p+1}...M_m\}$.

Below we explain the main part of our method, the *MSTA_With_Pivot* procedure.

3.1   Multiple Structural Alignment with Pivot

Our algorithm performs three types of multiple alignments and thus has three stages. The first type is the multiple structural alignment of contiguous frag-

---

[1] For the MSTA problem the term *partial* has two different meanings. First, only a subset of the molecules is aligned. Second, the structural core contains only subsets of the aligned molecule residues.

ments. Such solutions do not fully resolve the MSTA problem (*) defined above, but serve as an initial alignment for the global multiple structural alignment.

Stage 1. _Multiple Structure Fragment Alignment._ Let us define a fragment, $F_i^p(l)$ of molecule p, which starts at point i and has length l. We say that fragment $F_i^p(l)$ is ε-congruent to $F_j^k(l)$, $k \neq p$, if there exists a rigid 3-D transformation, T, that superimposes both fragments with $RMSD$ less than a predefined threshold ε (the default value is $3\mathring{A}$). The correspondence between the points is linear, i.e. point $i + t$, from molecule p, is matched with point $j + t$, from molecule k, where t moves between 0 and $l - 1$. Such an ε-congruent fragment pair is named $F_i^p F_j^k(l)$ and the function that measures the distance (similarity) between the fragments is denoted $RMSD_{opt}(F_i^p F_j^k(l))$, namely, $RMSD_{opt}(F_i^p F_j^k(l)) = min_T RMSD(F_i^p(l), T(F_j^k(l)))$.

If fragment $F_i^p(l)$ is ε-congruent to $F_j^k(l)$ and to $F_s^t(l)$, i.e. $RMSD_{opt}$ $(F_i^p F_j^k(l)) \leq \varepsilon$ and $RMSD_{opt}$ $(F_i^p F_s^t(l)) \leq \varepsilon$, then $RMSD_{opt}(F_s^t F_j^k(l)) \leq 2\varepsilon$ (it is true since the $RMSD_{opt}$ is a metric and, thus, satisfies the triangle inequality[11]). Though, the $RMSD_{opt}$ of 2ε between $F_s^t(l)$ and $F_j^k(l)$ is a theoretical worst case bound, in practice our results show much lower values. In experiments where $\varepsilon = 3\mathring{A}$, the average value of $RMSD_{opt}(F_s^t F_j^k(l))$ was about $2\mathring{A}$ with a small standard deviation (around 1.0). These values were consistent with different fragment lengths. This is not a surprise, since proteins are not randomly distributed structures in 3-D space, but rather well organized structural motifs built up of elementary sub-structures like α-helices or β-strands.

We define a multiple fragment alignment as a pivot fragment $F_i^p(l)$ and a set of fragments $S^F = \{F_{j_k}^k(l) : k \in \{(p+1)...m\}\}$ for which $RMSD_{opt}(F_i^p F_{j_k}^k(l)) \leq \varepsilon, \forall F_{j_k}^k \in S^F$. Notice, that the set S^F does not necessarily contain fragments from all the molecules $M_{p+1}...M_m$. Also, it may contain several fragments from the same molecule (e.g. several α-helices from some molecule could be matched with the same α-helix from M_p), and actually provides us with multiple solutions. Let us define a multiple fragment alignment, S_r^F, which contains fragments from exactly r different molecules. Thus, for this stage of the algorithm we state the problem as : _For a given integer κ, find the best κ multiple fragment alignments, S_r^F, for each r between 2 and $(m-p+1)$._ Saying "the best" we mean the longest. Next, we describe how do we detect these sets, S_r^F.

In the first step, given a pivot molecule M_p and the set of other molecules $S' = \{M_{p+1}...M_m\}$, the task is to find all congruent fragment pairs between the pivot molecule and each molecule from the set S'. To accomplish this task we can use an exact algorithm[2], but in order to achieve a favorable running time of the program we apply the same efficient method as in the _FlexProt_ algorithm [23]. At the end of this step we have a set of congruent fragment pairs, $S^{FP} = \{F_i^p F_j^k(l) : k \in \{(p + 1)...m\}, RMSD_{opt}(F_i^p F_j^k(l)) \leq \varepsilon\}$. The experimental complexity of this step is $O(|S| * n^2)$, where n is the size of the longest molecule.

---

[2] All ε-congruent fragments, $\{F_i^p F_j^k(l)\}$, can be obtained by an exhaustive verification in polynomial time.

Let us represent the above set by a 2 dimensional plot. The x-axis represents the sequence of the pivot molecule M_p. The y-axis is divided to bins, one bin for each molecule from the set $\{M_{p+1}...M_m\}$. The fragment $F_j^k(l)$, from the pair $F_i^p F_j^k(l)$, is plotted in bin $M(k)$ (y-axis) and aligned to $F_i^p(l)$, i.e. its projection onto the x-axis is exactly the set of the corresponding (according to the $F_i^p F_j^k(l)$ alignment) points of the $F_i^p(l)$. The order of the fragments inside the $M(k)$ bin (on the y-axis) is arbitrary. See Figure 1.

Drawing two vertical lines at points α and β defines a *cut* on the interval $[\alpha, \beta]$. A fragment belongs to the *cut* if the interval $[\alpha, \beta]$ lies inside the fragment, i.e. $F_i^p F_j^k(l)$ is in the *cut* if and only if $i \le \alpha$ and $\beta \le (i + l - 1)$. Since several fragments from the same molecule might participate in the *cut*, such a *cut* provides us with a set of multiple alignments. Choosing from the *cut* only one fragment for each molecule gives us some multiple alignment. The number of choices equals $\prod_i k_{M_i}$ where k_{M_i} is the number of fragments from molecule M_i in the *cut*. Thus, the number of possible multiple alignments (for the given *cut*) might grow exponentially with the number of molecules. This is the nature of the *multiple alignment* problem. We return to this problem later in *Stage 2* after we explain how to detect the *cuts*.

Cut Detection. From Figure 1 we can observe that it might be possible to extend a given *cut* to the left and to the right so that the *cut* contains the same fragments. Thus, we define a locally maximal cut $Cut[\alpha, \beta]$ as an interval $[\alpha, \beta]$ such that for any $\varepsilon > 0$, $Cut[\alpha - \varepsilon, \beta]$ and $Cut[\alpha, \beta + \varepsilon]$ contain different fragments than $Cut[\alpha, \beta]$. It is obvious that any $Cut[\alpha, \beta]$ starts at the beginning of a fragment and ends at the end of a (possibly, another) fragment. From now on, we use the terms *cut* and $Cut[\alpha, \beta]$ interchangeably.

All possible *cuts* can be detected efficiently by a simple algorithm which is similar to the sweeping technique from Computational Geometry [3]. The idea is to represent the projections on the x-axis of the fragment start(end)-points as events on the x-axis. The fragment left (right) end-point is the *start* (*end*) event. Starting from the first point (C_α atom) of the pivot molecule, M_p, we move along the x-axis (M_p sequence) with a vertical line and generate *cuts*. At every moment (i.e. position α of the vertical line) we remember the fragments that the vertical line passes through - let us call them $CFrags$. When a *start* event is encountered at position α we add the fragment $F_j^k(l)$ (which corresponds to some $F_\alpha^p(l)$) to the list of $CFrags$ and generate new *cuts*. We require that the newly encountered fragment, $F_j^k(l)$, participates in these new *cuts*. We generate the following *cuts*: $\{Cut[\alpha, \beta]\}$, where α is the current cutting point of the vertical line and $\beta \le (\alpha + l - 1)$, where β takes the values of the *end*-points, that are equal or smaller than the *end*-point of the $F_j^k(l)$, of fragments from the $CFrags$. Since $\beta \le (\alpha + l - 1)$, $F_j^k(l)$ participates in all new *cuts* (to make this step efficient we use the same "sweeping" technique, where the *end*-points of the fragments from the $CFrags$ constitute new events).

If there are several *start*-events at the same cutting point, than we first add newly discovered fragments to the list of $CFrags$ and then generate new *cuts*

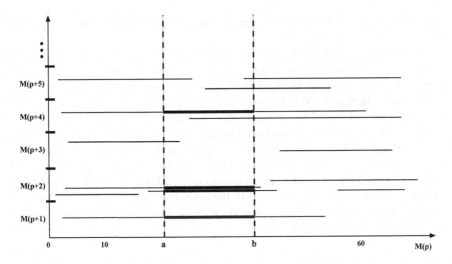

Fig. 1. *Cuts.* The x-axis is the sequence of M_p. Molecules $\{M_{p+1}...M_m\}$ are assigned into bins on the y-axis. Fragment pairs that include completely fragment $[\alpha, \beta]$ (shown in bold) are included into $Cut[\alpha, \beta]$.

($\{Cut[\alpha, \beta]\}$) where β is equal or smaller than the farthest *end*-point of the newly discovered fragments.

At an *end*-event we just remove the fragment from the $CFrags$ list. It is easy to see that the described algorithm generates all possible *cuts* according to the definition of $Cut[\alpha, \beta]$.

Stage 2. *Global Multiple Alignment.* Consider one of the previously detected *cuts* $Cut[\alpha, \beta]$. One of the possible approaches is to leave the *cuts* as is, i.e. not to choose the multiple alignment(s) from the set of possible ones of the specific *cut* (several fragments of the same molecule might be included in the *cut*). These complete *cuts* are included in the output of the program. Thus, an end-user can apply additional criteria to filter out the non-relevant fragments.

However, our goal is to detect the best multiple alignments based on the global structural similarity, as defined in (*). As we have already pointed out, this is a hard problem, so we provide only a heuristic solution.

In this step, we select from every *cut* only one fragment for each molecule. We aim to perform this selection, in a way that the resulting multiple alignment would give, possibly, the highest score. Namely, given a $Cut[\alpha, \beta]$ containing $\{F_i^p F_j^k(l) : i \leq \alpha, (i + l - 1) \geq \beta\}$, for each different k (for each molecule M_k) select the fragment (if there is more than one fragment from molecule M_k) so that it's transformation gives the largest, global, structural (pairwise) alignment with the pivot molecule (M_p). Let us explain this step in more detail. Given a fragment pair $F_i^p F_j^k(l)$ and the transformation T_{opt} that optimally superimposes $F_j^k(l)$ onto $F_i^p(l)$, we apply the transformation T_{opt} on molecule M_k. Now, when the two molecules are aligned, we calculate the size of their maximal structural alignment (for more details see section *Pairwise Correspondence* in the *Appendix*) which preserves ε-congruence.

At this stage of the algorithm every solution for the MSTA problem has a non ambiguous representation, i.e. each solution contains at most one representative from each molecule. Now, the task is to score these solutions based on the size of the multiple alignment. Notice, that the transformation for each molecule is now fixed thus we only need to calculate the *multiple correspondence* size. For the details of this procedure see section *Multiple Correspondence* of the *Appendix*.

To enlarge the obtained solutions we apply an iterative improvement procedure as follows. For each solution, after the *multiple correspondence* between the pivot molecule with the other molecules is established, we apply a rigid transformation that minimizes the RMSD between the matching points. Then, we compute the *multiple correspondence* once again and repeat this procedure (the default number of iterations is 3).

3.2   Ranking of the Solutions

When a common geometric core is detected, we compute the multiple *RMSD* (*mRMSD*) of the alignment. It is computed as an average of the *RMSD* values between the geometric core of the pivot molecule M_p with the corresponding geometric core of each molecule from the multiple alignment. Thus, solutions are grouped according to the number of aligned molecules and each group is sorted according to the size of the alignment and according to the *mRMSD*, giving priority to the alignment size.

Stage 3. *Bio-Core detection.* The solutions found in the previous stage are scored according to the global structural alignment. We can apply another scoring scheme, which requires that aligned points are of the same biological type (still, the points should be close enough in 3-D space). In our method the input points are positions of C_α atomic centers, so each point represents a specific amino acid (a user can choose other input atoms, for example C_β atoms). For instance, we can require residue identity matching, but it is, usually, too restrictive. We chose the following classification: *hydrophobic (Ala, Val, Ile, Leu, Met, Cys), polar/charged (Ser, Thr, Pro, Asn, Gln, Lys, Arg, His, Asp, Glu), aromatic (Phe, Tyr, Trp), glycine (Gly)*. Let us name this the *bio-core* classification.

Thus, at *Stage 3* of the algorithm, instead of applying a scoring scheme based on the largest common structural core, we apply scoring based on the *bio-core* classification. High scoring solutions from this stage of the algorithm are complementary to the solutions from *Stage 1* and *Stage 2*. The complexity of this stage is equivalent to the complexity of the *Stage 2*.

3.3   Complexity

Due to the lack of space, we do not go into details, but it can be shown that the experimental time complexity (based on the conducted experiments) of the algorithm is bounded by $O(m^2 * n^3)$, where m is the number of input molecules

Table 1. Pairwise structural alignment test. The protein pairs are classified as 'difficult' for structural analysis[6]. The alignments are performed by *VAST*[15], *Dali*[9], *CE* [25] and *MultiProt*. The information in this table, except for the *MultiProt* results, is taken from [25].

Molecule 1 (size)	Molecule 2 (size)	VAST S_{align}/rmsd	Dali S_{align}/rmsd	CE S_{align}/rmsd	MultiProt S_{align}/rmsd
1fxi:A(96)	1ubq(76)	48/2.1	-	-	48/1.8
1ten(89)	3hhr:B(195)	78/1.6	86/1.9	87/1.9	82/1.4
3hla:B(99)	2rhe(114)	-	63/2.5	85/3.5	73/2.5
2aza:A(129)	1paz(120)	74/2.2	-	85/2.9	86/2.4
1cew:I(108)	1mol:A(94)	71/1.9	81/2.3	69/1.9	75/1.9
1cid(177)	2rhe(114)	85/2.2	95/3.3	94/2.7	93/2.6
1crl(534)	1ede(310)	-	211/3.4	187/3.2	209/2.5
2sim(381)	1nsb:A(390)	284/3.8	286/3.8	264/3.0	291/2.9
1bge:B(159)	2gmf:A(121)	74/2.5	98/3.5	94/4.1	92/2.5
1tie(166)	4fgf(124)	82/1.7	108/2.0	116/2.9	102/2.3

S_{align} is the number of aligned atoms.

and n is the size of the longest molecule. The experimental complexity of *Stage-1* is $O(m*n^2)$. The complexity of *Stage-2,3* is $O(m^2*n^3)+n*O(m*n^2)$. Thus the overall algorithm time complexity is bounded by $O(m^2*n^3)$, though the actual running times are significantly lower (see *Results* section).

4   Results

Here we present experimental results performed by the program. The program output consists of 10 (changeable parameter) highest-scoring results for each number of molecules, i.e. if the number of input molecules is 15, then there are sets of results for 2, 3, .. 15 aligned molecules. Each result lists (1) 3-D rigid transformation for each aligned molecule, (2) matrix of aligned amino acids, (3) RMSD of the multiple alignment (mRMSD), calculated as described above. All experiments were performed on a Pentium©IV 1800 MHz processor with 1024MB internal memory.

Naturally, a good initial test for any multiple alignment method would be a set of "hard to detect" pairwise alignments. We applied our method to ten protein pairs tested by different pairwise structural alignment methods. The results are presented in Table 1. As it can be observed from the table our pairwise results are very competitive. The maximal running time (pair 1crl:534, 1ede:310) was less than 3 s.

Comparing to the results achieved by the *MUSTA* algorithm [12], our method achieved similar alignment results. The test includes the following cases:

- *Serpins* family - 7apiA, 8apiA, 1hleA, 1ovaA, 2achA, 9apiA, 1psi, 1atu, 1kct, 1athA, 1attA, 1antl, 2antl.
- *Serine Proteinase*: *Subtilases* family - 1cseE, 1sbnE, 1pekE, 3prk, 3tecE.

Table 2. Multiple structural alignment test. Comparison of *MUSTA* [12] and *Multi-Prot* algorithms.

Proteins	No. of Mols	Average Size	MUSTA S_{align}	MultiProt S_{align}	MultiProt run-time
Serpins	13	372	163	237	9m04s
7apiA, 8apiA, 1hleA, 1ovaA, 2achA, 9apiA,					
1psi, 1atu, 1kct, 1athA, 1attA, 1antl, 2antl					
Serine Proteinase	5	277	220	227	25s
1cseE, 1sbnE, 1pekE, 3prk, 3tecE					
Calcium-binding	6	140	31	36	9s
4cpv, 2scpA, 2sas, 1top, 1scmB, 3icb					
TIM-Barrels	7	391	40	44	3m12s
7timA, 1tml, 1btc, 4enl, 1pii, 6xia, 5rubA					
Helix-Bundle	10	140	27	27	2m10s
1flx, 1aep, 1bbhA, 1bgeB, 1le2, 1rcb,					
256bA, 2ccyA, 2hmzA, 3inkC					

S_{align} is the number of aligned atoms.

- *Calcium-binding*: *EF hand-like* superfamily. The protein 4cpv is from the *parvalbu-min* family; 2scpA, 2sas, 1top, and 1scmB from the *calmodulin-like* family; and 3icb from the *Cal-binding D9K* family.
- *TIM-Barrels*: proteins are taken from the 7 different superfamilies. *PDB codes*: 7timA, 1tml, 1btc, 4enl, 1pii, 6xia, 5rubA.
- *Helix-Bundle* proteins are taken from the 6 different superfamilies. *PDB codes*: 1flx, 1aep, 1bgeB, 1rcb, 3inkC, 1le2, 256bA, 2ccyA, 1bbhA, 2hmzA.

In all cases the size of the geometric core was at least the same (see Table 2). In addition, our method produced high-scoring partial alignments and run significantly faster (on the same computer).

In the following sections, we provide, along with the known results, new multiple protein structural alignments. We nickname the largest structural (bio) core to be *struct-core* (*bio-core*). The results are summarized in Table 3.

4.1 Globins

This family has been extensively studied in literature [2,28]. We compared our results to the ones obtained in [28]. In this paper the authors present results of the multiple structural alignment for the following 7 proteins: 5mbn, 1ecd, 2hbg, 2lh3, 2lhb, 4hhbA, 4hhbB. The largest geometrical core detected by their method [28] consists of 105 C_α atoms (or "corresponding landmarks" as they called it in the paper). Our program shows similar results. The size of the detected common *struct-core* varied between 92 C_α atoms ($\varepsilon = 3\mathring{A}$) to 109 C_α atoms ($\varepsilon = 4\mathring{A}$). The structural similarity is detected primarily between α-helices, while loop regions were left un-aligned. The detected *bio-core* is comparatively small. It ranged from 7 C_α atoms ($\varepsilon = 3\mathring{A}$) to 21 C_α atoms ($\varepsilon = 4\mathring{A}$). Running time was about 15 seconds.

Table 3. Multiple structural alignment results performed by *MultiProt*.

Proteins	No. of Mols	Average Size	Struct-Core S_{align}	run-time
Superhelix				14s
1lxa, 1qq0, 1xat, 2tdt, 1fwyA:252-328	5	205	61	
1lxa, 1qq0, 1xat, 2tdt	4	238	84	
1lxa, 1qq0, 2tdt	3	248	114	
1lxa.pdb, 1qq0.pdb	2	235	143	
Supersandwich				12s
1bgmI:731-1023, 1cb8A:336-599, 1oacA:301-724	3	360	118	
1cb8A:336-599, 1oacA:301-724	2	393	187	
Concanavalin				54s
2bqpA, 1gbg, 2galA, 1d2sA, 1sacA, 1a8d:1-247, 1kit:25-216, 2sli:81-276, 6cel, 1xnb	10	220	54	
1a8d:1-247, 1d2sA, 1gbg, 1sacA, 2galA, 2sli:81-276	6	194	75	
1a8d:1-247, 1gbg, 6cel	3	298	128	

S_{align} is the number of aligned atoms.

4.2   Superhelix

In this experiment we compare 5 proteins (1lxa, 1qq0, 1xat, 2tdt, 1fwy(A:252-328)) from the *Superfamily: Trimeric LpxA-like enzymes*. Each protein is taken from a different family. While the first 4 molecules are between 208 and 274 residue long, the last one, 1fwy(A:252-328), is a truncated form and has only 77 residues. Our algorithm detected multiple alignment between all 5 molecules with *struct-core* of size 61 with *mRMSD* 1.0 Å. The *bio-core* for this molecules consisted of 19 C_α-atoms. In this case, the detected highest ranked *bio-core* is NOT a sub-set of the highest ranked *struct-core* solution. The *struct-core* solution, which was ranked 8'th, corresponds to the highest ranked *bio-core*. Thus, one might argue that the *bio-core* scoring scheme might be more helpful for biological function prediction.

4 molecules (the first four) gave 88 (18) C_α-atoms in the *struct-core* (*bio-core*). Molecules 1lxa, 1qq0 and 2tdt gave 114 (27) C_α-atoms in the *struct-core* (*bio-core*) (also there are other combinations of the first four molecules presented in the solutions). Molecules 1lxa.pdb and 1qq0.pdb gave 143 (62) C_α-atoms in the *struct-core* (*bio-core*), as for the three molecules there are other molecule pairs that gave high similarities. The program running time was about 14 seconds. For details see Table 3.

4.3   Supersandwich

From the SCOP database [18] we selected the *Supersandwich* fold from the *All beta proteins* class. This fold contains three superfamilies. From each superfamily we selected one protein 1bgmI:731-1023, 1cb8A:336-599 and 1oacA:301-724. Our multiple alignment result contains 118 C_α atoms with *mRMSD* 2.21 Å. Protein 1bgmI:731-1023 has 17 strands, only two strands (914-921,894-901) are not in the alignment. This example demonstrates that our method does not totally depend on the order of the backbone chain. A number of strands were aligned in the opposite order. Below is a part of the alignment (notice the alignment between 1bgmI and 1oacA):

1bgmI: 738 739 740 741 746 747 748 749 750
1cb8A: 444 442 340 341 348 342 350 351 339
1oacA: 327 325 324 323 332 331 330 329 328

The *bio-core* of this multiple alignment contains 23 atoms, which is a sub-set of the largest detected *struct-core*. The program running time was about 12 seconds. For details see Table 3.

4.4   Concanavalin A-Like Lectins/Glucanases

In this experiment we selected from the SCOP database the *Concanavalin A-like lectins / glucanases* fold (sandwich; 12-14 strands in 2 sheets; complex topology). This fold contains only one *superfamily*, which includes 10 different families. We selected one protein from each family : 2bqpA, 1gbg, 2galA, 1d2sA, 1sacA, 1a8d:1-247, 1kit:25-216, 2sli:81-276, 6cel, 1xnb.

Aligning all 10 molecules results in the geometric core of size 54. Interestingly, protein 1a8d:1-247 participated in all alignments containing different numbers of molecules. 1a8d:1-247 has 5 (A, B, C, D, E) β-sheets. A, C and D create almost one β-sheet (let us call it S1), and so B and E (S2). In the alignment of all 10 molecules only β-sheet C was aligned well (from 7 β-strands 3 were aligned well and 2 received only small partial alignments) and β-sheet E obtained only a small partial alignment. Investigating multiple alignments containing less molecules, we notice, that the common core of β-sheet S1 increases and so of S2. See Figure 2 for a geometric core of 6 molecules - 1a8d:1-247, 1d2sA, 1gbg, 1sacA, 2galA, 2sli:81-276. The size of the *struct-core* is 75 C_α atoms with *mRMSD* 2.0 $\mathring{A}$. The *bio-core* size of these molecules is 9 atoms. The running time was 54 seconds. For details see Table 3.

4.5   Mixed Experiment

In this experiment, in order to show the power of our method, we included in the input set 18 proteins. 5 proteins from the *"Superhelix"* experiment (1lxa, 1qq0, 1xat, 2tdt, 1fwy(A:252-328)), 3 proteins from the *"Supersandwich"* experiment (1bgmI:731-1023, 1cb8A:336-599, 1oacA:301-724) and 10 proteins from the *"Concanavalin A-like lectins/glucanases"* experiment (2bqpA, 1gbg, 2galA, 1d2sA, 1sacA, 1a8d:1-247, 1kit:25-216, 2sli:81-276, 6cel, 1xnb).

It took only 8 minutes for the program to compare these 18 molecules. The multiple structural alignments for 3 molecules contained an alignment of the *"Supersandwich"* family. The results for 5 molecules contained the alignment of the *"Superhelix"* family. The results for 6 molecules contained the alignment of 6 proteins from *"Concanavalin A-like lectins/glucanases"* family (1a8d:1-247, 1d2sA, 1gbg, 1sacA, 2galA, 2sli:81-276), which alignment appears in Figure 2. Since both

(a)   **(b)**

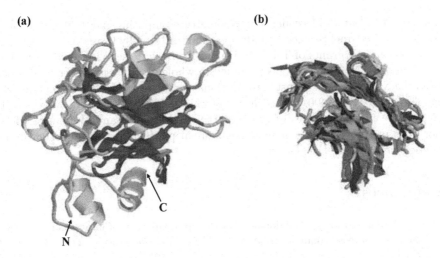

Fig. 2. . Concanavalin A-like lectins/glucanases. Structural core of 6 molecules - 1a8d:1-247, 1d2sA, 1gbg, 1sacA, 2galA, 2sli:81-276. Notice, the *two-sheet sandwitch* is conserved. (a) The backbone of molecule 1a8d:1-247 is shown complete in light green. The aligned core is in purple color. (b) Only the matched structural core is displayed, by assigning a different color to each molecule.

families *"Supersandwich"* and *"Concanavalin A-like lectins/glucanases"* contain a number of β-sheets, the result of 10-13 molecules contained the mixed alignments of proteins from these two families.

5   Conclusions and Future Work

We have presented a powerful tool for multiple structural alignment of protein molecules. Our method is extremely efficient and is suitable for comparison of up to tens of proteins. The experimental results have shown that the method is capable to detect the real partial multiple structural alignments, i.e. detection of common sub-structures and ability of distinguishing between the groups of structurally similar and dissimilar protein molecules.

Here we presented proteins as rigid molecules. A natural extension of multiple structural alignment is a *Flexible Multiple Structural Alignment*, where proteins are allowed to undergo hinge bending movements [24]. Further, we are going to implement our method on a parallel platform. This will allow to apply *MultiProt* on hundreds of proteins in practical running times. Applications of such a large-scale algorithm include, for example, analysis of protein populations where the number of structures is very big.

Acknowledgements

We thank Meir Fuchs for contribution of software to this project and we thank Hadar Benyaminy for biological suggestions. This research has been supported in part by the "Center of Excellence in Geometric Computing and its Applications" funded by the Israel Science Foundation (administered by the *Israel Academy of Sciences*) and by grants of Tel Aviv University Basic Research Foundation. The research of R. Nussinov has been funded in whole or in part with Federal funds from the National Cancer Institute, National Institutes of Health, under contract number NO1-CO-12400. The content of this publication does not necessarily reflect the view or policies of the Department of Health and Human Services, nor does mention of trade names, commercial products, or organization imply endorsement by the U.S. Government.

References

1. T. Akutsu and M. M. Halldorson. On the approximation of largest common subtrees and largest common point sets. *Theoretical Computer Science*, 233:33–50, 2000.
2. D. Bashford, C. Chothia, and A.M. Lesk. Determinants of a protein fold. Unique features of the globin amino acid sequences. *J Mol Biol.*, 196(1):199–216, 1987.
3. M. de Berg, M. van Kreveld, M. Overmars, and O. Schwarzkopf. *Computational Geometry - Algorithms and Applications*. Springer-Verlag, 2000.
4. Alon Efrat, Alon Itai, and Matthew J. Katz. Geometry helps in bottleneck matching and related problems. *Algorithmica*, 31(1):1–28, 2001.
5. I. Eidhammer, I. Jonassen, and WR. Taylor. Structure Comparison and Structure Patterns. *J Comput Biol.*, 7(5):685–716, 2000.
6. D. Fischer, A. Elofsson, D. Rice, and D. Eisenberg. Assessing the performance of fold recognition methods by means of a comprehensive benchmark. In L. Hunter and T. Klein, editors, *In Proc. Pacific Symposium on Biocomputing*, Singapore, 1996. World Scientific Press.
7. D. Fischer, S. L. Lin, H.J. Wolfson, and R. Nussinov. A Geometry-based Suite of Molecular Docking Processes. *J. Mol. Biol.*, 248:459–477, 1995.
8. D. Gusfield. *Algorithms on Strings, Trees, and Sequences : Computer Science and Computational Biology*. Cambridge University Press, 1997.
9. L. Holm and C. Sander. Protein structure comparison by alignment of distance matrices. *J. Mol. Biol.*, 233(1):123–38, 1993.
10. W. Kabsch. A solution for the best rotation to relate two sets of vectors. *Acta Crystallogr.*, A32:922–923, 1978.
11. TK. Kaindl and B. Steipe. Metric properties of the root-mean-square deviation of vector sets. *Acta Cryst.*, A53:809, 1997.
12. N. Leibowitz, Z.Y. Fligelman, R. Nussinov, and H.J. Wolfson. Automated multiple structure alignment and detection of a common substructural motif. *Proteins*, 43(3):235–45, 2001.
13. N. Leibowitz, R. Nussinov, and H.J. Wolfson. MUSTA-a general, efficient, automated method for multiple structure alignment and detection of common motifs: application to proteins. *J Comput Biol.*, 8(2):93–121, 2001.
14. C. Lemmen and T. Lengauer. Computational methods for the structural alignment of molecules. *J Comput Aided Mol Des*, 14(3):215–32, March 2000.
15. T. Madej, J. Gibrat, and S. Bryant. Threading a database of protein cores, 1995.
16. K. Mehlhorn. *The LEDA Platform of Combinatorial and Geometric Computing*. Cambridge University Press, 1999.

17. E.M. Mitchel, P.J. Artymiuk, D.W. Rice, and P. Willet. Use of Techniques Derived from Graph Theory to Compare Secondary Structure Motifs in Proteins. *J. Mol. Biol.*, 212:151–166, 1989.

18. A.G. Murzin, S.E. Brenner, T. Hubbard, and C. Chothia. SCOP: a structural classification of proteins database for the investigation of sequences and structures. *J. Mol. Biol.*, 247:536–540, 1995.

19. R. Nussinov and H.J. Wolfson. Efficient Detection of three-dimensional strcutral motifs in biological macromolecules by computer vision techniques. *Proc. Natl. Acad. Sci. USA*, 88:10495–10499, December 1991. Biophysics.

20. C.A. Orengo and W.R. Taylor. SSAP: Sequential Structure Alignment Program for Protein Structure Comparison. In R.F. Doolitle, editor, *Methods in Enzymology, Vol. 266*, pages 617–635. Academic Press, San Diego, 1996.

21. J.-R. Sack and J. Urrutia. *Handbook of Computational Geometry (chapter: Discrete Geometric Shapes: Matching, Interpolation, and Approximation)*. Elsevier, 2000.

22. B. Sandak, R. Nussinov, and H.J. Wolfson. An Automated Robotics-Based Technique for Biomolecular Docking and Matching allowing Hinge-Bending Motion. *Computer Applications in the Biosciences (CABIOS)*, 11:87–99, 1995.

23. M. Shatsky, Z.Y. Fligelman, R. Nussinov, and H.J Wolfson. Alignment of Flexible Protein Structures. In *8th International Conference on Intelligent Systems for Molecular Biology*, pages 329–343. The AAAI press, August 2000.

24. M. Shatsky, H.J Wolfson, and R. Nussinov. Flexible protein alignment and hinge detection. *Proteins: Structure, Function, and Genetics*, 48:242–256, 2002.

25. I. Shindyalov and P. Bourne. Protein structure alignment by incremental combinatorial extension (ce) of the optimal path. *Protein Engineering*, 11(9):739–747, 1998.

26. W. R. Taylor and C. A. Orengo. Protein structure alignment. *J. Mol. Biol.*, 208:1–22, 1989.

27. G. Vriend and C. Sander. Detection of Common Three-Dimensional Substructures in Proteins. *Proteins*, 11:52–58, 1991.

28. T.D. Wu, S.C. Schmidler, and T. Hastie. Regression analysis of multiple protein structures. *J Comput Biol*, 5(3):585–595, 1998.

Appendix

Pairwise Correspondence

Given two point sets in 3-D, A and B, and a threshold ε, find two equal size subsets A' and B' ($A' \subset A$, $B' \subset B$) satisfying the following criterion: there exist a one-to-one correspondence, f, between A' and B', such that $\forall a \in A'$, $\|a - f(a)\| \leq \varepsilon$. The task is to maximize the size of A' and B'. This problem can be solved by detecting maximal matching in a bipartite graph.

Let us construct the bipartite graph $G = (A \cup B, E)$. A graph edge $e = (u, v)$, $u \in A$ and $v \in B$, is created if and only if the distance between these points is less than ε. Now, the task is to select the maximal number of edges that satisfy the "matching" criterion - "A matching in a graph G is a subset of the edges of G such that no two share an end-point". This set of edges would give us a a one-to-one correspondence of maximal size. This is exactly the problem of finding "maximal cardinality matching in the bipartite graph" [16], which can be solved with time complexity $O(\sqrt{n} * m)$, where $n = |A| + |B|$ and $m = |E|$. In practice (for the case of protein molecules), a straightforward greedy method has been proven to work adequately well. It works as follows. An edge $e = (u, v)$ is added to the matching if and only if nodes u and v are not already matched. In other words the edge e should not violate the already created matching in the graph. The complexity of this greedy algorithm is only linear in the number of the graph edges $O(m)$. We show in the next paragraph that $m \in O(n)$.

A naive approach of building the "correspondence" graph $G = (A \cup B, E)$ takes $\Theta(n^2)$ steps, since every pair of nodes, one node from each part, is considered in order to decide, whether there is an edge between the nodes. To resolve this we have to find for each query point $u \in A$ all the points $v \in B$ within a ball of radius ε from u. Taking into consideration the geometrical constraints on atomic centers of protein molecules, this task can be accomplished in almost linear time $O(n)$.

We place one molecule, say A, on a three dimensional geometric grid, with grid size $max(\varepsilon, 3)$. Then, for each center of atom v of molecule B, which is just a point in 3-D space, we access the geometric grid and extract all atoms of A within radius ε from atom v. If ε is small enough, and it should be small since otherwise there is no meaning for alignment of two atoms, then we extract from the grid a constant number of atoms. If $\varepsilon = 3\text{Å}$, then in the ball of radius ε there are at most 2 or 3 C_α atoms. Therefore, the degree of each vertex in the graph is at most 3. It follows that the number of edges is linear in the number of nodes. Thus, the experimental complexity of building the graph G is (|Putting A onto a grid| + $|B|*$ |Number of atoms within radius ε|) $\in O(n)$.

To summarize, the complexity of solving the correspondence task using a greedy approach is linear in the number of atoms ($O(n)$), while using an exact algorithm of "Maximal Cardinality Matching in the Bipartite Graph" is $O(n^{3/2})$ (since $m \in O(n)$).

Multiple Correspondence

Given k point sets in 3-D, $A_1...A_k$, and a threshold ε, we are interested in k equal size ordered subsets $A'_i = \{a^i_1, ..., a^i_m\}$, $i = 1...k$, $A'_i \subset A_i$ satisfying the following criterion: $\forall i, j \in \{1...k\}$ and $\forall t \in \{1...m\}$ $||a^i_t - a^j_t|| \leq \varepsilon$. The task is to maximize m (the size of A'_i's).

In other words, we are looking for a maximal size set of k-tuples. Each k-tuple contains points, one from each set (molecule), that are close in 3-D space. Here we present a fast heuristic solution to this problem.

For each $i = 2...k$ solve Pairwise Correspondence for A_1 and A_i, as described above. The resulted correspondence-lists might contain different points from the set A_1. We are interested only in those that appear in all correspondence-lists. Thus we create a list of k-tuples, where each k-tuple contains - (1) a point p from A_1, which appeared in all correspondence-lists, (2) $k - 1$ points which are corresponding to p from Pairwise Correspondences.

A Hybrid Scoring Function
for Protein Multiple Alignment

Emily Rocke

University of Washington
ecrocke@cs.washington.edu

Abstract. Previous algorithms for motif discovery and protein alignment have used a variety of scoring functions, each specialized to find certain types of similarity in preference to others. Here we present a novel scoring function that combines the relative entropy score with a sensitivity to amino acid similarities, producing a score that is highly sensitive to the types of weakly-conserved patterns that are typically seen in proteins. We investigate the performance of the hybrid score compared to existing scoring functions. We conclude that the hybrid is more sensitive than previous protein scoring functions, both in the initial detection of a weakly conserved region of similarity, and given such a similarity, in the detection of weakly-conserved instances.

1 Introduction

Of the computations used to analyze the vast quantity of biological sequence data becoming available, a significant number involve detection or analysis of similarities between different DNA and protein sequences. This may take the form of sequence alignment (arranging two or more sequences so that corresponding residues are placed in alignment), pattern matching (searching a database for approximate instances of a given pattern), or pattern discovery (searching a database for a set of substrings that are similar to one another). In any of these cases, however, some score function must be established, to guide the progress of the algorithm and to measure whether the final result is interesting enough to report.

1.1 Tradeoffs

Any scoring function for protein multiple alignment has advantages and disadvantages. Some of the difference between scoring functions in use comes from different ideas of how to represent the underlying biology and the likelihoods of different evolutionary events.

However, another factor goes into the selection of a scoring function: for any given algorithm, some scoring functions dovetail well with the algorithm, allowing efficiency, and others are difficult to realize. For example, although one may believe that the biology is better represented by an exponential falloff in increase

R. Guigó and D. Gusfield (Eds.): WABI 2002, LNCS 2452, pp. 251–262, 2002.

of gap penalty with gap length, we typically use an affine gap penalty to perform Smith-Waterman alignment, precisely so that the score may be computed using an efficient dynamic program.

Another example of this phenomenon is the great prevalence of pattern discovery algorithms that find only ungapped patterns. Although biological reality includes many patterns whose instances vary in length, permission of gapped instances throws a confounding factor into many motif discovery algorithms and scoring schemata.

A third example, discussed below, is sensitivity to amino acid similarities in protein alignment.

1.2   Amino Acid Alignment

Protein sequences are made up of amino acids, some subsets of which are highly chemically similar, while others are chemically very different. Given two related proteins or protein segments, similar amino acids are more likely to have replaced one another than dissimilar ones.

The alignment of well-conserved protein regions does not absolutely require a sensitivity to this fact. Several programs, such as the Dialign algorithm for sequence alignment [15,14], or the Gibbs sampler for motif discovery [13], perform quite well in practice while only regarding exact matches between amino acids, as long as the conservation is sufficiently high.

However, there is compelling evidence that matching can be more sensitive if it takes amino acid similarity into account. Indeed, nearly any two-sequence alignment program will use amino acid similarities, in the form of the PAM or BLOSUM matrices, since the results improve significantly in their biological realism.

The main reason the same unanimity is not seen for multiple sequence alignment is that, for many multiple sequence scoring functions with otherwise desirable properties, it is unclear how to include amino acid similarities.

2   Previous Work

2.1   Two-Sequence Scoring Functions

In the case of evaluating a match between two sequences, the most popular scoring function is that used by the well-known Smith-Waterman / Needleman-Wunsch dynamic programming alignment algorithm [16,19]. The value of aligning any pair of residues depends on the two residues, independent of position. There are user-determined penalties for introducing a gap in a sequence and for extending it. The scoring function to maximize is the sum of this score over all positions.

Because of its position-independence, this score is very tractable. Each pairing of residues can be given a different score, such that the weight can be determined by empirical evidence as with the PAM [3] or BLOSUM [17,9] matrices. In addition, it lends itself to efficient dynamic programming, and the correspondence between scores and p-values has largely been solved (e.g. [11,4,12]). The same position independence also causes some of the flaws commonly observed in Smith-Waterman alignments: for example, long gaps are over-penalized, and if well-conserved segments are short compared to the overall sequence lengths the algorithm may not report them accurately.

Most software treating two-sequence alignment accepts these properties, working on faster algorithms to arrive at a high score rather than concentrating on the scoring function itself. For example, BLAST uses short ungapped matches to find candidate matches, but evaluates the candidates on the sum-of-similarities score against the query sequence.

An exception is the alignment program Dialign, which uses an alternative two-sequence scoring function called a *segment-to-segment* comparison [15,14]. The Dialign algorithm for two-sequence alignment uses dynamic programming to string together shorter ungapped matches. The ungapped matches are initially given a sum-of-similarities score, with a fixed value of 1 for each match and 0 for each mismatch, and then converted into a p-value before the second dynamic programming step.

Multiple-Sequence Equivalents. A straightforward way to extend this scoring function to multiple sequences is to use the sum-of-pairs score. This score adds together the pairwise similarity between each pair of sequences at each position, for a total of t^2 comparisons per column, if t is the number of sequences in the alignment. A similar approach chooses a single sequence as the template, and scores each position by taking the sum of similarities between each sequence and the template at that position. The template sequence may be one of the sequences being aligned, their consensus sequence, or another sequence that was determined externally. For example, MultiPipMaker [6], a DNA multiple alignment program, takes the first input sequence to be privileged and pairwise aligns the other sequences against it. In this case, the scoring function being optimized is the sum of similarities to the lead sequence.

Both scoring functions allow some of the same flexibility as the pairwise score: match scores can be weighted according to an arbitrary matrix and the score is easy to compute for most algorithms. However, in practice other scores can be more sensitive to patterns found in biological sequence.

2.2   Phylogeny-Based Score

One improvement to the sum-of-pairs score involves first placing the sequences to be aligned in a phylogenetic relationship, and then using any pairwise scoring function to greedily align the closest unprocessed neighbors. ClustalW [20], for example, first creates a phylogenetic tree using pairwise alignment distances, and then applies a sophisticated scoring function during the hierarchical alignment. The pairwise score is sensitive to secondary structure and evolutionary distance, giving the algorithm greater sensitivity.

This kind of scoring function is appropriate when a phylogenetic tree relates the candidate sequences, i.e., when aligning whole sequences (rather than pattern discovery or pattern matching) and when the sequences are closely enough related to determine an accurate phylogenetic relationship. In most contexts, a phylogenetic viewpoint is not useful for motif finding, except if the putative motif location is restricted to a family of phylogenetically related sequences (as in [1]).

In addition, even for those situations where a phylogeny applies, some other scoring function is usually necessary in the preprocessing stage to construct the tree. When sequences are distantly related, an algorithm like ClustalW might be improved by a more sensitive alignment algorithm during tree construction, to avoid mistakes in the phylogeny which will propagate to impact the accuracy of the alignment.

2.3   Segment-to-Segment Score

Dialign's "segment-to-segment" score extends to multiple dimensions by using the same p-value score as in the two-sequence case on ungapped pairwise matches, adding weight to a pairwise match if it is also matched in other sequences. The algorithm then uses a greedy heuristic to take the most significant ungapped matches first.

The method relies on there being enough significant ungapped matches to string together a correct alignment, which will be true only when the sequences are closely related. In section 7.1 we briefly discuss a method of extending the Dialign algorithm, using the scoring function proposed here as a subcomponent.

2.4   Relative Entropy

Another popular multiple sequence scoring function is the relative entropy score, as used by the Gibbs sampling algorithm introduced in [13]. This score rewards positions of the putative pattern that look very different from the background frequency by taking the ratio, for each residue in the column, of the frequency of the residue at that position of the pattern to the frequency of that residue in general. The logarithm of this ratio is summed over all residues in the column. This score is computed independently at each position, and added together to produce the score for the total pattern.

Mathematically, the score is expressed as

$$\sum_{\alpha \in \Sigma} p_\alpha \log(\frac{p_\alpha}{b_\alpha}),$$

where Σ is the alphabet, p_α is the fraction of the column made up of letter α, and b_α is the background frequency or prior probability, usually measured as the fraction of the database made up of letter α.

2.5   Benefits of Relative Entropy

The relative entropy score is an example of a scoring function that has not typically been considered in the context of gapped alignment or of including amino acid similarities. Although both restrictions would seem to limit the protein patterns an algorithm can find, the scoring function continues to be popular, particularly in the context of Gibbs sampling. It should be noted here that, although Gibbs sampling and relative entropy were proposed to be used together in [13], the Gibbs sampling algorithm can be used with any ungapped multiple-sequence scoring function.

The use of the relative entropy score, rather than, for example, a sum-of-pairs score which would allow accounting for amino acid similarities, is explained by its high sensitivity to subtle patterns. A great advantage of the score is that it serves as an approximate measure of the probability of an alignment, based on the priors predicted by an independent random background model. In practical terms, it means that the relative entropy score can recognize more subtle patterns than a score that counts differences, as long as the subtle pattern is significantly different from the background.

For this reason, the relative entropy score is an attractive candidate for extension to allow gaps and amino acid similarities.

3   Approach

The new method proposed in this paper is designed to combine the strengths of the relative entropy score with a sensitivity to amino acid similarities. In addition, as in our previous proposal for gapped Gibbs sampling ([18]), the relative entropy score is modified to allow for a gapped alignment.

In essence, the proposed scoring function adds weight to a residue if similar residues are in the same column. In order to describe the score more thoroughly, we begin with a description of the adaptation which allows the relative entropy score to deal with gaps, then describe how we allow sensitivity to amino acid similarities.

3.1   Adaptation to Gaps

The adaptation to permit gaps is little changed from that described in [18], in which we propose to extend Gibbs sampling to the gapped nucleotide alignment case. The most important addition to the regular relative entropy score is that in every column, in place of each blank, we substitute a fractional occurrence of every residue α equal to its background frequency b_α. The score is then computed as above, adding these fractional residues ($b_\alpha \times$ the number of blanks in the column) to the tally of p_α.

For example, suppose the alphabet consisted of letters A, B, C, and D, all occurring equally often in the database, and that the aligned column to be scored is "AA-ABA," where "-" represents a blank in the alignment. Then p_A would be 4.25 for the relative entropy computation, p_B would be 1.25, and C and D (neither of which appears explicitly in the column) would have $p_C = p_D = 0.25$.

The rationale behind this approach is that, with no remaining knowledge about a "missing" residue, i.e. gap, in an alignment, its possible values before deletion include any residue, and for lack of other information the best prior on this shape is the background frequencies.

More practically, this technique avoids the thorny problem of placing some kind of prior distribution on a gap. It has the pleasant side effect of penalizing gaps to a minor extent, although not sufficiently strongly to avoid the necessity for an additional gap penalty. A user-determined penalty, simply multiplied by the length of each gap, is subtracted from the combined score, although an affine gap penalty could also be used.

3.2 Amino Acid Similarities

Just as the idea of "fractional" residues is used to deal with gaps, a similar concept allows an accommodation of amino acid similarities. Suppose that for any pair of residues α_i and α_j, $(i \neq j)$, we have a weight $w_{i,j}$, $0 \leq w_{i,j} \leq 1$, which measures how often α_j is substituted for α_i in related sequences. The value $w_{i,j}$ in some way reflects to what extent the residue α_j *might actually have been* residue α_i before a substitution.

Then let $w_{i,j}$ be the "credit" that α_i gets for each pairing of α_i with α_j. This credit is multiplied by the fraction of possible pairings which are in fact a pairing of α_i with α_j. Finally, this credit is given to α_i as if α_i itself occurred those extra fractional times:

$$p^*_{\alpha_i} = p_{\alpha_i} + \sum_{j \neq i}(p_{\alpha_i} \times p_{\alpha_j} \times w_{i,j}).$$

The new measures p^*_α are then used in the relative entropy calculation in place of the p_α.

One way to look at this is that the number of occurrences of letter α_i is now taken to include an additional amount $w_{i,j}/t^2$ for each pair $\{\alpha_i, \alpha_j\}$ in the column out of $t(t-1)$ possible distinct pairs. In other words, if there are few literal instances of residue α_i, but many residues similar to it, then the computation would be as if α_i were more prevalent in the column, raising that component of the score.

Note that whereas before, with fractional residues only to fill the blanks in a column, $\sum_{\alpha \in \Sigma} p_\alpha$ must add to exactly 1, the new sum $\sum_{\alpha \in \Sigma} p^*_\alpha$ may be greater than 1, although less than 2. This is a peculiarity of the method. However, the property that the p_α sum to 1 is not critical to the function's use in measuring sequence similarity.

4 Implementation

4.1 Scoring Function

The scoring function is implemented as described above. For the weights $w_{i,j}$, we use the intermediate values from the computation of the BLOSUM62 matrix [10] to give an empirical value for the proportion of pairs involving α_i that use

α_j. This is a good estimate of $w_{i,j}$, the frequency with which α_j substitutes for α_i. However, any scoring system on amino acids could be substituted modularly in this function. A PAM matrix could be used, if preferred, or more interestingly a system such as Wu and Brutlag's ([21]), where here one would weight a pair of amino acids based on whether (and perhaps how strongly) they participate in a substitution group together.

The gap penalty used in all experiments is $1.5 \times$ (gap length). This is applied after normalizing the relative entropy variant to an approximate z-score (mean and standard deviation determined via random sampling). As is so often the case, this gap penalty is determined through trial and error. However, since all scores are normalized similarly, this gap penalty at least behaves consistently with all scoring functions tested and appears to remain appropriate at various pattern sizes and lengths.

4.2 Data

Two different protein motifs are used for these experiments, one for pattern discovery (finding the motif without advance knowledge) and one for pattern matching (finding instances of a given motif).

The input to each program consists of the isolated motifs, with added distractor sequence of unrelated protein, selected using the SWISS-PROT random entry option[1]. Using protein sequence relieves the necessity of providing a background model for random generation, and gives more realistic results. The reason to use unrelated sequence rather than the sequence surrounding the motif instances is that these surrounding protein sequences are likely to be related to each other, and so the algorithm will discover true patterns at many points in the sequence, making it hard to distinguish false positives from true positives.

DHFR. The tests of pattern discovery are run on 11 instances of the dihydrofolate reductase signature (Accession PS00075 on Prosite [7]), the set which was available from PDBsum ([5]) as occurring in PDB entries. This test motif was selected by eye from a small set of candidates as being a reasonable length, varying mildly in length (from 23 to 24 AA) but not so greatly that an ungapped method could not conceivably find them, and well-conserved but not so tightly as to make all scoring methods highly successful, which would make a distinction between them difficult.

The distractor sequence used for DHFR is 10,360 random AA, which is 40 times the aggregate length of the 11 planted DHFR motifs.

Zinc Finger. The tests of pattern matching are run on 100 instances of the C2H2-type zinc finger motif (Accession PS00028), plus an additional 12 instances used to specify the pattern, all downloaded from SwissPROT ([8]). These instances were chosen at random from the 3599 instances available at the time.

---

[1] http://us.expasy.org/sprot/get-random-entry.html

The zinc finger motif is particularly felicitous for this test because instances are so plentiful in the database that test cases illustrating performance are easy to construct, and because while the average motif is sufficiently strong for most relative entropy variants to find it (although never the sum-of-pairs score), there are a large number of anomalous instances. A good test is to see how often a scoring function can detect these without introducing too many false positives. This variety of anomalies includes gaps, insertions, and residue changes from the regular expression pattern describing the pattern. Several hundred of the C2H2 motifs cited do not follow the nominal regular expression (C.{2,4}C.{3}[LIVMFYWC].{8}H.{3,5}H) for the C2H2 zinc finger.

The twelve pattern-specifying zinc finger instances are pre-aligned with some human intervention to insure that the critical cysteine and histidine residues are aligned correctly. For the runs involving the ungapped algorithm, a different pattern specification must be used, in which the instances are aligned ungapped. For this alignment, the inner cysteine and histidine of each instance are aligned but the outer ones vary in position.

The distractor sequence for C2H2 is 23,620 AA, which is 10 times the aggregate length of the 100 planted zinc finger motifs.

4.3 Pattern Discovery

To test the effectiveness of the combined scoring function in detecting novel patterns, we use a hill-climbing search heuristic similar to a less randomized version of the Lawrence et al. [13] Gibbs sampling approach. This algorithm differs from typical Gibbs sampling for motifs in two aspects. The first is that it performs a dynamic programming match between the fixed pattern and the database, as described in [18], in order to permit gapped alignment of instances. The other important difference is that, while the instance it discards is chosen randomly, it selects the best of its choices instead of sampling from a probability distribution determined by the score. While not gaining the theoretical benefits that come with sampling (see, e.g., [2]), this algorithm's success or failure is easier to measure, and in practice, given repeated random restarts, it samples effectively from the high end of the set of scores with rapid convergence.

As a measure of each score's usefulness at finding novel motifs, we use the frequency with which a random restart results in discovering the planted motif.

The size parameters on the runs of the motif instructed it to find 10 motifs of size 20. This does not mesh exactly with the size (14) or number of instances (11) of the motif, in order to simulate a real-life usage, where one may not know precisely how many instances or what length motif to search for, but instead might choose round numbers at a reasonable order of magnitude for the problem.

4.4 Pattern Matching

Comparing the effectiveness of this scoring function to others for pattern matching is easier, since it requires no additional algorithm. Given a pattern, the scoring functions alone induce a score for each candidate sequence. We compare

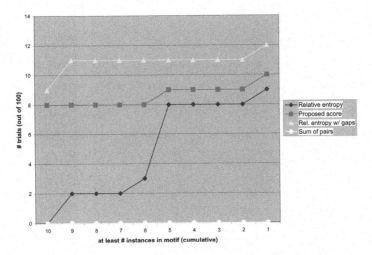

Fig. 1. Out of 100 hill-climbing runs, how many runs ended in instances of the planted pattern. Along the x-axis, the number of instances for "success" decreases; for the leftmost data, all 10 proposed instances must be motif instances.

the number of true positives (identified zinc fingers) and false positives (high-scoring regions that are not the planted zinc finger motifs) at various recognition score thresholds.

In this case, the pattern being searched for is precisely the length of the (aligned) predetermined zinc fingers.

5 Results

5.1 Pattern Discovery

The results for pattern discovery are shown in Figure 1. The difference between gapped relative entropy and the proposed score (gapped relative entropy with pairwise comparison weights) is minor: 8 versus 9 runs out of 100 found all 10 instances.

Note on the other hand that the ungapped relative entropy score never found as many as 10 of the 11 motif instances, and only in 3% of the runs found more than half of the instances. In real data, this could translate to a 3-4x increase in running time, or more importantly, to subtler patterns that the gapped scores barely find being permanently hidden to this scoring function.

The sum-of-pairs score is clearly poorly suited to this kind of heuristic sampling approach. This is likely because its score increases slowly as a pattern begins to appear, in contrast with a relative entropy score which rapidly recognizes a developing pattern and moves to add more instances.

5.2 Pattern Matching

Figure 2 shows the tradeoff between false positives and true positives at different score thresholds. In this case, it is the *ungapped* relative entropy score which

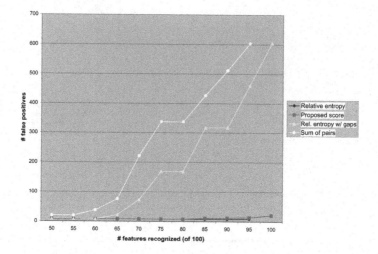

Fig. 2. At various score thresholds, how many of 100 zinc fingers were recognized (x-axis) versus how many other locations were falsely identified as zinc fingers (y-axis).

performs comparably to the proposed scoring function. The ungapped score has slightly fewer false positives than the proposed score. However, there is 1 out of the 100 zinc fingers which the ungapped score was unable to recognize, even after raising the threshold until 600 or more false positives were introduced, because of gaps in the inner portion.

The reason for the surprising performance of ungapped relative entropy is that, although the C2H2 pattern is variably sized, the center portion (a cysteine, 12 residues, and a histidine) is tightly conserved and usually, although not always, ungapped. Since a completely preserved column speaks loudly to a relative entropy score, the ungapped score in this case is relatively quick to recognize most instances.

Again, in this case, the sum of pairs score performs poorly, but the interesting thing here is the dramatic reversal of position of the gapped and ungapped relative entropy scores from the previous graph.

5.3 Repeatability

Both results shown appear to be robust, the relative orderings and magnitudes holding up to repeated experiments and minor changes to the data (such as regenerating the distractor sequence).

6 Discussion

The results demonstrate that adding amino acid similarities to a relative entropy score is a powerful tool for recognizing protein motifs and motif instances. Simpler scores that use amino acid similarities, such as a sum-of-pairs scores, are not viable; the power of the relative entropy score for multiple sequence alignment and comparison is important to preserve.

At first, it might appear unpromising that the proposed score was edged out in each experiment by a previous scoring function, in one case the common relative entropy score and in the other the score simply modified for gaps.

However, on the contrary, these results show compellingly that the extra power bestowed by recognizing beneficial amino acid combinations makes the score more robust to variation in the pattern being looked for and in the surrounding sequence. With the novel score, one need not decide in advance whether one wants a motif which is, like the zinc finger, still clear to an ungapped scoring function, or whether the motif in question is, like DHFR, multimodal enough in size that only a gapped scoring function can easily recognize that the differently-sized subsets are part of the same motif.

In addition, the new score's consistent high performance will become important in domains where neither previous score performs well. That such domains exist seems likely from the amount of separation between good and bad scores in the results. Although so far in our experiments no protein data has been seen to confound both the gapped and ungapped scores, the space of possible data types is large and notoriously difficult to sample.

7   Future Work

7.1   Sequence Alignment

Because of the arbitrary nature of the choice of affine gaps, and the error factor the penalty introduces over long sequences, it makes little sense to extend this scoring function directly to consider alignments of long, ill-conserved sequences. However, Dialign's segment-to-segment approach is an answer to this problem.

A reasonable way to align long, ill-conserved sequences, then, might be to discover short matches using the hill-climbing sampler and string them together, just as in [15,14], according to score and consistency of order.

An important future project is to implement this algorithm, in order to compare it to existing scoring functions and alignment software. If empirically successful, it would fill a compelling niche: the alignment of sequences sufficiently diverged that few short regions remain ungapped, making it impossible for a segment-to-segment algorithm using ungapped segments to find the correct alignment.

Acknowledgements

Thanks to Martin Tompa for much assistance, proofreading, and support; and to the anonymous referees for their helpful comments and advice.

This material is based on work funded in part by the National Science Foundation grant #DBI-9974498, and by the NIH under training grant #HG00035.

References

1. M. Blanchette, B. Schwikowski, and M. Tompa. Algorithms for phylogenetic footprinting. *J. Comp. Bio.*, 9(2):211–223, 2002.

2. K. S. Chan. Asymptotic behavior of the gibbs sampler. *J. Amer. Statist. Assoc.*, 88:320–326, 1993.

3. M. 0. Dayhoff, R. M. Schwartz, and B. C. Orcutt. A model of evolutionary change in proteins. In M. O. Dayhoff, editor, *Atlas of Protein Sequence and Structure*, volume 5, suppl. 3, pages 345–352. Natl. Biomed. Res. Found., Washington, 1978.

4. A. Dembo and S. Karlin. Strong limit theorems of empirical functionals for large exceedances of partial sums of iid variables. *Annals of Probability*, 19(4):1737–1755, 1991.

5. R. Laskowski et. al. Pdbsum. http://www.biochem.ucl.ac.uk/bsm/pdbsum/, 2002.

6. Schwartz et al. Pipmaker—a web server for aligning two genomic dna sequences. *Genome Research*, 10:577–586, April 2000.

7. ExPASy. Prosite. http://www.expasy.ch/prosite/, 2002. hosted by the Swiss Insitute of Bioinformatics.

8. ExPASy. Swiss-prot. http://www.expasy.ch/sprot/, 2002. hosted by the Swiss Insitute of Bioinformatics.

9. J. G. Henikoff and S. Henikoff. Using substitution probabilities to improve position-specific scoring matrices. *Comput. Appl. Biosci.*, 12(2):135–43, 1996.

10. S. Henikoff and J. G. Henikoff. Amino acid substitution matrices from protein blocks. *Proc. Natl. Acad. Sci. USA*, 89:10915–10919, 1992.

11. S. Karlin and S. F. Altschul. Methods for assessing the statistical significance of molecular sequence features by using general scoring schemes. *Proc. Natl Acad. Sci. USA*, 87:2264–2268, 1990.

12. S. Karlin and S. F. Altschul. Applications and statistics for multiple high-scoring segments in molecular sequences. *Proc. Natl Acad. Sci. USA*, 90:5873–5877, 1993.

13. C. E. Lawrence, S. F. Altschul, M. S. Boguski, J. S. Liu, A. F. Neuwald, and J. C. Wootton. Detecting subtle sequence signals: A gibbs sampling strategy for multiple alignment. *Science*, 262:208–214, 1993.

14. B. Morgenstern. Dialign 2: improvement of the segment-to-segment approach to multiple sequence alignment. *Bioinformatics*, 15:211–218, 1999.

15. B. Morgenstern, A. Dress, and T. Werner. Multiple dna and protein sequence alignment based on segment-to-segment comparison. *Proc. Natl. Acad. Sci. USA*, 93:12098–12103, 1996.

16. S. B. Needleman and C. D. Wunsch. A general method applicable to the search for similarities in the amino acid sequence of two proteins. *J. Mol. Biol.*, 48:443–453, 1970.

17. S. Pietrokovski, J. G. Henikoff, and S. Henikoff. The blocks database—a system for protein classification. *Nucl. Acids Res.*, 24(1):197–200, 1996.

18. E. Rocke and M. Tompa. An algorithm for finding novel gapped motifs in dna sequences. In *Proc. of the 2nd Annual International Conference on Computational Molecular Biology (RECOMB 1998)*, pages 228–233, March 1998.

19. T. F. Smith and M. S. Waterman. Identification of common molecular subsequences. *J. Mol. Biol.*, 147:195–197, 1981.

20. J. D. Thompson, D. G. Higgins, and T. J. Gibson. Clustal w: improving the sensitivity of progressive multiple sequence alignment through sequence weighting, position-specific gap penalties and weight matrix choice. *Nucl. Acids Res.*, 22:4673–4680, 1994.

21. T. D. Wu and D. L. Brutlag. Discovering empirically conserved amino acid substitution groups in databases of protein families. In *Proc. of the 4th International Conference on Intelligent Systems for Molecular Biology (ISMB 1996)*, pages 230–240, 1996.

Functional Consequences in Metabolic Pathways from Phylogenetic Profiles

Yonatan Bilu[1] and Michal Linial[2]

[1] School of Computer Sciences and Engineering
johnblue@cs.huji.ac.il
[2] Department of Biological Chemistry, Life Science Institute
The Hebrew University, Jerusalem 91904, Israel
Tel. (972) 2-658-5528, Fax. (972) 2-658-6448
michall@mail.ls.huji.ac.il

Abstract. Phylogenetic profiling is an ingenious computational method for identifying functional linkage between proteins. It is based on the assumption that a set of proteins involved in a common pathway or composing a structural complex tends to evolve in a correlated fashion. We present a rigorous statistical analysis of the success of this method in identifying proteins involved in the same metabolic pathway. We suggest several modifications for the use of phylogenetic profiles that improve the success measure significantly. The potential and the limitations of this method for protein functional predictions are discussed.

1   Introduction

Phylogenetic profiles were suggested in the seminal work of Pellegrini et al. [1] as a method for inferring functional linkage between proteins. The method is based on the assumption that a set of proteins involved in a common pathway or composing a structural complex tends to evolve in a correlated fashion. During evolution, such a set will tend to either be preserved, or to disappear, more or less in its entirety. Thus, the proteins in such a set are expected to be present in the same group of organisms.

The phylogenetic profile of a protein is a representation of this information - it is a list, or an indicator vector, of the organisms in which the protein is present. The prediction of this method is that proteins with similar profiles will tend to be functionally linked.

In [1], Pellegrini et al. use this method to identify functional linkage among *Escherichia coli* proteins. For each such protein, homologs are searched for within 16 fully sequenced genomes, mostly from bacteria. The phylogenetic profile of a protein is then defined as the indicator vector for the organisms in which a homolog was found. Two proteins are called "neighbors" if their Hamming distance is at most one (i.e., there is at most one organism in which one of them appears, and the other doesn't).

To analyze the accuracy of this method, Pellegrini et al. conduct three types of tests: First, three known proteins are examined in detail, and it is shown that

R. Guigó and D. Gusfield (Eds.): WABI 2002, LNCS 2452, pp. 263–276, 2002.

about half their neighbors have a known, related function. Second, several groups of proteins that are known to be related (sharing a common keyword in Swissprot [4], or appearing in the same class in EcoCyc [5]) are examined. It is shown that they have many more neighbors within the group than is expected at random. Finally, the Swissprot keyword annotations of each protein are compared with those of its neighbors. On the average, 18% of the neighboring keyword overlap those of the query protein, while only a 4% overlap is expected by choosing them at random.

Phylogenetic profiling was also used by Marcotte et al. in [2] as one of several methods to infer functional linkage between yeast (*S. cerevisiae*) proteins, and in [3] to identify cellular localization of proteins.

In this work we suggest a rigorous statistical analysis of the accuracy of this prediction method, and apply it to examine how well it identifies functional linkage among enzymes, induced by metabolic pathways. We show that when applied naively, predictions based on this method are no better than ones done at random. We suggest three modifications that greatly improve the predictive power.

The first is not to limit the profile to organisms with fully sequenced genomes. To the 40 organisms whose genomes were completely sequenced at the time this work started, we add 45 more - those with an abundance of proteins in Swissprot or TrEMBL [4]. Expanding the profile this way increases the proportion of correct prediction dramatically.

Next we suggest two improvements to the way a similarity between two profiles is measured. Rather than using Hamming distance in a straightforward manner, we modify it in two ways. The first is based on the number of organisms that are represented in the profile, defining a greater similarity score to larger profiles. The second is based on the phylogenetic distance between the organisms, considering a difference in profile stemming from similar organisms to be less severe than that originating from organisms that are far apart on the phylogenetic tree. Each of these modifications yields a further increase in the proportion of correct predictions, as many as 80% correct predictions when only a few predictions (15) are made.

We then investigate the reason for this surprisingly high number of correct predictions, and connect it to the distribution of correct predictions among the metabolic pathways that the give rise to them.

Finally, we test whether this method can be used to reconstruct entire pathways, or large parts of them. Even for known pathways, this could provide, for example, a useful supplemental computational tool in comparative analysis of the so-called "metabolic modules" in WIT [6]. It can also prove helpful in identifying the presence of pathway in a specific organism, which is often an interesting and non-trivial task (e.g. [7]).

A predicted connection between two proteins defines an edge over the graph of enzymes. We compare the connected components of this graph with over 100 maps of metabolic pathways, and find the correlation between them. Such a comparison resulted in an almost complete recovery of the "Aminoacyl-tRNA biosynthesis" map. For a few other maps, a small portion, yet bigger than expected at random, can also be identified.

We conclude that phylogenetic profiling, when applied properly, can be a powerful method for identifying functionally linked proteins, though additional data is needed to identify entire pathways or structural complexes.

2   Methods

2.1   Metabolic Pathway Maps

Our analysis focuses on the ability of the phylogenetic profiling method to identify proteins that take part in a common metabolic pathway. Information about such pathways was extracted from the KEGG [8] database. This database contains (among other things) information about 110 relevant metabolic and regulatory maps, which usually correspond to pathways. For each such pathway the enzymes involved in it (identified by their EC number) are listed, and links to related pathways are given. We define two enzymes to be functionally linked if there is some map that contains both of them. Our aim is therefore to identify these links.

2.2   Phylogenetic Profiles

Consider a set of proteins, P. Define a "profile organism" as a subset of P, and consider a set of profile organisms, O. The phylogenetic profile of a protein p is a 0-1 vector $(p_o)_{o \in O}$, with $p_o = 1 \iff p \in o$.

To construct the phylogenetic profile of a protein one first needs to choose the set of profiling organisms, and then to decide for each protein whether or not it appears in each of them.

The profiling organisms used in this work come from three sources. The first is those organisms whose genomes were completely sequenced at the time this work started. The other two are the fifty organisms with the largest number of proteins in Swissprot (release 39.15), and likewise in TrEMBL (release 16.3) [4]. These three lists were merged together, yielding a total of 85 profiling organisms.

Next, since our aim is to identify proteins involved in the same metabolic pathway, we consider only proteins that function as enzymes. We define an enzyme as appearing in a profile organism, if a protein from that organism is listed in either Swissprot or TrEMBL with the appropriate EC annotation.

Since Swissprot and TrEMBL are relied upon for defining the profile, and the reference for validation is the KEGG maps, the enzymes of interest are those 1015 enzymes that appear in both places. These enzymes appear in 110 KEGG maps. Of these, 7 include only one of these enzymes, thus, in effect, there are only 103 maps.

2.3   Statistical Analysis

To evaluate the predictive power of phylogenetic profiling in this context, two tests were conducted. In both, a defined number, n, of (ordered) enzyme pairs is picked, based on the similarity of the corresponding phylogenetic profiles. The underlying prediction is that both pair members are functionally linked - in our case, involved in the same metabolic pathway.

In the first test, the n pairs with the highest similarity score are picked. In the second test, the idea is to find, for each enzyme, other enzymes that are related to it. If n is a multiple of 1015 (the number of enzymes) then we simply pick for each enzyme the $n/1015$ enzymes with the most similar profile. Otherwise, we pick the most similar pairs, such that for each enzyme between $\lfloor \frac{n}{1015} \rfloor$ and $\lceil \frac{n}{1015} \rceil$ related enzymes are chosen.

The success ratio of either test is simply the number of correct predictions, that is, the number of pairs where both members appear in a common KEGG map, divided by the total number of predictions (n).

We call the first comparison the "overall test", and the second one, the "individual test". Note that since the individual test is more restrictive, one would expect the success ratio of this test to be lower. Note also, that usually the success ratio to decline as the number of predictions grows.

The KEGG maps define 44,756 functionally linked enzyme pairs. Dividing this by the total number of enzyme pairs in our data set, we get (by linearity of expectation) that the expected success ratio of either test is about 4.3% (if the pairs are chosen at random).

2.4   Similarity Measure

A crucial point for either of the two methods is to define a similarity measure for phylogenetic profiles. Indeed, the focus of this work is to define an appropriate similarity measure. The obvious way to do this is to base the similarity measure on the norm of the distance between the profile vectors. That is, if p and q are two profile vectors, their similarity measure can be defined as $-\|p\text{-}q\|$. Call this similarity measure the "norm similarity". Note that since p and q are 0-1 vectors, and we are only interested in the order of similarities, any L_p norm will lead to the same predictions.

We consider two modifications of this similarity measure. The first is motivated by the notion that profiles with a larger support are more "trustworthy". This seems especially true when organisms, whose genomes are incompletely sequenced, are used for profiling (see the Discussion section for more details), and the incomplete annotation of enzymes in Swissprot is relied upon. We would like to penalize the similarity of proteins in a way that differentiates, and gives a severe penalty to, profiles with a small support, while being relatively transparent when the profile's support is large. Thus we use an exponential function. Define the "support similarity" between p and q to be:

$$-|p - q| - k(e^{-c|p|} + e^{-c|q|}),$$

where $|x|$ is the L_1 norm of x (the size of its support), and k and c are normalization factors. The largest possible L_1 distance between two enzymes is the number of profile organisms - 85 in our case. We wanted the penalty factor to be no more than half that. Thus, we wanted a c such that for a profile of size 1 the exponent part of the penalty factor is nearly 1 (higher than 0.9), and k that is no more than 21.25. This is rounded up to obtain $c=0.1$ and $k=20$.

The second modification to the "norm similarity" is based on a coarser version of phylogenetic profile. That is, instead of defining the profile over individual organisms, one may define them over families of organisms. More generally put, instead of basing the profile on the leaves of the taxonomy tree, one can use nodes that are higher up in the tree.

Each protein in Swissprot and TrEMBL is associated (through the OC field) with a branch of the taxonomy tree - the one leading to the organism in which the protein is found. Define the "level l profile" of a protein as its profile vectors over the nodes in the taxonomy tree that are l levels above the leaves.

For example, consider 4 profile organisms: Rat, Mouse, Human and Drosophila. The branches in the taxonomy tree that correspond to these organisms (respectively) are:

1. *Eukaryota; Metazoa; Chordata; Craniata; Vertebrata; Euteleostomi; Mammalia; Eutheria; Rodentia; Sciurognathi; Muridae; Murinae; Rattus.*

2. *Eukaryota; Metazoa; Chordata; Craniata; Vertebrata; Euteleostomi; Mammalia; Eutheria; Rodentia; Sciurognathi; Muridae; Murinae; Mus.*

3. *Eukaryota; Metazoa; Chordata; Craniata; Vertebrata; Euteleostomi; Mammalia; Eutheria; Primates; Catarrhini; Hominidae; Homo.*

4. *Eukaryota; Metazoa; Arthropoda; Tracheata; Hexapoda; Insecta; Pterygota; Neoptera; Endopterygota; Diptera; Brachycera; Muscomorpha; Ephydroidea; Drosophilidae; Drosophila.*

Consider a protein appearing in Mouse and Human. Its regular profile is [0 1 1 0]. In constructing the level 1 profile, the Rat and Mouse coordinates are merged (both are *Murinae*), thus the level 1 profile is [1 1 0]. Note that with the definition above, the level 5 profile is also [1 1 0], since going up 5 levels in the first 2 organisms brings us to *Eutheria*, while in the third to *Mammalia*. To deal with this problem, the definition can be altered as follows - two coordinates are merged if going up l levels in the corresponding branches of both organisms leads to a common ancestor.

In what follows we consider the sum of the "norm similarity" scores for the regular profiles and for the "level 3" profiles, and refer to it as the "taxonomy similarity" score.

2.5   Reconstructing Entire Maps

A more presumptuous goal than simply identifying two functionally linked proteins, is to identify entire sets of proteins that are all related. In our case, this means identifying an entire metabolic pathway.

A natural way to go about this is to cluster together proteins with similar profiles, and predict that each cluster corresponds to a pathway. A simple clustering algorithm, so called single linkage clustering, is implied by the predictions described above. Consider a graph where the vertices correspond to proteins.

Choosing a pair of proteins defines an edge in this graph, connecting the vertices corresponding to the two pair members.

Thus, either of the two methods discussed in the previous subsections defines a graph with n (directed) edges. A (weakly) connected component in this graph defines a cluster of proteins. To evaluate the correlation between these clusters and the original maps, each cluster is compared with all the maps, and is considered to correspond with the map with which it has the largest overlap.

2.6   Random Simulations

To evaluate the significance of the success ratio of both tests we use a random simulation. 1000 random symmetric matrices are generated. For each value of n (the number of predictions made), the top n (off diagonal) entries are chosen. The corresponding row and column define for each such entry a random pair. The highest success ratio achieved by any of the 1000 random matrices is then recorded. This gives us an estimate of where the graph of p-value <0.001 crosses. In other words, similarity measures achieving a higher graph are significantly (with p-value <0.001) better than random.

Random simulation is also used to estimate the expected overlap between a cluster defined by the single-linkage clustering, and a metabolic map. The same number of proteins as in the cluster is chosen at random, and the best fit with any of the maps is found. This is averaged over 1000 simulations as an estimate of the expected overlap size.

3   Results

Following the procedure used in [1], one would consider profiles over organisms whose genomes are fully sequenced, and use the "norm similarity". Doing so yields rather disappointing results with the "overall test" (see subsection 2.3 for a definition of the statistical tests). As can be seen in Figure 1, the success ratio is even lower than would be expected at random (4.3%). Interestingly, the more restrictive and, arguably, less natural, "individual test" is more successful (Figure 1), achieving a success ratio that is significantly higher than random.

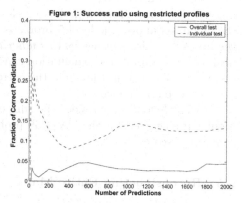

Fig. 1.   Success ratio of the two statistical tests, when applied to profiles over full-sequenced organisms.

Expanding the profile over the entire set of profile organisms improves the results dramatically (Figure 2). For the "overall test", expanding the profile means the difference between a seemingly random result, and one that is significantly

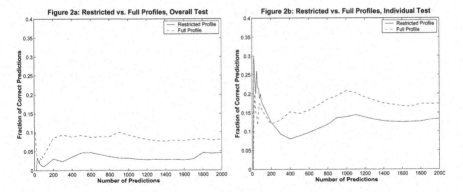

Fig. 2. Comparison of the success ratio of the two statistical tests, when applied to profiles restricted to organisms with fully sequenced genomes, and to profiles over the entire "profile organisms" set.

better. For the "individual test", for $n < 200$, the success ratio actually worsens, but beyond that, expanding the profiles improves the success ratio by a factor of about 1.5.

To appreciate the significance of these results, see Figure 3. The figure depicts the best results obtained, for each value of n, by any of 1000 random similarity scores. Thus the success ratio described by this graph is that of p-value significance of 0.001 above random. Next, we examine the effect of choosing the alternative similarity scores discussed above. Figure 4a shows a drastic improvement in the success ratio of the "overall test" when the "support similarity" score is employed, and a more moderate one in the success ratio of the "individual test". Also, the graphs

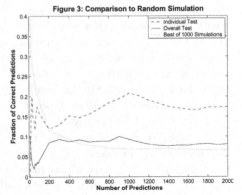

Fig. 3. Success ratio of both tests, compared with the best result achieved in any of 1000 random simulations.

display the behavior we expect - descending as n grows, and better results for the "overall test".

To check the robustness of the values chosen for the parameters k and c, other values were tested. For k in the range of 10-50 the results are about the same. For values outside this range (5 and 55 were checked) they somewhat worsen for $n > 1500$. For $c=0.2$ the success ratio is the same for the "individual test" and varies by about 5% for the "overall test", being lower for $n < 300$, and higher for larger values (data not shown).

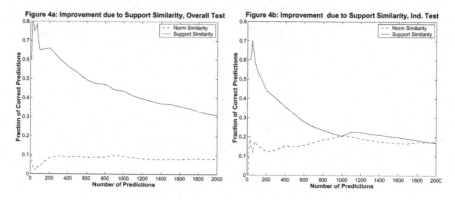

Fig. 4. Comparison of the "Support Similarity" and "Norm Similarity", with respect to the success ratio of both tests. Profiles are defined over the entire set of organisms.

Employing the "taxonomy similarity" score also leads to some improvement, though a small one. Indeed, for the "overall test" the results are about the same as those of the "norm similarity" (not shown). For the "individual test", as is seen in Figure 5, there is no improvement for small values of n, but a distinct one, albeit small, for large values. Similar results are obtained for other values of profiling level ($l=1,2,3,5$ were tested).

A preliminary question to the aforementioned problem of reconstructing entire maps, is which maps underline the successful predictions. In other words, do the identified functionally linked pairs all belong to a small set of metabolic pathways, or are they widespread among them? To some extent, the "individual test" is designed to be biased towards the latter, since in this test predictions are made for (at least) n different proteins.

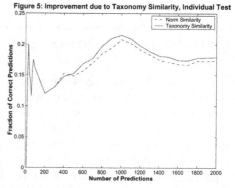

Fig. 5. Comparison of the "Taxonomy Similarity" and "Norm Similarity".

We examine the predictions based on the "support similarity" score, which were shown to perform best. Figure 6 shows the distribution of successful predictions among the maps in which the predicted pair appears, for $n=500$. As is seen in Figure 6a, most of the successful predictions in the "overall test" are of pairs of enzymes involved in the Aminoacyl-tRNA biosynthesis (henceforth "AA-tRNA") map. This map is also a major contributor to the success ratio of the "individual test", as seen in Figure 6b. Figure 7 shows the proportion of successful predictions based on pairs of enzymes involved in the AA-tRNA map for the various values of n. We defer the discussion of this map to the next section.

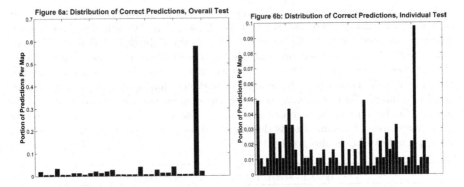

Fig. 6. Distribution of the number of correct predictions, among the maps to which these predictions correspond. The tallest bar in both diagrams indicates the Aminoacyl-tRNA biosynthesis map.

An inherent problem to the phylogenetic profiling method is that there is no way to tell apart two pathways that have evolved in the same set of organisms, since their proteins will have the same profile. Assuming that some of the "False Positive" results are because of this, we broadened the definition of when two proteins are truly linked, to include two proteins in related pathways. This indeed increases the accuracy of the method, but also the chance that such a linkage is found at random. As before, results of both tests are significantly above random, but it is inconclusive

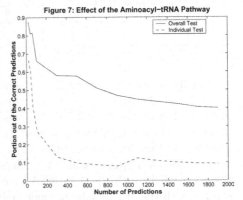

Fig. 7. The proportion of correct predictions of linkage due to the Aminoacyl-tRNA pathway, out of all correct predictions.

whether they show a relative improvement over the restricted definition of functional linkage, and require further study (data not shown).

A different way to handle "False Positives" is to try and weed them out. In higher organisms, we might assume that proteins involved in the same pathway will have to be present in the same tissue. Thus, this data was extracted from Swissprot for each protein. However, for most enzymes (940) at least one Swissprot entry did not contain the name of the tissue. Conversely, among the 55 enzymes that were localized this way, we identified a few that originated from different tissues, but were involved in the same pathway. Thus, we conclude that for this setting, this kind of data is not abundant and accurate enough to be useful. However, when looking for linkage between proteins from a specific organism (as in [1]) such considerations may be useful. Similarly, one can look at the organelles in which a protein is present, but as with tissue data, the data examined was not sufficient.

Table 1. The correlation between KEGG maps and the clusters generated for $n=300$, containing at least 15 enzymes (see text for details). The left side describes clusters based on the predictions of the "overall" test, and the right those based on the "individual test". The first column on each side lists the number of enzymes in the cluster, the next two list the name of the map with which it has the highest correlation and its size. The fourth column lists the number of enzymes that appear both in the cluster and in the map. The last column gives the expected size of the best overlap, had the proteins in the clusters been chosen at random. The value $n=300$ was chosen so that clusters are about the same size as the maps. Larger clusters have greater overlap, but the expected overlap is also greater.

Table 1 - Overlap of KEGG maps with clusters

"Overall" clustering					"Individual" clustering				
Cluster Size	Map Name	Map Size	Over lap	Expec- ted	Cluster Size	Map Name	Map Size	Over lap	Expec- ted
18	Aminoacyl-tRNA biosynthesis	21	16	2.71	21	Aminoacyl-tRNA biosynthesis	21	14	2.98
15	Phenylalanine metabolism	42	5	2.45	20	Peptideglycan biosynthesis	11	4	2.88
					21	Histidine metabolism	29	4	2.98

Finally, we are ready to tackle our most ambitious goal, that of reconstructing entire metabolic pathways. As might be expected from the high representation of enzymes from the AA-tRNA map, basing the clustering on the "overall test", leads to a cluster that includes most of the enzymes from this map. (See the previous section for a description of the clustering.) As seen in Table 1, for $n=300$ this clustering contains two clusters with at least 15 enzymes. One of these fits nicely with the AA-tRNA map, while the other has a less impressive overlap with the "Phenylalanine metabolism" map. The clustering based on the "individual test" leads to similar results.

4   Discussion

Given a set of proteins and a set of profile organisms, employing the phylogenetic profiling methods to infer functional linkage requires tackling two problems. The first is to decide which proteins appear in which organisms, and the second is to define a similarity measure between profiles. When analyzing the validation of the result, a third problem arises - that of deciding which pairs of proteins are truly linked. Note that the problems are not independent of each other - defining when two proteins from different organisms are "the same" might depend on the type of functional linkage sought.

In [1], two proteins are considered "the same" if the BLAST E-score of their sequence similarity is low enough, and, as part of the validation tests, proteins are considered to be "truly" linked if they share a common keyword in Swissprot. While this is plausible, it introduces another level of complication, and an additional source of possible errors. BLAST might miss two proteins which should be considered the same, or match together proteins that shouldn't (this is addressed in [3] by incorporating the E-score into the profile). Two proteins sharing a common Swissprot keyword are, in all likelihood, functionally related, but not necessarily in the sense that is relevant to phylogenetic profiles. Keywords such as Structural protein, Phosphorylation, Glycoprotein, Repeat, Alternative splicing, Disease and many more are not informative for phylogenetic profile tests. Even more specific keywords, such as Erythrocyte maturation, Diabetes or Magnesium provide little relevant information. That is, at least for some keywords, it is not clear that one can assume that proteins so tagged have indeed co-evolved.

In this sense, the setting described here is quite appropriate for testing phylogenetic profiles. Because enzymes are tagged with EC numbers, we solve part of the first problem mentioned above. Two proteins sharing the same EC annotation are indeed "the same" enzyme for our purpose. On the other hand, as is implied by the results, the EC annotation is far from complete, and thus we might erroneously conclude that some enzymes do not appear in some organisms. In other words, we expect some false negatives but no false positives. This is similar to taking a very conservative threshold when using BLAST, as discussed above.

However, since we are looking at enzymes, the metabolic pathway maps give a complete picture of which proteins are linked. This is also exactly the type of linkage that phylogenetic profiles are thought to identify - pathways and structural complexes. With these two problems out of the way, this work focuses on the problem of defining an appropriate similarity score.

The first modification suggested is not to restrict profiles to organisms with fully sequenced genomes. The rationale behind this is simple - by restricting the profiles we implicitly assume that they are exactly the same in the coordinates of the organisms that are not included. Generally, this need not be the case, and thus it is probably better not to ignore information regarding genomes that are largely sequenced.

The problem with including such organisms is, of course, that when the organism does not appear to have a certain protein, it might be because that protein was not yet identified. Our initial hope was that in the setting described here, this would not be an acute problem, as it is not likely that, among the profile organisms used, there are many enzymes that are yet undiscovered, but do participate in a metabolic pathway listed in KEGG. The results depicted in Figure 1 imply that if such enzymes are known, they might still lack the EC annotation in Swissprot.

Still, including organism whose genomes are not fully sequenced introduces an asymmetry between the 1's and the 0's in the phylogenetic profile. While a 1 means that the protein appears in the relevant organism, a 0 may mean either that it doesn't, or that it wasn't discovered yet (or, in this setting, that it wasn't annotated with an EC number). Thus we suggest the "support measure", where we consider two profiles more similar if the support of their profiles is bigger.

The first reason to justify this is this asymmetry. A larger number of 1's means a higher degree of certainty about the profile, making it more "trustworthy".

Yet, even when restricting the profile to fully sequenced genomes, this modification might be useful, as is illustrated by the following example:

Consider two pairs of profiles:

$[1\ 0\ 0\ 0\ 0\ 0\ 0]$, $[0\ 1\ 0\ 0\ 0\ 0\ 0]$ and $[1\ 1\ 1\ 1\ 1\ 1\ 0]$, $[0\ 1\ 1\ 1\ 1\ 1\ 1]$.

In both cases the Hamming distance between the two proteins is 2. However, the first pair seems to represent two proteins that are specific to two different organisms, and thus, probably unrelated. By contrast, the second pair, appears in 5 common organisms, and therefore is likely to be linked.

Now, to understand the motivation behind the "taxonomy similarity", consider what two organisms the first two coordinates in the above example represent. If these are, for instance, Human and *E. Coli*, it further supports the notion that the proteins corresponding to the first pair are each specific to the organism in which it is found. If, on the other hand, the first two coordinates correspond to Mouse and Rat, we might suspect that the two proteins are related, as they serve some function peculiar to rodents. The fact that we do not see both proteins in both rodents might be due to our incomplete knowledge of their genomes, or because we failed to identify the appropriate homolog.

This brings us to the final point regarding the incorporation of incomplete genomes. To some extent, when using sequence similarity to construct phylogenetic profiles, there is an inherent uncertainty regarding the 0 entries, and, to a lesser degree, the 1 entries as well. While, with a conservative E-score threshold, BLAST matches do usually indicate that the matched proteins are "the same", the converse is less often true. Namely, it is not uncommon for BLAST to miss two proteins that may be considered "the same", in the sense of their role within a structural complex or pathway (see [9]). Thus, the arguments made above may also be valid for profiles constructed from fully sequenced genomes.

The results described in the previous section indeed support the three suggested modifications. For the "overall test", which seems the more natural one, using restricted profiles leads to results that are not better than random. So in this sense, in this setting, the difference between using the restricted and full profiles seem to be the difference between success and failure. For that test, using the "support measure", instead of simply the distance between the two vectors, increases the ratio of correct prediction, for small values of n, by a factor of more than 8.

It is surprising that such a high success ratio (80% for small n) is achieved. As described in the previous section, the vast majority of these successful predictions are due to a single map - that of Aminoacyl-tRNA biosynthesis. This map is highly conserved, and is needed in practically all organisms. Thus, it is expected that it would be recognized, especially with the similarity score that "prefers" proteins that appear in many organisms. In this sense, the high success ratio of this scoring method for small values of n is likely to be biased favorably, and will probably be lower in a setting where the sought functional linkage is rarely widespread among the profiling organisms.

It is somewhat disappointing that this is the only map that was reconstructed reliably, as its enzymes do not really form a pathway, and being common to all organisms, it is probably the "easiest" one to recognize. Also, the fact that when restricting the profiles to fully-sequenced genomes this map did not yield a high success ratio, implies to the noisy nature of the profiles.

The poor reconstruction of most maps might also be due to the clustering algorithm. It suggests a hard clustering - each enzyme is put in one cluster, while the KEGG maps induce a soft clustering - an enzyme may appear in more than one pathway. Its underlying assumption is that functional linkage is a transitive relation (if A is linked to B, and B to C, then A and C are linked), which does not hold for the KEGG maps, for the same reason. Better clustering algorithms, using one of the suggested similarity measures, will hopefully lead to better results.

5   Conclusion and Future Work

This work suggests an appropriate setting for testing the phylogenetic profiling method. In this setting it is shown that it is advantageous to include profile organisms whose genomes are largely, though not fully, sequenced. It is also shown, that incorporating the "trustworthiness" of a profile, and considering the phylogenetic distance between the organisms on which two profiles differ, leads to better predictions.

We are currently working to better understand the biology behind the statistics. Namely, to analyze which pathways lead to the correct predictions, which connections are missed, especially in comparison to the restricted profiles. Similarly, it is interesting to try and categorize the organisms behind the profiles that lead to these correct predictions.

Another task is to understand the reason for seemingly incorrect predictions. Can the method be improved to avoid them, or do they contain a biological truth that is not apparent in our data? As mentioned before, our first attempt at this - adding the relations between pathways - leads to inconclusive results, requiring a careful analysis.

Even with perfect data, this method will inevitably link together proteins that appear in the same organisms, yet belong to different pathways or structural complexes. Can we use other types of data to tell them apart? Data about the tissue or organelle origin of enzymes is currently insufficient for this, but other sources of data, such as gene expression levels or protein-protein interactions, may prove better suited.

Finally, an important question is the generality of the results. Are the claims supported by the results here, valid in other settings? For example, the "support similarity" which seems to work very well here, is probably favorably biased due to pathways common to many organisms. Would it still outperform other similarity scores when trying to identify structural complexes?

A most appropriate Database is COG [10], which in effect lists the phylogenetic profiles of many proteins, and relates them to pathways. Another direction, which we completely ignored here, is that of identifying protein complexes. The same kind of predictions might be verified against the DIP [11] instead of KEGG maps. For annotation, another well curated source is the GO [12] database, which might be combined to give better a-priori identification of which proteins are "the same". These, and similar databases, can be used to further validate and develop the phylogenetic profiling method.

References

1. Pellegrini, M., Marcotte, E. M., Thompson, M. J., Eisenberg, D. and Yeates, T. O.: Assigning protein functions by comparative genome analysis: protein phylogenetic profiles. Proc. Natl. Acad. Sci. U.S.A. **96** (1999) 4285-4288.
2. Marcotte, E. M., Matteo, P., Thompson, M. J., Yeates, T. O. and Eisenberg, D.: A combined algorithm for genome-wide prediction of function. Nature **402** (1999) 83-86.
3. Marcotte, E. M., Xenarios, I., van Der Bliek, A. M. and Eisenberg, D.: Localizing proteins in the cell from their phylogenetic profiles. Proc. Natl. Acad. Sci. U.S.A. **97** (2000) 12115-12120.
4. Bairoch A. and Apweiler R.: The SWISS-PROT protein sequence database and its supplement TrEMBL in 2000.{Nucleic Acids Res. **28** (2000) 45-48.
   http://www.expasy.ch/sprot/sprot-top.html
5. Karp, P.D., Riley, M., Saier, M., Paulsen, I.T., Paley, S. and Pellegrini-Toole, A.: EcoCyc: Electronic Encyclopedia of E. coli Genes and Metabolism. Nucleic Acids Res. **28** (2000) 55-58.
   http://ecocyc.pangeasystems.com/ecocyc/ecocyc.html
6. Overbeek, R., Larsen, N., Pusch, G. D., D'Souza, M., Selkov, E. Jr, Kyrpides, N., Fonstein, M., Maltsev, N. and Selkov, E.: WIT: integrated system for high-throughput genome sequence analysis and metabolic reconstruction.Nucleic Acids Res. **28** (2000) 123-125.
   http://wit.mcs.anl.gov/WIT2/
7. Roberts, F., Roberts, C. W., Johnson, J. J., Kyle, D. E., Krell, T., Coggins, J. R., Coombs, G. H., Milhous, W. K., Tzipori, S., Ferguson, D. J., Chakrabarti, D. and McLeod R.: Evidence for the shikimate pathway in apicomplexan parasites. Nature **393** (1998) 801-805.
8. Ogata, H., Goto, S., Sato, K., Fujibuchi, W., Bono, H., and Kanehisa, M.: KEGG: Kyoto Encyclopedia of Genes and Genomes. Nucleic Acids Res. **27** (1999) 29-34
   http://www.genome.ad.jp/kegg/
9. Devos, D. and Valencia, A.: Practical Limits of Function Prediction. (2000) Proteins **41** (2000) 98-107.
10. Tatusov, R.L., Natale, D.A., Garkavtsev, I.V., Tatusova, T.A., Shankavaram, U.T., Rao, B.S., Kiryutin, B., Galperin, M.Y., Fedorova, N.D. and Koonin, E.V.: The COG database: new developments in phylogenetic classification of proteins from complete genomes. Nucleic Acids Res **29** (2001) 22-28.
    http://www.ncbi.nlm.nih.gov/COG/
11. Xenarios I, Fernandez E, Salwinski L, Duan XJ, Thompson MJ, Marcotte EM, Eisenberg D.: DIP: The Database of Interacting Proteins: 2001 update. NAR **29** (2001) 239-241.
    http://dip.doe-mbi.ucla.edu/
12. The Gene Ontology Consortium. 2000: Gene Ontology: tool for the unification of biology. Nature Genetics **25** (2000) 25-29.

Finding Founder Sequences
from a Set of Recombinants

Esko Ukkonen[*]

Department of Computer Science, P.O Box 26 (Teollisuuskatu 23)
FIN-00014 University of Helsinki, Finland
ukkonen@cs.helsinki.fi

Abstract. Inspired by the current interest in the so–called haplotype blocks we consider a related, complementary problem abstracted from the following scenario. We are given the DNA or SNP sequences of a sample of individuals from a (human) population. The population is assumed to have evolved as an isolate, founded some generations ago by a relatively small number of settlers. Then the sequences in our given sample should be a result of recombinations of the corresponding sequences of the founders, possibly corrupted by (rare) point mutations. We are interested in finding plausible reconstructions of the sequences of the founders. Formulated as a combinatorial string problem, one has to find a set of *founder sequences* such that given sequences can be composed from fragments taken from the corresponding locations of the founder sequences. The solution can be optimized for example with respect to the number of founders or the number of crossovers. We give polynomial–time algorithms for some special cases as well as a general solution by dynamic programming.

1 Introduction

Currently, DNA sequences and haplotyped SNP sequences are available in rapidly growing amounts. Therefore it is realistic to consider a situation in which we are given DNA sequences or dense SNP sequences of a sample of individuals of the same species, such as humans. From the sequence sample, one can attempt to uncover the structure of the variation of the sequences. The findings can be, for example, so–called haplotype blocks [1, 9, 2, 6] or phylogenies, e.g. [3].

If our given sequences are from individuals sampled from an isolated population we can hope to see a quite clear structure. Such a population has been evolved as an isolate that was originally founded by a relatively small number of settlers, called the founders, perhaps only a few generations ago. There can even be several such bottleneck periods in the history of the sampled population. In such a case the sequences in our given sample should be a result of iterated recombinations of the corresponding sequences of the founders. In addition, some corruption by point mutations is possible. This is, however, rare and therefore will be ignored in our present study.

[*] Supported by the Academy of Finland under grant 22584.

R. Guigó and D. Gusfield (Eds.): WABI 2002, LNCS 2452, pp. 277–286, 2002.

We want to build plausible reconstructions of the sequences of the founders from the given sequences. Seeing potential founder sequences and crossover points associated with them can be useful for example in the analysis of the possible haplotype blocks and in the study of genetic linkage in general. The given sequences are strings of equal length. The task is, according to our simple abstract formulation, to find a set of *founder sequences* such that the given sequences can be composed of fragments taken from the corresponding locations of the founder sequences.

Two optimality criteria for the solution will be studied. It is of interest to estimate the size of the founder population as well as the number of generations after the founders, as reflected by the fragmentation of the founder sequences (see e.g. [7]). Therefore we consider the problems of minimizing the size of the founder set as well as finding founder sets that explain the given sequences in smallest number of crossovers. It is immediate that these are contradictory goals for optimization.

A general solution method based on dynamic programming will be developed. In some special cases the method is shown to run in polynomial time.

In the literature, Hudson and Kaplan [4] give an algorithm for computing a lower bound for the number of recombinations but they do not try to reconstruct the sequences. Kececioglu and Gusfield [5] consider the problem of reconstructing step–by–step history of recombinations and hence the approach has more explicit phylogenetic nature than adopted here. Zhang et al [10] observed that haplotype blocking and related problems can optimally be solved by dynamic programming. Here we will use similar algorithms as building blocks of our methods.

The paper is organized as follows. The sequence reconstruction problem is formalized in Section 2. A convenient equivalent formulation as a coloring problem is also given. In Section 3 we develop a technique for eliminating crossovers in the reconstruction and analyze its behavior in some special cases. Section 4 gives a general dynamic programming algorithm which aims at finding long recombination fragments and a small number of founder sequences.

2   The Founder Sequence Reconstruction Problem

Let the set of *recombinants* $\mathcal{C} = \{C_1, C_2, \ldots, C_m\} \subseteq \Sigma^n$ be a set of sequences of length n over alphabet Σ. Each C_i is written as $C_i = c_{i1}c_{i2}\ldots c_{in}$. In a typical application the recombinants are haplotyped SNP sequences in alphabet $\Sigma = \{0, 1\}$ where the two symbols encode the two most common alleles of each SNP. In the case of DNA, we have $\Sigma = \{A, C, G, T\}$, of course.

A set $\mathcal{F} = \{F_1, \ldots, F_k\} \subseteq (\Sigma \cup \{-\})^n$ is called a *founder set* of $\mathcal{C}$ and each F_i is a *founder sequence* if $\mathcal{C}$ has a *parse* in terms of $\mathcal{F}$. This is the case if each $C_i \in \mathcal{C}$ has a decomposition into non–empty *fragments* f_{ih} such that $C_i = f_{i1}f_{i2}\cdots f_{ip_i}$ and each f_{ih} occurs in some $F \in \mathcal{F}$ exactly in the same location as in C_i. That is, one can write $F = \alpha f_{ih}\beta$ where sequence α is of length $|\alpha| = |f_{i1}\cdots f_{ih-1}|$ and $|\beta| = |f_{ih+1}\cdots f_{ip_i}|$. We always assume that a parse is *reduced* in the sense that two successive fragments f_{ih} and f_{ih+1} are always from different elements of $\mathcal{F}$. The extra symbol '−' can occur in the founder sequences but not in the recombinant sequences of $\mathcal{C}$. This symbol indicates positions in the founder sequences that are not reconstructed; hence some founders can be fragmented. Similarly, our given recombinants $\mathcal{C}$ can in practice have missing and uncertain data encoded with specific symbols but we omit these issues here. Also note that an underlying assumption of our developments is the so–called equal crossing over in which the sequence fragments exchanged in a recombination event are of equal length, and hence the fragments retain their location in the sequences.

In a (reduced) parse of $\mathcal{C}$, a recombinant C_i is said to have a *crossover point* at j if the parse of C_i is $C_i = f_1 \cdots f_h f_{h+1} \cdots f_{p_i}$ and $|f_1 \cdots f_h| = j - 1$. Hence C_i has $p_i - 1$ crossover points and $\mathcal{C}$ has $\sum p_i - m$ such points. Other interesting parameters of the parse are the *average fragment length* $\lambda_{ave} = mn / \sum_i p_i$ and the *minimum fragment length* $\lambda_{min} = \min_{ih}\{|f_{ih}|\}$.

Example 1. The recombinant set

$$
\begin{array}{l}
0\,0\,1\,0\,0\,0\,0\,1\,0\,0\,1\,1 \\
1\,1\,1\,1\,1\,1\,1\,0\,0\,1\,1\,0 \\
0\,0\,1\,0\,1\,1\,1\,0\,1\,1\,0 \\
1\,1\,1\,1\,0\,0\,1\,0\,0\,0\,1\,1
\end{array}
$$

has a founder set

$$
\begin{array}{l}
0\,0\,1\,0\,1\,1\,1\,0\,0\,0\,1\,1 \\
1\,1\,1\,1\,0\,0\,0\,1\,0\,1\,1\,0
\end{array}
$$

giving for example a parse

$$
\begin{array}{l}
\underline{0\,0\,1\,0}\,0\,0\,0\,1\,0\,\underline{0\,1\,1} \\
1\,1\,1\,1\,\underline{1\,1\,1}\,0\,0\,1\,1\,0 \\
\underline{0\,0\,1\,0\,1\,1}\,1\,0\,1\,1\,0 \\
1\,1\,1\,1\,0\,0\,\underline{1\,0\,0\,0\,1\,1}
\end{array}
$$

where the fragments from the first founder are underlined. This parse has average fragment length $\lambda_{ave} = 4.8$ and minimum fragments length $\lambda_{min} = 4$. Note that the second crossover point of the first and the second sequence which is now at location 9 could be at 10 as well. The resulting parse would still have $\lambda_{ave} = 4.8$ but $\lambda_{min} = 3$. On the other hand, if we want to get a parse with $\lambda_{min} > 4$ a founder set of at least size 4 is needed. Then even the recombinants themselves would do as founders in this case, and the analysis does not uncover any interesting structure.

Moreover,

$$
\begin{array}{l}
1\,1\,1\,1\,1\,1\,1\,1\,1\,1\,1\,1 \\
0\,0\,0\,0\,0\,0\,0\,0\,0\,0\,0\,0
\end{array}
$$

is also a founder set. The resulting parse (there is in fact only one parse in this case) has $\lambda_{ave} = 48/22 \approx 2.2$ and $\lambda_{min} = 1$. □

A founder set $\mathcal{F}$ 'explains' the recombinants $\mathcal{C}$ via a parse. Our problem is to reconstruct from $\mathcal{C}$ an $\mathcal{F}$ that is small in size but also gives a 'good' parse for $\mathcal{C}$. As parses with large λ_{ave} and λ_{min} are considered good (they capture more of the 'linkage' in the recombinants), we want simultaneously to minimize $|\mathcal{F}|$ and to maximize λ_{ave} and λ_{min}. As illustrated by Example 1, these are contradictory goals.

This trade–off of optimization goals leads to the following two versions of our reconstruction problem.

- The *minimum founder set problem* is, given a fragment length bound L, to construct a smallest possible founder set $\mathcal{F}$ such that $\mathcal{C}$ has in terms of $\mathcal{F}$ a parse for which $\lambda_{ave} \geq L$ or $\lambda_{min} \geq L$.
- The *maximum fragment length problem* is, given a bound M for the number of founders, to find largest λ_{ave} and/or λ_{min} that are possible in a parse of $\mathcal{C}$ in terms of a founder set $\mathcal{F}$ of size at most M.

Note that maximizing λ_{ave} is equivalent to minimizing the number of crossover points in the parse.

It is illuminating to formulate the founder reconstruction as a coloring problem. Let assume that each of the k founders has a unique *color* from $\{1, \ldots, k\}$. A parse gives these colors to the symbols of the recombinants: the symbols in a fragment will get the color of the founder which the fragment belongs to in the parse. This coloring of $\mathcal{C}$ has the property that whenever two symbols c_{ij} and $c_{i'j}$ taken from the same location of different recombinants are different, then their colors must also differ. This is because the parse would map two symbols with the same color on the same symbol of the founder. Colorings with this property are called *consistent*. A coloring has a crossover whenever two adjacent symbols of the same recombinant have a different color. On the other hand, if a consistent coloring of $\mathcal{C}$ is given, it immediately gives a founder set in terms of which the coloring is a parse. A founder sequence with color h is built using the symbols of the recombinants with the color h. The jth location of the founder gets the h colored symbol from the column j of $\mathcal{C}$ if the column has this color, and otherwise it gets '$-$'. The fragments implied by a coloring consist of blocks of equally colored adjacent symbols of a recombinant.

The reconstruction problem is thus equivalent to the problem of finding consistent colorings of the recombinants. The coloring should have some desired property such as a small number of colors or crossover points.

3   Segmentation Based Algorithms

3.1   Finding the Segments

We start with observing that the minimum founder set problem becomes easy if $L = 1$, i.e., if fragments of any length are allowed. Let namely μ be the maximum number of different elements of Σ that can occur in the same position of the sequences in $\mathcal{C}$. That is,

$$\mu = \max_j |\{c_{ij} \mid 1 \leq i \leq m\}|.$$

It follows that an $\mathcal{F}$ for $\mathcal{C}$ must be at least of size μ because $\mathcal{F}$ must offer μ different symbols for a certain location. However, the size μ also suffices if $L = 1$: take μ colors and consistently color with these colors the symbols on each column of $\mathcal{C}$, and then build $\mathcal{F}$ from this coloring as explained above. The corresponding parse can have one–symbol long fragments. So we have:

Theorem 1. *If $L = 1$, the minimum founder set problem has optimal solution of size $\mu \leq |\Sigma|$.* $\square$

The construction of Theorem 1 can be generalized for blocks of adjacent symbols of recombinants. A *segment* $\mathcal{C}[j, k]$ of $\mathcal{C}$ is a set consisting of the subsequences $c_{ij} \cdots c_{ik}$ of recombinants from location j to location k. The segment has *length* $k - j + 1$; its size, i.e., the number of different sequences in the segment, is denoted $|\mathcal{C}[j, k]|$. A *segmentation* of $\mathcal{C}$ consists of any collection of disjoint segments that cover the whole $\mathcal{C}$.

The current interest in the so–called haplotype blocks aims at finding from dense SNP haplotype data segments and segmentations whose internal variation (the size of the segments) is smaller than expected. Once a scoring function to evaluate each potential segment is fixed and the score is based only on the internal properties of the segment, one can find by dynamic programming a globally optimal segmentation. Recently, Zhang et al [10] gave such an algorithm to find an optimal segmentation that has a minimum total number of SNPs to uniquely distinguish at least α percent of haplotypes within each segment. This method also improved the results of [9] which were obtained using a greedy algorithm.

Our goal is to find a segmentation that would give a parse with $\lambda_{min} \geq L$ for some fixed L. To this end, we construct a segmentation such that each segment has length $\geq L$, and the maximum size of the segments is minimized. Let S_L denote the set of all segmentations of $\mathcal{C}$ such that each segment is of length $\geq L$. We want to compute

$$M(n) = \min_{s \in S_L} \max\{|\mathcal{C}[j, k]| \mid \mathcal{C}[j, k] \in s\}.$$

Then $M(n)$ can be obtained by evaluating $M(L), M(L+1), \ldots, M(n)$ from

$$\begin{cases} M(L) = |\mathcal{C}[1, L]| \\ M(j) = \min_{h \leq j-L} \max\{M(h), |\mathcal{C}[h+1, j]|\}, j = L+1, \ldots, n \end{cases} \quad (1)$$

The resulting dynamic programming algorithm needs to evaluate $|\mathcal{C}[j', j]|$ for $j' = j - L, j - L - 1, \ldots$. This can be dome by reading all sequences of $\mathcal{C}$, starting from location j and proceeding to the left and constructing a tree, level by level. The number of leaves at level that corresponds to location j' gives $|\mathcal{C}[j', j]|$. As constructing a new level takes time $O(m)$, we certainly get all necessary size values for a particular $M(j)$ in time $O(nm)$. Repeating for different j gives time bound $O(n^2 m)$.

Once we have $M(n)$, we can construct from the corresponding segmentation (this can be found by the usual trace–back after $M(n)$ has been evaluated) a founder set $\mathcal{F}$ of size $M(n)$. The segmentation gives a parse of $\mathcal{C}$ in terms of $\mathcal{F}$ consisting of fragments of length $\geq L$, hence $\lambda_{min} \geq L$.

Recursions similar to (1) can be used for the other variants of the segmentation problem. For example, to find segmentation such that $\lambda_{ave} \geq L$ and the size of the largest segment (i.e., the size of the founder set) is minimized, first observe that then the number of segments must be $\leq n/L$. Denoting by S all segmentations with at most n/L segments, we want to evaluate

$$M(n, n/L) = \min_{s \in S} \max\{|\mathcal{C}[j, k]| \mid \mathcal{C}[j, k] \in s\}.$$

This leads to recursion

$$\begin{cases} M(j, 1) = |\mathcal{C}[1, j]| \\ M(j, r) = \min_{h<j} \max\{M(h, r-1), |\mathcal{C}[h+1, j]|\} \end{cases} \tag{2}$$

where $j = 1, \ldots, n$ and $r = 2, \ldots, n/L$. Evaluation by dynamic programming is possible in time $O(n^3 m)$.

A still simpler greedy algorithm finds a segmentation with smallest number of segments such that each segment has size $\leq M$ where M is a given bound. One only needs first to find the largest k such that $|\mathcal{C}[1, k]| \leq M$, then the largest k' such that $|\mathcal{C}[k+1, k']| \leq M$, and so on until the whole $\mathcal{C}$ has been segmented. The minimality of this segmentation is easy to see. The running time becomes $O(nm)$.

This subsection will be concluded with the following example that shows that founder sets constructed from a segmentation with the above simple method are not necessarily optimal for the given constraint.

Example 2. Let $L = 2$ and consider recombinants

$$\begin{matrix}
1 & 1 & 1 & 0 & 0 & 0 & 0 & 0 & 0 & 0 \\
1 & 1 & 1 & 1 & 0 & 0 & 0 & 0 & 0 & 0 \\
1 & 1 & 1 & 1 & 1 & 0 & 0 & 0 & 0 & 0 \\
0 & 0 & 0 & 1 & 1 & 1 & 0 & 0 & 0 & 0 \\
0 & 0 & 0 & 0 & 1 & 1 & 1 & 0 & 0 & 0 \\
0 & 0 & 0 & 0 & 0 & 1 & 1 & 1 & 1 & 1 \\
0 & 0 & 0 & 0 & 0 & 0 & 1 & 1 & 1 & 1 \\
0 & 0 & 0 & 0 & 0 & 0 & 0 & 1 & 1 & 1
\end{matrix}$$

From (1) we get an optimal segmentation with segment lengths 4, 2, and 4

$$\begin{matrix}
1 & 1 & 1 & 0|0 & 0|0 & 0 & 0 & 0 \\
1 & 1 & 1 & 1|0 & 0|0 & 0 & 0 & 0 \\
1 & 1 & 1 & 1|1 & 0|0 & 0 & 0 & 0 \\
0 & 0 & 0 & 1|1 & 1|0 & 0 & 0 & 0 \\
0 & 0 & 0 & 0|1 & 1|1 & 0 & 0 & 0 \\
0 & 0 & 0 & 0|0 & 1|1 & 1 & 1 & 1 \\
0 & 0 & 0 & 0|0 & 0|1 & 1 & 1 & 1 \\
0 & 0 & 0 & 0|0 & 0|0 & 1 & 1 & 1
\end{matrix} \tag{3}$$

The middle segment has the largest size 4. Hence the corresponding founder set is of size 4, and using the segment boundaries as crossover points we get $\lambda_{ave} = 80/24 \approx 3.3$ and $\lambda_{min} = 2$.

On the other hand, sequences

$$1\ 1\ 1\ 1\ 1\ 1\ 1\ 1\ 1\ 1$$
$$0\ 0\ 0\ 0\ 0\ 0\ 0\ 0\ 0\ 0$$

are also a founder set, giving an obvious parse with 18 fragments, the shortest of them being of length 3. This founder set of size 2 thus has $\lambda_{ave} = 80/18 \approx 4.5$ and $\lambda_{min} = 3$. $\square$

3.2   Colorings with Minimum Crossovers

The parse suggested by a segmentation can be improved using the idea that we should color the fragments of adjacent segments such that the colors match as well as possible. If the color is the same across a segment boundary on some recombinant, then there is actually no crossover and the two fragments can be joined to make a longer one.

Removing the maximum possible number of potential crossover points from the segmentation (3) of Example 2 will give

$$
\begin{array}{l}
1\ 1\ 1\ 0|0\ 0\ 0\ 0\ 0\ 0 \\
1\ 1\ 1\ 1|0\ 0\ 0\ 0\ 0\ 0 \\
1\ 1\ 1\ 1\ 1\ 0|0\ 0\ 0\ 0 \\
0\ 0\ 0\ 1\ 1\ 1|0\ 0\ 0\ 0 \\
0\ 0\ 0\ 0|1\ 1\ 1\ 0\ 0\ 0 \\
0\ 0\ 0\ 0|0\ 1\ 1\ 1\ 1\ 1 \\
0\ 0\ 0\ 0\ 0\ 0|1\ 1\ 1\ 1 \\
0\ 0\ 0\ 0\ 0\ 0|0\ 1\ 1\ 1
\end{array}
\qquad (4)
$$

From this we get founder sequences

$$
\begin{array}{l}
1\ 1\ 1\ 0\ 0\ 1\ 1\ 1\ 1\ 1 \\
1\ 1\ 1\ 1\ 1\ 0\ 0\ 1\ 1\ 1 \\
0\ 0\ 0\ 1\ 1\ 1\ 1\ 0\ 0\ 0 \\
0\ 0\ 0\ 0\ 0\ 0\ 0\ 0\ 0\ 0
\end{array}
$$

The parse indicated by the fragmentation (4) has $\lambda_{ave} = 80/16 = 5$ and $\lambda_{min} = 4$.

We now describe the illustrated technique to eliminate a number of crossover points from a segmentation s. Let P_r denote the equivalence partition of the recombinants that is induced by the equality of the fragments in the rth segment. Hence i and i' belong to the same class of P_r if and only if C_i and $C_{i'}$ are equal in the rth segment. We assume that P_r has always M classes where M is the size of the largest segment (add empty classes to P_r if needed).

An M–coloring (a coloring with M colors) of $\mathcal{C}$ is called *consistent with segmentation s* if all crossovers of the coloring are on the segment boundaries of

s and *strictly consistent* with s if it additionally is consistent with the partition P_r of each segment r. We restrict here on strictly consistent M–colorings and minimize the crossovers in this class.

For the minimization, the color of each P_r should be selected properly. Assume that we have fixed the colors of P_r and must next color P_{r+1}. If we give to a class $Y \in P_{r+1}$ the color of $X \in P_r$, we write $\phi(X) = Y$. This way each consistent coloring of P_{r+1} induces a one–to–one *color propagation mapping* ϕ from P_r to P_{r+1}. The best among them can be found with bipartite matching techniques as follows.

If $\phi(X) = Y$ then recombinants $i \in X \cap Y$ do not have a crossover at the boundary between segments r and $r+1$ while all $i \in X - Y$ must have a crossover. Hence selecting $\phi(X) = Y$ eliminates $w(X, Y) = |X \cap Y|$ crossovers.

To find the ϕ that eliminates the largest number of crossovers, denoted $W(P_r, P_{r+1})$, we form a complete weighted bipartite graph with node sets P_r and P_{r+1} and with weights $w(X, Y)$ for each $X \in P_r$, $Y \in Y_{r+1}$. The coloring is propagated optimally by the one–to–one mapping from P_r to P_{r+1} that corresponds to the maximum weight matching in this graph. The matching can be found by standard 'Hungarian method' in time $O(M^3)$; [8]. The weight of this matching equals $W(P_r, P_{r+1})$. This many crossovers are eliminated and hence $U(P_r, P_{r+1}) = m - W(P_r, P_{r+1})$ crossovers are left.

Note that this minimization works for *any* partitions. Hence we summarize it in a general form:

Theorem 2. *A color propagation mapping between two partitions P and P' of the recombinants that gives smallest possible number $U(P, P')$ of crossovers can be constructed in time $O(M^3)$ where M is the size of the partitions.* □

The minimization is repeated on each segment boundary to get a global coloring. The running time (omitting the segmentation time) becomes $O(n(m + M^3))$.

If the size of each segment is M and hence each P_r has M non–empty classes then this coloring has smallest possible number of crossovers among M–color parses whose crossovers are on the predetermined segment boundaries. This is because all such M–colorings must be strictly consistent with the segmentation and the color propagation can be done optimally by the above technique.

The case $M = 2$ is of special interest. We want to construct two founders and minimize the crossovers. Consider segmentation whose every segment has length 1, i.e., every column of $\mathcal{C}$ is its own segment. The size of each such segment must be at most 2, otherwise we cannot succeed. Remove all (uninformative) segments of size 1. All 2–colorings of the remaining segments must be strictly consistent, hence the color propagation can be done optimally with respect to the segmentation. But this segmentation does not exclude any crossover points, so we have the global optimum. Finally let the uninformative segments to inherit the coloring of their closest informative neighbor. This does not increase the number of crossovers.

We have obtained:

Theorem 3. *The maximum average fragment length problem can be solved in time $O(mn)$ for founder sets of size 2.*   □

4   General Solution

The segmentation approach of the previous section leads to polynomial time algorithms but the results are suboptimal as we resorted to crossover points given by segmentations and to strictly consistent colorings. For completeness, we give here dynamic programming algorithms for the general case.

Given bound M for the size of $\mathcal{F}$, a parse and the corresponding $\mathcal{F}$ with largest possible λ_{ave} can be constructed as follows. In effect we use again the segmentation in which every column is its own segment, hence no crossover points are excluded. For any column j, the partition P_j that gives the optimum is not known. Therefore we tabulate the minimum number of crossovers in the initial segment of $\mathcal{C}$ up to column j for each possible partition P_j. Then similar values for column $j+1$ can be evaluated using the algorithm of Theorem 2.

More formally, let $\mathcal{P}_j$ denote the set of all consistent M–partitions of the rows of $\mathcal{C}$ at column j. Such partitions have M classes, some of them possibly empty, and they are subpartitions of the partition induced by the equality of the symbols on column j. For any two M–partitions P and P' of successive columns we get by the method of Theorem 2 the smallest number $U(P, P')$ of crossovers between the columns when partitions P and P' are used.

Denote by $S(j, P)$ the minimum possible number of crossovers in M–colorings of $\mathcal{C}[1, j]$ that have partition $P \in \mathcal{P}_j$ at column j. Then we have $S(1, P) = 0$ for all $P \in \mathcal{P}_1$ and

$$S(j, P) = \min_{P' \in \mathcal{P}_{j-1}} \{S(j-1, P') + U(P', P)\} \tag{5}$$

for all $P \in \mathcal{P}_j$ and $j = 2, \ldots, n$. Then $S(n) = \min_{P \in \mathcal{P}_n} S(n, P)$ is the global minimum and the corresponding parse has $\lambda_{ave} = mn/(S(n, P) + m)$.

Evaluating (5) by dynamic programming gives an algorithm with running time proportional to nN where N denotes the size of the largest $\mathcal{P}_j$. Bound N explodes when m and M grow but the method can possibly be feasible for small parameter values.

The problem of minimizing the number of founders under the requirement $\lambda_{ave} \geq L$ can be solved by reducing to (5). Make a binary search over $1, \ldots, m$ to find the smallest M such that the corresponding λ_{ave}, as evaluated using (5), is at least L. This works correctly as λ_{ave} grows monotonically with M.

5   Conclusion

We described algorithms for the inversion of recombination. The approach was combinatorial ignoring several aspects necessary to consider in practical application of these methods.

An interesting candidate for further developments is to look at recombination history in this framework.

References

1. M. Daly, J. Rioux, S. Schaffner, T. Hudson, and E. Lander. High–resolution haplotype structure in the human genome. *Nature Genetics* 29: 229–232, 2001.
2. S. B. Gabriel, S. F. Schaffner, H. Ngyen, J. M. Moore et al. The structure of haplotype blocks in the human genome. *Science* 296: 2225–2229, 2002.
3. D. Gusfield. Haplotyping as perfect phylogeny: Conceptual framework and efficient solutions. *RECOMB 2002*, 166–175.
4. R. R. Hudson and N. L. Kaplan. Statistical properties of the number of recombination events in the history of a sample of DNA sequences. *Genetics* 111: 147–164, 1985.
5. J. Kececioglu and D. Gusfield. Reconstructing a history of recombinations from a set of sequences. *Discrete Applied Mathematics* 88: 239–260, 1998.
6. H. Mannila, M. Koivisto, M. Perola, L. Peltonen, E. Ukkonen et al. A method to find and compare the strength of haplotype block boundaries. In preparation.
7. M. S. McPeek and A. Strahs. Assessment of linkage disequilibrium by the decay of haplotype sharing, with application to fixed–scale genetic mapping. *Am.J. Hum. Genet.* 65: 858–875, 1999.
8. C. H. Papadimitriou and K. Steiglitz. *Combinatorial Optimization: Algorithms and Complexity.* Dover Publications 1998.
9. N. Patil, A. J. Berno, D. A. Hinds, W. A. Barrett et al. Blocks of limited haplotype diversity revealed by high–resolution scanning of human chromosome 21. *Science* 294: 1719–1723, 2001.
10. K. Zhang, M. Deng, T. Chen, M. S. Waterman and F. Sun. A dynamic programming algorithm for haplotype block partition. *PNAS* 99: 7335–7339, 2002.

Estimating the Deviation
from a Molecular Clock

Luay Nakhleh[1], Usman Roshan[1], Lisa Vawter[2], and Tandy Warnow[1]

[1] Department of Computer Sciences, University of Texas, Austin, TX 78712
{nakhleh,usman,tandy}@cs.utexas.edu
[2] Informatics Department, Aventis Pharmaceuticals, Bridgewater, NJ 08807-0800
lisa.vawter@aventis.com

Abstract. We address the problem of estimating the degree to which
the evolutionary history of a set of molecular sequences violates a strong
molecular clock hypothesis. We quantify this deviation formally, by defining the "stretch" of a model tree with respect to the underlying ultrametric tree (indicated by time). We then define the "minimum stretch"
of a dataset for a tree and show how this can be computed optimally
in polynomial time. We also present a polynomial-time algorithm for
computing a lower bound on the stretch of a given dataset for any tree.
We then explore the performance of standard techniques in systematics for estimating the deviation of a dataset from a molecular clock. We
show that standard methods, whether based upon maximum parsimony
or maximum likelihood, can return infeasible values (i.e. values for the
stretch which cannot be realized on a tree), and often under-estimate
the true stretch. This suggests that current approximations of the degree to which data sets deviate from a molecular clock may significantly
underestimate these deviations. We conclude with some suggestions for
further research.

1 Introduction

A phylogenetic tree is a rooted tree in which leaves represent a given set of taxa
(species, DNA sequences, etc.), and internal nodes represent ancestral taxa. The
inference of phylogenetic trees plays a role in many aspects of biological research,
including drug design, the understanding of human migrations and the origins
of life.

Most phylogenetic methods produce unrooted trees or produce rooted trees
with unreliable roots. Although in some applications, the topology of an unrooted
phylogenetic tree is sufficient, for most applications, a rooted tree is desirable. For
example, in the famous *African Eve* study, it was assumed that the location of
the root of the tree was reliable and that location was used to infer that humans
evolved from African ancestors. Yet rooting a phylogenetic tree is often difficult.
The techniques thought by systematists to be most reliable use an outgroup (a
taxon or taxa thought to attach to the true tree by an edge off of the root).
However, if the outgroup is more closely related to some ingroup taxa than to

R. Guigó and D. Gusfield (Eds.): WABI 2002, LNCS 2452, pp. 287–299, 2002.

others, it is not an outgroup. Also, if the outgroup is too distantly-related to the ingroup, it may be difficult to reconstruct the location of its attachment to the remainder of the phylogeny. This is because repeated character state changes over time obscure relationship between the ingroup and outgroup.

Alternate approaches for locating the root of a phylogenetic tree assume that the data set is evolving via a (rough) molecular clock. The molecular clock hypothesis asserts that the expected number of times that a randomly chosen character will change in t time units is (roughly) proportional to t. The assumption that a strong molecular clock underlies the data can be tested through the log-likelihood ratio test [4], but there are no tests for estimating the degree of deviation from a strong molecular clock. Furthermore, the log-likelihood ratio test is computationally intensive if used appropriately with an exact method for finding the maximum likelihood tree (a computationally intractable problem).

The accuracy of the molecular clock hypothesis is of broad interest to biologists (see [19,2,8,14,13,20,5,1]). One reason for this interest is that data sets that conform closely to a molecular clock hypothesis can be used to predict times at which speciation (or gene duplication) events occurred, thus allowing comparison with the fossil record and enabling a more fine-grained analysis of the processes underlying evolution of the genes or other taxa.

Here, we present a formal definition of the deviation from the molecular clock of a data set on a tree, which we call the *stretch*. We then present two $O(n^2)$ algorithms: the first computes the optimal stretch of a given tree for a given data set; the second computes a tree with the optimal stretch for a given dataset. Furthermore, we describe common methods that biologists use for computing the deviation of a dataset from the molecular clock and provide empirical evidence which shows that the values obtained via these methods may be infeasible.

2   Background and Definitions

We define the terms from the biological literature that are used in this paper.

Definition 1. *A* **phylogenetic tree** *for a set S of taxa is a rooted tree whose leaves are labeled by the taxa in S and whose internal nodes represent the (hypothetical) ancestors of the leaf taxa.*

Phylogenetic trees represent the evolutionary history of sets of taxa (genes, species, etc.). The strong molecular clock assumption is the assumption that the taxa under consideration evolved at equal rates from a common ancestor (at the root of the tree), such that the number of evolutionary events on every root-to-leaf path in the model tree will tend to be approximately equal. More formally, a strong molecular clock assumption implies the following: If one weights the edges of a model tree by the expected number of times a randomly chosen site changes on an edge of that tree, then the lengths of all root-to-leaf paths in that model tree are the same.

Definition 2. *An edge-weighted tree (with no outgroup) T is called* **ultrametric** *if it can be rooted in such a way that the lengths of all the root-leaf paths in* *the tree are equal. (A tree with an outgroup requires additionally that the root lie on the edge between the outgroup and ingroup for that tree to be called* **ultrametric.** *An ultrametric distance matrix D is the matrix of leaf-to-leaf distances for an ultrametric tree.)*

One method for estimating the edge weights of a tree is to let $t(e)$ equal the time indicated by the edge e, so that the weight on edge e is given by $\lambda(e) = p \cdot t(e)$, where p is the expected number of events per time unit. Similarly, if $t_{i,j}$ is the time since i and j diverged from their most recent common ancestor (i.e. $2t_{i,j} = \sum_{e \in P_{ij}} t(e)$, where $P_{i,j}$ is the path in T between i and j), then $D_{ij} = 2pt_{i,j}$ is an ultrametric matrix.

Note that given a rooted ultrametric tree (and hence its matrix U), we can assign "heights" to the nodes of the tree as follows. Given a node v, let i and j be two leaves whose most recent common ancestor is v; set $height(v) = U_{i,j}$. By construction, this is well-defined, and if v and w are nodes in T such that v is on the path from w towards the root (i.e., $v > w$), then $height(v) \geq height(w)$. Hence, an alternative way of defining an ultrametric matrix is as follows:

Definition 3. *A matrix D is an ultrametric matrix if there exists a rooted tree T with function $height(v) \geq 0$ defined for each node $v \in V(T)$, (height$(v) = D_{i,j}$, where v is the most recent common ancestor of i and j) such that whenever $v > w$ (i.e., v is on the path from w to the root of T) we have height$(v) >$ height(w).*

Given a rooted tree T with heights assigned to the nodes, we can then compute the distance $D_{i,j}$ between any two leaves i and j as follows:

$$D_{i,j} = height(lca_T(i,j)),$$

where $lca_T(i,j)$ denotes the most recent common ancestor of i and j. For most stochastic models of evolution it is possible to estimate model distances in a statistically consistent manner (see [10]). This means that estimates of model distances converge to the model distances as the lengths of the sequences generated on the model tree increase.

This convergence has the following consequence for estimating deviation from a molecular clock hypothesis: suppose we are given sequences generated by a stochastic process operating on a model tree. We then apply a statistically consistent estimator for the pairwise distances, thus obtaining a matrix $\{d_{ij}\}$ of distances. If the matrix d is exactly correct (i.e., d is the model distance matrix), then times can be set at the internal nodes, thus defining an ultrametric matrix U, so that $\lambda_{i,j} = U_{i,j}$. However, when the estimates d_{ij} are not exactly correct, we will, instead, seek to minimize the following quantity:

Definition 4. *We define the* **stretch** *of an ultrametric matrix U with respect to a dissimilarity matrix d as follows:*

$$Stretch_d(U) = max\{\frac{d_{ij}}{U_{ij}}, \frac{U_{ij}}{d_{ij}}\}.$$

$Stretch_d(U)$ is thus an estimate of how much the evolutionary rate deviates from the molecular clock on the tree. For example, if the speed-up or slow-down on each edge is bounded between c and $1/c$ (for some positive constant c), then $Stretch_d(U) \leq c$. Furthermore, although our matrix d of estimated pairwise distances will not generally be exactly correct, for long sequences they will be close to the model distances, and so the value computed in this way will be a reasonable estimate of a lower bound of the degree of speed-up or slow down in the model tree.

The relationship between the stretch of the ultrametric matrix U with respect to the corrected distance matrix d and the deviation of the rates of change for the dataset from a strict molecular clock is thus straightforward. If the molecular clock hypothesis applies to the dataset, then as the sequence length increases, it will be possible to label the internal nodes of the model tree so that the value computed by this formula is close to 1. On the other hand, if we cannot label internal nodes so as to obtain an ultrametric matrix U so that $Stretch_d(U)$ is close to 1, then we might suspect that the molecular clock hypothesis does not hold on the dataset (and furthermore, the magnitude of $\min_U\{Stretch_d(U)\}$ will allow us to assess the degree to which it fails to hold).

This discussion suggests two computational problems:

- *Problem 1: Min Stretch Tree.* The input is an $n \times n$ dissimilarity matrix d, and the objective is to find an ultrametric matrix U with a minimum $Stretch_d(U)$ among all possible ultrametric matrices. We call this $Stretch(d)$:

$$Stretch(d) = \min_U\{Stretch_d(U)\}$$

- *Problem 2: Min Stretch Fixed Tree T.* The input is an $n \times n$ dissimilarity matrix d and a rooted tree T. Our goal is to find an ultrametric assignment of heights to the nodes of the tree T, thus defining an ultrametric matrix U, so that U minimizes $Stretch_d(U)$. We call this $Stretch_d(T)$.

The first problem is of interest because the minimum stretch obtained for any ultrametric tree is by necessity a lower bound on the stretch of the model tree on the matrix of estimated pairwise distances. The second problem arises when we use techniques such as maximum parsimony, maximum likelihood (see [7] for details), and neighbor joining [16] to infer trees from biomolecular sequence datasets.

In this paper, we show that both these problems can be solved exactly in polynomial time, using techniques from [3]. We solve the first problem through the use of the general algorithm given in [3], as we show in Section 3. We solve the second problem by a modification to another algorithm in [3], as we show in Section 4. Both algorithms run in $O(n^2)$ time, i.e., linear in the input size.

3   Finding the Stretch when the Topology Is not Fixed

In [3], Farach *et al.* described an $O(n^2)$ algorithm for finding optimal ultrametric trees with respect to an input distance matrix. We use this algorithm in order to solve the optimal stretch problem for the case where the tree is not given. We will describe the general problem they address, and show how our first issue is a special case of their general problem. Consequently, their $O(n^2)$ algorithm solves this problem.

Definition 5. THE GENERAL ULTRAMETRIC OPTIMIZATION PROBLEM:

- **Input:** *A distance matrix M and two functions $f(x, \varepsilon)$ and $g(x, \varepsilon)$ which take two real arguments such that both f and g are monotone non-decreasing on their first argument, f is monotone non-increasing on ε and g is monotone non-decreasing on ε.*
- **Output:** *The smallest ε such that there exists an ultrametric matrix U in which for all (i, j), $f(M[i, j], \varepsilon) \le U_{ij} \le g(M[i, j], \varepsilon)$.*

We show how our problem can be stated in these terms by defining the functions $f(x, \varepsilon)$ and $g(x, \varepsilon)$ appropriately.

Because our goal is to find an ultrametric matrix U with the minimum stretch, we want to minimize the value of the following:

$$max\{\frac{M[i,j]}{U_{ij}}, \frac{U_{ij}}{M[i,j]}\}.$$

In other words, we want to find the minimum value of ε, such that there exists an ultrametric matrix U satisfying

$$\max\{\frac{M[i,j]}{U_{ij}}, \frac{U_{ij}}{M[i,j]}\} \le \varepsilon.$$

We solve for U_{ij} and obtain the following:

$$\frac{M[i,j]}{U_{ij}} \le \varepsilon \text{ and } \frac{U_{ij}}{M[i,j]} \le \varepsilon$$

which is equivalent to

$$\frac{M[i,j]}{\varepsilon} \le U_{ij} \le \varepsilon M[i,j].$$

The problem is reduced now to the following:

Given a distance matrix M, find the smallest ε such that there exists an ultrametric matrix U so that for all i, j, $\frac{M[i,j]}{\varepsilon} \le U_{ij} \le \varepsilon M[i, j]$. In other words, we wish to solve the General Ultrametric Optimization problem, with $f(x, \varepsilon) = \frac{x}{\varepsilon}$ and $g(x, \varepsilon) = \varepsilon x$.

Hence, we have:

Theorem 1. *We can solve the Min Stretch Tree problem in $O(n^2)$ time, using the algorithm in [3].*

The algorithm in [3] is therefore useful directly in solving our first problem. As we will show, the techniques in that algorithm are also useful for solving our second problem.

4   Finding the Stretch when the Topology Is Fixed

In this section we describe a polynomial time algorithm for solving the problem of finding the minimum stretch of a fixed tree. More formally, given a tree topology T and a distance matrix M, defined on the leaves of T, the algorithm finds an ultrametric assignments of heights to the nodes of T, thus defining an ultrametric matrix U, so that U minimizes $Stretch_M(U)$.

Lemma 1. *For a given tree T, and upper and lower matrices, M_l and M_h, there exists an ultrametric assignment of heights to the internal nodes of T if the following condition holds for every internal node $v \in T$:*

$$min \; M_h[i,j] \; \geq \; max \; M_l[p,q],$$

where i, j, p, and q are leaves of T such that $lca(i,j) = v$, and $v \geq lca(p,q)$.

Proof. Given the tree T, M_l and M_h, we let the height of node $v \in T$ be

$$height(v) \; = \; max \; M_l[p,q],$$

where p and q are leaves of T such that $v \geq lca(p,q)$.

To complete the proof, we show the following two properties of the height function as we defined it:

- If $v > w$ then $height(v) \geq height(w)$. Since $v > w$, it follows that $S = \{(i,j) : i,j \text{ are leaves below } w\} \subseteq \{(i,j) : i,j \text{ are leaves below } v\} = S'$. Therefore, $height(v) = max \; M_l[p,q] \geq max \; M_l[i,j] = height(w)$, where $(p,q) \in S$ and $(i,j) \in S'$.
- $M_l[i,j] \leq U_{ij} \leq M_h[i,j]$ for all i and j. This follows from the definition of the height function, the fact that $U_{ij} = height(v)$, where $v = lca(i,j)$, and the given assumption.

Theorem 2. *Given a tree T, and a distance matrix M, we can find their deviation from ultrametricity in polynomial time.*

Proof. For each internal node $v \in T$, we define $p(v) = \{max \; M[i,j] : i,j \text{ are below } v\}$, and $q(v) = \{min \; M[i,j] : lca(i,j) = v\}$.

Using Lemma 1, we want to find the "tightest" M_h and M_l such that for every internal node $v \in T$, the following holds:

$$min \; M_h[i,j] \geq max \; M_l[a,b],$$

where i, j, a, and b are leaves of T, such that $lca(i,j) = v$, and $lca(a,b) \leq v$.

Using the same monotone functions $f(x, \varepsilon)$ and $g(x, \varepsilon)$, that we defined before, for each node $v \in T$, we find the smallest $\varepsilon(v)$ such that

$$\varepsilon(v) \cdot q(v) \geq \frac{p(v)}{\varepsilon(v)}$$

Solving for $\varepsilon(v)$, we obtain

$$\varepsilon(v) \geq \sqrt{\frac{p(v)}{q(v)}}$$

To obtain the minimum value of $\varepsilon(v)$, we choose $\varepsilon(v) = \sqrt{\frac{p(v)}{q(v)}}$. The deviation of the distance matrix M, given the tree topology T, is the maximum value of $\varepsilon(v)$, where v is any internal node in the tree T.

The algorithm we have described runs in $O(n^2)$ time. We can compute $lca(i,j)$ for all pairs of leaves in $O(n^2)$ time by first doing a linear-time pre-processing of the tree followed by constant time calculations for each pair of leaves (See [6]). Therefore, by processing the distance matrix once, we can compute $q(v)$ for each v. At the same time, for each node v, we initialize the value of $p(v)$ to $max\, M[i,j]$, where i and j are below v. Then, in a bottom-up fashion, we update $p(v)$ to $max\{p(v), p(u), p(w)\}$, where u and w are the two children of v. Once $p(v)$ and $q(v)$ are computed, finding $\varepsilon(v)$ takes constant time, and finding the maximum among them takes linear time. Therefore, the algorithm takes $O(n^2)$ time.

5 Simulation Study

5.1 Introduction

The general design of our simulation study is as follows. We implemented several techniques for estimating the stretch of a given dataset: our own technique for obtaining a lower bound on the stretch of the model tree on a dataset when the topology is not given (described in Section 3); the technique for the fixed-tree case given in Section 4; the two techniques biologists use (described below, in Section 5.3). We applied these techniques to a number of datasets obtained by simulating DNA sequence evolution down model trees, under the K2P+Gamma model.

5.2 Model Trees

We used the *r8s* software [17] to produce a number of random birth-death trees with a strong molecular clock under the K2P+ gamma [9] model. Hence as the sequence length increases, the stretch on these datasets on the true tree will tend to 1. To obtain trees that deviate from a molecular clock, we multiplied each edge in the tree by e^x, where x is a random number drawn uniformly from the range $[-\ln c, \ln c]$. We used six different values of for c: 1, 3, 5, 8, 15, and 25. The expected value of the scaling factor on an edge is $\frac{c-1/c}{2\ln c}$, so the expected deviations are moderate even for the values of c that we examined. The height of the trees generated by *r8s* is 1. To obtain trees with additional heights, we scaled those trees by factors of 0.25, 1, and 4. We set $\alpha = 1$ and ts/tv ratio equals 2 for the K2P+Gamma evolution model. We examined trees with 20, 40 and 80 taxa, and used sequences of lengths 100, 200, 400 and 800.

5.3 Biological Methods for Estimating the Stretch

Our experimental study examines the accuracy of two methods used by systematists for estimating the stretch of a dataset. The two methods have the same basic structure. First, they obtain an estimate of the phylogenetic tree using either Maximum Parsimony (MP) or Maximum Likelihood (ML). Parsimony and likelihood methods produce not only topologies but also edge lengths. For example, with MP, the edge lengths are the Hamming Distances on each edge, in an optimal labeling of the internal nodes of the tree so as to minimize the sum of the Hamming Distances. In ML, the edge lengths represent the expected number of times a random site changes on the edge. Given either method for defining edge lengths, we can then define distances between nodes v in the tree and leaves i. We indicate such a distance by the notation $d(v, i)$. These trees are then rooted using some technique (for example, the "mid-point rooting" technique, whereby the midpoint of the longest path in the tree is identified as the root of the tree). Then, the stretch of the rooted edge-weighted tree T ($w(e)$ denotes the weight of edge e in T) is computed as follows:

$$Dev(T, w) = \max\{\sqrt{\tfrac{d(v,i)}{d(v,j)}} : v \text{ is a node in } T \text{ and } i \text{ and } j \text{ are leaves below } v\}$$

where $d(v, i) = \sum_e w(e)$, where e is an edge on the path from node v to leaf i.

We denote by $Deviation_\Phi(T)$ the stretch of tree T as computed by the previous formula when the branch lengths are estimated using criterion Φ. In this paper we consider $Deviation_{MP}(T)$ and $Deviation_{ML}(T)$, using PAUP* 4.0 [18] to assign branch lengths. In addition, we also consider $Deviation_{Model}(T)$, where the model branch lengths are used.

Note that this way of estimating the stretch of a tree with respect to an input does not verify that the internal nodes can be assigned heights so that the resultant values are feasible solutions to the stretch problem. One objective of this study was to determine if these calculations produce feasible solutions. For the same reason, we do not call the resultant value the *stretch* of the tree with respect to the estimated distances, but rather the *deviation* of the tree with respect to the estimated distances.

There are several places where this technique can err: in particular, in obtaining a good estimate of the rooted tree, and then in assigning edge lengths. We have simplified the problem by studying the accuracy of these methods assuming that the rooted model tree is given to the methods; hence, we only use the methods to infer branch lengths and not to also find the best tree. We then compare the estimated values for the stretch obtained by those methods against the lower bound for the rooted model tree.

5.4   Simulations

We used the program Seq-Gen [15] to randomly generate a DNA sequence for the root and evolve it through the tree under the K2P+Gamma model. We calculated K2P+Gamma distances appropriately for the model (see [10]). We then applied the algorithm in Section 3 to compute the tree with the optimal stretch (and hence the optimal stretch). We also applied our algorithm (Section 4) to the dataset on the model topology, as well as the other techniques where the MP and ML branch length estimates of the model topology were computed.

In order to obtain statistically robust results, we used a number of *runs*, each composed of a number of *trials* (a trial is a single comparison), computed the mean for each run, and studied the mean over the runs of these events. This method enables us to obtain estimates of the mean that are closely grouped around the true value. This method was recommended by McGeoch [11] and Moret [12]. In our experiments, each run is composed of 30 trials. A new tree is generated for each run, and in each trial a new set of sequences is evolved on the tree.

6   Results and Analysis

In this section we report on the results of the experimental studies that we carried out according to the description in Section 5. We examine the performance of the following methods for estimating the stretch:

1. The minimum stretch of the dataset: $Stretch(d)$, where d is the distance matrix of the dataset on the model tree.
2. The stretch of the ultrametric model tree (i.e. the model tree before we deviate the branch lengths away from ultrametricity) with respect to the model branch lengths obtained after deviation from ultrametricity. Thus, this is $Stretch_D(U)$ where U is the ultrametric matrix underlying the model tree, and D is the additive matrix of the model tree).
3. The minimum stretch of the dataset on the rooted model tree topology: $Stretch_d(T)$, where T is the model topology.
4. The deviation of the rooted model tree topology with respect to MP branch lengths: $Deviation_{MP}(T)$.
5. The deviation of the rooted model tree topology with respect to model branch lengths: $Deviation_{Model}(T)$.
6. The deviation of the rooted model tree topology with respect to ML branch lengths: $Deviation_{ML}(T)$.

(Recall that the deviation of the rooted model tree is calculated using a midpoint rooting technique, and hence does not produce a solution that is guaranteed to be feasible for the stretch problem.)

The values plotted in the figures are the mean of 30 runs for each experimental setting.

Because the values of $Deviation_{ML}(T)$ were so large (sometimes in the thousands); see Figure 3, we did not plot those values in the graphs since they almost always either give infeasible solutions or gross overestimates of the stretch values. Figures 2(a), 1(c), 1(d) and 1(a) show clearly that the values of the $Stretch(d)$ and $Deviation_{Model}(T)$ are equal when the the model tree is ultrametric. However, as the deviation from the molecular clock increases, we see that the discrepancy between those two values grows. Therefore, even if we had a method that could estimate the branch lengths of a tree accurately, the midpoint rooting with MP or ML trees computes values that are far from the values of the true stretch of the tree.

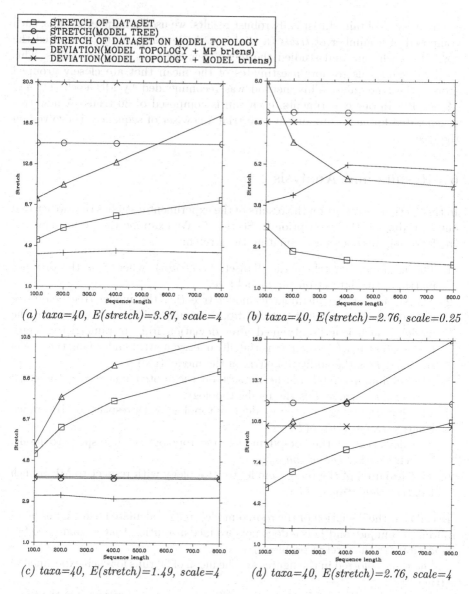

(a) *taxa=40, E(stretch)=3.87, scale=4*

(b) *taxa=40, E(stretch)=2.76, scale=0.25*

(c) *taxa=40, E(stretch)=1.49, scale=4*

(d) *taxa=40, E(stretch)=2.76, scale=4*

Fig. 1.

By definition, $Stretch(d) \leq Stretch_d(T)$ for all trees T, and Figures 1 and 2 demonstrate this observation empirically. Figures 2(a) and 2(b) demonstrate that our methods "converge" to the true stretch value on ultrametric datasets, as the sequence length increases. In these two figures, we notice that the two curves corresponding to the values of $Stretch(d)$ and $Stretch_D(U)$ go to 1 as the sequence length increases.

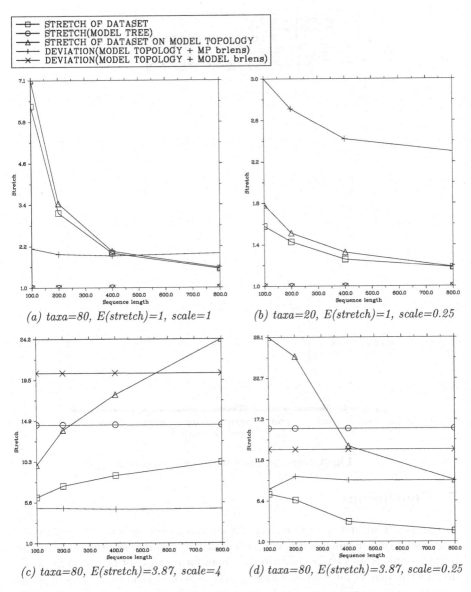

(a) taxa=80, E(stretch)=1, scale=1

(b) taxa=20, E(stretch)=1, scale=0.25

(c) taxa=80, E(stretch)=3.87, scale=4

(d) taxa=80, E(stretch)=3.87, scale=0.25

Fig. 2.

Figures 1(a), 1(b), 1(c) and 2(c) show that $Deviation_{MP}(T)$ are sometimes inconsistent with the definition of the stretch, since we see the values computed by this method are lower than $Stretch(d)$. Figure 2(d) shows the value computed by the same method is greater than the stretch of the dataset, but is lower than the stretch of the dataset on the same topology $(Stretch_D(U))$, which is an inconsistent result. This means that the values computed by $Deviation_{MP}(T)$ are sometimes infeasible.

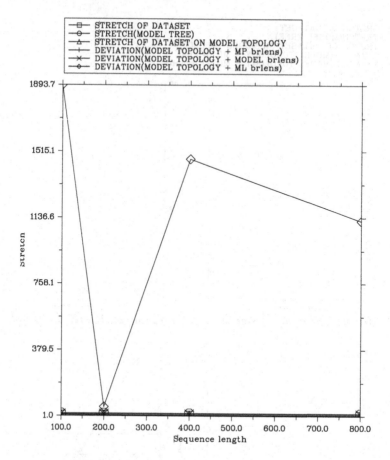

Fig. 3. taxa=20, E(stretch)=3.87, scale=0.25

7   Conclusions

In this paper we defined the concept of stretch as the amount of deviation of a dataset from ultrametricity. We presented two theoretical results in this paper: the first is an $O(n^2)$ algorithm for computing the optimal stretch of any ultrametric matrix for a given dataset, and the second is an $O(n^2)$ algorithm for computing the optimal stretch of a fixed tree with respect to a given dataset.

The experimental study presented here is surprising, and shows that the two standard methods (MP and ML) used by systematists to estimate the degree to which a dataset deviates from the strong molecular clock hypothesis are quite faulty. Both can produce estimates that are not feasible (i.e. there are no heights for the nodes of a tree that would produce such values), and the ML method in particular can produce enormous values, clearly much larger than are required. More generally, our study suggests that accurate estimates of the deviation from the molecular clock may be beyond what can be inferred using existing stochastic models of evolution.

References

1. F. Bossuyt and M.C. Milinkovitch. Amphibians as indicators of early tertiary "Out-of-India" dispersal of vertebrates. *Science*, 292:93–95, 2001.
2. V. Dvornyk, O. Vinogradova, and E. Nevo. Long-term microclimatic stress causes rapid adaptive raditiona of kaiABC clock gene family in a cyanobacterium Nostoc linckia, from 'evolution canyons' I and II, Israel. *Proc. National Academy of Sciences (USA)*, 99(4):2082–2087, 2002.
3. M. Farach, S. Kannan, and T. Warnow. A robust model for finding optimal evolutionary trees. *Algorithmica*, 13(1):155–179, 1995.
4. N. Goldman. Statistical tests of models of DNA substitution. *J. Mol. Evol.*, 36:182–198, 1993.
5. X. Gu and W-H. Li. Estimation of evolutionary distances under stationary and nonstationary models of nucleotide substitution. *Proc. Natl. Acad. Sci. (USA)*, 95:5899–5905, 1998.
6. D. Gusfield. *Algorithms on strings, trees, and sequences.* Cambridge University Press, 1997.
7. D.M. Hillis, C.Moritz, and B.K. Mable. *Molecular Systematics.* Sinauer Associates, Sunderland, MA, 1996.
8. S. Y. Kawashita, G. F. Sanson, O. Fernandez, B. Zingales, and M. R. Briones. Maximum-likelihood divergence date estimates based on rRNA gene sequences suggest two scenarios of trypanosoma crazi intrapsecific evolution. *Mol. Biol. Evol.*, 18(12):2250–2259, 2001.
9. M. Kimura. A simple method for estimating evolutional rates of base substitutions through comparative studies of nucleotide sequences. *J. Mol. Evol.*, 16:111–120, 1980.
10. W.-H. Li. *Molecular Evolution.* Sinauer Assoc., 1997.
11. C.C. McGeoch. Analyzing algorithms by simulation: variance reduction techniques and simulation speedups. *ACM Comp. Surveys*, 24:195–212, 1992.
12. B.M.E. Moret. Towards a discipline of experimental algorithmics. In *Proc. 5th DIMACS Challenge.* available at www.cs.unm.edu/moret/dimacs.ps.
13. M. Nei, P. Xu, and G. Glazko. Estimation of divergence times from multiprotein sequences for a few mammalian species and several distantly related organisms. *Proc. Natl. Acad. Sci. (USA)*, 98(5):2497–2502.
14. M. Nikaido, K. Kawai, Y. Cao, M. Harada, S. Tomita, N. Okada, and M. Hasegawa. Maximum likelihood analysis of the complete mitochondirial genomes of eutherians and a reevaluation of the phylogeny of bats and insectivores. *J. Mol. Evol.*, 53(4-5):508–516, 2001.
15. A. Rambaut and N.C. Grassly. Seq-gen: An application for the Monte Carlo simulation of dna sequence evolution along phylogenetic trees. *Comp. Applic. Biosci.*, 13:235–238, 1997.
16. N. Saitou and M. Nei. The neighbor-joining method: A new method for reconstructing phylogenetic trees. *Mol. Biol. Evol.*, 4:406–425, 1987.
17. Michael Sanderson. available from http://loco.ucdavis.edu/r8s/r8s.html.
18. D. Swofford. PAUP*: Phylogenetic analysis using parsimony (and other methods), version 4.0. 1996.
19. L. Vawter and W.M. Brown. Nuclear and mitochondrial DNA comparisons reveal extreme rate variation in the molecular clock. *Science*, 234(4773):194–196, 1986.
20. Z. Yang, I.J. Lauder, and H.J. Lin. Molecular evolution of the hepatitis B virus genome. *J. Mol. Evol.*, 41(5):587–596, 1995.

Exploring the Set of All Minimal Sequences of Reversals – An Application to Test the Replication-Directed Reversal Hypothesis

Yasmine Ajana[1], Jean-François Lefebvre[1],
Elisabeth R.M. Tillier[2], and Nadia El-Mabrouk[1]

[1] Département d'informatique et de recherche opérationnelle (DIRO)
Université de Montréal, Québec, Canada
{ajanay,lefebvrj,mabrouk}@iro.umontreal.ca
[2] Clinical Genomics Centre, University Health Network
101 College street, MBRC 5-501, Toronto, Ontario M5G1L7, Canada
e.tillier@utoronto.ca

Abstract. Given two genomes, the problem of sorting by reversals is to explain the evolution of these genomes from a common ancestor by a minimal sequence of reversals. The Hannenhalli and Pevzner (HP) algorithm [8] gives the reversal distance and outputs one possible sequence of reversals. However, there is usually a very large set of such minimal solutions. To really understand the mechanism of reversals, it is important to have access to that set of minimal solutions. We develop a new method that allows the user to choose one or several solutions, based on different criteria. In particular, it can be used to sort genomes by weighted reversals. This requires a characterization of all "safe" reversals, as defined in the HP theory. We describe a procedure that outputs the set of all safe reversals at each step of the sorting procedure in time $O(n^3)$, and we show how to characterize a large set of such reversals in a more efficient way. We also describe a linear algorithm allowing to generate a random genome of a given reversal distance. We use our methods to verify the hypothesis that, in bacteria, most reversals act on segments surrounding one of the two endpoints of the replication axis [12].

1   Introduction

The availability of complete genome sequence data from a multitude of bacterial and archaeal species now enables evolutionary relationships to be inferred from entire genetic material. The algorithmic study of comparative genomics has focused on inferring the minimal number of genomic mutations (genome rearrangements) necessary to explain the observed differences in gene orders. The rearrangement operations that have been mostly considered by authors are inversion (reversal) of a chromosomal segment [8,10,1], transposition of a segment from one site to another in one chromosome [2,13], and reciprocal translocation in the case of multi-chromosomal genomes [7]. In particular, the exact polynomial

R. Guigó and D. Gusfield (Eds.): WABI 2002, LNCS 2452, pp. 300–315, 2002.

algorithm of Hannenhalli and Pevzner (hereafter "HP") for sorting signed permutations by reversals [8] was a breakthrough for the formal analysis of genome rearrangement. Moreover, they were able to extend their approach to the analysis of both reversals and reciprocal translocations at the same time [9], and we further generalized the method to account for insertions and losses of fragments of chromosomes [6].

The HP algorithm allows to compute the reversal distance between two genomes and output one minimal sequence of reversals. However, there is usually a very large set of such minimal solutions. To really understand the mechanism of reversals, it is important to have access, not only to one, but to a large set of such solutions. This allows to choose one or several solutions on specific criteria, as for example the length of the segments involved in the reversals, the cut-sites, or the position of the reversals with respect to the replication axis. So far, the only existing algorithm that allows to weight the different rearrangement operations and to choose one solution of minimal cost has been the *Derange* heuristic [5]. It approximates the optimal distance by minimizing a weighted sum of reversals, transpositions and transversions. As only some specific reversals are allowed, sorting by reversals only is not always possible with *Derange*.

We develop a new method for sorting signed permutations by reversals, that allows the user to choose one or several solutions, based on different criteria. In particular, it can be used to sort genomes by weighted reversals. The only constraint is minimality: the solutions are all minimal sequences of reversals. This minimality constraint is justified by the fact that the complete set of solutions can be huge and impossible to explore in its entirety. To give an example, in the case of the genomes *C.pneumoniae*/*C.trachomatis* of reversal distance $d = 422$, there are an average of 661 safe reversals (reversals leading to the reversal distance), at each step of the sorting procedure (422 steps). In *Derange*, the solution space explored was also restricted to a very small number of reversals.

Our new algorithm is based on the HP method. After clearing the hurdles, each step of the HP algorithm consists in choosing one safe reversal [8]. The problem of finding a safe reversal has been considered by many authors [3,10,8]. Bergeron [3] formulated an elementary criterion for a reversal to be safe. Recently, Siepel [11] enumerated all sorting reversals, in particular all reversals that clear the hurdles. As the number of hurdles is usually close to 0, we do not focus on hurdles in this paper. The key idea is to begin by clearing the hurdles, and then, at each step, generate all safe reversals, sort them with the user's criteria, and choose the first one in that order.

The paper is organized as follows. In Section 2, we begin by introducing the problem and the HP graph and results. In Section 3, we characterize a large set of safe reversals, and describe an efficient algorithm allowing to generate them. We then show, in Section 4, how to generate all safe reversals at each step of the sorting procedure in time $O(n^3)$. We then develop a new method for generating all minimal sequences of reversals. This is used to describe a branch-and-bound method for sorting permutations by weighted reversals. Section 5 gives an algorithm that generates a random genome of a given reversal distance. These algorithms are finally used, in Section 6, to test the replication-directed reversal hypothesis in bacteria. By comparing orthologous gene positions in several pairs of bacterial genomes, Tillier and Collins [12] provided evidence that a substantial proportion of reversals results from recombination sites that are determined by the positions of the replication forks. Our results are in agreement with this hypothesis for reasonably shuffled genomes.

2   Preliminaries

This work will be situated in the context of circular genomes. Formally, given a set $\mathcal{B}$ of genes, a genome is a circular string of signed ($+$ or $-$) elements of $\mathcal{B}$, no gene appearing more than once. Though we will write a genome as a linear string, it is understood that the first gene is adjacent to the right of the last gene. For a string $X = x_1 x_2 \cdots x_r$, denote by $-X$ the **reverse** string $-x_r - x_{r-1} \cdots - x_1$. A **reversal** (or inversion) transforms some proper substring of a genome into its reverse. By the size of a reversal, we mean the length of the reversed segment.

Given two genomes G, H on the same set of genes $\mathcal{B}$, the HP algorithm computes the minimal number of reversals (reversal distance) required to transform G to H, and outputs one minimal sequence of reversals. Without loss of generality, we assume that H is the identity permutation, i.e. $H = +1 + 2 + 3 \cdots + n$. We denote by $R(G)$ the reversal distance between G and H.

The HP algorithm is based on a bicolored **breakpoint graph** $B(G)$, defined as follows. If gene x_i in G has positive sign, replace it by the pair $x_i^t x_i^h$, and if it is negative, by $x_i^h x_i^t$. The vertices of $B(G)$ are just the x_i^t and the x_i^h for all x_i of $\mathcal{B}$. Any two vertices that are adjacent in G, other than x_i^t and x_i^h from the same x_i, are connected by a black edge, and any two adjacent in H, by a gray edge (Figure 1.a). This graph decomposes into a set of $c(G)$ disjoint color-alternating cycles. By the **size of a cycle** we mean the number of black edges it contains.

We say that a **reversal R is determined by the two black edges** $e = (a, b)$ and $f = (c, d)$ if R reverses the segment of G bounded by the vertices b and c. For example, in Figure 1.a, the reversal determined by the black edges $(4^t, 6^t)$ and $(3^h, 8^t)$ reverses the segment $< +6 + 5 - 7 + 3 >$. Similarly, the **reversal determined by a gray edge** g is the one determined by the two black edges adjacent to g. For example, the reversal determined by the gray edge $(4^h, 5^t)$ reverses the segment $< -4 + 6 >$, i.e. creates the segment $< -6 + 4 >$.

Let e and f be two black edges of the same cycle C. Think of both e and f as being oriented clockwise. Each one of them induces an orientation for C. If the orientation induced by e and f coincide, we say that the two edges are **convergent**; otherwise they are **divergent**. For example, in Figure 1.a, $(4^t, 6^t), (3^h, 8^t)$ diverge, while $(4^t, 6^t), (5^h, 7^t)$ converge. We also say that a gray edge g is **oriented** if it connects two divergent black edges, and **unoriented** otherwise. For example, $(6^h, 7^t)$ is oriented, while $(6^t, 5^h)$ is unoriented.

Lemma 1. ([8]) *Let R be a reversal determined by the two black edges e and f. 1. If e and f belong to different cycles, then the number of cycles of the breakpoint graph decreases by 1 after performing the reversal R. We say that R*

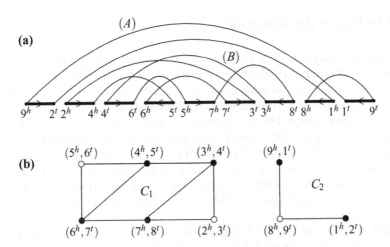

Fig. 1. (a) The breakpoint graph $B(G)$ of $G = +2-4+6+5-7+3+8-1+9$. A possible orientation of the black edges is specified. The graph has three cycles, all of them being oriented, and two oriented components (A) and (B). **(b)** The overlap graph $OV(G)$, containing two components C_1, C_2. The oriented vertices are the black ones.

*is a **bad reversal**; 2. If e and f belong to the same cycle and converge, then the number of cycles of the graph remains unchanged after R; 3. If e and f belong to the same cycle and diverge, then the number of cycles increases by 1 after R. We say that R is a **good reversal**.*

The number of cycles of the breakpoint graph is maximized when $G = H$. Therefore, the problem of minimizing the number of reversals reduces to the one of increasing the number of cycles as fast as possible, that is of performing as many good reversals as possible. This depends on the decomposition of the graph into its components. To explain this decomposition, we need to introduce some basic definitions. A cycle C is **oriented** if it contains at least one oriented gray edge, and **unoriented** otherwise. A **component** of $B(G)$ is a maximal subset of crossing cycles. An **oriented component** has at least one oriented cycle; otherwise it is an **unoriented component** (see Figure 1.a). Finally, a **safe reversal** is a good reversal that does not create any new unoriented component.

HP proved that an oriented component can be **solved**, i.e transformed to a set of cycles of size 1, by a sequence of safe reversals. As for unoriented components, some of them can still be solved by safe reversals, whereas others, called **hurdles**, require bad reversals (see [8] for a more detailed description of hurdles).

Finally, the minimal number of reversals necessary to transform G into H is given by: $R(G) = n - c(G) + h(G) + fr(G)$ **(HP-formulae)**
where n is the size of G (number of genes), $c(G)$ the number of cycles of $B(G)$, $h(G)$ its number of hurdles, and $fr(G) = 1$ or 0 depending on the set of hurdles.

3   Finding Safe Reversals

The bottleneck of the HP theory is to solve the oriented components by performing exclusively safe reversals. Different methods have been developed [8] [4,10,3], one the most efficient and simple one being that presented in [3]. We use their notations and concepts to generate a set of safe reversals.

The **overlap graph** $OV(G)$ is defined as follows: the vertices of $OV(G)$ are all the gray edges of $B(G)$, and two vertices v_1, v_2 are connected by an edge iff v_1, v_2 are two intersecting edges in $B(G)$ (Figure 1.b). For any vertex v, the **reversal** v will refer to the one determined by the gray edge v, and the reversal orientation will be the orientation of the vertex v. Bergeron proved that a vertex of $OV(G)$ is oriented iff it has an odd number of adjacent edges. She also proved:

- **Fact 1 ([3]):** If one performs the reversal v, the effect on $OV(G)$ is to complement the subgraph formed by the vertex v and its adjacent vertices.
- **Fact 2 ([3]):** If one performs the reversal v, each vertex adjacent to v changes its orientation.

Proposition 1. ([3]). *Let the* **score** *of a vertex v of $OV(G)$ be the number of oriented vertices in the overlap graph resulting from the reversal v. Then, an oriented reversal of maximal score is safe.*

From the results given above, an efficient bit vector algorithm for solving each oriented component C of $OV(G)$ has been developed [3]. To do so, each vertex v of C is represented by a bit vector $\mathbf{v}$ whose wth coordinate is 1 iff v is adjacent to vertex w. Then, C is represented by the bit matrix containing as many lines as the number of vertices of the component. Moreover, a vector p is used to store the orientation of each vertex of C (1 if oriented and 0 otherwise), and a vector s is used to store the score for each v. Bergeron showed that the matrix is easily updated after performing one safe reversal.

But proposition 1 usually allows to output only one safe reversal. Here, we give more general criteria allowing to characterize a large subset of safe reversals. Due to the lack of space, we don't give the proof of the following proposition.

Proposition 2. *Let v be an oriented reversal. If v is unsafe, then there exists an oriented vertex w adjacent to v such that the three following conditions are verified:*

1. *There exists at least one vertex adjacent to w but not to v, or adjacent to v but not to w.*
2. *For any unoriented vertex u adjacent to v, u is adjacent to w.*
3. *For any oriented vertex u adjacent to w, u is adjacent to v.*

Sufficient conditions for safe reversals are obtained by taking the negative of Proposition 2. These conditions allow to output a large subset of safe reversals in a very efficient way, by using exclusively bit wise operations.

```
Algorithm safe-reversals (C):
1.    For each oriented vertex v of C Do
2.         v̄ = v & ¬p; bool=true;
3.         For each oriented vertex w adjacent to v Do
4.              Z = (w & ¬v)|(v & ¬w);
5.              V ← v̄ & ¬w;
6.              w̄ ← w & p;
7.              W ← w̄ & ¬v;
8.              If V = 0 and W = 0 and Z ≠ 0 Then
9.                   bool = false; exit;
10.             End of if
11.        End of for
12.        If bool=true Then v is safe Else v is unsafe;
13.        End of if
14. End of for
```

This is done by *Algorithm safe-reversals (C)*, that outputs the subset of safe reversals for an oriented component C. In this algorithm, **p** is the orientation vector defined in [3], & is the *and* bit wise operation, | the *or* bit wise operation and ¬ the *not* bit wise operation.

Algorithm safe-reversals has at most k^2 steps, where k is the number of vertices of the component C. As the procedure stops as soon as an appropriate w is found, the number of steps is usually lower than k^2. The number of words needed to represent each bit vector is $\lceil \frac{k}{\omega} \rceil$, where ω is the word size. Therefore, the time complexity of the algorithm applied to a component C of k vertices is $O(\lceil \frac{k}{\omega} \rceil k^2)$.

4   Exploring the Set of All Minimal Sequences of Reversals

We first consider the problem of generating all minimal sequences of reversals for a breakpoint graph that does not contain any hurdle. We then explore all the possible ways to solve the hurdles with the HP method. Finally, we use these results to develop a new heuristic for sorting permutations by weighted reversals.

4.1   Generating All Minimal Solutions

Given a graph with only oriented components, a **minimal solution** is a minimal sequence of reversals transforming the graph into a set of cycles of size 1.

Lemma 2. *A sequence of reversals is a minimal solution iff it contains exclusively safe reversals.*

Proof. It follows directly from the HP theory [8] □

From the above lemma, the problem of generating all minimal solutions reduces to the problem of generating all the safe reversals at each step of the sorting procedure. Notice that a reversal on one component of the graph does not affect the other components. Thus, if the order of reversals is not important, we can solve each oriented component independently. Now, given an oriented component C, to find all safe reversals, one should check each pair of divergent black edges of the breakpoint graph. For each divergent pair (e, f), the resulting breakpoint graph should be decomposed into its components. The reversal determined by (e, f) is safe iff all the components are oriented. Checking all the components of a graph takes $O(n)$ time [10]. Now, as there are at most $O(n^2)$ pairs of divergent black edges in the breakpoint graph, the total time required to find all safe reversals at each step of the sorting procedure is in $O(n^3)$.

4.2   Hurdles

If the breakpoint graph contains hurdles, then they have to be transformed to oriented components. This is done by merging and cutting hurdles in a particular way, described in the HP procedure for clearing hurdles [8]. This procedure allows some flexibility in choosing hurdles to be merged and cut, and reversals to be performed. More precisely, hurdles can be cleared in several optimal ways: **1.** non-consecutive hurdles can be paired in all possible ways; **2.** to merge a pair of hurdles, one black edge is arbitrarily chosen in each of the two hurdles; **3.** to cut a hurdle, two black edges are arbitrarily chosen from the same cycle.

Therefore, if one is interested in finding all possible sequences of reversals transforming genome G into H, all ways of clearing hurdles should be considered. To limit the number of solutions, specific constraints on the reversals can be established. For example, if we are interested in finding the shortest reversals, then the chosen black edges for hurdle merging or hurdle cutting are the closest ones, and the only remaining flexibility is in the way to pair the hurdles.

However, notice that the probability for a component to be oriented is higher than its probability to be unoriented. Indeed, with the hypothesis that a vertex has the same probability to be oriented or unoriented, and that the vertices are two-two independent, the probability for a component of size k to be unoriented is $\frac{1}{2^k}$. Moreover, as a hurdle is a special case of an unoriented component, the probability for a component to be a hurdle is even lower. In practice, the unoriented components are almost non-existent (see Table 1).

4.3   Sorting Signed Permutations by Weighted Reversals

Understanding the mechanism of rearrangements, and in particular when and how reversals occur, is still an open question in biology. In order to answer this question and to test different hypothesis, the HP algorithm giving the reversal distance and one possible solution is not sufficient. So far, the only existing algorithm that explored different solutions has been *Derange* [5]. It approximates

Table 1. The breakpoint graph structure for various pairwise comparisons of bacterial genomes. Orthologous genes have been identified by Tillier and Collins [12]. Column 2 gives the number of genes considered, column 3 the total number of components of the graph, and column 4 the number of hurdles among these components.

Bacterial Genomes	Genome size (Nb of common genes)	Number of components	Number of hurdles
C.pneumoniae/C.trachomatis	732	70	0
V.cholera/E.coli	1252	76	1
M.leprae/M.tuberculosis	1295	212	0
C.jejuni/H.pylori	706	9	0
S.pyogenes/S.aureus	438	3	0
U.urealyticum/M.pneumoniae	278	6	0

the optimal distance by minimizing a weighted sum of reversals, transpositions and transversions. At each step, the algorithm considers only rearrangements that reduce the number of breakpoints. The method is a depth-first search with limited look-ahead and branch-and-bound. As the only operations that are allowed by *Derange* are those that reduce the number of breakpoints, in some cases sorting by reversals only is impossible and transpositions can not be avoided. This is always the case when the overlap graph contains hurdles, as an optimal sequence of reversals should contain at least one bad reversal.

Here, we develop a new depth-first branch-and-bound heuristic for weighted reversals. As for *Derange* algorithm, we do not consider all possible reversals for each node in the search tree, which make the algorithm a heuristic. Rather, we choose a set of reversals leading to the reversal distance.

The general method is as follows: we first clear the hurdles to end up with oriented components. We then solve the oriented components by considering the complete set of safe reversals. In other words, the algorithm has two major steps:

1. If the graph contains hurdles, perform a set of reversals clearing the hurdles.
2. Solve each oriented component independently, by choosing safe reversals.

The depth-first branch-and-bound method is used for the second step. For each component, all its safe reversals are ordered in decreasing order of their weight. Once a first solution is obtained (a sequence of reversals solving the oriented component), its cost (sum of its reversals weights) becomes the current bound. Then we backtrack, undoing the most recent reversal, and trying the next-best reversal, as long as one exists. If not, we undo the most recent move at the previous level. After each backtrack step, we again proceed until we end up with a partial solution of cost higher than the current best solution, or we completely solve the current component. In that case, if the new solution has a cost lower than the current bound, then the bound is replaced by this cost.

If the components of the overlap graph are small, the search tree is small and the branch-and-bound algorithm is efficient. However, even with our restricted set of reversals, the search becomes quickly huge and unmanageable. For example in the case of large genomes with a high rate of rearrangements, the number of components is close to 1, and the search tree size can be on the order of n^r, where n is the size of the genome, and r the reversal distance. To avoid such situation, the weighting function should allow a large prune of the search tree. This is not the case for a weighting proportional to the length of the reversed segment. Indeed, a long reversal that is avoided at the beginning of a path, is usually postponed and performed farther in the path. One way to improve the branch-and-bound is to select a small number of safe reversals at each step. This restriction has been considered in *Derange*.

5   Generating a Random Genome of a Given Reversal Distance

To test various hypothesis on reversals, one should compare the results obtained for a real genome G with those obtained with a random genome. For example, in the case of the replication-directed reversal hypothesis, we are interested in weighting the reversals according to their position with respect to the replication axis. Now, suppose we find a solution of minimal weight with $R(G)$ reversals. To evaluate the distribution of the obtained reversals, we should compare it with the one obtained with a random genome of the same length and the same reversal distance. The problem is, therefore, to generate a random genome of reversal distance $R(G)$. This can not be done by simply starting with the identity permutation and performing $R(G)$ random reversals. Indeed, we usually end up with a genome that has a reversal distance significantly lower than $R(G)$.

Consider the HP-formulae (Section 2). As the number of hurdles is usually very close to 0, the dominant variable in this formulae is the number of cycles of the graph. In other words, suppose for now that we ignore the hurdles. Then, $R(G) = n - c(G)$, where n is the size of G and $c(G)$ the number of cycles in the breakpoint graph $B(G)$. Therefore, the problem of generating a random genome of size n and reversal distance $R(G)$ can be reduced to that of constructing a random breakpoint graph of size n (number of black edges), having $c(G)$ cycles. To do so, we start with the **partial graph** $B'(G)$, that is the breakpoint graph $B(G)$ without the gray edges, and we complete it with gray edges by *Algorithm construct-cycles* below. In this algorithm, a **free vertex** is a vertex not already adjacent to a gray edge. At the beginning of the algorithm, all vertices are free. At any step of the algorithm, the graph is constituted of a set of alternating paths and cycles. A gray edge is said to **close a path** if it links the two free vertices of the path. We denote by p the number of paths and c the number of cycles of the current graph. At the beginning of the algorithm, $p = n$ and $c = 0$. At the end, we should have $p = 0$ and $c = n - R(G)$.

Lemma 3. Algorithm construct-cycles *constructs a graph with $R(G)$ cycles.*

The problem is now to use the graph constructed by *Algorithm construct-cycles* to generate a random genome of reversal distance $R(G)$. Notice that the correspondence is not straightforward. Indeed, the gray edges constructed by the

---

Algorithm construct-cycles:
1. $p = n$, $c = 0$;
2. **While** $c + p \neq n - R(G)$ **Do**
3.   **If** $c = n - R(G) - 1$ and $p > 1$ **Then**
4.   Choose randomly two free vertices belonging to two different paths, and link them with a gray edge;
5.   **Else**
6.   Choose randomly two free vertices and link them with a gray edge;
7.   **End of if**
8.   update p and c;
9. **End of while**
10. Close each path individually;

---

algorithm do not necessarily give rise to a unique circular genome containing all the genes of G. Instead, we can end up with a set of circular fragments, where a fragment contains a subset of the genes of G. The problem is then to link these fragments without changing the number of cycles of the graph.

Let F_1, F_2 be two circular fragments obtained with *Algorithm construct-cycles*. Let a_1, b_1 be two adjacent genes in F_1, and a_2, b_2 two adjacent genes in F_2. As a_1, b_1 are connected by a gray edge, they belong to one cycle of the graph. The same holds for a_2, b_2. Therefore, a_1, b_1, a_2, b_2 are in the same cycle or in two different cycles. The (a) part of Figure 2 illustrates one (over two) possible situation with one cycle. The (b) part of Figure 2 illustrates one (over four) possible situation with two cycles. In that case, the other endpoints of the gray edges should be considered. There are different situations depending on the distribution of the 8 vertices in 2, 3 or 4 fragments. Figure 2.b represents the case of 4 fragments. The right part of Figure 2 shows that it is possible to destroy the "bad" gray edges, and construct new ones, in such a way that the total number of cycles is not modified. All the other situations are treated in a similar way, and are not detailed here.

After the correction described above, we end up with a circular genome represented by a breakpoint graph with $c = n - R(G)$ cycles. If the graph does not contain any hurdle, then the genome has the reversal distance $R(G)$. Otherwise, we run the algorithm until we obtain a graph without hurdles. Usually, a unique run of the algorithm is sufficient to obtain an appropriate random genome.

6   Verifying the Replication-Directed Reversal Hypothesis

Tillier and Collins [12] have determined plots for the relative positions of orthologous pairs of genes in pairwise comparisons of bacterial genomes. In most cases, a high number of orthologous are located at the same relative position and orientation in both genomes. But the most interesting fact is that the other genes

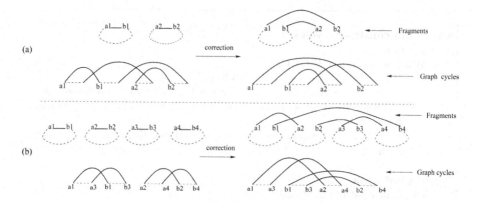

Fig. 2. Thick lines represent gray edges and thin lines are alternating paths of black and gray edges, ending with black edges. Circles represent circular fragments. Left is the situation after Algorithm construct-cycles, and right is the correction to decrease the number of fragments without changing the number of cycles. (a) One possible situation with one cycle. (b) One possible situation with two cycles.

that have moved are not randomly distributed throughout the genome. Instead, they form a perpendicular diagonal line that intersects the first line at approximately the position of the termination of replication. Moreover, the genes in these orthologous pairs have opposite signs. Of course, the rate of recombination varies from one pairwise comparison to another, and the diagonal is more or less clear depending on the plot. These observations have led to the hypothesis that apparent transpositions are actually resulting from pairs of consecutive reversals at different sites. Moreover, the segments involved in these reversals are not randomly distributed in the genome. Rather, they are surrounding one of the two endpoints of the replication axis, and this endpoint is approximately situated in the middle of the segment. We refer to this hypothesis as the replication-directed reversal hypothesis. Figure 3 is an example of a reversal across the replication axis. The size of a reversal is the length $l = l_1 + l_2$ of the segment involved in the reversal, where l_1 (respectively l_2) is the length of the sub-segment on the left side (respectively right side) of the replication axis endpoint. The lengths are represented as a percentage of the total size of the genome, in term of nucleotide number. Notice that if $l > 50\%$, the complementary reversal, obtained by taking the complementary segment in the genome, gives rise to the same permutation. In particular, a reversal containing the two endpoints of the replication axis can be replaced by the complementary one that does not contain any of these two endpoints. Therefore, we consider only the reversals for which $l \leq 50\%$.

We define the asymmetric coefficient $C = \frac{|l_1 - l_2|}{l}$. If $C = 0$, the reversal is perfectly symmetric, and if $C = 1$, the reversed segment does not cross the replication axis.

Method: We ran our algorithm on various pairs of bacterial genomes. For one of them, *C.pneumoniae/C.trachomatis*, the two diagonals are clear in the plot

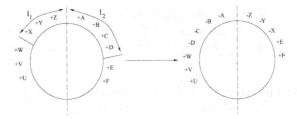

Fig. 3. A reversal surrounding one of the two endpoints of the replication axis. l_1 (resp. l_2) is the length of the sub-segment on the left side (resp. right side) of the replication axis endpoint. A "valid" reversal is a reversal such that $\frac{|l_1-l_2|}{l_1+l_2}$ is close to 0.

(relative positions of orthologous pairs of genes), while for the others, they are hardly visible [12]. We used simplified data: the genes present in only one genome have not been considered. This simplification is possible because it does not alter the two diagonals on the plots. However, we considered the real nucleotide distance between pairs of genes.

As no evidence for any particular weighting is established, we first selected, at each step of the algorithm, one random reversal from the set of all safe reversals. In that way, running the algorithm several times gives rise to several possible solutions. Let S be the set of all reversals obtained by running our algorithm t times. To verify the hypothesis, it is not sufficient to count the reversals that have an asymmetric coefficient higher than a threshold value, and compare it with the value obtained for a random genome. Indeed, in real data, short reversals are predominant. And as the probability of having a reversal with an asymmetric coefficient close to 0 depends on the reversal size (a short reversal has a high probability of being asymmetric), the size of the reversals should be considered in our experiments. Therefore, we plotted the reversals of the set S depending on their size, and compared them with the theoretical results described below.

Consider a random permutation of n genes, and suppose they are uniformly distributed on the genome. Then, the size of a reversal can be viewed as the number of genes in its segment. For a reversal R of length l, let X be the random variable whose value is equal to the asymmetric coefficient C of R.

Lemma 4. *The mean and variance of the random variable X are respectively:*

$$E[X] \sim 1 - \frac{l}{n}, \quad Var[X] \sim 1 - \frac{4(s^2-1)}{3ns} - \frac{(n-s)^2}{n^2}$$

Proof. We want to compute the value of $P\{X = C\}$, for $0 \le C \le 1$.

The total number of possible reversals of length l is n if $l < \frac{n}{2}$, and $\frac{n}{2}$ if $l = \frac{n}{2}$. Consider the number of possible reversals of length l depending on C:

1. **C = 0 :** The number of possible reversals is 2 if $l \le \frac{n}{2}$, and 1 if $l = \frac{n}{2}$. Thus, in both cases, $P\{X = 0\} = \frac{2}{n}$.
2. **0 < C < 1 :** The number of possible reversals is 4 if $l < \frac{n}{2}$, and 2 if $l = \frac{n}{2}$. Thus, in both cases, $P\{X = C\} = \frac{4}{n}$.

3. $C = 1$: If $l < \frac{n}{2}$, the number of possible reversals is $2(\frac{n}{2} - l + 1) = n - 2l + 2$, and $P\{X = 1\} = 1 - 2\frac{s-1}{n}$; if $l = \frac{n}{2}$, there is only 1 possible reversal, and thus $P\{X = 1\} = \frac{2}{n}$. But for $l = \frac{n}{2}$, $1 - 2\frac{s-1}{n} = \frac{2}{n}$. Thus, in both cases, $P\{X = 1\} = 1 - 2\frac{s-1}{n}$.

The set of all possible values of C other than 0 and 1 is $\{\frac{2i}{l}, \text{ for } 1 \le i \le \frac{l-2}{2}\}$ if l is even, and it is $\{\frac{2i-1}{l}, \text{ for } 1 \le i \le \frac{l-1}{2}\}$ if l is odd.
$E[X] = \sum_{0 \le C \le 1} C P\{X = C\} = 1 - 2\frac{s-1}{n} + \frac{4}{n}\sum_{0 < C < 1} C$. By adding the different values of C we obtain: **1.** If l is even, $E[X] = 1 - \frac{l}{n}$; **2.** If l is odd, $E[X] = 1 - \frac{l}{n} + \frac{1}{nl}$.
$Var(X) = E[X^2] - (E[X])^2$ with $E[X^2] = \sum_{0 \le C \le 1} C^2 P\{X = C\}$. By adding the different values of C we find $E[X^2] = 1 - \frac{4(l^2-1)}{3nl}$ for any l. Therefore:
1. If l is even, $Var[X] = 1 - \frac{4(l^2-1)}{3nl} - \frac{(n-l)^2}{n^2}$. **2.** If l is odd, $Var[X] = 1 - \frac{4(l^2-1)}{3nl} - (\frac{n-l}{n} + \frac{1}{nl})^2$.
As $\frac{1}{nl}$ is negligible in comparison to the other terms, we omit it in both $E[X]$ and $Var(X)$. The results are then the same for l even and l odd $\square$

From Lemma 4, it follows that the function $f(l) = E[X]$ is a straight line. It is represented by a gray line in the diagrams of Figures 4. Notice that this theoretical line is obtained by assuming that the probability of picking a reversal of size l is the same for any l. However, as our algorithm only chooses sorting reversals (reversals leading to the reversal distance), and as these reversals are chosen on specific cycles of the graph, the assumption of equiprobability on reversals does not hold for the solutions of our algorithm. To have a better idea about this bias, we ran our algorithm several times on random permutations. We plotted the reversals and obtained a second straight line falling below the theoretical line. This is represented by a black line in the diagrams of Figures 4.

Finally, we weighted the safe reversals by their asymmetric coefficients, and we used our branch-and-bound to find a solution of minimal weight. Due to the fact that the reversal distances are large and that the weighting function does not permit an extensive pruning of the search tree, we had to limit the search to three safe reversals at most at each step of the algorithm.

Figure 5 shows the results obtained for two pairwise comparisons. To evaluate these results, we compared them with those obtained for a random genome of the same length and reversal distance. These random genomes were generated by the algorithm described in Section 5.

Discussion: In [12], plots for the relative positions of orthologous pairs of genes in pairwise comparisons of bacterial genomes have been determined. Depending on the rate of recombination, the diagonals are more or less visible. For example, the diagonals are clearly identifiable for *C. pneu./ C.trach.*, hardly visible for *L.lactis/S.pyo.*, and completely invisible for *C.jejuni/H.pylori*.

The first diagram of Figure 4 shows the curve for *C. pneu./C.trach.*. Most points, except the ones for very short reversals, are far below both theoretical

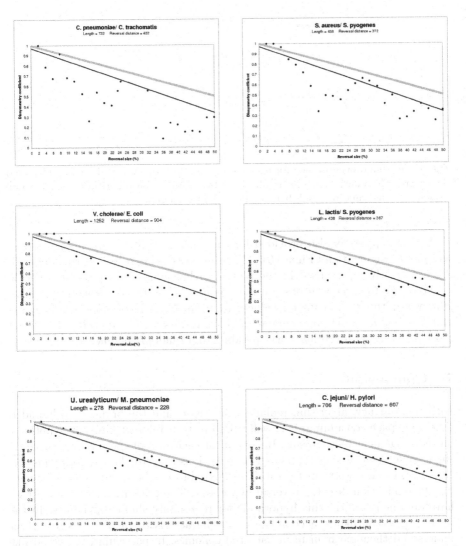

Fig. 4. The diagrams obtained for pairwise comparisons of bacterial genomes. The horizontal axis is the reversal size, written as a percentage of the total length of the genome. The vertical axis it the asymmetric coefficient. Dots represent the reversals obtained by our algorithm. The gray line represents the theoretical mean, and its width is the theoretical variance. The black line is the one obtained by running our algorithm on random permutations. For each pairwise comparison, the reversal distance and the length of the genomes (number of common genes) are given.

lines. The replication-directed reversal hypothesis is strongly validated. This is further supported by the diagram of Figure 5, where all points are far below the line corresponding to a random genome of the same reversal distance.

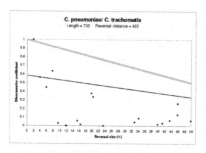 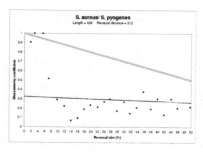

Fig. 5. The diagrams obtained for two pairwise comparisons by our branch-and-bound algorithm. The black thin line is obtained by running the algorithm on a random sequence of the same reversal distance than the tested genomes.

Strong support is also provided for the pairwise *S.aureus/S. pyogenes* (second plot of Figure 4 and Figure 5). This is an interesting result, as the diagonals are not clearly represented in the plot of orthologous genes positions. For *V. Cholerae/E. Coli*, we still observe most points below the two theoretical lines. However, nothing can really be deduced from the diagrams of *L.lactis/S.pyogenes* and *U.urealyticum/M. pneumoniae*, and the results for *C.jejuni/H.pylori* fit almost perfectly with the line corresponding to a random genome.

7 Conclusion

Although our new algorithm remains a heuristic, and the branch-and-bound is time and space consuming, it is well suited to test different kinds of hypothesis on rearrangement mechanisms. Indeed, as at each step of the sorting procedure the complete set of safe reversals is known, it is easy to choose, from this set, the reversals that we want to test. This flexibility has been very useful for testing the replication-directed reversal hypothesis. The preliminary results that we have obtained confirm this hypothesis for reasonably shuffled genomes. In the future, we would like to account for other characteristics of reversals to further understand their mechanism in bacterial genomes. In particular, we would like to account for deleted genes and gene families.

The key idea of our method is to take advantage of an exact parsimony approach to develop an efficient general heuristic. This can be extended to other rearrangement operations. In particular, a new heuristic for weighted reversals and transpositions based on the same idea can be envisaged.

References

1. D.A. Bader, B.M.E. Moret, and M. Yan. A linear-time algorithm for computing inv. dist. between signed perm. *J. Comput. Biol.*, 8(5):483-491, 2001.
2. V. Bafna and P.A. Pevzner. Sorting by transpositions. *SIAM Journal on Discrete Mathematics*, 11(2):224-240, 1998.

3. A. Bergeron. A very elementary presentation of the Hannenhalli-Pevzner theory. In *CPM'01*, LNCS, pages 106-117. Springer Verlag, 2001.

4. P. Berman and S. Hannenhalli. Fast sorting by reversals. In *CPM'96*, LNCS, pages 168- 185. Springer-Verlag, 1996.

5. M. Blanchette, T. Kunisawa, and D. Sankoff. Parametric genome rearrangement. *Gene-Combis*, 172:11-17, 1996.

6. N. El-Mabrouk. Sorting signed permutations by reversals and insertions/deletions of contiguous segments. *to appear in J. Disc. Algo.*, 2000.

7. S. Hannenhalli. Polynomial-time algorithm for computing translocation distance between genomes. In *CPM*, LNCS 937, pages 162–176. Springer, 1995.

8. S. Hannenhalli and P.A. Pevzner. Transforming cabbage into turnip. In *Proc. 27th Annu. ACM-SIAM Symp. Theory of Comp.*, pages 178–189, 1995.

9. S. Hannenhalli and P.A. Pevzner. Transforming men into mice. In *Proc. of the IEEE 36th Annu. Symp. on Found. of Comp. Sci.*, pages 581–592, 1995.

10. H. Kaplan, R. Shamir, and R.E. Tarjan. Faster and simpler algo. for sorting sign. perm. by rev. In *Proc. 8th Annu. ACM-SIAM Symp. Disc. Algo.*, 1997.

11. A. C. Siepel. An algorithm to find all sorting reversals. In *RECOMB'02*, pages 281-290, Washington, 2002. ACM.

12. E.R.M. Tillier and R.A. Collins. Genome rearrangement by replication-directed translocation. *Nature Genetics*, 26:195 - 197, October 2000.

13. M. E. Walter, Z. Dias, and J. Meidanis. Reversal and transposition distance of linear chromosomes. In *SPIRE '98*, 1998.

Approximating the Expected Number of Inversions Given the Number of Breakpoints

Niklas Eriksen

Dept. of Mathematics
Royal Institute of Technology
SE-100 44 Stockholm, Sweden
niklas@math.kth.se
http://www.math.kth.se/~niklas

Abstract. We look at a problem with motivation from computational biology: Given the number of breakpoints in a permutation (representing a gene sequence), compute the expected number of inversions that have occurred. For this problem, we obtain an analytic approximation that is correct within a percent or two. For the inverse problem, computing the expected number of breakpoints after any number of inversions, we obtain an analytic approximation with an error of less than a hundredth of a breakpoint.

1 Introduction

For about a decade, mathematicians and computer scientists have been studying the problem of inferring evolutionary distances from gene order. We are given two permutations (of a gene sequence) and want to calculate the evolutionary distance between them. The most common distance studied is the shortest edit distance: the least number of operations needed to transform one of the permutations into the other, given a set of allowed operations. The operations primarily considered are inversions, transpositions and inverted transpositions. For an overview, see Pevzner's book [11].

The most prominent result in this field is the Hannenhalli-Pevzner inversion distance [10]. They provide a closed formula for the minimal edit distance of *signed* permutations, a formula which can be computed in linear time [1]. Other results worth mentioning are the NP-hardness of computing the inversion distance for unsigned permutations (Caprara, [3]) and the $(1 + \varepsilon)$-approximation of the combined inversion and transposition distance, giving weight 2 to transpositions (Eriksen, [6]).

Although incorporating both inversions and transpositions makes the model more realistic, it seems that the corresponding distance functions usually do not differ much [2,8]. Thus, the inversion edit distance seems to correspond quite well to the evolutionary edit distance, even though this is not *a priori* obvious.

However, there is another problem with using the Hannenhalli-Pevzner formula for the shortest edit distance using inversions: For permutations that are quite disparate, the shortest edit distance is much shorter than the expected

R. Guigó and D. Gusfield (Eds.): WABI 2002, LNCS 2452, pp. 316–330, 2002.

edit distance. The reason is that as the distance between the permutations increases, so does the probability that the application of yet another inversion will not increase the distance. Obtaining the true evolutionary distance is of course impossible, but it will usually be closer to the expected value of the edit distance than to the the shortest edit distance. Therefore we want to study the following problem.

Problem 1. Given the number of breakpoints between two signed, circular, permutations, compute the expected number of inversions giving rise to this number of breakpoints.

For unsigned permutations, this problem has been solved by Caprara and Lancia [4]. One easily realizes that if gene g_2 is not adjacent to gene g_1, then the probability that a random inversion will place g_2 next to g_1 is the same for all such permutations. From this observation, the expected number of breakpoints after i random inversions is not hard to come by. This contrasts with the signed case, where the sign of g_2 will affect the probability.

For signed permutations, the problem has been attacked independently in two sequences of papers. One sequence includes papers by Sankoff and Blanchette [12] and Wang [13]. They have calculated transition matrices for a Markov chain, using which the expected number of breakpoints, given that i random inversions have been applied, can be computed. By inverting this, the expected number of inversions can be computed. This is a fair solution in practice, but it does not give an analytical expression and it is slow for large problems.

The other sequence contains papers by Eriksson et al. [9] and Eriksen [7]. They have looked at the similar problem of computing the expected number of pairs of elements in the wrong order (inversion number) given that t random **adjacent** transpositions have been applied to an ordinary (unsigned) permutation. Again, the purpose is to invert the result. This is an analogue of the above case and the study was initiated to raise ideas on how to solve the inversion problem. In the latter paper, Markov chain transition matrices for this problem are briefly investigated and the conclusion is drawn that an approximation of the expected inversion number can be found if we can compute the two largest eigenvalues of the transition matrix. This raises the question if it is possible to compute the largest eigenvalues of the matrices found by Sankoff, Blanchette and Wang.

We will show that the answer is yes. In fact, we can compute most of the eigenvalues of these transition matrices and sufficient information about the eigenvectors to get a very good approximation of the expected number of inversions, given b breakpoints:

$$i_{\mathrm{appr}}(b) = \frac{\log\left(1 - \frac{b}{n(1-\frac{1}{2n-2})}\right)}{\log\left(1 - \frac{2}{n}\right)}.$$

In this paper, we will derive this formula. We then take a look at some adjustments that can be made to improve it. The formula above is the inverse of an approximation of the expected number of breakpoints after i random inversions. If we drop the demand that the formula for the expected number of breakpoints should be analytically invertible, then we can provide a significantly better formula, with an error that rarely exceeds 0.01 breakpoints.

2 Preliminaries

We are using signed, circular permutations as a model for bacterial genomes. Each gene corresponds to a unique element. The two possible orientations of a gene corresponds to the sign (+ or −) of the corresponding element in the permutation.

An **inversion** is an operation on a permutation which takes out a segment of consecutive elements and insert it at the same place in reverse order, altering the sign of all elements in the segment.

Example 1. Let $\pi = [g_1 \ -g_4 \ g_3 \ -g_6 \ g_5 \ g_2]$ be our permutation. It is understood that g_1 follows directly after g_2, since the permutation is circular. If we invert the segment $[g_3 \ -g_6 \ g_5]$ in π, we get $\pi' = [g_1 \ -g_4 \ -g_5 \ g_6 \ -g_3 \ g_2]$. Had we inverted the segment $[g_2 \ g_1 \ -g_4]$ (the complement of the previous segment), the result would have been the same.

A measure of the difference between two permutations π_1 and π_2 is the number of **breakpoints** between them. There is a breakpoint between two adjacent genes $g_i \ g_j$ in π_1 if π_1 contains neither of the segments $[g_i \ g_j]$ or $[-g_j \ -g_i]$.

Example 2. The permutations $\pi_1 = [g_1 \ -g_4 \ g_3 \ -g_6 \ g_5 \ g_2]$ and $\pi_2 = [g_1 \ g_4 \ -g_5 \ g_6 \ -g_3 \ g_2]$ are separated by three breakpoints. In π_1 these are between g_1 and $-g_4$, between $-g_4$ and g_3, and between g_5 and g_2.

3 Obtaining the Formula

We will in this paper consider Markov chains for circular genomes of length n. At each step in the process, an inversion is chosen at random from a uniform distribution, and the inversion is applied to the genome. The states in the process correspond to the position of the gene g_2 as follows. We fix the first element g_1 and consider the various places where the element g_2 can be located, relative to g_1. Each such position (with orientation) is considered a state in the Markov process. This makes $2n-2$ states, since there are $n-1$ positions and two possible orientations at each position.

The transition matrix for this process was presented in 1999 by Sankoff and Blanchette [12] and it was generalized to include transpositions and inverted transpositions in a 2001 WABI paper [13] by Li-San Wang.

Theorem 1. *(Sankoff and Blanchette [12] and Wang [13]) Consider the Markov process where the states corresponds to the positions of g_2, with a possible minus sign which signals that g_2 is reversed. The states are ordered as $\{2, -2, 3, -3, \ldots,$*

$n, -n\}$. *At each step an inversion is chosen at random from a uniform distribution. Then the transition matrix is* $\frac{1}{\binom{n}{2}}M_n$, *where n is the length of the genome, and $M_n = (m_{ij})$ is given by*

$$m_{ij} = \begin{cases} \min\{|u| - 1, |v| - 1, n + 1 - |u|, n + 1 - |v|\}, & \text{if } uv < 0; \\ 0, & \text{if } u \neq v, uv > 0; \\ \binom{|u|-1}{2} + \binom{n+1-|u|}{2}, & \text{otherwise.} \end{cases}$$

Here $u = (-1)^{i+1}\left(\lceil\frac{i}{2}\rceil + 1\right)$ *and* $v = (-1)^{j+1}\left(\lceil\frac{j}{2}\rceil + 1\right)$, *that is, u and v are the (signed) positions of the states that corresponds to row i and column j, respectively.*

The proof is straightforward — just count the number of inversions that move g_2 from position u to position v.

Example 3. Let $n = 4$. At position $i = 3, j = 4$ in M_4, we have $u = 3$ and $v = -3$. Thus, $m_{34} = \min\{3 - 1, 3 - 1, 5 - 3, 5 - 3\} = 2$. The entire matrix is given by

$$M_4 = \begin{pmatrix} 3 & 1 & 0 & 1 & 0 & 1 \\ 1 & 3 & 1 & 0 & 1 & 0 \\ 0 & 1 & 2 & 2 & 0 & 1 \\ 1 & 0 & 2 & 2 & 1 & 0 \\ 0 & 1 & 0 & 1 & 3 & 1 \\ 1 & 0 & 1 & 0 & 1 & 3 \end{pmatrix}$$

The transition matrix M_n can be used to calculate the estimated number of breakpoints in a genome, given that i inversions have been applied. The reason is that the entry at $(1,1)$ of M_n^i gives the probability p that g_2, after i inversions, is positioned just after g_1, where it does not produce a breakpoint. The probability of a breakpoint is the same after any gene, so the expected number of breakpoints after i inversions is $n(1 - p)$. This is the same as

$$n\left(1 - \frac{\bar{e}_1 M_n^i \bar{e}_1^T}{\binom{n}{2}^i}\right),$$

where $\bar{e}_1 = (1, 0, 0, \ldots, 0)$.

Now, M_n is a real symmetric matrix, so we can diagonalize it as $M_n = V_n D_n V_n^T$, where D_n is a diagonal matrix with the eigenvalues of M_n on the diagonal, and V_n is the orthogonal matrix of eigenvectors. We write $\bar{v}_n = \bar{e}_1 V_n$. The expected number of breakpoints after i inversions, $b(i)$, is then given by

$$b(i) = n\left(1 - \frac{\bar{e}_1 M_n^i \bar{e}_1^T}{\binom{n}{2}^i}\right) = n\left(1 - \frac{\bar{e}_1 V_n D_n^i V_n^T \bar{e}_1^T}{\binom{n}{2}^i}\right) = n\left(1 - \frac{\bar{v}_n D_n^i \bar{v}_n^T}{\binom{n}{2}^i}\right).$$

This analysis proves our first result.

Theorem 2. *Let $\bar{v}_n = (v_1, v_2, \ldots, v_{2n-2}) = \bar{e}_1 V_n$ and let $\lambda_1 \geq \lambda_2 \geq \ldots \geq \lambda_{2n-2}$ be the eigenvalues of M_n. Then the expected number of breakpoints after i random inversions in a genome with n genes can be written as*

$$b(i) = n \left(1 - \frac{\sum_{j=1}^{2n-2} v_j^2 \lambda_j^i}{\binom{n}{2}^i} \right),$$

where $\sum v_j^2 = 1$.

Calculating how fast the expected number of breakpoints approaches the obvious limit $n \left(1 - \frac{1}{2n-2} \right)$ primarily amounts to calculating the eigenvalues of M_n. We will prove the following result, which contains the most important information about the eigenvalues.

Theorem 3. *Let M_n, $n \geq 2$, be the matrix described above and $\lambda_1 \geq \lambda_2 \geq \ldots \geq \lambda_{2n-2}$ its eigenvalues. Then $\lambda_1 = \binom{n}{2}$ and $\lambda_2 = \lambda_3 = \lambda_4 = \binom{n-1}{2}$. The coefficient v_1^2 of λ_1 in the sum above will be $\frac{1}{2n-2}$ and the sum of the coefficients of $\lambda_2 = \lambda_3 = \lambda_4$ will be $\frac{3}{4} - \frac{1}{2n-2}$. The smallest eigenvalue λ_{2n-2} is greater than or equal to $-\lfloor \frac{n}{2} \rfloor$.*

In the appendix, we have gathered information on some of the other eigenvalues and their coefficients.

If we set $v_2^2 = 1 - v_1^2$ and $v_j = 0$ for all $j \geq 3$ in Theorem 3, we get the following approximation.

Corollary 1. *The expected number of breakpoints given i inversions, $b(i)$, can be approximated by*

$$b_{appr}(i) = n \left(1 - \frac{1}{2n-2} \right) \left[1 - \left(1 - \frac{2}{n} \right)^i \right]$$

such that

$$\lim_{i \to \infty} (b(i) - b_{appr}(i)) = 0.$$

By taking the inverse of this map, we obtain an approximation of the expected number of inversions, given that we observe b breakpoints.

Corollary 2. *The expected number of inversions given b breakpoints, $i(b)$, can be approximated by*

$$i_{appr}(b) = \frac{\log \left(1 - \frac{b}{n(1 - \frac{1}{2n-2})} \right)}{\log \left(1 - \frac{2}{n} \right)}.$$

The quality of these approximations will be investigated in the next section. We now give a concrete example of what is going on, followed by the proof of Theorem 3.

Example 4. For $n = 4$, the matrix M_n looks like

$$M_4 = \begin{pmatrix} 3 & 1 & 0 & 1 & 0 & 1 \\ 1 & 3 & 1 & 0 & 1 & 0 \\ 0 & 1 & 2 & 2 & 0 & 1 \\ 1 & 0 & 2 & 2 & 1 & 0 \\ 0 & 1 & 0 & 1 & 3 & 1 \\ 1 & 0 & 1 & 0 & 1 & 3 \end{pmatrix}$$

and has eigenvalues $\{6, 3, 3, 3, 2, -1\}$. It is obvious that both 6 and 3 are eigenvalues, since the row sums are all 6 (hence $(1, 1, 1, 1, 1, 1)$ is an eigenvector with eigenvalue 6) and subtracting 3 from the diagonal would make the rows 1 and 5 equal, as well as rows 2 and 6. If we diagonalize $M_4 = V_4 D_4 V_4^T$, we get

$$V_4 = \begin{pmatrix} -0.6132 & -0.1687 & -0.4230 & 0.4082 & 0.2887 & 0.4082 \\ 0.2835 & 0.1893 & -0.6835 & -0.4082 & -0.2887 & 0.4082 \\ 0.2516 & -0.4719 & 0.2175 & -0.4082 & 0.5774 & 0.4082 \\ 0.2516 & -0.4719 & 0.2175 & 0.4082 & -0.5774 & 0.4082 \\ 0.3615 & 0.6406 & 0.2055 & 0.4082 & 0.2887 & 0.4082 \\ -0.5351 & 0.2826 & 0.4660 & -0.4082 & -0.2887 & 0.4082 \end{pmatrix}$$

and

$$D_4 = \begin{pmatrix} 3 & 0 & 0 & 0 & 0 & 0 \\ 0 & 3 & 0 & 0 & 0 & 0 \\ 0 & 0 & 3 & 0 & 0 & 0 \\ 0 & 0 & 0 & 2 & 0 & 0 \\ 0 & 0 & 0 & 0 & -1 & 0 \\ 0 & 0 & 0 & 0 & 0 & 6 \end{pmatrix}$$

From this, we can calculate

$$\bar{v}_n = \bar{e}_1 V_n = (-0.6132, \ -0.1687, \ -0.4230, \ 0.4082, \ 0.2887, \ 0.4082),$$

and thus

$$b(i) = 4 \left(1 - \frac{0.1667 \cdot 6^i + 0.5833 \cdot 3^i + 0.1667 \cdot 2^i + 0.0833 \cdot (-1)^i}{6^i} \right).$$

Our approximation (from Corollary 1) yields

$$b_{\text{appr}}(i) = \frac{10}{3} \left(1 - \frac{1}{2^i} \right),$$

the inverse of which is (from Corollary 2)

$$i_{\text{appr}}(b) = \frac{\log \left(1 - \frac{3b}{10} \right)}{\log \frac{1}{2}}.$$

We also have

$$
M_5 = \begin{pmatrix}
6 & 1 & 0 & 1 & 0 & 1 & 0 & 1 \\
1 & 6 & 1 & 0 & 1 & 0 & 1 & 0 \\
0 & 1 & 4 & 2 & 0 & 2 & 0 & 1 \\
1 & 0 & 2 & 4 & 2 & 0 & 1 & 0 \\
0 & 1 & 0 & 2 & 4 & 2 & 0 & 1 \\
1 & 0 & 2 & 0 & 2 & 4 & 1 & 0 \\
0 & 1 & 0 & 1 & 0 & 1 & 6 & 1 \\
1 & 0 & 1 & 0 & 1 & 0 & 1 & 6
\end{pmatrix}
$$

with eigenvalues $\{10,\ 6,\ 6,\ 6,\ 4.8284,\ 4,\ 4,\ -0.8284\}$ and

$$
M_6 = \begin{pmatrix}
10 & 1 & 0 & 1 & 0 & 1 & 0 & 1 & 0 & 1 \\
1 & 10 & 1 & 0 & 1 & 0 & 1 & 0 & 1 & 0 \\
0 & 1 & 7 & 2 & 0 & 2 & 0 & 2 & 0 & 1 \\
1 & 0 & 2 & 7 & 2 & 0 & 2 & 0 & 1 & 0 \\
0 & 1 & 0 & 2 & 6 & 3 & 0 & 2 & 0 & 1 \\
1 & 0 & 2 & 0 & 3 & 6 & 2 & 0 & 1 & 0 \\
0 & 1 & 0 & 2 & 0 & 2 & 7 & 2 & 0 & 1 \\
1 & 0 & 2 & 0 & 2 & 0 & 2 & 7 & 1 & 0 \\
0 & 1 & 0 & 1 & 0 & 1 & 0 & 1 & 10 & 1 \\
1 & 0 & 1 & 0 & 1 & 0 & 1 & 0 & 1 & 10
\end{pmatrix}
$$

with eigenvalues $\{15,\ 10,\ 10,\ 10,\ 8.7392,\ 7,\ 7,\ 7,\ 5.7759,\ -0.5151\}$. It is clear that in these examples, both $\binom{n}{2}$ and $\binom{n-1}{2}$ are eigenvalues of M_n. Also, it turns out that the eigenvalues of M_6 are related to the eigenvalues of M_4. In fact, if we write

$$
M_6 = \begin{pmatrix}
6 & 0 & 0 & 0 & 0 & 0 & 0 & 0 & 0 & 0 \\
0 & 6 & 0 & 0 & 0 & 0 & 0 & 0 & 0 & 0 \\
0 & 0 & 3 & 1 & 0 & 1 & 0 & 1 & 0 & 0 \\
0 & 0 & 1 & 3 & 1 & 0 & 1 & 0 & 0 & 0 \\
0 & 0 & 0 & 1 & 2 & 2 & 0 & 1 & 0 & 0 \\
0 & 0 & 1 & 0 & 2 & 2 & 1 & 0 & 0 & 0 \\
0 & 0 & 0 & 1 & 0 & 1 & 3 & 1 & 0 & 0 \\
0 & 0 & 1 & 0 & 1 & 0 & 1 & 3 & 0 & 0 \\
0 & 0 & 0 & 0 & 0 & 0 & 0 & 0 & 6 & 0 \\
0 & 0 & 0 & 0 & 0 & 0 & 0 & 0 & 0 & 6
\end{pmatrix} + \begin{pmatrix}
0 & 1 & 0 & 1 & 0 & 1 & 0 & 1 & 0 & 1 \\
1 & 0 & 1 & 0 & 1 & 0 & 1 & 0 & 1 & 0 \\
0 & 1 & 0 & 1 & 0 & 1 & 0 & 1 & 0 & 1 \\
1 & 0 & 1 & 0 & 1 & 0 & 1 & 0 & 1 & 0 \\
0 & 1 & 0 & 1 & 0 & 1 & 0 & 1 & 0 & 1 \\
1 & 0 & 1 & 0 & 1 & 0 & 1 & 0 & 1 & 0 \\
0 & 1 & 0 & 1 & 0 & 1 & 0 & 1 & 0 & 1 \\
1 & 0 & 1 & 0 & 1 & 0 & 1 & 0 & 1 & 0 \\
0 & 1 & 0 & 1 & 0 & 1 & 0 & 1 & 0 & 1 \\
1 & 0 & 1 & 0 & 1 & 0 & 1 & 0 & 1 & 0
\end{pmatrix} + 4I = B_6 + A_6 + 4I,
$$

then A_6 has eigenvalues $\{5, 0, 0, 0, 0, 0, 0, 0, 0, -5\}$, $4I$ has only 4 as eigenvalue and B_6 has eigenvalues $\{6, 6, 6, 6, 6, 3, 3, 3, 2, -1\}$. If we compare the latter with the eigenvalues of M_4, we see that apart from the first four eigenvalues, we now have the same eigenvalues as for M_4. This comes as no surprise, since if we remove the first and last two rows and columns, we get a matrix that is exactly M_4. We will show below that we can always write $M_n = B_n + A_n + (n-2)I$, where B_n is a block matrix containing M_{n-2} in a similar fashion as B_6 contains M_4.

Lemma 1. *Let* $A_n = (a_{ij})$ *be a quadratic matrix of size* $2n - 2$ *with entries given by*

$$a_{ij} = \begin{cases} 1, & \text{if } i+j \text{ is odd;} \\ 0, & \text{otherwise.} \end{cases}$$

Then the eigenvalues are $n-1$ *with multiplicity* 1, 0 *with multiplicity* $2n-4$ *and* $-(n-1)$ *with multiplicity* 1.

Proof. Since A_n is real and symmetric, it has $2n - 2$ eigenvalues and an orthonormal set of equally many eigenvectors. It is easy to see that $(1, 1, \ldots, 1)$ is an eigenvector with eigenvalue $n - 1$ and that $(1, -1, 1, -1, \ldots, 1, -1)$ is an eigenvector with eigenvalue $-(n - 1)$. Clearly the rank of A_n is 2, and hence all remaining eigenvalues must be 0.

We are now able to prove Theorem 3.

Proof. (of Theorem 3) It is easy to see that $\binom{n}{2}$ and $\binom{n-1}{2}$ always are eigenvalues of M_n, since the common row sum equals the number of inversions on a genome with n genes, which is $\binom{n}{2}$, and since the second and last rows are equal, except for the term $\binom{n-1}{2}$ on the diagonal. However, we also need to show that there are no other eigenvalues as large as these.

We will use induction. Since our claim is true for M_4 and M_5 in the example above (and anyone can check that it also holds for M_2 and M_3), we can concentrate on the inductive step.

Consider M_n and define $B_n = M_n - A_n - (n - 2)I$, where A_n is given in the previous lemma and I is the identity matrix. Let $C_n = (c_{ij})$ be the matrix B_n with the first and last two rows and columns removed. C_n will have the same size as $M_{n-2} = (m_{ij})$ and we shall see that these matrices are in fact identical. For $i + j$ odd, we have

$$c_{ij} = \min\{\lceil \tfrac{i+2}{2} \rceil - 1, \lceil \tfrac{j+2}{2} \rceil - 1, n + 1 - \lceil \tfrac{i+2}{2} \rceil, n + 1 - \lceil \tfrac{j+2}{2} \rceil\} - 1$$

$$= \min\{\lceil \tfrac{i}{2} \rceil - 1, \lceil \tfrac{j}{2} \rceil - 1, (n - 2) + 1 - \lceil \tfrac{i}{2} \rceil, (n - 2) + 1 - \lceil \tfrac{j}{2} \rceil\} = m_{ij},$$

and for $i + j$ even, with $i \neq j$, we have

$$c_{ij} = 0 = m_{ij}.$$

Finally, on the main diagonal, we have

$$c_{ii} = \binom{\lceil \tfrac{i+2}{2} \rceil - 1}{2} + \binom{n + 1 - \lceil \tfrac{i+2}{2} \rceil}{2} - (n - 2)$$

$$= \binom{\lceil \tfrac{i}{2} \rceil - 1}{1} + \binom{\lceil \tfrac{i}{2} \rceil - 1}{2} + \binom{n - 1 - \lceil \tfrac{i}{2} \rceil}{1} + \binom{n - 1 - \lceil \tfrac{i}{2} \rceil}{2} - (n - 2)$$

$$= n - 2 + m_{ii} - (n - 2) = m_{ii}.$$

Thus, with the only entries on the first and last two rows of B_n being four copies of $\binom{n-1}{2} - (n - 2) = \binom{n-2}{2}$ on the diagonal, the eigenvalues of B_n consists of these four copies of $\binom{n-2}{2}$ and all the eigenvalues of M_{n-2}. Since the greatest eigenvalue of M_{n-2} is $\binom{n-2}{2}$, this is also the greatest eigenvalue of B_n.

We are now in the position to estimate the eigenvalues of M_n. If we let $\lambda_i(A)$ be the ith greatest eigenvalue of any matrix A, then it is known (see for instance [5], p. 52) that

$$\lambda_{1+i+j}(A+B) \leq \lambda_{1+i}(A) + \lambda_{1+j}(B).$$

We will apply this inequality to find the eigenvalues of $M_n = A_n + B_n + (n-2)I_n$. We know the eigenvalues of A_n from Lemma 1 and B_n by the induction hypothesis. Hence, we get

$$\lambda_1(M_n) \leq \lambda_1(A_n) + \lambda_1(B_n) + \lambda_1((n-2)I)$$
$$= (n-1) + \binom{n-2}{2} + (n-2) = \binom{n}{2}$$

and

$$\lambda_2(M_n) \leq \lambda_2(A_n) + \lambda_1(B_n) + \lambda_1((n-2)I)$$
$$= 0 + \binom{n-2}{2} + (n-2) = \binom{n-1}{2}.$$

Since we know that these are in fact eigenvalues, the inequalities are actually equalities.

So far, we have shown that $\lambda_1 = \binom{n}{2}$ and $\lambda_2 = \binom{n-1}{2}$. In order to see that $\binom{n-1}{2}$ is an eigenvalue with multiplicity at least 3, we need to check that the three linearly independent vectors $\overline{w}_1 = (1,0,0,\ldots,0,-1,0)$, $\overline{w}_2 = (0,1,0,\ldots,0,0,-1)$ and $\overline{w}_3 = (\frac{n-3}{2}, \frac{n-3}{2}, -1, -1, \ldots, -1, \frac{n-3}{2}, \frac{n-3}{2})$ all are eigenvectors of M_n with eigenvalue $\binom{n-1}{2}$. It is clear for the two first, and for $\overline{w}_3$, multiplying it with M_n give the entries

$$\frac{n-3}{2}\binom{n-1}{2} + 2\frac{n-3}{2} - \frac{2n-2-4}{2} = \frac{n-3}{2}\binom{n-1}{2}$$

for the first and last two positions, and

$$2\frac{n-3}{2} - \left(\binom{n}{2} - 2\right) = \binom{n-1}{1} - \binom{n}{2} = -\binom{n-1}{2}$$

for the other entries. Thus, $M_n\overline{w}_3^T = \binom{n-1}{2}\overline{w}_3^T$.

We now turn to the coefficients of these eigenvalues in the sum giving $b(i)$. The coefficients are given by the first element in the normalized eigenvectors. For $\lambda_1 = \binom{n}{2}$, the first element is $v_1 = \frac{1}{\sqrt{2n-2}}$ and thus gives the coefficient $v_1^2 = \frac{1}{2n-2}$. For the eigenvalue $\lambda_2 = \lambda_3 = \lambda_4 = \binom{n-1}{2}$, we get

$$v_2^2 + v_3^2 + v_4^2 = \frac{1}{2} + 0 + \frac{\left(\frac{n-3}{2}\right)^2}{4\left(\frac{n-3}{2}\right)^2 + 2n - 6} = \frac{1}{2} + \frac{(n-3)}{4((n-3)+2)} = \frac{3}{4} - \frac{1}{2n-2}.$$

Finally, turning to the smallest eigenvalue, we can use

$$\lambda_{m-i-j}(A+B) \geq \lambda_{m-i}(A) + \lambda_{m-j}(B),$$

where $m \times m$ is the common size of A and B, to show that the smallest eigenvalue is greater than or equal to $-\lfloor \frac{n}{2} \rfloor$. This holds for M_2 and M_3, and by the same procedure as above, we find that

$$\lambda_{2n-2}(A+B) \geq \lambda_{2n-2}(A) + \lambda_{2n-2}(B) + \lambda_{2n-2}((n-2)I)$$

$$\geq -(n-1) - \lfloor \frac{n-2}{2} \rfloor + (n-2) = -\lfloor \frac{n}{2} \rfloor.$$

This ends the proof of Theorem 3.

4   Analyzing and Improving the Formula

We will now leave the analytical trail and look at how well these approximations behave in practice. Based on abundant observations (every $n \leq 100$), we believe that the largest eigenvalues are distributed as follows.

Conjecture 1. Let M_n be the scaled transition matrix studied in the previous section and $\lambda_1 \geq \lambda_2 \geq \ldots \geq \lambda_{2n-2}$ its eigenvalues. Then $\lambda_6 = \lambda_7 = \lambda_8 = \binom{n-2}{2}$ for $n \geq 6$.

Since we know the four greatest eigenvalues, this conjecture implies that there is only one unknown eigenvalue λ_5 larger than $\binom{n-2}{2}$. Knowledge of this eigenvalue and its coefficient v_5^2 would give a better approximation than the one found above. We now take a closer look at these parameters.

First we look at the unknown coefficients. We know that the sum of all coefficients is 1 and that, according to Theorem 3, the coefficients of the eigenvalues $\binom{n}{2}$ and $\binom{n-1}{2}$ sum to $\frac{3}{4}$. For the remaining $\frac{1}{4}$, numerical calculations for $n \leq 100$ indicate that almost everything has been given to λ_5. We see in Figure 1, where the coefficients has been plotted for $n \leq 40$ that this coefficient fast approaches $\frac{1}{4}$. The sum of the remaining coefficients has been plotted in Figure 2. We see that for $n = 40$, their sum is less than 0.001. We can neglect this without worries.

Supported by this, we now propose an improved approximation of the expected number of breakpoints, given i inversion. By setting $v_5^2 = \frac{1}{4}$ and writing $\lambda_5 = \binom{n-1}{2} - \varepsilon(n)$, we get

$$b_{\text{appr2}}(i) = n \left[1 - \frac{1}{2n-2} - \left(\frac{3}{4} - \frac{1}{2n-2} \right) \left(1 - \frac{2}{n} \right)^i - \frac{1}{4} \left(\frac{\binom{n-1}{2} - \varepsilon(n)}{\binom{n}{2}} \right)^i \right]$$

$$\approx n \left(1 - \frac{1}{2n-2} \right) \left(1 - \left(1 - \frac{2}{n} \right)^i \right) + \frac{i\varepsilon(n)}{2n-2} \left(1 - \frac{2}{n} \right)^{i-1}.$$

The final approximation was obtained by including only the first two terms in the binomial expansion of the last term. From this approximation of $b_{\text{appr2}}(i)$, we find that the error of $b_{\text{appr}}(i)$ is approximately $\frac{i\varepsilon(n)}{2n-2} \left(1 - \frac{2}{n} \right)^{i-1}$.

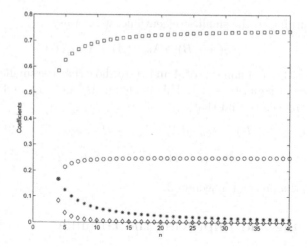

Fig. 1. How the coefficients evolve with increasing n. The stars correspond to the eigenvalue $\binom{n}{2}$, the squares to $\binom{n-1}{2}$, the circles to the eigenvalue just below $\binom{n-1}{2}$ and the diamonds to the sum of all other coefficients.

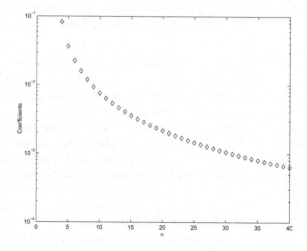

Fig. 2. The coefficients of the eigenvalues we neglect. These tend to zero quite rapidly as n increases.

The usefulness of this improved approximation, which equals our first if we put $\varepsilon(n) = 0$, depends on our ability to calculate $\varepsilon(n)$. We have plotted $\varepsilon(n)$ as a function of n in Figure 3. We find (left graph) that for $40 \leq n \leq 100$, we can approximate $\varepsilon(n)$ with, for instance, $\varepsilon(n)_{\mathrm{appr}} = 1.7 + 0.0016n$, and for larger n, $\varepsilon(n) = 2$ seems to be an as good approximation as one could hope for.

A comparison between the quality of our approximations can be found in Figures 4 and 5. The true values have, of course, been taken from the Markov chain. We have plotted the error depending on i for the following values of n:

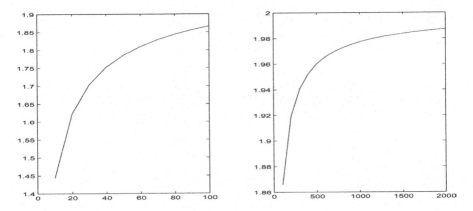

Fig. 3. Two graphs of the function $\varepsilon(n)$. We have plotted this function for different ranges of n in order to make its behavior clear for small n as well as for large n. For small n, it increases from less than $\frac{3}{2}$ to about 2, but for large n it stays fairly constant, just below 2.

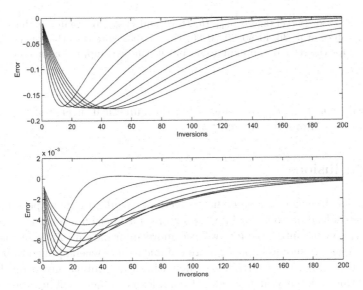

Fig. 4. The error of $b_{\mathrm{appr}}(i)$ (top) and $b_{\mathrm{appr2}}(i)$ for these n: $30, 40, 50, 60, 70, 80, 90$ and 100. The latter is one or two orders of magnitude lower.

$30, 40, 50, 60, 70, 80, 90$ and 100. We see that for the approximations of $b(i)$, the first approximation $b_{\mathrm{appr}}(i)$ has an error that stays below 0.2 breakpoints, which is fairly good. With the second approximation, $b_{\mathrm{appr2}}(i)$, the error is well below 0.01 breakpoints.

For the approximations of $i(b)$, we see that when b approaches its upper limit, the error increases. This is bound to happen, since the slope of $b(i)$ decreases towards zero for large i. Still, the error percentage is not to high, even for the

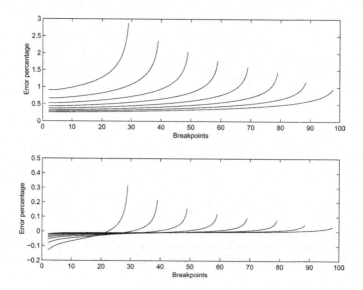

Fig. 5. The error percentage of $i_{appr}(b)$ (top) and $i_{appr2}(b)$. Again, we have used the following values of n: 30, 40, 50, 60, 70, 80, 90 and 100. Note that there is no analytical expression for the second approximation, but it can be computed numerically from its inverse.

analytical expression $i_{appr}(b)$, and using the numerical inverse of $b_{appr2}(i)$, the error vanishes in practice for most b.

5   Conclusions

Caprara and Lancia considered the problem of calculating the expected number of inversions leading to b breakpoints for *unsigned* permutations. By taking the sign of the genes into account, we use more information. Thus, using signed permutations, we should hope for more reliable results when applied to biological problems. In this case, contrary to calculating the minimal inversion distance, we gain this reliability at the cost of complexity. Where solving the unsigned case amounts to calculating the expected number of breakpoints in a random permutation ($n - 2$ by linearity of expectation values) and the probability that a random inversion creates or destroys the adjacency of two genes, the signed case requires the calculation of many permutation-specific probabilities and the calculation of the eigenvalues of a matrix containing these probabilities. In other words: for signed permutations we calculate the eigenvalues of a $(2n-2) \times (2n-2)$-matrix, for unsigned permutations the corresponding matrix has size 1×1.

It is a truly remarkable property of the transition matrices M_n that most of their eigenvalues can be calculated without much effort. This insight provides us with the means to compute the expected number of inversions giving rise to b breakpoints. The error in this calculation is most certainly negligible, compared to the standard deviation. For the future, calculating this standard deviation seems to be an important problem. Also, it would be interesting to see if the information not used for the problem considered in this paper, for example the cycle structure of the permutations, can be used to obtain an even more realistic distance measure.

Acknowledgements

I wish to thank my advisor Kimmo Eriksson for valuable comments during the preparation of this paper.

Niklas Eriksen was supported by the Swedish Research Council and the Swedish Foundation for Strategic Research.

A   Further Information on the Spectrum of M_n

There is some information known about the eigenvalues (and their coefficients) of M_n, which do no affect the analysis above. We will present this information here.

Theorem 4. *In addition to the eigenvalues found in Theorem 3, M_n has eigenvalues $\binom{n-2}{2}+\binom{2}{2}, \binom{n-3}{2}+\binom{3}{2}, \ldots, \binom{n-\lceil\frac{n-2}{2}\rceil}{2}+\binom{\lceil\frac{n-2}{2}\rceil}{2}$ with multiplicity 3 (if n is odd, $\binom{n-\lceil\frac{n-2}{2}\rceil}{2}+\binom{\lceil\frac{n-2}{2}\rceil}{2}$ has multiplicity 2). The coefficients of these eigenvalues are 0.*

Proof. We saw in the proof for Theorem 3 that $\overline{w}_1 = (1,0,0,\ldots,0,-1,0)$, $\overline{w}_2 = (0,1,0,\ldots,0,0,-1)$ and $\overline{w}_3 = (\frac{n-3}{2},\frac{n-3}{2},-1,-1,\ldots,-1,\frac{n-3}{2},\frac{n-3}{2})$ are eigenvectors of M_n with eigenvalue $\binom{n-1}{2}$. In fact, if we add two zeros at the front and at the back of the eigenvector $\overline{w}_i$ of M_{n-2}, we get a corresponding eigenvector for the eigenvalue $\binom{n-2}{2}+\binom{2}{2}$ in M_n. Using this inductively, it is easy to show that the eigenvalues $\binom{n-k}{2}+\binom{k}{2}$ for $2 \leq k \leq \lceil\frac{n-2}{2}\rceil$ have three-dimensional eigenspaces. The exception is, as mentioned, $\binom{n-\lceil\frac{n-2}{2}\rceil}{2}+\binom{\lceil\frac{n-2}{2}\rceil}{2}$ for odd n, which only has a two-dimensional eigenspace, since the eigenvector that would correspond to $\overline{w}_3$ then equals the sum of the other two eigenvectors.

Turning to their coefficients, we just found that all eigenvectors of $\binom{n-k}{2}+\binom{k}{2}$, $2 \leq k \leq \lceil\frac{n-2}{2}\rceil$ has a zero in the first position. Hence, the coefficients are also zero.

References

1. Bader, D. A., Moret, B. M. E., Yan, M.: A Linear-Time Algorithm for Computing Inversion Distance Between Signed Permutations with an Experimental Study. Journal of Computational Biology, **8**, 5 (2001), 483–491
2. Blanchette, M., Kunisawa, T., Sankoff, D.: Parametric genome rearrangement. Gene **172** (1996), GC 11–17

3. Caprara, A.: Sorting permutations by reversals and Eulerian cycle decompositions. SIAM Journal of Discrete Mathematics **12** (1999), 91–110
4. Caprara, A., Lancia, G.: Experimental and statistical analysis of sorting by reversals. Sankoff and Nadeau (eds.), Comparative Genomics (2000), 171–183
5. Cvetković, D. M., Doob, M., Sachs, H.: Spectra of Graphs. Johann Ambrosius Barth Verlag, Heidelberg, 1995
6. Eriksen, N.: $(1 + \varepsilon)$-Approximation of Sorting by Reversals and Transpositions. Algorithms in Bioinformatics, Proceedings of WABI 2001, LNCS 2149, 227–237
7. Eriksen, N.: Expected number of inversions after a sequence of random adjacent transpositions — an exact expression. Preprint
8. Eriksen, N., Dalevi, D., Andersson, S. G. E., Eriksson, K.: Gene order rearrangements with Derange: weights and reliability. Preprint
9. Eriksson, H., Eriksson, K., Sjöstrand, J.: Expected inversion number after k adjacent transpositions Proceedings of Formal Power Series and Algebraic Combinatorics 2000, Springer Verlag, 677–685
10. Hannenhalli, S., Pevzner, P.: Transforming cabbage into turnip (polynomial algorithm for sorting signed permutations with reversals). Proceedings of the 27th Annual ACM Symposium on the Theory of Computing (1995), 178-189
11. Pevzner, P.: Computational Molecular Biology: An Algorithmic Approach. The MIT Press, Cambridge, MA 2000
12. Sankoff, D., Blanchette, M.: Probability models for genome rearrangements and linear invariants for phylogenetic inference. Proceedings of RECOMB 1999, 302–309
13. Wang, L.-S.: Exact-IEBP: A New Technique for Estimating Evolutionary Distances between Whole Genomes. Algorithms in Bioinformatics, Proceedings of WABI 2001, LNCS 2149, 175–188

Invited Lecture –
Accelerating Smith-Waterman Searches

Gene Myers[1] and Richard Durbin[2]

[1] Celera Genomics, Rockville, MD, USA
Gene.Myers@celera.com
[2] Sanger Centre, Hinxton Hall, Cambridgeshire, UK

Abstract. Searching a database for a local alignment to a query under a typical scoring scheme such as PAM120 or BLOSUM62, is a computation that has resisted algorithmic improvement due to its basis in dynamic programming and the weak nature of the signals being searched for. In a query preprocessing step, a set of tables can be built that permit one to (a) eliminate a large fraction of the dynamic programming matrix from consideration, and (b) to compute several steps of the remainder with a single table lookup. While this result is not an asymptotic improvement over the original Smith-Waterman algorithm, its complexity is characterized in terms of some sparse features of the matrix and it does yields 3.33 factor improvement over the basic algorithm in practice.

1   Introduction

Consider the problem of searching a database of protein sequences looking for those that are similar to a given query sequence. By similar we mean that the query and the target sequence in question have a local alignment between a substring of each that scores above a specifiable threshold under a specifiable alignment scoring scheme. The first algorithm for this problem was given by Smith and Waterman [1] whose names have now become synonymous with the search. A Smith-Waterman search is considered a *full-sensitivity* search as it is guaranteed to find all local alignments above a threshold, whereas the popular heuristics BLAST [2] and FASTA [3] miss some matches.

Techniques for speeding up Smith-Waterman searches have primarily involved appeals to hardware parallelization. A coarse-grained parallelization is achieved by evenly partitioning the database to be searched and assigning each partition to a processor of a MIMD machine or network of workstations [4]. Fine-grained parallelization of the dynamic programming algorithm using SIMD supercomputers or systolic arrays has also been reported and involves assigning an array of processing elements to the cells on an anti-diagonal of the dynamic programming matrix [5]. In this paper, we explore software acceleration using the approach of building a set of tables that allow us (1) to perform several steps of the underlying dynamic programming computation in a single lookup, and (2) to more fully exploit sparsity, often eliminating over 90% of the entries because they are unproductive (i.e. cannot be part of a local alignment above threshold).

R. Guigó and D. Gusfield (Eds.): WABI 2002, LNCS 2452, pp. 331–342, 2002.

The impetus for this work stemmed from one of the authors earlier work on the application of the 4-Russians techniques to approximate matching under unit-cost distance measures [6]. We were interested in whether precomputed tables could be used to accelerate Smith-Waterman searches. It was immediately clear that while the 4-Russians method could be applied, the tables would be horrendously large and ineffective. In our application, the query is relatively small (length 100-1000) in comparison to the database it is to be compared against (length 10's of millions or more). We therefore looked at building large tables specific to a query as a prelude to the search. Basically, we found that one can eliminate many short paths that cannot contribute to a match by building tables that only bring to the fore paths that are positive over the first k_S target characters and have a positive extension over the next k_E characters. We show how to build these tables in Section 3 and present empirical results that show that 95.5% of the matrix is eliminated from consideration when $k_S = k_E = 4$, the largest possible values of the parameters given that the tables grow exponentially in these parameters. In Section 4, we develop a table that allows one to look up the values attainable k_J target chars forward of a given matrix entry, thus permitting the database to be processed k_J characters at a time. The best overall empirical results were obtained with $k_J = 3$, where a 3.33 speed factor over the basic dynamic programming algorithm is achieved.

2   Preliminaries

Our problem is formally as follows. We are given:

1. an alphabet Σ over which sequences are composed,
2. a $|\Sigma| \times |\Sigma|$ scoring matrix S giving the score, $S[a][b]$, of aligning a with b,
3. a uniform gap penalty $g > 0$,
4. a query sequence $Query = q_1 q_2 \ldots q_P$ of P letters over alphabet Σ,
5. a target sequence $Target = t_1 t_2 \ldots t_N$ of N letters over Σ, and
6. a threshold $\tau > 0$.

The problem is to determine if there is an alignment between a substring of $Query$ and a substring of $Target$ whose score is not less than threshold τ.

To reduce the length of forthcoming recurrences, we will let let T_j^{-a} denote $Target[j - (a-1) \ldots j]$, the a symbols of the target ending at the jth, and T_j^b denote $Target[j + 1 \ldots j + b]$, the b symbols following it.

We assume that the reader is familiar with the *edit graph* formulation [7] of sequence comparison where the problem of finding a high-scoring local alignment between $Query$ and $Target$ is mapped to one of finding a high-scoring path in an edge-weighted, directed graph whose vertices are arranged in a $(P+1) \times (N+1)$ array as illustrated in Figure 1. Briefly reviewing its construction, each diagonal substitution edge from $(i-1, j-1)$ to (i, j) models aligning q_i and t_j and has weight or score $S[q_i][t_j]$, each horizontal deletion edge from $(i-1, j)$ to (i, j) models leaving q_i unaligned and has weight $-g$, and each vertical insertion edge from $(i, j-1)$ to (i, j) models leaving t_j unaligned, also with weight $-g$. There is then

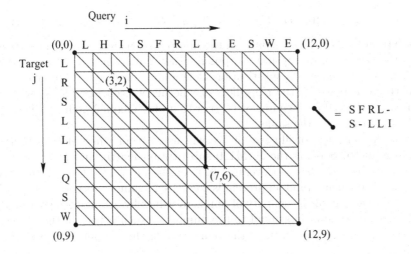

Fig. 1. A sample edit graph and local alignment.

a one-to-one correspondence between local alignments of say $Query[g \ldots i]$ and $Target[h \ldots j]$ and paths from $(g{-}1, h{-}1)$ to (i, j) where the score of the alignment equals the weight or score of the path. Thus, in the edit graph framework, our problem is to determine if there is a path in the graph whose score is τ or greater.

The first observation is that it suffices to limit our attention to *prefix-positive* paths. A prefix of a path is a sub-path consisting of some initial sequence of its edges. A path is prefix-positive if the score of every non-empty prefix of the path is positive. Given an arbitrary path of positive score τ, one obtains a prefix-positive sub-path of score τ or greater by eliminating the prefix (if any) with the most negative score. Thus if there is a path of score τ or greater in the edit graph of $Query$ and $Target$ then there is a prefix-positive path of score τ or greater. While we will not use it here, one could go on to define suffix-positive paths and argue that it suffices to consider only paths that are both prefix- and suffix-positive.

Now consider the sequence of edges in a prefix-positive path. Let S denote a substitution edge, I an insertion edge, and D a deletion edge. Then the sequence of edge types must be in the regular language $S(I|S|D)^*$ as the first edge must be positive and only substitution edges score so. We may equivalently write this as $SD^*((S|I)D^*)^*$ which in plain words asserts that a path can be partitioned into sub-paths such that each subpath is a substitution or insertion followed by a possibly empty series of deletions, save for the first which must begin with a substitution. We define the *length* of a path to be the number of substitution and insertion edges within it, or equivalently, to be the number of target symbols in its corresponding alignment. Thus a path from (g, h) to (i, j) has length $j{-}h$.

The basic dynamic programming algorithm for Smith-Waterman searches involves computing $C(i, j)$, the score of a heaviest or maximal path ending at vertex (i, j). It is not difficult to observe the following recurrence [1]:

$$C(i,j) = \max\{0, C(i-1,j-1) + S[q_i][t_j], C(i-1,j)-g, C(i,j-1)-g\}$$

The 0 term models the possibility of the null-path that begins and ends at (i,j), the other three terms capture the possibilities of reaching (i,j) by a maximal path whose last edge is a substitution, deletion, or insertion, respectively. Note that the recurrence is true regardless of whether we consider $C(i,j)$ to be the score of a maximal path ending at (i,j) or that of a maximal prefix-positive path to (i,j) as these are the same.

For the purposes of illustration we will use, except where noted otherwise, the example of S being the PAM120 substitution matrix and $g = 12$. In this case, one observes that, on average, 62% of the $(P+1) \times (N+1)$ matrix of C-values is 0 when a random protein is compared against another[1]. It is possible to take advantage of the sparsity of non-zero values by arranging a computation which spends time proportional to the number of such entries in the C-matrix. We sketch such an approach in the remainder of this section as a preliminary exercise.

Consider a set p of index-value pairs $\{(i,v) : i \in [0,P]\}$ where each index is to a column of the edit graph. In what follows, the pairs will model the column and score of a path ending in a given row of the edit graph. We call such collections *path sets*. The following operators on path sets will be used to succinctly describe forthcoming algorithms. Given two path sets p and q let their merged union, $\cup^{max}$ be

$$p \cup^{max} q = \{(i, \max u) : (i,u) \in p \cup q\}$$

That is, for index-value pairs in p and q with the same index, keep the pair with the greatest value. Furthermore for a path set p and symbol $a \in \Sigma$, let

$$Advance(p,a) = \{(i, v-g) : (i,v) \in p\} \ \cup^{max} \ \{(i+1, v + S[q_{i+1}][a]) : (i,v) \in p\}$$

Informally, $Advance(p,a)$ extends the paths in p with a substitution or insertion edge presuming the next target sequence letter is a. To capture the effect of extending by a (possibly empty) sequence of deletion edges, let

$$Delete_\tau(p) = \cup_{max}\{(i+\Delta, v-\Delta g) : (i,v) \in p \text{ and } v-\Delta g > \tau\}$$

Finally, the path set $Start(a)$ will define the set of paths that begin with a positive substitution edge on target letter a:

$$Start(a) = \{(i,v) : v = S[q_i][a] > 0\}$$

The value of $Start(a)$ is computed in an $O(P|\Sigma|)$ preprocessing step for each value of $a \in \Sigma$ and stored in a table for use during the search proper.

With these definition in hand we now turn to the problem of computing the non-zero values of the C-matrix row by row. Let Pos_j be the path set

---

[1] We assumed an average length of 300 residues and generated each "random" protein via a series of Bernoulli trials that selected residues with the frequency with which they are found in a recent download of Genbank.

$\{(i, v) : v = C(i, j) > 0\}$ that models the non-zero values in row j. A positive entry in row j of the edit graph is the value of a prefix-positive path that is either (a) a maximal prefix-positive path to row $j-1$ followed by an insertion or substitution edge to row j followed by a possibly empty sequence of deletion edges in row j, or (b) a positive weight substitution edge from row $j-1$ to row j followed by a possibly empty sequence of deletion edges in row j. It thus directly follows that

$$Pos_j = \begin{cases} Delete_0(Advance(Pos_{j-1}, t_j) \cup^{max} Start(t_j)) & \text{if } j \geq 1 \\ \emptyset & \text{if } j < 1 \end{cases}$$

Thus we have a computable recurrence for Pos_j in terms of Pos_{j-1}. All that remains is to examine how efficiently the computation can be done.

Once a table of $Start$ values has been precomputed, one can perform any number of searches against targets in a database in $O(N + Pos)$ search time per target using the recurrence immediately above, where $Pos = \Sigma_j |Pos_j|$ is the number of non-zero entries in the C-matrix. In the practical case of comparison under PAM120 scores with a uniform gap penalty of 12, this simple algorithm results in only 38% of the matrix being computed on average. Nonetheless, the algorithm takes 201 seconds on average to search a 300 residue query against a 30million residue database, whereas a careful coding of the basic dynamic programming algorithm performs the search in only 186 seconds[2]. This is because the increased logic required to compute each entry in the sparse algorithm requires about 2.74 times longer than in the basic algorithm, making our sparse algorithm take about 7% more time on average.

3   The Start and Extension Tables

Entries that are zero in the dynamic programming matrix are clearly "unproductive" in that a prefix-positive path does not pass through such a vertex. In this section we lever this basic idea further by building tables that help us avoid expending effort on additional unproductive vertices. The size of our tables and the degree to which they screen the matrix will be determined by two small parameters k_S and k_E which in practice might typically be set from 1 to 4, but no higher . Let the length of a path or alignment be the number of symbols in the target subsequence it aligns. Consider a vertex x in row j of the edit graph of $Query$ vs. $Target$ and the following definitions:

1. x is a *seed* if there exists a prefix-positive path to x that is of length k_S, i.e., that starts in row $j - k_S$.
2. x is *startable* if there exists a positive path to x whose prefix of length k_S is a seed.

---

[2] All timing experiments were performed on a T22 IBM laptop with 256Mb of memory and a 1Ghz Pentium III processor with 256Kb cache, running SuSE Linux 7.1. Our implementation of the dynamic programming algorithm is a straight implementation of the earlier recurrence with care taken to load and store from row vectors only once, and with a scoring matrix $Q[a][i] = S[a][q_i]$ tailored to the query.

3. x is *extendable* if there exists a positive path through x to row $j + k_E$ whose suffix of length k_E is prefix positive, i.e., it has a prefix-positive "extension" of length k_E beyond x.
4. x is *productive* if it is both startable and extendable.

Let *Seed*, *Startable*, *Extensible*, and *Productive* be the sets of vertices that are seeds, are startable, are extendable, and are productive, respectively. Note that *Seed* is a proper subset of *Startable* as the suffix of a long prefix-positive path that spans the last k_S rows need not be prefix-positive. In this section we show how to build tables for a given query that with an $O(1)$ lookup deliver the startable vertices for any target substring of length k_S, and indicate if a given vertex is extendable given any ensuring target substring of length k_E. Using these tables allows an algorithm that takes time to the number of productive vertices for a given problem instance.

3.1   Start Trimming

Limiting the dynamic programming algorithm to the startable vertices requires a table $Start(w)$ where w ranges over the $|\Sigma|^{k_S}$ possible sequences of length k_S. The entry for a given sequence w is defined as follows:

$$Start(w) = \{ \ (i,v) : \exists \text{ prefix-positive path from row 0 to vertex } (i, k_S) \text{ in}$$
$$\text{the edit graph of } Query \text{ vs. } w, \text{ and } v \text{ is the score}$$
$$\text{of a maximal such path } \}$$

The entire table may be computed efficiently using the simple sparse algorithm of the previous section for figuring the path set for each entry w and by factoring the computation of entries that share a common prefix as detailed in the following recurrence:

$$Start(\varepsilon) = \{(i,0) : i \in [0, P]\}$$
$$Start(wa) = Delete_0(Advance(Start(w), a))$$

Effectively $Start(w)$ is computed for every sequence of length k_S or less, with the path set for wa being computed incrementally from that for w. This reduces the worst-case time for computing the entire table $Start$, $O(P|\Sigma|^{k_S})$, by a factor of k_S over the brute-force algorithm. Let α be the expected percentage, $|Seed|/(N+1)(P+1)$, of vertices that are seed vertices in a comparison, for example, as reported in Table 1. It follows by definition that the expected size of the path set for an entry of the $Start$ table is αP and thus that the entire table occupies $O(\alpha P|\Sigma|^{k_S})$ space in expectation. With care, only 4 bytes are required for each index-value pair in a path set and the table can be realized as an array of pointers to sequentially allocated path sets giving a concrete estimate of $(4\alpha P+4)|\Sigma|^{k_S}$ bytes on a 32-bit address machine.

With the $Start$ table in hand, one can easily limit the dynamic programming computation of $Query$ vs. $Target$ to the startable vertices in each row. Let $Stable_j$ be the path set $\{(i,v) : (i,j) \text{ is a startable vertex and } v = C(i,j)\}$. It

Table 1. Sparsity of Table-Computed Vertex Sets.

k_S	Startable	Seed
1	37.8%	22.1%
2	24.1%	11.2%
3	17.2%	6.9%
4	12.7%	4.6%

k_E	Extensible
1	23.4%
2	15.6%
3	12.7%
4	10.3%

Productive		k_S			
		1	2	3	4
k_E	1	23.4%	16.1%	11.8%	9.2%
	2	15.6%	11.3%	8.5%	6.9%
	3	12.7%	9.4%	6.9%	5.3%
	4	10.3%	7.9%	6.6%	4.5%

follows directly from the definition of a startable vertex and the structure of prefix-positive paths that:

$$Stable_j = \begin{cases} Delete_0(Advance(Stable_{j-1}, t_j) \cup^{max} Start(T_j^{-k_S}) & \text{if } j \geq k_S \\ \emptyset & \text{if } j < k_S \end{cases}$$

Incrementally converting each successive subsequence $T_j^{-k_S}$ of the target into an integer index into the *Start* table takes constant time, so the above recurrence outlines an $O(N + |Startable|)$ algorithm for detecting local alignments at or above threshold τ.

3.2   Extension Trimming

We now turn to the development of a table that eliminates vertices that are not extendable. For a given sequence w of length k_E and a column index in the range $[0, P]$, let:

$$Extend(i, w) = min\{ \, drop(p) : p \text{ is a path from } (i, 0) \text{ to row } k_E \text{ in the edit graph of } Query \text{ vs. } w \, \}$$

where $drop(p) = - min \{ \, score(q) : q \text{ is a prefix of } p \, \} \geq 0.$

That is, $Extend(i, w)$ gives the least decrease in score that will be encountered along a path from column i when the next k_S target symbols are w. Thus (i, j) is an extendable vertex if and only if $C(i, j) > Extend(i, T_j^{k_E})$.

Computing the *Extend* table involves the following variation of the basic dynamic programming algorithm. Let $E_w[0 \ldots P]$ be a $(P+1)$-vector of numbers for which $E_w[i] = -Extend(i, w)$. It is a straightforward exercise in induction to verify the following recurrences:

$$E_\varepsilon[i] = 0$$
$$E_{aw}[i] = min(0, max(E_w[i+1] + S[q_{i+1}][a], E_w[i] - g, E_{aw}[i+1] - g))$$

As for the *Start* table, a factor of k_E in efficiency is gained by factoring the computation of $Extend(i, w)$ for different w but this time for sequences sharing a common *suffix*. Working "backwards" by extending suffixes is required in this case as we are interested in optimal paths *starting* in a given column as opposed to ending in one. Overall the computation of the *Extend* table takes $O(P|\Sigma|^{k_E})$ time and the table occupies the same amount of space. In practice, the values in the *Extend* table are small enough to fit in a single byte so that the table occupies exactly $(P+1)|\Sigma|^{k_E}$ bytes.

Once the *Extend* table is computed it can be used to prune vertices that are not extendable during the search computation. The point at which to do so is during the extension of path sets along deletion edges. To this end define:

$$Delete_E(p, w) = \cup^{max}\{(i+\Delta, v-\Delta g) : (i, v) \in p \text{ and } v-\Delta g > Extend(i+\Delta, w)\}.$$

Since $Extend(i, w) \geq 0$ for all i and w it follows that $Delete_E(p, w) \subseteq Delete_0(p)$ for all p and w. One must trim vertices that are not extendable during the deletion extension of an index-value pair, as opposed to afterwards, in order to avoid potentially wasting time on a series of vertices that are all not extendable along a chain of deletion edges. Stopping at the first vertex that is not extendable in such a chain is correct as $Extend(i, w) - g \leq Extend(i + 1, w)$ implies that all vertices reached from it by virtue of a deletion are also not extendable.

Let $Prod_j$ be the path set $\{(i, v) : (i, j) \text{ is a productive vertex and } v = C(i, j)\}$. Then by definition it follows that:

$$Prod_j = \begin{cases} Delete_E(Advance(Prod_{j-1}, t_j) \cup^{max} Start(T_j^{-k_S}), T_j^{k_E}) & \text{if } j \geq k_S \\ \emptyset & \text{if } j < k_S \end{cases}$$

Computing the indices into the *Start* and *Extend* tables can easily be done incrementally at an overhead of $O(N)$ time over the course of scanning *Target*. The time to compute $Prod_j$ from $Prod_{j-1}$ is as follows. *Advance* takes time $O(|Prod_{j-1}|)$. The merge of this result with the *Start* table path set takes time proportional to the sum of the lengths of the two lists, i.e., $O(|Prod_{j-1}| + |Seed_j|)$, where $Seed_j$ is the set of seed vertices in row j. Finally computing $Delete_E$ takes no longer than the sum of the size of the path set given to it and the size of the path set that results, i.e., $O(|Prod_{j-1}| + |Seed_j| + |Prod_j|)$. Thus the overall time to search *Target* is $O(N + \Sigma_j|Prod_j| + \Sigma_j|Seed_j|) = O(N + |Productive| + |Seed|)$ as claimed at the outset.

A subtlety ignored to this point is that alignments of score τ or more can end on non-productive vertices when τ is sufficiently small, specifically when $\tau \leq max(\sigma(k_S - 1), g(k_E - 1))$, where σ is the largest entry in the scoring matrix S. This happens in two ways: when the alignment is of length less than k_S and so does not give rise to a seed vertex, and when the last vertex of the alignment is followed by such negative scoring edges that all extensions drop to 0 in less than k_E rows. While the algorithm can be modified to detect these cases at no extra cost, in this abstract, we simply note that to create such instances τ has to be set to an unrealistically small value. For example, for our running

Table 2. Times and space for sparse SW-searches on queries of length 300 and a database of 100,000 300 residue entries.

Start k_S	Preprocessing Time(secs)	Space (Kb)
1	.00027	5
2	.0019	62
3	.025	892
4	.38	13,390

Extend k_E	Preprocessing Time(secs)	Space (Kb)
1	.00053	7
2	.0081	159
3	.19	3,650
4	4.3	83,952

	Search Time(secs)	k_S 1	2	3	4	
	off	201	127	96	81	vs. Dynamic
k_E 1		177	115	94	82	Programming:
2		132	91	79	72	187
3		118	92	87	78	

example of PAM120 and $g = 12$, with $k_S = k_E = 3$, τ would have to be 24 or less, and typically users never set τ below 50.

Table 2 shows the time and space to produce the tables, and the search time for given choices of k_S and k_E. Time and space for both tables increase geometrically with k, with the factor for the *Start* table being somewhat less because the number of seeds per entry is also decreasing geometrically. For running time, one notes that times generally decrease as the k parameters increase save that when k_E goes from 2 to 3, times actually increase in columns for larger k_S, so that the best overall time is obtained with $k_S = 4$ and $k_E = 2$. The reason for this is due to memory caching behavior on todays hardware. As processor speeds have gotten ever faster, memory speeds have not, so that for todays machines any memory cache miss typically incurs slowdowns of more than a factor of 10 over accesses that are in cache. As the *Start* and *Extension* tables get larger, it becomes more and more likely that an access into the table will not be in the cache as the access pattern to these tables is more or less random. Thus the observed behavior: even though only 1/20th of the d.p. matrix is being computed for $k_S = 4$ and $k_E = 2$, only a factor of 2.6 is gained in time over the naive algorithm. Nonetheless, this is not an inconsequential performance gain.

4   A Table-Driven Scheme for Dynamic Programming

In addition to restricting the SW computation to productive vertices, one can further develop a "jump" table that captures the effect of *Advance* and *Delete* over $k_J > 0$ rows. For each sequence of k_J potential row labels w and each column position i, an entry $Jump(i, w)$ gives the change in score that can be obtained at each column position $k \geq i$. That is, $Jump(i, w) = \{(k, u) :$ the maximal path from $(0, i)$ to (k_j, k) is u in the edit graph of w versus $Query\}$.

As formulated the table $Jump$ is unmanageably large at $O(\Sigma^{k_J} P^2)$ space. However, this is greatly improved by realizing that one need only record those (k, u) for which $u > -(\tau - 1)$. One will never jump from a vertex of score τ or more as the algorithm can quit as soon as it encounters such a vertex. Jumping from a vertex of score less than τ with a change of $-(\tau - 1)$ or more leads to a non-positive entry and so need not be considered. We can thus formulate the computations of the $Jump$ table as:

$$Jump(i, \varepsilon) = Delete_{-(\tau-1)}(\{(i, 0)\})$$
$$Jump(i, wa) = Advance(Delete_{-(\tau-1)}(Jump(i, w)), a)$$

If we truly wish to compute only critical entries on every other k_J^{th} row of the dynamic programming matrix, then care must be taken to model all possible seed paths that can lead to the next row to be computed. For example, if $k_J = 2$, so that only every even row is actually computed, and $k_S = 4$, then one might miss seeds that end in an odd row. The required fix is to modify the $Start(w)$ table so that it models all seeds of length $k_S - (k_J - 1)$ to k_S that align to a suffix of w. While this decreases the sparsity of the table, the loss is offset by the gain in being able to skip rows. So the $Start$ table is now defined by the recurrence:

$$Start(\varepsilon) = \{(i, 0) : i \in [0, P]\}$$

$$Start(wa) = \begin{cases} Delete_0(Advance(Start(w), a)) \cup^{max} Start(\varepsilon) & \text{if } |w| < k_J - 1 \\ Delete_0(Advance(Start(w), a)) & \text{otherwise} \end{cases}$$

Unlike the algorithms of the proceeding sections, the tactic of letting the $Jump$ table handle deletions implies that the set of entries, $Cand_j$, explicitly computed for every k_J^{th} row is a subset of the productive entries in the row and a superset of the entries not producible by deletions. To compute $Cand_j$, one first applies the $Jump$ table to the entries of $Cand_{j-k_J}$ and then merges these with all the seeds of lengths between $k_S - (k_J - 1)$ and k_S ending on row j. Finally, this aggregate set of entries on row j are checked for extendability. Putting this together formally in recurrences:

$$Cand_j = Trim_E(JUMP(Cand_{j-k_J}, T_j^{-k_J}) \cup^{max} Start'(T_j^{-k_S}), T_j^{k_E})$$

$$\text{where } JUMP(p, w) = \{(i+k, v+u) : \begin{matrix} (i, v) \in p \text{ and } v+u > 0 \\ \text{and } (k, u) \in Jump(i, w) \end{matrix}\}$$

$$\text{and } Trim_E(p, w) = \{(i, v) \in p : v > Extend(i, w)\}$$

Efficiently implementing $JUMP$ involves organizing the index-value pairs for a given entry $Jump(i, w)$ as a list ordered on the value of the pairs, not their index. Thus when jumping from an entry (i, v) one processes the offsets (k, u) in order until $v + u \leq 0$ at which point no further elements of the list need be considered.

The remaining detail is to modify the check for paths scoring τ or more. One should, of course examine every entry in $Cand_j$, but since one is skipping $k_J - 1$

Table 3. Times and space for sparse, jump-based SW-searches on queries of length 300 and a database of 100,000 300 residue entries.

Jump' k_J	Preprocessing Time(secs)	Space (Kb)
1	0.0031	32
2	0.098	1,175
3	2.18	35,786

Seeds		k_S		
k_J	1	2	3	4
1	22.1%	11.2%	6.9%	4.6%
2		27.3%	15.4%	9.7%
3			31.9%	18.4%

Start' (Kb)		k_S		
k_J	1	2	3	4
1	5	62	880	14,020
2		156	1,949	28,356
3			3,966	52,677

Search Time (secs)

$k_J = 1$		k_S		
k_E	1	2	3	4
1	130	88	76	63
2	102	76	75	62
3	112	92	88	79
4	102	89	85	78

$k_J = 2$		k_S	
k_E	2	3	4
1	134	110	94
2	126	107	87
3	89	80	72
4	78	72	62

$k_J = 3$	k_S	
k_E	3	4
1	102	85
2	84	71
3	94	72
4	64	56

rows, it is necessary to check if there is an extension of each of these that has score over τ in the skipped rows. To this end let $Peak(i, w)$ be the maximum score achieved on all possible paths starting at $(i, 0)$ in the matrix of $Query$ versus w. Like the analysis for the $Extend$ table, except easier, $Peak(i, w) = P_w[i]$ where:

$$P_\varepsilon[i] = 0$$
$$P_{aw}[i] = max(0, P_w[i + 1] + S[q_{i+1}][a], P_w[i] - g, P_{aw}[i + 1] - g))$$

To test for paths of score τ or greater, it suffices to check if $v + Peak(i, T_j^{k_J-1}) \geq \tau$ for each (i, v) in $Cand_j$. The total space for the $Peak$ table is $O(|\Sigma|^{k_J-1}P)$.

It is no longer possible to give a statement of the complexity of the resulting algorithm in simple terms as $JUMP(p, w)$ potentially involves enumerating a position of the underlying dynamic programming matrix several times. It simply becomes an empirical question as to whether in practice this inefficiency is offset by the fact that one no longer needs to consider deletion sequences and that one skips rows when $k_J > 1$. In Table 3 one sees that the time and space for producing the jump table increase geometrically with the highest reasonable value of k_J being 3. In the second row of tables, one sees that for every row skipped there is a corresponding increase in the number of vertices that are seeds and in the size of the $Start$ tables. If $Seed(s, j)$ is the fraction of seeds for $k_S = s$ and $k_J = j$, then one sees that $Seed(s, j) \leq \Sigma_{t=s-j+1}^s Seed(t, 1)$ as a seed for a given k_S is often a seed for $k_S - 1$. Running time is a complex function of the three k parameters due to the stochastic nature of competition for the cache as the parameters increase. In particular, increasing k_J doesn't win, *unless* one is running with large values of k_S and k_E. The basic insight is that the cost of the table lookups per row is such that skipping rows only wins if the number of entries per row is made small enough with the use of the *Start* and *Extension* tables. Overall, on our test machine, the best performance was achieved with $k_S = k_E = 4$ and $k_J = 3$. At 56 seconds for the search, this is 3.33 times faster than the our implementation of the original Smith-Waterman algorithm.

References

1. Smith, T.F., Waterman, M.S.: Identification of common molecular sequences. J. Mol. Biol 147 (1981) 195-197
2. Altschul, S.F., Gish, W., Miller W., Myers, E.W., Lipman, D.J.: A basic local alignment search tool. J. Mol. Biol. 215 (1990) 403-410
3. Lipman, D.J., and Pearson, W.R.: Rapid and sensitive protein similarity searches. Science 227 (1985) 1435-1441
4. Huang, X.: A space-efficient parallel sequence comparison algorithm for a message-passing multi-processor. Int. J. Parallel Prog. 18, 3 (1989) 223-239
5. Lopresti, D.P.: P-NAC: A systolic array for comparing nucleic acid sequences. Computer 20, 7 (1987) 98-99
6. S. Wu, U. Manber, and E. Myers. A sub-quadratic algorithm for approximate limited expression matching. *Algorithmica* 15, 1 (1996), 50-67.
7. Myers, E.W.: An O(ND) difference algorithm and its variations. Algorithmica 1, 2 (1985) 251-266

Sequence-Length Requirements
for Phylogenetic Methods

Bernard M.E. Moret[1], Usman Roshan[2], and Tandy Warnow[2]

[1] Department of Computer Science, University of New Mexico
Albuquerque, NM 87131
moret@cs.unm.edu
[2] Department of Computer Sciences, University of Texas, Austin, TX 78712
{usman,tandy}@cs.utexas.edu

Abstract. We study the sequence lengths required by neighbor-joining, greedy parsimony, and a phylogenetic reconstruction method ($DCM_{NJ}+MP$) based on disk-covering and the maximum parsimony criterion. We use extensive simulations based on random birth-death trees, with controlled deviations from ultrametricity, to collect data on the scaling of sequence-length requirements for each of the three methods as a function of the number of taxa, the rate of evolution on the tree, and the deviation from ultrametricity. Our experiments show that $DCM_{NJ}+MP$ has consistently lower sequence-length requirements than the other two methods when trees of high topological accuracy are desired, although all methods require much longer sequences as the deviation from ultrametricity or the height of the tree grows. Our study has significant implications for large-scale phylogenetic reconstruction (where sequence-length requirements are a crucial factor), but also for future performance analyses in phylogenetics (since deviations from ultrametricity are proving pivotal).

1 Introduction

The inference of phylogenetic trees is basic to many research problems in biology and is assuming more importance as the usefulness of an evolutionary perspective is becoming more widely understood.

The inference of very large phylogenetic trees, such as might be needed for an attempt on the "Tree of Life," is a big computational challenge. Indeed, such large-scale phylogenetic analysis is currently beyond our reach: the most popular and accurate approaches are based on hard optimization problems such as maximum parsimony and maximum likelihood. Even heuristics for these optimization problems sometimes take weeks to find local optima; some datasets have been analyzed for years without a definite global optimum being found.

Polynomial-time methods would seem to offer a reasonable alternative, since they can be run to completion even on large datasets. Recent simulation studies have examined the performance of fast methods (specifically, the distance-based method neighbor-joining [21] and a simple greedy heuristic for the maximum

R. Guigó and D. Gusfield (Eds.): WABI 2002, LNCS 2452, pp. 343–356, 2002.

parsimony problem) and have shown that both methods might perform rather well (with respect to topological accuracy), even on very large model trees.

These studies have focused on trees of low diameter that also obey the strong molecular clock hypothesis (so that the expected number of evolutionary events is roughly proportional to time). Our study seeks to determine whether these, and other, polynomial-time methods really do scale well with increasing number of taxa under more difficult conditions—when the model trees have large diameters and do not obey the molecular clock hypothesis.

Our paper is organized as follows. In Section 2 we review earlier studies and the issues they addressed. In Section 3, we define our terms and present the basic concepts. In Section 4 we outline our experimental design and in Section 5 we report the results of our experiments. We conclude with some remarks on the implications of our work.

2   Background

The *sequence-length requirement* of a method is the sequence length that this method needs in order to reconstruct the true tree topology with high probability. Earlier studies established analytical upper bounds on the sequence-length requirements of various methods, including the popular *neighbor-joining (NJ)* [21] method. These studies showed that standard methods, such as NJ, recover the true tree with high probability from sequences of lengths that are exponential in the evolutionary diameter of the true tree. Based upon these studies, we defined a parameterization of model trees in which the longest and shortest edge lengths are fixed [6,7], so that the sequence-length requirement of a method can be expressed as a function of the number, n, of taxa. This parameterization led us to define *fast-converging* methods, methods that recover the true tree with high probability from sequences of lengths bounded by a polynomial in n once f and g, the minimum and maximum edge lengths, are bounded. Several fast-converging methods were subsequently developed [4,5,11,24]. We and others analyzed the sequence-length requirements of standard methods, such as NJ, under the assumptions that f and g are fixed. These studies [1,7] showed that NJ and many other methods can recover the true tree with high probability when given sequences of lengths bounded by a function that grows exponentially in n.

We then performed studies on a different parameterization of the model tree space, where we fixed the evolutionary diameter of the tree and let the number of taxa vary [17]. This parameterization, suggested to us by J. Huelsenbeck, allowed us to examine the differential performance of methods with respect to "taxon-sampling" strategies [10]. In these studies, we evaluated the performance of NJ and $DCM_{NJ}+MP$ on simulated data, obtained from random birth-death trees with bounded deviation from ultrametricity. We found that $DCM_{NJ}+MP$ consistently dominated NJ throughout the parameter space; we also found that the difference in performance increased as the number of taxa increased or the evolutionary rate increased.

In previous studies [15] we had studied the sequence-length requirements for NJ, Weighbor [3], Greedy MP, and $DCM_{NJ}+MP$ as a function of the numbers of taxa, with fixed height and deviation. We had found that $DCM_{NJ}+MP$ requires shorter sequences than the other four methods when trees of high topological accuracy are desired. However, we had used datasets of modest size and had not explored a very large range of evolutionary rates and deviations from ultrametricity. In this paper we remedy these limitations.

3 Basics

3.1 Simulation Study

Simulation studies are the standard technique used in phylogenetic performance studies [9,10,14]. In a simulation study, a DNA sequence at the root of the model tree is evolved down the tree under some assumed stochastic model of evolution. This process generates a set of sequences at the leaves of the tree. The sequences are then given to phylogenetic reconstruction methods, with each method producing a tree for the set of sequences. These reconstructed trees are then compared against the model tree for topological accuracy. The process is repeated many times in order to obtain a statistically significant test of the performance of the methods under these conditions.

3.2 Model Trees

We used random birth-death trees as model trees for our experiments. They were generated using the program r8s [22], where we specified the option to generate birth-death trees using a backward Yule (coalescent) process with waiting times taken from an exponential distribution. An edge e in a tree generated by this process has length λ_e, a value that indicates the expected number of times a random site changes on the edge. Trees produced by this process are by construction ultrametric (that is, the path length from the root to each leaf is identical). Furthermore, a random site is expected to change once on any root-to-leaf path; that is, the (evolutionary) height of the tree is 1.

In our experiments we scale the edge lengths of the trees to give different heights. To scale the tree to height h, we multiply the length of each edge of T by h. We also modify the edge lengths as follows in order to deviate the tree away from ultrametricity. First we pick a deviation factor, c; then, for each edge $e \in T$, we choose a number, x, uniformly at random from the interval $[-\lg(c), \lg(c)]$. We then multiply the branch length λ_e by e^x. For deviation factor c, the expected deviation is $(c - 1/c)/2lnc$, with a variance of $((c^2 - \frac{1}{c^2}) - 2(c - \frac{1}{c}))/4 \ln c$. For instance, for an expected deviation of 1.36, the standard deviation has the rather high value of 1.01.

3.3 DNA Sequence Evolution

We conducted our experiments under the Kimura 2-parameter [13] + Gamma [25] model of DNA sequence evolution, one of the standard models for studying the performance of phylogenetic reconstruction methods. In this model a site evolves down the tree under the Markov assumption and all site substitutions are partitioned into two classes: *transitions*, which substitute a purine for a purine; and *transversions*, which substitute a pyrimidine for a pyrimidine. This model has a parameter to indicate the transition/transversion ratio; in our experiments, we set it to 2 (the standard setting).

We assume that the sites have different rates of evolution drawn from a known distribution. In our experiments we use the gamma distribution with shape parameter α (which we set to 1, the standard setting), which is the inverse of the coefficient of variation of the substitution rate.

3.4   Measures of Accuracy

Since the trees returned by each of the three methods are binary, we use the *Robinson-Foulds* (RF) distance [20], which is defined as follows. Every edge e in a leaf-labeled tree T defines a bipartition, π_e, of the leaves (induced by the deletion of e); the tree T is uniquely encoded by the set $C(T) = \{\pi_e : e \in E(T)\}$, where $E(T)$ is the set of all internal edges of T. If T is a model tree and T' is the tree obtained by a phylogenetic reconstruction method, then the set of *False Positives* is $C(T') - C(T)$ and the set of *False Negatives* is $C(T) - C(T')$. The RF distance is then the average of the number of false positives and the number of false negatives. We plot the *RF rates* in our figures, which are obtained by dividing the RF distance by $n - 3$, the number of internal edges in a binary tree on n leaves. Thus the RF rate varies between 0 and 1 (or 0% and 100%). Rates below 5% are quite good, while rates above 25% are unacceptably large. We focus our attention on the sequence lengths needed to get at least 75% accuracy.

3.5   Phylogenetic Reconstruction Methods

Neighbor-Joining. Neighbor-Joining (NJ) [21] is one of the most popular polynomial time distance-based methods [23]. The basic technique is as follows: first, a matrix of estimated distances between every pair of sequences is computed. (This step is standardized for each model of evolution, and takes $O(kn^2)$ time, where k is the sequence length and n the number of taxa.) The NJ method then uses a type of agglomerative clustering to construct a tree from this matrix of distances. The NJ method is provably statistically consistent under most models of evolution examined in the phylogenetics literature, and in particular under the K2P model that we use.

Greedy Maximum Parsimony. Maximum Parsimony is an NP-hard optimization problem for phylogenetic tree reconstruction [8]. Given a set of aligned sequences (in our case, DNA sequences), the objective is to reconstruct a tree with leaves labelled by the input sequences and internal nodes labelled by other sequences so as to minimize the total "length" of the tree. This length is calculated by summing up the Hamming distances on the edges. Although the MP problem is NP-hard, many effective heuristics (most based on iterative local refinement) are offered in popular software packages. Earlier studies [19,2] have suggested that a very simple greedy heuristic for MP might produce as accurate a reconstructed tree as the best polynomial-time methods. Our study tests this hypothesis by explicitly looking at this greedy heuristic. The greedy heuristic begins by randomly shuffling the sequences and building an optimal tree on the first four sequences. Each successive sequence is inserted into the current tree, so as to minimize the total length of the resulting, larger tree. This method takes $O(kn^2)$ time, since finding the best place to insert the next sequence takes $O(kn)$ time, where k is the sequence length and n is the number of taxa.

DCM$_{NJ}$+MP. The $DCM_{NJ}+MP$ method was developed by us in a series of papers [16]. In our previous simulation studies, $DCM_{NJ}+MP$ outperformed, in terms of topological accuracy, both DCM^*_{NJ} (a provably fast-converging method of which $DCM_{NJ}+MP$ is a variant) and NJ. Let d_{ij} represent the distance between taxa i and j; $DCM_{NJ}+MP$ operates in two phases.

- *Phase 1:* For each choice of threshold $q \in \{d_{ij}\}$, compute a binary tree T_q, by using the Disk-Covering Method from [11], followed by a heuristic for refining the resultant tree into a binary tree. Let $\mathcal{T} = \{T_q : q \in \{d_{ij}\}\}$. (For details of Phase I, the reader is referred to [11].)
- *Phase 2:* Select the tree from $\mathcal{T}$ that optimizes the maximum parsimony criterion.

If we consider all $\binom{n}{2}$ values for the threshold q in Phase 1, then $DCM_{NJ}+MP$ takes $O(n^6)$ time; however, if we consider only a fixed number p of thresholds, it only takes $O(pn^4)$ time. In our experiments we use $p = 10$ (and $p = 5$ for 800 taxa), so that the running time of $DCM_{NJ}+MP$ is $O(n^4)$. Experiments (not shown here) indicate that choosing 10 thresholds does not reduce the performance of the method significantly (in terms of topological accuracy).

4   Experimental Design

We explore a broad portion of the parameter space in order to understand the factors that affect topological accuracy of the three reconstruction methods under study. In particular, we are interested in the type of datasets that might be needed in order to attempt large-scale phylogenetic analyses that span distantly related taxa, as will be needed for the "Tree of Life." Such datasets will have large heights and large numbers of taxa. Therefore we have created random model trees with heights that vary from quite small (0.05) to quite large (up to 4), and with a number of taxa ranging from 100 to 800. We have also explored the effect of deviations from the molecular clock, by using model trees that obey the molecular clock (expected deviation 1) as well as model trees with significant violations of the molecular clock (expected deviation 2.87). In order to keep the experiment tractable, we limited our analysis to datasets in which the sequence lengths are bounded by 16,000; this has the additional benefit of not exceeding by too much what we might expect to have available in a phylogenetic analysis in the coming years. In contrast the trees in [2] have smaller heights (up to a maximum of 0.5) and smaller expected deviations with less variance (their actual deviation is truncated if it exceeds a preset value), providing much milder experimental conditions than ours. The trees studied in [19] are even simpler, being ultrametric (no deviation from molecular clock) with a very small height of 0.15.

For the two distance-based methods (NJ and $DCM_{NJ}+MP$), we computed evolutionary distances appropriately for the K2P+gamma model with our settings for α and the transition/transversion ratio. Because we explore performance on datasets with high rates of evolution, we had to handle cases in which the standard distance correction cannot be applied because of saturation. In such a case, we used the technique developed in [12], called "fix-factor 1:" distances that cannot be set using the distance-correction formula are simply assigned the largest corrected distance in the matrix.

We used seqgen [18] to generate the sequences, and PAUP* 4.0 to construct Greedy MP trees as described. The software for the basic DCM_{NJ} was written by D. Huson. We generated the rest of the software (a combination of C++ programs and Perl scripts) explicitly for these experiments. For job management across the cluster and public laboratory machines, we used the Condor software package.

5   Experimental Results

We present a cross-section of our results in three major categories. All address the sequence lengths required by each reconstruction method to reach prescribed levels of topological accuracy on at least 80% of the reconstructed trees as a function of three major parameters, namely the deviation from ultrametricity, the height of the tree, and the number of taxa. We compute the sequence-length requirements to achieve a given accuracy as follows. For each method and each collection of parameter values (number of taxa, height, and deviation), we perform 40 runs to obtain 40 FN rates at each sequence length of the input dataset. We then use linear interpolation on each of the 40 curves of FN rates vs. sequence length to derive the sequence-length requirement at a given accuracy level. (Some curves may not reach the desired accuracy level even with sequences of length 16,000, in which case we exclude those runs.) We then average the computed sequence lengths to obtain the average sequence-length requirement at the given settings. Points not plotted in our curves indicate that less than 80% of the trees returned by the method did not have the required topological accuracy even at 16000 characters.

In all cases the relative requirements of the three methods follow the same ordering: $DCM_{NJ}+MP$ requires the least, followed by NJ, with greedy MP a distant third. However, even the best of the three, $DCM_{NJ}+MP$, demonstrates a very sharp, exponential rise in its requirements as the deviation from ultrametricity increases—and the worst of the three, Greedy MP, cannot be used at all with a deviation larger than 1 or with a height larger than 0.5. Similarly sharp rises are also evident when the sequence-length requirements are plotted as a function of the height of the trees. (In contrast, increases in the number of taxa cause only mild increases in the sequence length requirements, even some decreases in the case of the greedy MP method.)

5.1   Requirements as a Function of Deviation

Figure 1 shows plots of the sequence-length requirements of the three methods for various levels of accuracy as a function of the deviation. Most notable is the very sharp and fast rise in the requirements of all three methods as the deviation from ultrametricity increases. The greedy MP approach suffers so much from such increases that we could not obtain even 80% accuracy at expected deviations larger than 1.36. $DCM_{NJ}+MP$ shows the best behavior among the three methods, even though it requires unrealistically long sequences (significantly longer than 16,000 characters) at high levels of accuracy for expected deviations above 2.

5.2   Requirements as a Function of Height

Figure 2 parallels Figure 1 in all respects, but this time the variable is the height of the tree, with the expected deviation being held to a mild 1.36. The rises in sequence-length requirements are not quite as sharp as those observed in Figure 1, but remain severe enough that accuracy levels beyond 85% are not achievable for expected deviations above 2. Once again, the best method is clearly $DCM_{NJ}+MP$ and greedy MP is clearly inadequate. Figure 3 repeats the experiment with a larger expected deviation of 2.02. At this deviation value, greedy MP cannot guarantee 80% accuracy even with strings of length 16,000, so the plots only show curves for NJ and $DCM_{NJ}+MP$, the latter again proving best. But note that the rise is now sharper, comparable to that of Figure 1. Taken together, these two figures suggest that the product of height and deviation is a very strong predictor of sequence-length requirements and that these requirements grow exponentially as a function of that product. The number of taxa does not play a significant role, although, for the larger height, we note that increasing the number of taxa decreases sequence-length requirements—a behavior that we attribute to the gradual disappearance of long tree edges under the influence of better taxon sampling.

5.3   Requirements as a Function of the Number of Taxa

Figure 4 shows the sequence-length requirements as a function of the number of taxa over a significant range (up to 800 taxa) at four different levels of accuracy for our three methods. The tree height here is quite small and the deviation modest, so that the three methods place only modest demands on sequence length for all but the highest level of accuracy. (Note that parts (a) and (b) of the figure use a different vertical scale from that used in parts (c) and (d).) At high accuracy (90%), NJ shows distinctly worse behavior than the other two methods,

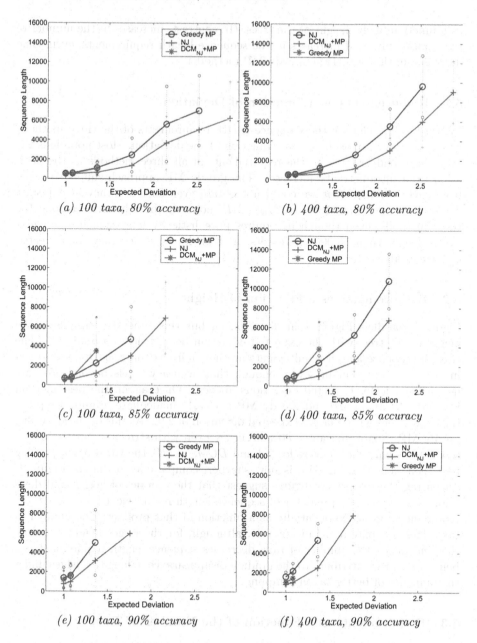

Fig. 1. Sequence-length requirements as a function of the deviation of the tree, for a height of 0.5, two numbers of taxa, and three levels of accuracy.

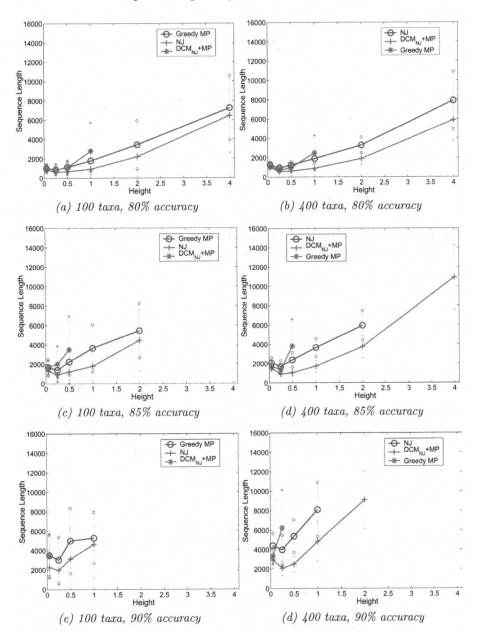

Fig. 2. Sequence-length requirements as a function of the height of the tree, for an expected deviation of 1.36, two numbers of taxa, and three levels of accuracy.

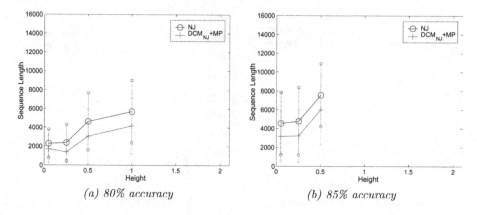

Fig. 3. Sequence-length requirements as a function of the height of the tree, for an expected deviation of 2.02, 100 taxa, and two levels of accuracy.

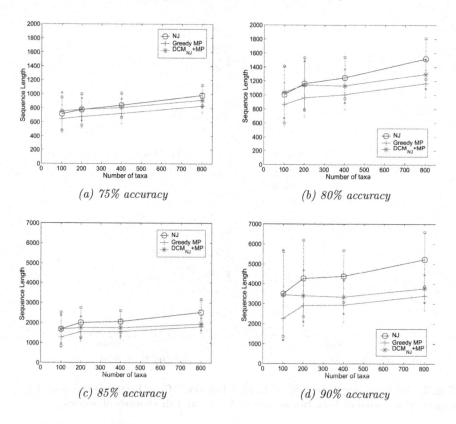

Fig. 4. Sequence-length requirements as a function of the number of taxa, for an expected deviation of 1.36, a height of 0.05, and four levels of accuracy.

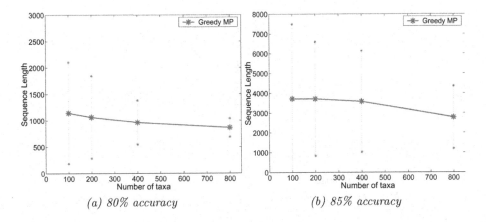

(a) 80% accuracy   *(b) 85% accuracy*

Fig. 5. Sequence-length requirements for Greedy MP (60 runs) as a function of the number of taxa, for an expected deviation of 1.36, a height of 0.5, and two levels of accuracy.

with nearly twice the sequence-length requirements. Figure 5 focuses on just the greedy MP method for trees with larger height. On such trees, initial length requirements are quite high even at 80% accuracy, but they clearly *decrease* as the number of taxa increases. This curious behavior can be explained by taxon sampling: as we increase the number of taxa for a fixed tree height and deviation, we decrease the expected (and the maximum) edge length in the model tree. Breaking up long edges in this manner avoids one of the known pitfalls for MP: its sensitivity to the simultaneous presence of both short and long edges.

Figure 6 gives a different view on our data. It shows how $DCM_{NJ}+MP$ and NJ are affected by the number of taxa as well as by the height of the model tree. Note that the curves are not stacked in order of tree heights because Figures 2 and 3 show that the sequence-length requirements decrease from height 0.05 to 0.25 and then start to increase gradually. The juxtaposition of the graphs clearly demonstrate the superiority of $DCM_{NJ}+MP$ over NJ: the curves for the former are both lower and flatter than those for the latter—that is, $DCM_{NJ}+MP$ can attain the same accuracy with significantly shorter sequences than does NJ and the growth rate in the length of the required sequences is negligible for $DCM_{NJ}+MP$ whereas it is clearly visible for NJ.

6   Conclusions and Future Work

6.1   Summary

We have explored the performance of three polynomial-time phylogenetic reconstruction methods, of which two (greedy maximum parsimony and neighbor-joining) are well known and widely used. In contrast to earlier studies, we have found that neighbor-joining and greedy maximum parsimony have very different

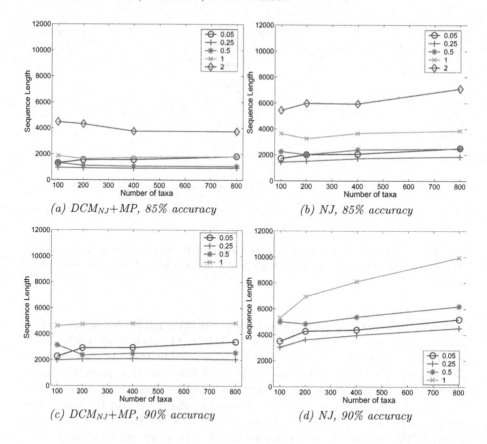

(a) $DCM_{NJ}+MP$, 85% accuracy

(b) NJ, 85% accuracy

(c) $DCM_{NJ}+MP$, 90% accuracy

(d) NJ, 90% accuracy

Fig. 6. Sequence-length requirements (for at least 70% of trees) as a function of the number of taxa, for an expected deviation of 1.36, various heights, and two levels of accuracy, for each of $DCM_{NJ}+MP$ and NJ.

performance on large trees and that both suffer substantially in the presence of significant deviations from ultrametricity. The third method, $DCM_{NJ}+MP$, also suffers when the model tree deviates significantly from ultrametricity, but it is significantly more robust than either NJ or greedy MP. Indeed, for large trees (containing more than 100 taxa) with large evolutionary diameters (where a site changes more than once across the tree), even small deviations from ultrametricity can be catastrophic for greedy MP and difficult for NJ.

Several inferences can be drawn from our data. First, it is clearly not the case that greedy MP and NJ are equivalent: while they both respond poorly to increasing heights and deviations from ultrametricity, greedy MP has much the worse reaction. Secondly, neither of them has good performance on large trees with high diameters—both require more characters (longer sequences) than are likely to be easily obtained. Finally, the newer method, $DCM_{NJ}+MP$, clearly outperforms both others and does so by a significant margin on large trees with high rates of evolution.

6.2   Future Work

A natural question is whether a better, albeit slower, heuristic for maximum parsimony might demonstrate better behavior. Other polynomial-time methods could also be explored in this same experimental setting, and some might outperform our new method $DCM_{NJ}+MP$. Maximum-likelihood methods, while typically slower, should also be compared to the maximum-parsimony methods; to date, next to nothing is known about their sequence-length requirements.

Also of interest is the impact of various aspects of our experimental setup on the results. For example, a different way of deviating a model tree from the molecular clock might be more easily tolerated by these reconstruction methods—and might also be biologically more realistic. The choice of the specific model of site evolution (K2P) can also be tested, to see whether it has a significant impact on the relative performance of these (and other) methods.

Acknowledgements

This work is supported by the National Science Foundation under grants ACI 00-81404 (Moret), EIA 99-85991 (Warnow), EIA 01-13095 (Moret), EIA 01-21377 (Moret), EIA 01-21680 (Warnow), and DEB 01-20709 (Moret and Warnow), and by the David and Lucile Packard Foundation (Warnow).

References

1. K. Atteson. The performance of the neighbor-joining methods of phylogenetic reconstruction. *Algorithmica*, 25:251–278, 1999.
2. O.R.P. Bininda-Emonds, S.G. Brady, J. Kim, and M.J. Sanderson. Scaling of accuracy in extremely large phylogenetic trees. In *Proc. 6th Pacific Symp. Biocomputing PSB 2002*, pages 547–558. World Scientific Pub., 2001.
3. W. J. Bruno, N. Socci, and A. L. Halpern. Weighted neighbor joining: A likelihood-based approach to distance-based phylogeny reconstruction. *Mol. Biol. Evol.*, 17(1):189–197, 2000.
4. M. Csűrös. Fast recovery of evolutionary trees with thousands of nodes. To appear in RECOMB 01, 2001.
5. M. Csűrös and M. Y. Kao. Recovering evolutionary trees through harmonic greedy triplets. *Proceedings of ACM-SIAM Symposium on Discrete Algorithms (SODA 99)*, pages 261–270, 1999.
6. P. L. Erdős, M. Steel, L. Székély, and T. Warnow. A few logs suffice to build almost all trees– I. *Random Structures and Algorithms*, 14:153–184, 1997.
7. P. L. Erdős, M. Steel, L. Székély, and T. Warnow. A few logs suffice to build almost all trees– II. *Theor. Comp. Sci.*, 221:77–118, 1999.
8. L. R. Foulds and R. L. Graham. The Steiner problem in phylogeny is NP-complete. *Advances in Applied Mathematics*, 3:43–49, 1982.
9. J. Huelsenbeck. Performance of phylogenetic methods in simulation. *Syst. Biol.*, 44:17–48, 1995.
10. J. Huelsenbeck and D. Hillis. Success of phylogenetic methods in the four-taxon case. *Syst. Biol.*, 42:247–264, 1993.

11. D. Huson, S. Nettles, and T. Warnow. Disk-covering, a fast-converging method for phylogenetic tree reconstruction. *Comput. Biol.*, 6:369–386, 1999.

12. D. Huson, K. A. Smith, and T. Warnow. Correcting large distances for phylogenetic reconstruction. In *Proceedings of the 3rd Workshop on Algorithms Engineering (WAE)*, 1999. London, England.

13. M. Kimura. A simple method for estimating evolutionary rates of base substitutions through comparative studies of nucleotide sequences. *J. Mol. Evol.*, 16:111–120, 1980.

14. K. Kuhner and J. Felsenstein. A simulation comparison of phylogeny algorithms under equal and unequal evolutionary rates. *Mol. Biol. Evol.*, 11:459–468, 1994.

15. L. Nakhleh, B.M.E. Moret, U. Roshan, K. St. John, and T. Warnow. The accuracy of fast phylogenetic methods for large datasets. In *Proc. 7th Pacific Symp. Biocomputing PSB 2002*, pages 211–222. World Scientific Pub., 2002.

16. L. Nakhleh, U. Roshan, K. St. John, J. Sun, and T. Warnow. Designing fast converging phylogenetic methods. In *Proc. 9th Int'l Conf. on Intelligent Systems for Molecular Biology (ISMB01)*, volume 17 of *Bioinformatics*, pages S190–S198. Oxford U. Press, 2001.

17. L. Nakhleh, U. Roshan, K. St. John, J. Sun, and T. Warnow. The performance of phylogenetic methods on trees of bounded diameter. In O. Gascuel and B.M.E. Moret, editors, *Proc. 1st Int'l Workshop Algorithms in Bioinformatics (WABI'01)*, pages 214–226. Springer-Verlag, 2001.

18. A. Rambaut and N. C. Grassly. Seq-gen: An application for the Monte Carlo simulation of DNA sequence evolution along phylogenetic trees. *Comp. Appl. Biosci.*, 13:235–238, 1997.

19. B. Rannala, J. P. Huelsenbeck, Z. Yang, and R. Nielsen. Taxon sampling and the accuracy of large phylogenies. *Syst. Biol.*, 47(4):702–719, 1998.

20. D. F. Robinson and L. R. Foulds. Comparison of phylogenetic trees. *Mathematical Biosciences*, 53:131–147, 1981.

21. N. Saitou and M. Nei. The neighbor-joining method: A new method for reconstructing phylogenetic trees. *Mol. Biol. Evol.*, 4:406–425, 1987.

22. M.J. Sanderson. r8s software package. Available from http://ginger.ucdavis.edu/r8s/.

23. M.J. Sanderson, B.G. Baldwin, G. Bharathan, C.S. Campbell, D. Ferguson, J.M. Porter, C. Von Dohlen, M.F. Wojciechowski, and M.J. Donoghue. The growth of phylogenetic information and the need for a phylogenetic database. *Systematic Biology*, 42:562–568, 1993.

24. T. Warnow, B. Moret, and K. St. John. Absolute convergence: true trees from short sequences. *Proceedings of ACM-SIAM Symposium on Discrete Algorithms (SODA 01)*, pages 186–195, 2001.

25. Z. Yang. Maximum likelihood estimation of phylogeny from DNA sequences when substitution rates differ over sites. *Mol. Biol. Evol.*, 10:1396–1401, 1993.

Fast and Accurate Phylogeny
Reconstruction Algorithms Based on
the Minimum-Evolution Principle

Richard Desper[1] and Olivier Gascuel[2]

[1] National Center for Biotechnology Information, National Library of Medicine, NIH
Bethesda, MD USA
`desper@ncbi.nlm.nih.gov`
[2] Dept. Informatique Fondamentale et Applications
LIRMM, Montpellier, France
`gascuel@lirmm.fr`
`http://www.lirmm.fr/~w3ifa/MAAS/`

Abstract. This paper investigates the standard ordinary least-squares version [24] and the balanced version [20] of the minimum evolution principle. For the standard version, we provide a greedy construction algorithm (GME) which produces a starting topology in $O(n^2)$ time, and a tree swapping algorithm (FASTNNI) that searches for the best topology using nearest neighbor interchanges (NNIs), where the cost of doing p NNIs is $O(n^2 + pn)$, i.e. $O(n^2)$ in practice because p is always much smaller than n. The combination of GME and FASTNNI produces trees which are fairly close to Neighbor Joining (NJ) trees in terms of topological accuracy, especially with large trees. We also provide two closely related algorithms for the balanced version, called BME and BNNI, respectively. BME requires $O(n^2 \times \mathrm{diam}(T))$ operations to build the starting tree, and BNNI is in $O(n^2 + pn \times \mathrm{diam}(T))$, where $\mathrm{diam}(T)$ is the topological diameter of the output tree. In the usual Yule-Harding distribution on phylogenetic trees, the diameter expectation is in $\log(n)$, so our algorithms are in practice faster that NJ. Furthermore, BNNI combined with BME (or GME) has better topological accuracy than NJ, BIONJ and WEIGHBOR (all three in $O(n^3)$), especially with large trees.

Minimum evolution (ME) was proposed by several authors as a basic principle of phylogenetic inference [28]. Given the matrix of pairwise evolutionary distances between the taxa being studied, this principle involves first estimating the length of any given topology and then selecting the topology with shortest length. Numerous variants exist, depending on how the edge lengths are estimated and how the tree length is calculated from these edge lengths [14]. Several definitions of tree length have been proposed, differing from one another by the treatment of negative edge lengths. The most common solution [24,25] simply defines the tree length as the sum of all edge lengths, regardless of whether they are positive or negative. Edge lengths are usually estimated within the least-squares framework. When all distance estimates are supposed to be independent and to have the same variance, we use ordinary least-squares (OLS).

R. Guigó and D. Gusfield (Eds.): WABI 2002, LNCS 2452, pp. 357–374, 2002.

The weighted least-squares corresponds to the case were distance estimates are independent but (possibly) with different variances, while the generalized least-squares approach does not impose any restriction and is able to take benefit from the covariances of the distance estimates. Distance estimates obtained from sequences do not have the same variance, large distances are much more variable than the shortest ones [11], and are mutually dependent when they share a common history (or path) in the true phylogeny [19]. Therefore, to estimate edge lengths from evolutionary distances, using generalized least-squares is theoretically superior to using weighted least-squares, which is in turn more appropriate than ordinary least-squares [4].

The ME principle has been shown to be statistically consistent when combined with ordinary least-squares [7,24]. However, surprisingly (and unfortunately) we recently demonstrated [14] that the combination of this principle with weighted least-squares might be inconsistent in some (likely unrealistic) cases, while the combination with generalized least-squares is deeply inconsistent and cannot be used for phylogenetic inference.

The ME principle, combined with a form of OLS length estimation, constitutes the basis of the popular Saitou and Nei's [25] Neighbor Joining (NJ) algorithm, which greedily refines a star tree by agglomerating taxa pairs in order to minimize the tree length at each step. The traditional approach to using ME is then to start with the NJ tree, and then do a topological search from that starting point. The first stage requires $O(n^3)$, where n is the number of taxa, while the current implementations [24] of the second are in $O(pn^3)$, where p is the number of swaps performed by the program. Another approach, available in FITCH 3.6 [10], is to greedily construct a tree by iterative addition of taxa to a growing tree, and then to perform swaps from this initial tree. But the time complexity is $O(n^4)$ or more.

This paper further investigates the ME principle. First, we demonstrate that its combination with OLS, even when not fully optimal in terms of topological accuracy, has the great advantage to lead to very fast $O(n^2)$ algorithms, able to build huge trees as envisaged in biodiversity studies. Second, we show that a new version of this principle, first introduced by Pauplin [20] to simplify tree length computation, is more appropriate than the OLS version. In this new version, sibling subtrees have equal weight, as opposed to the standard unweighted OLS, where all taxa have the same weight so that the weight of a subtree is equal to the number of its taxa. Algorithms to deal with this new "balanced" version are faster than NJ, although not as fast as their OLS counter-part, and have a better topological accuracy than NJ, but also than BIONJ [12] and WEIGHBOR [2]. The rest of this paper is organized as follows: we first provide the notation, definitions and formulae, we describe the tree swapping algorithms for both OLS and balanced ME versions, we explain how these algorithms are modified to greedily construct a tree by iterative taxa addition, we provide simulation results to illustrate the gain in topological accuracy and run times, and we then conclude by a brief discussion. We have omitted some of the details of the algorithms and some mathematical proofs, which will appear in the journal version of the paper.

1   Notation, Definitions, and Formulae

A tree is made of nodes (or vertices) and of edges. Among the nodes we distinguish the internal (or ancestral) nodes and the leaves (or taxa), The leaves are denoted as i, j or k, the internal nodes as u, v or w, while an edge e is defined by a pair of nodes and a length $l(e)$. We shall be considering various length assignments of the same underlying shape. In this case, we shall use the word "topology", while "tree" will be reserved for an instance of a topology with given edge lengths associated. We use the letter $\mathcal{T}$ to refer to a topology and T to refer to a tree. A tree is also made of subtrees typically denoted as A, B, C or D. For the sake of simplicity, we shall use the same notation for the subtrees and for the sets of taxa they contain. Accordingly, T also represents the set of taxa being studied, and n is the number of these taxa. Moreover, we shall use lowercase letters, e.g. a, b, c or d, to represent the subtree roots. If A and B are two disjoint subtrees, with roots a and b respectively, we'll say that A and B are distant-k subtrees if there are k edges in the path from a to b.

Δ is the matrix of pairwise evolutionary distance estimates, with Δ_{ij} being the distance between taxa i and j. Let A and B be two non-intersecting subtrees from a tree T. We define the average distance between A and B as:

$$\Delta_{A|B} = \frac{1}{|A||B|} \sum_{i \in A, j \in B} \Delta_{ij}. \tag{1}$$

$\Delta_{A|B}$ may also be defined recursively as:

- If A and B are singleton sets, i.e. $A = \{a\}$ and $B = \{b\}$, then $\Delta_{A|B} = \Delta_{ab}$,
- Else, without loss of generality let $B = B_1 \cup B_2$, so we then have

$$\Delta_{A|B} = \frac{|B_1|}{|B|} \Delta_{A|B_1} + \frac{|B_2|}{|B|} \Delta_{A|B_2}. \tag{2}$$

It is easily seen that Equations (1) and (2) are equivalent. Equation (2) follows the notion that the weight of a subtree is proportional to the number of its taxa. Thus every taxon has the same weight, and the same holds for the distances as shown by Equation (1). Thus, this average is said to be unweighted (Sneath and Sokal 1973). It must be noted that the unweighted average distance between subtrees does not depend on their topologies, but only of the taxa they contain.

Δ^T is the distance induced by the tree T, i.e., Δ_{ij}^T is equal to the length of the path connecting i to j in T, for every taxon pair (i, j). Given a topology $\mathcal{T}$ and a distance matrix Δ, the OLS edge length estimation produces the tree T with topology $\mathcal{T}$ minimizing the sum of squares

$$\sum_{i, j \in T} (\Delta_{ij}^T - \Delta_{ij})^2.$$

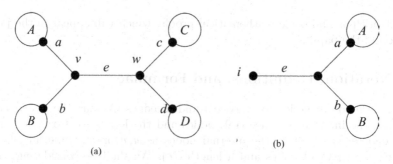

Fig. 1. Corner subtrees used to estimate the length of e, (a) for e an internal edge; (b) for e an external edge

Vach [29], Rzhetsky and Nei [24] and others showed analytical formulae for the proper OLS edge length estimation, as functions of the average distances. Suppose e is an internal edge of T, with the four subtrees A, B, C and D defined as depicted in Figure 1(a). Then, the OLS length estimate of e is equal to:

$$l(e) = \frac{1}{2} \left[\lambda(\Delta_{A|C} + \Delta_{B|D}) + (1 - \lambda)(\Delta_{A|D} + \Delta_{B|C}) - (\Delta_{A|B} + \Delta_{C|D}) \right], \quad (3)$$

where

$$\lambda = \frac{|A||D| + |B||C|}{(|A| + |B|)(|C| + |D|)}.$$

Suppose e is an external edge, with i, A and B as represented in Figure 1(b). Then we have:

$$l(e) = \frac{1}{2}(\Delta_{A|i} + \Delta_{B|i} - \Delta_{A|B}). \quad (4)$$

Following Saitou and Nei [25] and Rzhetsky and Nei [24], we define the tree length $l(T)$ of T to be the sum of the edge lengths of T. The OLS minimum evolution tree is then that tree with topology $\mathcal{T}$ minimizing $l(T)$, where T has the OLS edge length estimates for $\mathcal{T}$, and $\mathcal{T}$ ranges over all possible tree topologies for the taxa being studied.

Now, suppose that we are interested in the length of the tree T shown in Figure 1(a), depending on the configuration of the corner subtrees. We then have (proof deferred until journal publication):

$$l(T) = \frac{1}{2} \left[\lambda(\Delta_{A|C} + \Delta_{B|D}) + (1 - \lambda)(\Delta_{A|D} + \Delta_{B|C}) + \Delta_{A|B} + \Delta_{C|D} \right] \quad (5)$$
$$+ l(A) + l(B) + l(C) + l(D) - \Delta_{a|A} - \Delta_{b|B} - \Delta_{c|C} - \Delta_{d|D},$$

where λ is defined as in Equation (3). The advantage of Equation (5) is that the lengths $l(A), l(B), l(C)$ and $l(D)$ of the corner subtrees as well as the average root/leaf distances, $\Delta_{a|A}, \Delta_{b|B}, \Delta_{c|C}$ and $\Delta_{d|D}$, do not depend of the configuration of A, B, C and D around e. Exchanging B and C or B and D might change the length of the five edges shown in Figure 2(a), and then the length of T, but not the lengths of A, B, C and D. This simply comes from the fact that the edge e is within the corner subtrees associated to any of the edges of A, B, C and D. As we shall see in the next section, this property is of great help in designing fast OLS tree swapping algorithms.

Let us now turn our attention toward the balanced version of minimum evolution, as defined by Pauplin [20]. The tree length definition is the same. Formulae for edge length estimates are identical to Equations (3) and (4), with λ replaced by $1/2$, but using a different definition of the average distance between subtrees, a definition which depends on the topology under consideration. Letting $\mathcal{T}$ be this topology, the balanced average distance between two non-intersecting subtrees A and B is then recursively defined by:

- If A and B are singleton sets, i.e. $A = \{a\}$ and $B = \{b\}$, then $\Delta^{\mathcal{T}}_{A|B} = \Delta_{ab}$,
- Else, without loss of generality let $B = B_1 \cup B_2$. We then have

$$\Delta^{\mathcal{T}}_{A|B} = \frac{1}{2}(\Delta^{\mathcal{T}}_{A|B_1} + \Delta^{\mathcal{T}}_{A|B_2}). \tag{6}$$

The change from Equation (2) is that the sibling subtrees B_1 and B_2 now have equal weight, regardless of the number of taxa they contain. Thus taxa do not have the same influence depending on whether they belong to a large clade or are isolated, which can be seen as consistent in the phylogenetic inference context [26]. Moreover, a computer simulation comparing variants of the NJ algorithm [13] showed that this "balanced" (or "weighted") approach is more appropriate than the unweighted one for reconstructing phylogenies with evolutionary distances estimated from sequences. Therefore, it was tempting to test the performance of this balanced version of the minimum evolution principle.

Unfortunately, this new version does not have all of the good properties as the OLS version: the edge length estimates given by Equations (3) and (4) now depend on the topology of the corner subtrees, simply because the weighted average distances between these subtrees depend on their topologies. As we shall see, this makes the algorithms more complicated and more expensive in computing time than with OLS.

However, the same tree length Equation (5) holds with Δ being replaced by $\Delta^{\mathcal{T}}$ and λ by $1/2$, and, fortunately, we still have the good property that tree lengths $l(A), l(B), l(C)$ and $l(D)$ as well as average root/leaves distances $\Delta^{\mathcal{T}}_{a|A}, \Delta^{\mathcal{T}}_{b|B}, \Delta^{\mathcal{T}}_{c|C}$ and $\Delta^{\mathcal{T}}_{d|D}$ remain unchanged when B and C or B and D are swapped. Edge lengths within the corner subtrees may change when performing a swap, but their (balanced) sums remain identical (proof deferred until journal publication).

2   FASTNNI & BNNI Tree Swapping Algorithms

This paper presents two algorithmic schemes for phylogenetic inference. The first constructs an initial tree by the stepwise addition of taxa to a growing tree, while the second improves this tree by performing local rearrangements (or swapping) of subtrees. Both follow a greedy approach and seek, at each step, to improve the tree according to the minimum evolution criterion. This approach does not guarantee that the global optimum will be reached, but only a local optimum. However, this kind of approach has proven to be effective in many optimization problems [5], and we shall see that further optimizing the minimum evolution criterion would not yield significant improvement in terms of topological accuracy. Moreover, such a combination of heuristic and optimization algorithms is used in numerous phylogenetic reconstruction methods, for example those from the PHYLIP package [10].

In this section, we consider *tree swapping* to minimize the minimum evolution criterion, when edge lengths are assigned to the topology either by OLS or the balanced version. The algorithm iteratively exchanges subtrees of an initial tree, in order to minimize the length of the tree as described in Equation (5). There are many possible definitions of "subtree exchange". Since the number of combinations is high, we usually only consider exchanging neighboring subtrees, and, at least initially, we restrict ourselves to the exchange of subtrees separated by three edges, for example B and C in Figure 1(a). Such a procedure is called a "nearest neighbor interchange", since exchanging subtrees separated by one or two edges does not yield any modification of the initial tree. We adopted this approach because it allows a fast algorithm, and it is sufficient to reach a good topological accuracy.

2.1   The FASTNNI Tree Swapping Algorithm

Assume that OLS edge lengths are used, and that the swap between B and C in Figure 1(a) is considered. Let T be the initial tree and T' the swapped tree. According to Equation (5) and our remarks, we have:

$$L(T) - L(T') = \frac{1}{2} \left[(\lambda - 1)(\Delta_{A|C} + \Delta_{B|D}) - (\lambda' - 1)(\Delta_{A|B} + \Delta_{C|D}) \right.$$
$$\left. - (\lambda - \lambda')(\Delta_{A|D} + \Delta_{B|C}) \right], \tag{7}$$

where λ is as defined in Section 1, and

$$\lambda' = \frac{|A||D| + |B||C|}{(|A| + |C|)(|B| + |D|)}.$$

If $L(T) - L(T') > 0$, T is not optimal, and we greedily choose the tree swap which corresponds to the largest difference between $L(T)$ and $L(T')$ (there are 2 such possible swaps for each internal edge of T). Moreover, assuming that the average distances between the corner subtrees have already been computed, $L(T) - L(T')$ can be obtained in $O(1)$ time via Equation (7). Instead of computing the average distances between the corner subtrees (which change when swaps are realized), we compute the average distances between every pair of non-intersecting subtrees. This takes place before evaluating the swaps and requires $O(n^2)$ time, using an algorithm that is based on the recursive Equation (2) and is completely explained in the journal version of the paper. The whole algorithm can be summarized as follows:

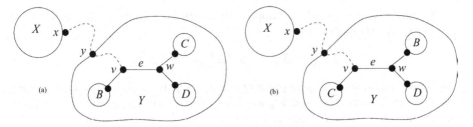

Fig. 2. T' is obtained from T by swapping B and C

1. Precompute the average distances between non-intersecting subtrees;
2. Run over all internal edges and select the best swap using Equation (7);
3. If the best swap is not relevant $(L(T) - L(T') \leq 0)$ stop and return the current tree, else perform the swap, compute the average distances between the newly created subtrees $(A \cup C$ and $B \cup D$ in our example above) and the other non-intersecting subtrees using Equation (2), and go to Step 2.

Step 1 requires $O(n^2)$ time, Step 2 requires $O(n)$ time and Step 3 also requires $O(n)$ time because the total number of new averages is $O(n)$. Thus, the total complexity of the algorithm is $O(n^2 + pn)$, where p is the number of swaps performed. In practice, p is much smaller than n, as we shall see in Section 4, so this algorithm has a practical time complexity of $O(n^2)$. It is very fast, able to improve trees with thousands of taxa, and we called it FASTNNI (Fast Nearest Neighbor Interchanges).

Rzhetsky and Nei [24] describe a procedure that requires $O(n^2)$ to compute every edge length. In one NNI, five edge lengths are changed, so evaluating a single swap is in $O(n^2)$, searching for the best swap in $O(n^3)$, and their whole procedure in $O(pn^3)$. This can be improved using Bryant and Waddell [3] results, but the implementation in the PAUP environment of these ideas is still in progress (David Bryant, personal communication). In any case, our $O(n^2)$ complexity is optimal since it is identical to the data size.

2.2   The BNNI Tree Swapping Algorithm

The BNNI algorithm is modeled after FASTNNI, with a simple change to each step. Step 1 is changed by replacing Equation (2) with Equation (6) to calculate subtree average distances. Step 2 is changed in the balanced setting by using the equation:

$$L(T) - L(T') = \frac{1}{4} \left[(\Delta_{A|B}^{\mathcal{T}} + \Delta_{C|D}^{\mathcal{T}}) - (\Delta_{A|C}^{\mathcal{T}} + \Delta_{B|D}^{\mathcal{T}}) \right]$$

instead of Equation (7). There is a major difference within Step 3, where average distances between subtrees are updated. Consider the swap of subtrees B and C as depicted in Figure 2. To maintain an accurate table of average distances,

BNNI must recalculate $\Delta_{X|Y}^{T'}$. Suppose there are $l-2$ edges in the path from y to u in T. We then compute

$$\Delta_{X|Y}^{T'} = \Delta_{X|Y}^{T} + 2^{-l}\Delta_{X|C}^{T} - 2^{-l}\Delta_{X|B}^{T}.$$

We'll need to do such a recomputation for every pair (x,y) in the same subtree (here A, see Figure 1(a) and Figure 2) of the tree swapping step. Since there are at most $n \times \operatorname{diam}(T)$ such pairs for any tree T, the time complexity of the updating step is $O(n \times \operatorname{diam}(T))$. The upper bound is worst when T is a chain. In this case, the diameter is proportional to n and both the number of distances to update and the bound are proportional to n^2. However, the diameter is usually much lower. Assuming, as usual, a Yule-Harding speciation process [15,30], the expected diameter is $O(\log(n))$ [8], which implies an average complexity of the updating step in $O(n \log(n))$. Other (e.g. uniform) distributions on phylogenetic trees are discussed in [1,18], with expected diameter at most $O(\sqrt{n})$. Therefore, the total time complexity of the algorithm is $O(pn^2)$ in the worst case, and $O(n^2 + pn \log(n))$ (or $O(n^2 + pn\sqrt{n})$) in practice. As we shall see, BNNI is a bit slower than FASTNNI, but is still able to improve thousand-taxa trees.

Neither FASTNNI nor BNNI explicitly computes the edge lengths until all tree swaps are completed. This is done in $O(n)$ time by using the edge length formulae (3) and (4) for FASTNNI, while for BNNI, we use

$$l(e) = \Delta_{A\cup B|C\cup D}^{T} - \frac{1}{2}(\Delta_{A|B}^{T} + \Delta_{C|D}^{T})$$

and

$$l(e) = \frac{1}{2}(\Delta_{i|A}^{T} + \Delta_{i|B}^{T} - \Delta_{A|B}^{T}),$$

for the external and internal edges, respectively (see Figure 1 for the notation).

3   The GME and BME Addition Algorithm

The tree swapping algorithms of Section 2 are very fast at moving from an initial topology to a local minimum for their respective minimum evolution criterion. Rzhetsky and Nei [23] proposed the Neighbor-Joining tree as a good initial starting point. However, Neighbor-Joining requires $O(n^3)$ computations, more than FASTNNI or BNNI would require. In this section, we will consider an insertion algorithmic scheme, which iteratively adds taxa to partial trees by using tree swapping techniques similar to those from Section 2 to find the optimal insertion point. This simple method provides an $O(n^2)$ (resp. $O(n^2 \times \operatorname{diam}(T))$ algorithm for finding a good initial topology for the OLS (resp. balanced) version of minimum evolution.

3.1   The GME Greedy Addition Algorithm

Given an ordering on the taxa, denoted as $(1, 2, 3, \ldots, n)$, for $k = 4$ to n we create a tree T_k on the taxa set $(1, 2, 3, \ldots, k)$. We do this by testing each edge

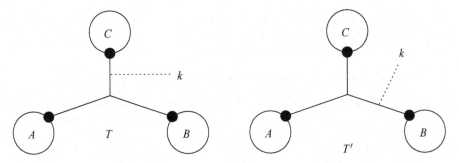

Fig. 3. T' is obtained from T by swapping A and k

of T_{k-1} as a possible insertion point for k, and the different insertion points are compared by the minimum evolution criterion. Inserting k on any edge of T_{k-1} removes that edge, changes the length of every other already existing edge, and requires the computation of the length of the three newly created edges. For each edge $e \in T_{k-1}$, define T_e to be tree created by inserting k along e, and set $c(e)$ to be the OLS size of T_e. Computing $c(e)$ for each e would seem to be computationally expensive. However, a much simpler approach using tree swapping exists to determine the best insertion point.

Consider the tree T of Figure 3, where k is inserted between subtrees C and $A \cup B$, and assume that we have the length $l(T) = L$ of this new tree. If we knew the size of T, we could quickly calculate the size of T', using a special version of Equation (5):

$$l(T') = L + \frac{1}{2} \left[(\lambda - \lambda')(\Delta_{k|A} + \Delta_{B|C}) + (\lambda' - 1)(\Delta_{A|B} + \Delta_{k|C}) \right.$$
$$\left. + (1 - \lambda)(\Delta_{A|C} + \Delta_{k|B}) \right]. \tag{8}$$

In this situation,

$$\lambda = \frac{|A| + |B||C|}{(|A| + |B|)(|C| + 1)},$$

and

$$\lambda' = \frac{|A| + |B||C|}{(|A| + |C|)(|B| + 1)}.$$

Suppose we were to first root the tree at some leaf r, let $e_0 = (r, s)$ be the edge incident to r, and L to be $c(e_0)$. We could then calculate $c(e_1)$ and $c(e_2)$ where e_1 and e_2 are the two edges incident to e_0, and in fact we could find $c(e)$ for each edge $e \in T_{k-1}$ merely by repeated applications of Equation (8). To simplify matters, we could write $c(e) = L + f(e)$. Because we only seek to determine the best insertion point, we need not calculate the actual value of L, as it is sufficient to minimize $f(e)$ with $f(e_0) = 0$.

To carry out the $(2k - 5)$ calculations of Equation (8), we need only pre-calculate

1. all average distances $\Delta_{k|S}$ between k and any subtree S from T_{k-1};

2. all average distances between subtrees of T_{k-1} separated by two edges, for example A and B in Figure 3;
3. the number of leaves of every subtree.

The algorithm can be summarized as follows:

- For $k = 3$, initialize the matrix of average distances between distant-2 subtrees and the array counting the number of taxa per subtree. Form T_3 with leaf set $\{1, 2, 3\}$.
- For $k = 4$ to n,
  1. Compute all $\Delta_{k|S}$ average distances;
  2. Starting from an initial edge e_0 of T_{k-1}, set $f(e_0) = 0$, and recursively search each edge e to obtain $f(e)$ from Equation (8).
  3. Select the best edge by minimizing f, insert k on that edge to form T_k, and update the average distance between every pair of distant-2 subtrees as well as the number of taxa per subtree.
- Return T_n.

To achieve Step 1, we recursively apply Equation (2), which requires $O(k)$ computing time. Step 2 is also done in $O(k)$ time, as explained above. Finally, to update the average distance between any pair A, B of distant-2 subtrees, if k is inserted in the subtree A,

$$\Delta_{\{k\} \cup A|B} = \frac{1}{1 + |A|} \Delta_{k|B} + \frac{|A|}{1 + |A|} \Delta_{A|B}. \qquad (9)$$

Step 3 is also done in $O(k)$ time, because there are $O(k)$ pairs of distant-2 subtrees, and all the quantities in the right-hand side of Equation (9) have already been computed. So we build T_k from T_{k-1} in $O(k)$ computational time, and thus the entire computational cost of the construction of T, as we sum over k, is $O(n^2)$. This is much faster than NJ-like algorithms which require $O(n^3)$ operations, and the FITCH [9] program which requires $O(n^4)$ operations. As we shall see, this allows trees with 4,000 taxa to be constructed in few minutes, while this task requires more than an hour for NJ, and is essentially impossible for FITCH on any existing and imaginable machine. This algorithm is called GME (Greedy Minimum Evolution algorithm).

3.2   The BME Addition Algorithm

This algorithm is very similar to GME. Equation (8) simplifies into:

$$L(T') = L + \frac{1}{4} \left[(\Delta^{\mathcal{T}}_{A|C} + \Delta^{\mathcal{T}}_{k|B}) - (\Delta^{\mathcal{T}}_{A|B} + \Delta^{\mathcal{T}}_{k|C}) \right]. \qquad (10)$$

Step 1 is identical and provides all $\Delta^{\mathcal{T}}_{k|S}$ distances by recursively applying Equation (6). Step 2 simply uses Equation (10) instead of Equation (8) to calculate the relative cost of each possible insertion point. The main difference is the updating performed in Step 3. Equation (10) only requires the balanced average

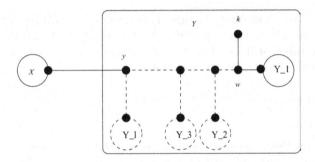

Fig. 4. Calculating balanced average $\Delta_{A|B}^{\mathcal{T}}$ when k is inserted into A

distances between distant-2 subtrees, but to iteratively update these distances, we use (and update) a data structure that contains all distances between every pair of non-intersecting subtrees (as for FASTNNI).

When k is inserted into T_{k-1}, we must calculate the average $\Delta_{A|B}^{\mathcal{T}}$ for any subtree A of T_k which contains k, and any subtree B disjoint from A. We can enumerate all such pairs by considering their respective roots. Let a be the root of A and b the root of B. Regardless of the position of k, any node of T_k could serve as the root b of B. Then, considering a fixed b, any node in the path from b to k could serve as the root a. Thus there are $O(k \times \operatorname{diam}(T_k))$ such pairs.

Given such a pair, A, B, let us consider how we may quickly calculate $\Delta_{A|B}^{\mathcal{T}}$ from known quantities. Consider the situation as depicted in Figure 4. Suppose k is inserted by creating a new node w which pushes the subtree A_1 farther away from B. Suppose there are $(l-1)$ edges in the path from w to a, and the subtrees branching off this path are, in order from w to a, $A_2, A_3, \ldots, A_l$. Then

$$\Delta_{A \cup k|B}^{\mathcal{T}} = 2^{-l}(\Delta_{k|B}^{\mathcal{T}} + \Delta_{A_1|B}^{\mathcal{T}}) + \sum_{i=2}^{l} 2^{-(l+1-i)} \Delta_{A_i|B}^{\mathcal{T}}.$$

However, we already know the value of

$$\Delta_{A|B}^{\mathcal{T}} = 2^{-(l-1)} \Delta_{A_1|B}^{\mathcal{T}} + \sum_{i=2}^{l} 2^{-(l+1-i)} \Delta_{A_i|B}^{\mathcal{T}}.$$

Thus

$$\Delta_{A \cup k|B}^{\mathcal{T}} = \Delta_{A|B}^{\mathcal{T}} + 2^{-l}(\Delta_{k|B}^{\mathcal{T}} - \Delta_{A_1|B}^{\mathcal{T}}).$$

The upper bound on the number of pairs is worst when T_k is a chain, with k inserted at one end. In this case, the diameter is proportional to k and both the number of distances to update and the bound are proportional to k^2. However, the diameter is usually much lower. We repeat the arguments of Section 2.2: if we assume a Yule-Harding speciation process, the expected diameter is $O(\log(k))$ [8], which implies an average complexity of the updating step in $O(k \log(k))$, while a uniform distribution on phylogenetic trees yields an expected diameter of $O(\sqrt{k})$.

Therefore, the complexity of the whole insertion algorithm is $O(n^3)$ in the worst case, and $O(n^2 \log(n))$ (or $O(n^2\sqrt{n})$) in practice. As we shall see in the next section, this is still less than NJ and allows trees with thousands of taxa to be constructed within a reasonable amount of time.

4   Results

4.1   Protocol

We used simulations based on random trees. Parameter values were chosen so as to cover the features of most real data sets, as revealed by the compilation of the phylogenies published in the journal *Systematic Biology* during the last few years. This approach induces much smaller contrasts between the tested methods than those based on model trees (e.g., [2,12]). Indeed, model trees are generally used to emphasize a given property of the studied methods, for example their performance when the molecular clock is strongly violated. Thus, model trees are often extreme and their use tends to produce strong and possibly misleading differences between the tested methods. On the other hand, random trees allow comparisons with a large variety of tree shapes and evolutionary rates, and provide a synthetic and more realistic view of the average performances.

We used 24- and 96-taxon trees, and 2000 trees per size. For each of these trees, a true phylogeny, denoted as T, was first generated using the stochastic speciation process described by Kuhner and Felsenstein [17], which corresponds to the usual Yule-Harding distribution on trees [15,30]. Using this generating process makes T ultrametric (or molecular clock-like). This hypothesis does not hold in most biological data sets, so we created a deviation from the molecular clock. Each edge length of T was multiplied by $1.0 + \mu X$, where X followed the standard exponential distribution ($P(X > \eta) = e^{-\eta}$) and μ was a tuning factor to adjust the deviation from molecular clock; μ was set to 0.8 with 24 taxa and to 0.6 with 96 taxa. The average ratio between the mutation rate in the fastest evolving lineage and the rate in the slowest evolving lineage was then equal to about 2.0 with both tree sizes. With 24 taxa, the smallest value (among 2000) of this ratio was equal to about 1.2 and the largest to 5.0 (1.0 corresponds to the strict molecular clock), while the standard deviation was approximately 0.5. With 96 taxa, the extreme values became 1.3 and 3.5, while the standard deviation was 0.33.

These (2×2000) trees were then rescaled to obtain "slow", "moderate" and "fast" evolutionary rates. With 24 taxa, the branch length expectation was set to 0.03, 0.06 and 0.15 mutations per site, for the slow, moderate and fast conditions, respectively. With 96 taxa, we had 0.02, 0.04 and 0.10 mutations per site, respectively. For both tree sizes, the average maximum pairwise divergence was of about 0.2, 0.4 and 1.0 substitutions per site, with a standard deviation of about 0.07, 0.14 and 0.35 for 24 taxa, and of about 0.04, 0.08 and 0.20 for 96 taxa. These values are in good accordance with real data sets. The maximum pairwise divergence is rarely above 1.0 due to the fact that multiple alignment from highly divergent sequences is simply impossible. Moreover, with such distance value, any correction formula, e.g. Kimura's (1980), becomes very doubtful, due to our ignorance of the real substitution process and to the fact that the larger the distance the higher the gap between estimates obtained from different formulae. The medium condition ($\sim$0.4) corresponds to the most favorable practical setting, while in the slow condition (from $\sim$0.1 to $\sim$0.3) the phylogenetic signal is only slightly perturbed by multiple substitutions, but it can be too low, with some short branches being not supported by any substitution.

SeqGen [21] was used to generate the sequences. For each tree T (among $3 \times 2 \times 2000$), these sequences were obtained by simulating an evolving process along T according to the Kimura [16] two parameter model with a transition/transversion ratio of 2.0. The sequence length was set to 500 sites. Finally, DNADIST from the PHYLIP package [10] was used to compute the pairwise distance matrices, assuming the Kimura model with known transition/transversion ratio. The data files are available on our web page (see title page).

Every inferred tree, denoted as $\widehat{T}$, was compared to the true phylogeny T (i.e., that used to generate the sequences and then the distance matrix) with a topological distance equivalent to Robinson and Foulds' [22]. This distance is defined by the proportion of internal edges (or bipartitions) which are found in one tree and not in the other one. This distance varies between 0.0 (both topologies are identical) and 1.0 (they do not share any internal edge). The results were averaged over the 2000 test sets for each tree size and evolutionary rate. Finally, to compare the various methods to NJ, we measured the relative error reductions $(P_M - P_{NJ})/P_{NJ}$, where M is any tested method different from NJ and P_X is the average topological distance between $\widehat{T}$ and T when using method X. This measure highlights the performance difference between the related, distance-based, tested methods.

4.2 Phylogeny Estimation Algorithm Comparison

We used a variety of different algorithms to try to reconstruct the original tree, given the matrix of estimated distances. We used the NJ [25] program from the PAUP package [27], with and without the BIONJ [12] flag; WEIGHBOR version 1.2 [2], available at http://www.t10.lanl.gov/billb/weighbor/, the FITCH [9] program from the PHYLIP package [10]; the Harmonic Greedy Triplet (HGT/FP) program, provided by Miklos Csűrös [6]; GME and BME. Also, we used output topologies from other programs as input for FASTNNI and BNNI. GME, BME, FASTNNI, and BNNI will be available in the near future at our web page, or via ftp at ftp://ncbi.nlm.nih.gov/~pub/desper/ME.

We also measured how far from the true phylogeny one gets with NNIs. This served as a measure of the limitation of each of the minimum evolution frameworks, as well as a performance index for evaluating our algorithms. Ordinarily, the (OLS or balanced) minimum evolution criterion will not, in fact, observe a minimum value at the true phylogeny. So, starting from the true phylogeny and running FASTNNI or BNNI, we end up with a tree with a significant proportion of false edges. When this proportion is high, the corresponding criterion can be seen as poor regarding topological accuracy. This proportion represents the best possible topological accuracy that can be achieved by optimizing the considered

Table 1. Topological accuracy for 24-taxon trees at various rates of evolution

	slow rate			moderate rate		
	w/o NNIs	FASTNNI	BNNI	w/o NNIs	FASTNNI	BNNI
True Tree		.109 -1.6%	.104 -6.2%		.092 3.7%	.083 -5.8%
FITCH	.109 -1.9%	.113 2.0%	.107 -3.4%	.085 -4.9%	.094 6.0%	.085 -4.0%
WEIGHBOR	.109 -1.8%	.112 1.7%	.107 -3.0%	.085 -4.3%	.094 6.2%	.085 -4.0%
BIONJ	.111 -0.3%	.113 2.0%	.107 -3.6%	.087 -2.0%	.094 6.5%	.085 -4.2%
NJ	.111 0%	.113 2.0%	.107 -3.5%	.088 0%	.094 6.6%	.085 -4.0%
BME	.118 7.1%	.113 1.9%	.107 -2.8%	.100 13%	.094 6.3%	.084 -4.9%
GME	.122 10%	.113 2.1%	.107 -3.4%	.107 21%	.095 7.1%	.084 -4.8%
HGT/FP	.334 202%	.112 1.1%	.107 -2.9%	.326 268%	.095 7.5%	.088 -0.2%

	fast rate		
	w/o NNIs	FASTNNI	BNNI
True Tree		.088 6.5%	.076 -8.3%
FITCH	.076 -7.8%	.090 8.8%	.077 -7.0%
WEIGHBOR	.077 -6.8%	.089 8.2%	.077 -6.9%
BIONJ	.079 -3.6%	.090 9.0%	.077 -6.6%
NJ	.082 0%	.090 9.1%	.077 -6.9%
BME	.098 19%	.090 9.1%	.076 -7.1%
GME	.105 28%	.090 9.8%	.076 -7.6%
HGT/FP	.329 300%	.090 9.8%	.083 0.8%

criterion, since we would not expect any algorithm optimizing this criterion in the whole tree space to find a tree closer from the true phylogeny than the tree that is obtained by "optimizing" the true phylogeny itself.

Results are displayed in Table 1 and Table 2. The first number indicates the average topological distance between the inferred tree and the true phylogeny. The second number provides the percentage of relative difference in topological distance between the method considered and NJ (see above); the more negative this percentage, the better the method was relative to NJ.

The performances of the basic algorithms (1st columns) are strongly correlated with the number of computations that they perform. Both $O(n^2)$ algorithms are clearly worse than NJ. BME (in $O(n^2 \log(n))$) is still worse than NJ, but becomes very close with 96 taxa, which indicates the strength of the balanced minimum evolution framework. BIONJ, in $O(n^3)$ like NJ and having identical computing times, is slightly better than NJ in all conditions, while WEIGHBOR, also in $O(n^3)$ but requiring complex numerical calculations, is better than BIONJ. Finally, FITCH, which is in $O(n^4)$, is the best with 24 taxa, but was simply impossible to evaluate with 96 taxa (see the computing times below).

After FASTNNI (2nd columns), we observe that the output topology does not depend much on the input topology. Even the deeply flawed HGT becomes close to NJ, which indicates the strength of NNIs. However, except for the true phylogeny in some cases, this output is worse than NJ. This confirms our previous results [13], which indicated that the OLS version of minimum evolution is reasonable but not excellent for phylogenetic inference. But this phenomenon

Table 2. Topological accuracy for 96-taxon trees at various rates of evolution

	slow rate			moderate rate		
	w/o NNIs	FASTNNI	BNNI	w/o NNIs	FASTNNI	BNNI
True Tree		.172 -5.6%	.167 -8.8%		.132 -3.0%	.115 -15.4%
WEIGHBOR	.178 -2.5%	.181 -0.7%	.173 -5.2%	.129 -5.4%	.137 0.5%	.118 -13.0%
BIONJ	.180 -0.9%	.182 -0.3%	.173 -5.1%	.134 -1.9%	.138 1.3%	.118 -13.0%
NJ	.183 0%	.182 -0.2%	.173 -5.2%	.136 0%	.139 1.8%	.119 -12.9%
BME	.186 1.9%	.181 -0.6%	.173 -5.3%	.137 1.0%	.138 1.1%	.118 -13.2%
GME	.199 8.8%	.183 0.3%	.173 -5.3%	.158 16%	.140 2.7%	.118 -13.2%
HGT/FP	.512 185%	.185 1.5%	.175 -4.3%	.480 253%	.143 5.2%	.123 -.9.3%

	fast rate		
	w/o NNIs	FASTNNI	BNNI
True Tree		.115 0.6%	.088 -23.4%
WEIGHBOR	.103 -10%	.119 3.8%	.091 -21.0%
BIONJ	.112 -2.5%	.121 5.1%	.090 -21.7%
NJ	.115 0%	.121 5.5%	.090 -21.3%
BME	.117 1.8%	.120 4.4%	.090 -21.4%
GME	.144 25%	.122 6.3%	.091 -21.1%
HGT/FP	.465 306%	.126 9.4%	.098 -14.7%

is much more visible with 24 than with 96 taxa, so we expect the very fast GME+FASTNNI $O(n^2)$ combination to be equivalent to NJ with large n.

After BNNI (3rd columns), all initial topologies become better than NJ. With 24 taxa, the performance is equivalent to that of FITCH, while with 96 taxa the results are far better than those of WEIGHBOR, especially in the Fast condition where BNNI improves NJ by 21%, against 10% for WEIGHBOR. These results are somewhat unexpected, since BNNI has a low $O(n^2 + pn \log(n))$ average time complexity. Moreover, as explained above, we do not expect very high contrast between any inference method and NJ, due to the fact that our data sets represent a large variety of realistic trees and conditions, but do not contain extreme trees, notably concerning the maximum pairwise divergence. First experiments indicate that maximum parsimony is close to BNNI in the Slow and Moderate conditions, but worse in the Fast condition, the contrast between these methods being in the same range as those reported in Table 1 and 2. The combination of BNNI with BME or GME can thus be seen as remarkably efficient and accurate. Moreover, regarding the true phylogeny after BNNI, it appears that little gain could be expected by further optimizing this criterion, since our simple optimization approach is already close to these upper values.

4.3   Computing Times

To compare the actual running speeds of these algorithms, we tested them on a Sun Enterprise E4500/E5500, with ten 400 MHz processors and 7 GB of memory, running the Solaris 8 operating system. Table 3 summarizes the average computational times (in hours:minutes:seconds) required by the various programs to

Table 3. Average computational time (HH:MM:SS) for reconstruction algorithms

Algorithm	24 taxa	96 taxa	1000 taxa	4000 taxa
GME	.02522	.07088	8.366	3:11.01
GME + FASTNNI	.02590	.08268	9.855	3:58.79
GME + BNNI	.02625	.08416	11.339	6:02.10
HGT/FP	.02518	.13491	13.808	3:33.14
BME	.02534	.08475	19.231	12:14.99
BME + BNNI	.02557	.08809	19.784	12:47.45
NJ	.06304	.16278	21.250	20:55.89
BIONJ	.06528	.16287	21.440	20:55.64
WEIGHBOR	.42439	26.88176	*****	*****
FITCH	4.37446	*****	*****	*****

Table 4. Average number of NNIs performed

Algorithm	24 taxa	96 taxa	1000 taxa	4000 taxa
GME + FASTNNI	1.244	8.446	44.9	336.50
GME + BNNI	1.446	11.177	59.1	343.75
BME + BNNI	1.070	6.933	29.1	116.25

build phylogenetic trees. The leftmost two columns were averaged over two thousand 24- and 96-taxon trees, the third over ten 1000-taxon trees, and the final column over four 4000-taxon trees. Stars indicate entries where the algorithm was deemed to be feasible

FITCH would take approximately 25 minutes to make one 96-taxon tree, a speed which makes it impractical for real applications, because they often include bootstrapping which requires the construction of a large number of trees. As WEIGHBOR's running time increased more than 60-fold when moving from 24-taxon trees to 96-taxon trees, we judged it infeasible to run WEIGHBOR on even one 1000-taxon tree.

Table 3 confirms that all of our algorithms are faster than NJ, with the improvement becoming notably impressive when thousands of taxa are dealt with. Furthermore, we suspect that implementation refinements such as those used in PAUP's NJ could be used to make our algorithms still much faster.

Table 4 contains the number of NNIs performed by each of the three combinations which appear in Table 3. Not surprisingly, the largest number of NNIs was consistently required when the initial topology was made to minimize the OLS ME criterion, but NNIs were chosen to minimize the balanced ME criterion (i.e., GME + BNNI). This table shows the overall superiority of the BME tree over the GME tree, when combined with BNNI. In all of the cases considered, the average number of NNIs considered for each value of n was considerably less than n itself.

5   Discussion

We have presented a new distance-based approach to phylogeny estimation which is faster than most distance algorithms currently in use. The current most popular fast algorithm is Neighbor-Joining, an $O(n^3)$ algorithm. Our greedy ordinary least-squares minimum evolution tree construction algorithm (GME) runs at $O(n^2)$, the size of the input matrix. Although the GME tree is not quite as accurate as the NJ tree, it is a good starting point for nearest neighbor interchanges (NNIs). The combination of GME and FASTNNI, which achieves NNIs according to the ordinary least-squares criterion, also in $O(n^2)$ time, has a topological accuracy close to that of NJ, especially with large numbers of taxa and moderate or slow rates.

However, the balanced minimum evolution framework appears much more appropriate for phylogenetic inference than the ordinary least-squares version. This is likely due to the fact that it gives less weight to the topologically long distances, i.e. those containing numerous edges, while the ordinary least-squares method puts the same confidence on each distance, regardless of its length. Even when the usual and topological lengths are different, they are strongly correlated. The balanced minimum evolution framework is thus conceptually closer to weighted least-squares [11], which is more appropriate than ordinary least-squares for evolutionary distances estimated from sequences.

The balanced NNI algorithm (BNNI) achieves a good level of performance, superior to that of NJ, BIONJ and WEIGHBOR. BNNI is an $O(n^2 + np \times \text{diam}(T))$ algorithm, where $\text{diam}(T)$ is the diameter of the inferred tree, and p the number of swaps performed. With $p \times \text{diam}(T) = O(n)$ for most data sets, the combination of GME and BNNI effectively gives us an $O(n^2)$ algorithm with high topological accuracy.

Acknowledgements

Special thanks go to Stephane Guindon, who generated the simulated data sets used in Section 4.

References

1. Aldous, D.J.: Stochastic models and descriptive statistics for phylogenetic trees from Yule to today. Statist. Sci. **16** (2001) 23–34
2. Bruno, W.J., Socci, N.D., Halpern, A.L.: Weighted neighbor joining: A likelihood-based approach to distance-based phylogeny reconstruction. Mol. Biol. Evol. **17** (2000) 189–197
3. Bryant, D., Waddell, P.: Rapid evaluation of least-squares and minimum-evolution criteria on phylogenetic trees. Mol. Biol. Evol. **15** (1998) 1346–1359
4. Bulmer, M.: Use of the method of generalized least squares in reconstructing phylogenies from sequence data. Mol. Biol. Evol. **8** (1991) 868–883
5. Cormen, T.H., Leiserson, C.E., Rivest, R.L.: Introduction to algorithms. MIT Press, Cambridge, MA (2000)
6. Csűrös, M.: Fast recovery of evolutionary trees with thousands of nodes. Journal of Computational Biology **9** (2002) 277–297
7. Denis, F., Gascuel, O.: On the consistency of the minimum evolution principle of phylogenetic inference. Discr. Appl. Math. **In press** (2002)

8. Erdös, P.L., Steel, M., Székély, L., Warnow, T.: A few logs suffice to build (almost) all trees: Part II. Theo. Comp. Sci. **221** (1999) 77–118

9. Felsenstein, J.: An alternating least-squares approach to inferring phylogenies from pairwise distances. Syst. Biol. **46** (1997) 101–111

10. Felsenstein, J.: PHYLIP — Phylogeny Inference Package (Version 3.2). Cladistics **5** (1989) 164–166

11. Fitch, W.M., Margoliash, E.: Construction of phylogenetic trees. Science **155** (1967) 279–284

12. Gascuel, O.: BIONJ: an improved version of the NJ algorithm based on a simple model of sequence data. Mol. Biol. Evol. **14** (1997) 685–695

13. Gascuel, O.: On the optimization principle in phylogenetic analysis and the minimum-evolution criterion. Mol. Biol. Evol. **17** (2000) 401–405

14. Gascuel, O., Bryant, D., Denis, F.: Strengths and limitations of the minimum evolution principle. Syst. Biol. **50** (2001) 621–627

15. Harding, E.: The probabilities of rooted tree-shapes generated by random bifurcation. Adv. Appl. Probab. **3** (1971) 44–77

16. Kimura, M.: A simple method for estimating evolutionary rates of base substitutions through comparative studies of nucleotide sequences. J. Mol. Evol. **16** (1980) 111,120

17. Kuhner, M.K., Felsenstein, J.: A simulation comparison of phylogeny algorithms under equal and unequal rates. Mol. Biol. Evol. **11** (1994) 459–468

18. McKenzie, A., Steel, M.: Distributions of cherries for two models of trees. Math. Biosci. **164** (2000) 81–92

19. Nei, M., Jin, L.: Variances of the average numbers of nucleotide substitutions within and between populations. Mol. Biol. Evol. **6** (1989) 290–300

20. Pauplin, Y.: Direct calculation of a tree length using a distance matrix. J. Mol. Evol. **51** (2000) 41–47

21. Rambaut, A., Grassly, N.C.: Seq-Gen: An application for the Monte Carlo simulation of DNA sequence evolution along phylogenetic trees. Computer Applications in the Biosciences **13** (1997) 235–238

22. Robinson, D., Foulds, L.: Comparison of phylogenetic trees. Math. Biosci. **53** (1981) 131–147

23. Rzhetsky, A., Nei, M.: A simple method for estimating and testing minimum-evolution trees. Mol. Biol. Evol. **9** (1992) 945–967

24. Rzhetsky, A., Nei, M.: Theoretical foundation of the minimum-evolution method of phylogenetic inference. Mol. Biol. Evol. **10** (1993) 1073–1095

25. Saitou, N., Nei, M.: The neighbor-joining method: A new method for reconstructing phylogenetic trees. Mol. Biol. Evol. **4** (1987) 406–425

26. Sneath, P.H.A., Sokal, R.R. In: Numerical Taxonomy. W.K. Freeman and Company, San Francisco (1973) 230–234

27. Swofford, D.: PAUP—Phylogenetic Analysis Using Parsimony (and other methods), Version 4.0 (1996)

28. Swofford, D.L., Olsen, G.J., Waddell, P.J., Hillis, D.M.: Phylogenetic inference. In Hillis, D., Moritz, C., Mable, B., eds.: Molecular Systematics. Sinauer, Sunderland, MA (1996) 407–514

29. Vach, W.: Least squares approximation of addititve trees. In Opitz, O., ed.: Conceptual and numerical analysis of data. Springer-Verlag, Berlin (1989) 230–238

30. Yule, G.: A mathematical theory of evolution, based on the conclusions of Dr. J. C. Willis. Philos. Trans. Roy. Soc. London Ser. B, Biological Sciences **213** (1925) 21–87

NeighborNet: An Agglomerative Method for the Construction of Planar Phylogenetic Networks

David Bryant[1] and Vincent Moulton[2]

[1] McGill Centre for Bioinformatics, 3775 University St, Montréal, Québec, H3A 2B4
bryant@math.mcgill.ca
http://www.math.mcgill.ca/bryant
[2] Linnaeus Center for Bioinformatics, Uppsala University
Box 598, 751 24 Uppsala, Sweden
vincent.moulton@lcb.uu.se

Abstract. We introduce NeighborNet, a network construction and data representation method that combines aspects of the neighbor joining (NJ) and SPLITSTREE. Like NJ, NeighborNet uses agglomeration: taxa are combined into progressively larger and larger overlapping clusters. Like SPLITSTREE, NeighborNet constructs networks rather than trees, and so can be used to represent multiple phylogenetic hypotheses simultaneously, or to detect complex evolutionary processes like recombination, lateral transfer and hybridization. NeighborNet tends to produce networks that are substantially more resolved than those made with SPLITSTREE. The method is efficient ($O(n^3)$ time) and is well suited for the preliminary analyses of complex phylogenetic data. We report results of three case studies: one based on mitochondrial gene order data from early branching eukaryotes, another based on nuclear sequence data from New Zealand alpine buttercups (*Ranunculi*), and a third on poorly corrected synthetic data.

Keywords: Networks, Phylogenetics, Hybridization, SplitsTree, Neighbor-Joining

1 Introduction

As the importance of complex evolutionary processes such as recombination, hybridization, lateral gene transfer and gene conversion becomes more evident, biologists are showing a growing interest in using networks as a tool for aiding phylogenetic analysis. Often, a single tree is simply not the most accurate way to represent the underlying evolutionary relationships. Networks can provide a useful alternative to trees, for they can allow the simultaneous representation of multiple trees, or at least give extra insight into the structure of data set in question.

Phylogenetic networks are typically applied during the early stages of data analysis. A network provides a *representation* of the data rather than a phylogenetic inference. It can indicate whether or not the data is substantially tree-like and give evidence for possible reticulation or hybridization events. Hypotheses

R. Guigó and D. Gusfield (Eds.): WABI 2002, LNCS 2452, pp. 375–391, 2002.

suggested by the network can then be tested directly using more detailed phylogenetic analysis and specialized biochemical methods (e.g. DNA fingerprinting or chromosome painting). Data representation tools, as distinct from phylogenetic inference tools, are especially important when there is no well established model of evolution, since the use of phylogenetic inference methods based on incorrect models, or even the assumption that the data is tree-like, can lead to positively misleading conclusions.

Although there are a few programs available for constructing networks (most notably the popular SPLITSTREE package [28]), few use agglomeration, a technique that has paid dividends in the tree building business (with methods such as neighbor-joining [33] etc.). It is natural to look for a method which constructs networks using the agglomeration principle.

NeighborNet iteratively selects pairs of taxa to group together. However, unlike tree building, it does not agglomerate these pairs immediately but rather at a later stage when it agglomerates pairs of pairs which share one node in common. In this way, NeighborNet generates a (weighted) *circular split system* rather than a hierarchy or a tree, which can subsequently be represented by a planar *split graph*. In these graphs, bipartitions or *splits* of the taxa are represented by classes of parallel lines, and conflicting signals or incompatibilities appear as boxes. An important advantage of the agglomerative approach is that NeighborNet tends to produce more resolved networks than those produced by split decomposition, especially when the number of taxa grows.

The NeighborNet algorithm runs in $O(n^3)$ time when applied to n taxa—the same time complexity as NJ. One is thus able to apply computationally intensive validation techniques such as bootstrapping. The resulting split graph can be generated, displayed, and manipulated using the user-friendly software package SPLITSTREE [28].

NeighborNet is designed to be a data representation tool for situations when standard phylogenetic inference methods may not be appropriate. We have chosen three such situations to illustrate the method. The results we present in this abstract are preliminary: a comprehensive experimental study is currently underway.

First we analyze mitochondrial gene order data for early branching eukaryotes [35]. This is difficult data for phylogenetic analysis. The divergences are ancient and there appears to be a great deal of noise in the signal. Furthermore, the physical processes behind gene order evolution are not well understood, particularly for highly diverged genomes, making it difficult to create realistic synthetic data.

The second data set, in contrast, involves sequences that have evolved recently and rapidly. We analyze nuclear ITS sequence data taken from a group of 46 New Zealand alpine buttercups (refer to [29] for a description of the sequences and accession numbers). As with many plants, the evolutionary history of the buttercups is not tree-like. There is substantial evidence of hybridization events. There are also systematic difficulties associated with the sequence data: polymorphism, polyploidy, multiple alleles, interbreeding between species, and concerted evolution all confuse standard phylogenetic methods.

The third data set was generated on a randomly selected model tree. To attempt to simulate wrong model choice we used uncorrected distances to construct both the NJ tree and NeighborNet. Both NJ and NeighborNet produced similar networks, both capturing most of the clades of the original tree. However, in the places where NJ gave an incorrect resolution NeighborNet indicated an ambiguous resolution.

In all of the case studies, NeighborNet was able to quickly construct a detailed network representation of the data. It was able to detect weak and conflicting phylogenetic signals in the data, and recover the major accepted groupings, including several recently established groups.

NeighborNet has been implemented in C++ and incorporated into the SPLITsTREE package. It will be available for download at http://www.math.mcgill.ca/bryant/.

2   Relation to Existing Methodologies

2.1   The Success of NJ

Neighbor-joining (NJ) is one of the most popular, and most successful, methods for constructing phylogenies from distance data. The algorithm belongs to the agglomerative family of tree methods, a family that includes the classical linkage methods, Sattath and Tversky's ADDTREE algorithm [36], and the average linkage UPGMA method [37].

For some time, the success of NJ proved to be an enigma for computational biologists, particularly given the all-important lack of approximation bounds. The first explicit performance bound was provided by Atteson [1], who showed that NJ would return the true tree provided that the observed distance is sufficiently close to the true evolutionary distance. The bound is tight, but also holds for many other phylogenetic methods [19], many of which perform poorly in practice.

A better explanation for the success of NJ is provided in a series of papers by Gascuel [22,23,24]. Gascuel analyses the statistical properties of the NJ method and agglomerative methods in general, showing that the edge length and distance matrix reduction formulae in NJ give low variance estimators. Indeed, the accuracy of the distance estimates between internal nodes improves as nodes are agglomerated. Accuracy can be improved further by incorporating minimum variance optimization directly into the reduction formulae, as with BIONJ. [23]

2.2   Network Methods

The idea of representing evolutionary processes by networks as opposed to trees has a long history, though it was not until recent years that the use of networks has become more common. The programs SPLITSTREE [28] and T-REX [31] both compute networks from distances: SPLITSTREE computes *split graphs* which are derived from a canonical *split decomposition* of the given distance, whereas T-REX [31] computes *reticulograms* by first computing a phylogeny and subsequently "growing" a network by adding in branches which minimize a certain least-squares loss function [32]. In contrast, the program NETWORK [5] constructs *phylogenetic networks* directly from characters, using variants of the median construction [4].

Among other applications, SPLITSTREE has been used to analyze viral data [15], plant hybridization [29], and the evolution of manuscripts [6]. T-REX has been used to construct spatially-constrained representations of fish dispersal and for analysis of homoplasy in primates [32]. NETWORK has been mainly used for intra-specific data such as human population data [4,5] (based on mitochondrial sequences).

NeighborNet complements these programs. Unlike T-REX, NeighborNet is not restricted by having to compute an initial phylogeny, which could cause problems if the the data is not tree-like. NeighborNet tends to produce more resolved networks than SPLITSTREE. Split decomposition is quite conservative – it only represents splits of the taxa with positive *isolation index*, a quantity which is computed by minimizing a certain quartet index over all quartets which "cross" the split in question. Since it only takes one "bad" quartet to reduce the isolation index, SPLITSTREE tends to under estimate edge lengths, and so the split graph resolution tends to degrade for larger data-sets [26].

2.3 Pyramidal Clustering

A network method that deserves special mention is the pyramidal clustering technique developed by Diday [14], Betrand [8] and others (refer to Aude *et al.* [2] for a survey and application of the method to the analysis of protein sequences.) A pyramid can be seen as a dual structure to circular splits in the same sense that hierarchies are dual to unrooted trees.

Like NeighborNet, the pyramidal clustering schemes are agglomerative. There are, however, important differences. First, during the agglomeration step of pyramid construction, one is forced to assign an order to the nodes being combined [14]. With NeighborNet, the order rises naturally from the agglomeration process and one does not need to enforce some arbitrary ordering during combination.

Secondly, pyramidal clustering methods are suitable for phylogenetic analysis only when we have some kind of molecular clock assumption. They form clusters directly from similarity or distance data, and are adversely affected by unequal rates across lineages or long external edges. To extend an analogy, pyramidal clustering is to UPGMA what NeighborNet is to NJ, and many of the theoretical advantages of NJ over UPGMA apply to NeighborNet over pyramidal clustering.

3 Circular Split Systems, Circular Distances and Split Graphs

3.1 Compatible Splits and Trees

Let X denote a finite set of n taxa. A *split* $S = A|B = B|A$ is a bipartition $\{A, B\}$ of X, and any collection of splits of X is called a *split system*. We say that two splits $S_1 := A|B$ and $S_2 := C|D$ of X are *compatible* if one (and hence precisely one) of the four intersections $A \cap C$, $A \cap D$, $B \cap C$ and $B \cap D$ is empty; otherwise S_1, S_2 are *incompatible*. A split system is compatible if every pair of splits within it is compatible.

As is well-known, compatible split systems on X are in 1-1 correspondence with unweighted phylogenies (e.g. see [7]). Using this correspondence, we can consider a phylogeny as a weighted compatible split system, where each split is weighted by the length of the corresponding branch in the phylogeny. The distance or divergence between two taxa in a tree is equal to the length of the unique path connecting them, which in turn equals the sum of the weights of the splits separating them. A distance matrix arising in this way is called *additive* and has a simple characterization [11].

3.2   Circular Splits and Circular Distances

We now consider split systems which can contain incompatible pairs of splits, and that, like phylogenies, can be represented exactly by a planar graph. A *circular ordering* $\theta = (x_1, x_2, \ldots, x_n, x_{n+1} = x_1)$ of X is a bijection between X and the vertices of a convex n-gon in the plane such that x_i and x_{i+1} are consecutive vertices of the n-gon, $1 \leq i \leq n$. For each i, j such that $1 \leq i \neq j \leq n$ we define the split $S[x_i, x_j] = S[x_i, x_j, \theta]$ to be the bipartition consisting of the set $\{x_i, x_{i+1}, \ldots, x_{j-1}\}$ and its complement (see Figure 1), and we let $\mathcal{S}(\theta)$ denote the split system containing all of the splits $S[x_i, x_j]$ with $1 \leq i \neq j \leq n$. Clearly $\mathcal{S}(\theta)$ contains $\binom{n}{2}$ elements. An arbitrary split system $\mathcal{S}$ of X is *circular* if there is a circular ordering θ of X with $\mathcal{S}$ contained in $\mathcal{S}(\theta)$. Thus, when $\mathcal{S}$ is circular $|\mathcal{S}| \leq \binom{n}{2}$; when equality holds we call $\mathcal{S}$ *binary*. Circular split systems were introduced in [3] and further studied in, for example, [12,13,17,20]. They generalize compatible split systems because every system of compatible splits is circular.

We also consider weighted circular split systems, that is, split systems in which we associate some weight to each split corresponding to "branch length". The distance between two taxa in a weighted circular split system is equal to the sum of the weights of the splits that separate them. Distance matrices arising this way are called *circular distances* and can be recognized in linear time [13].

3.3   Split Graphs

In general, to represent split systems containing incompatible subsystems it can be necessary to employ non-planar graphs (usually subgraphs of hypercubes) – see e.g. [17]. However, if the split system is circular it is possible to represent it by a *planar split graph* [18]. These graphs represent splits by parallel classes of edges. Weights are represented by assigning to each edge a length proportional to the weight of the corresponding split. If the split system is compatible the split graph is precisely the corresponding phylogeny. The split graph corresponding to a given weighted circular split system can be computed using the program SPLITSTREE [28], which accepts split systems in NEXUS format [30].

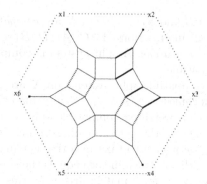

Fig. 1. The circular split system $\mathcal{S}(\theta)$ for the circular ordering $\theta = (x_1, x_2, \ldots, x_6, x_1)$ of the set $\{x_1, \ldots, x_6\}$. The edges in the split graph corresponding to $S[x_2, x_4]$ are marked with solid lines, while those corresponding to $S[x_1, x_4]$ are marked with dotted lines.

Significantly, the number of edges and vertices in the split graph for circular splits is polynomial in the number of taxa. This does not hold for general split systems. The split graph of a binary circular split system $\mathcal{S}(\theta)$ contains $O(n^4)$ vertices and edges [16]. In practice, the removal of zero weight splits gives graphs with far fewer vertices and edges.

Split graphs provide a representation of distance data in terms of implicit groupings in the data. However they do not necessarily provide a direct representation of evolutionary history: one should not assume that internal nodes correspond to ancestral sequences of taxa [38]. To use networks to depict evolutionary history we require a specific model for the part of the signal that is not treelike, for example, models involving data heterogeneity, lateral transfer, or duplications.

4   The NeighborNet Method

NJ, UPGMA, and the linkage tree algorithms all follow the same general scheme. We start with one node for each taxon. At each iteration, a pair of nodes is selected and replaced by a new composite node. The procedure continues until only two (or three) nodes remain, at which point we reverse the process to construct a tree or hierarchy.

NeighborNet works along the same lines, with one important difference. When we select a pair of nodes we do not combine and replace them immediately. Instead we wait until a node has been paired up a second time. We replace the three linked nodes with two linked nodes, reduce the distance matrix, and proceed to the next iteration. This simple change in the agglomerative framework generates circular split systems, instead of hierarchies or trees.

Within the agglomerative framework, the NeighborNet method is defined by the three following elements:

Fig. 2. Generic situation for the selection criterion

1. The criterion used to select the nodes to merge.
2. The reduction of the distance matrix after agglomeration.
3. The estimation of edge lengths (split coefficients).

We discuss each of these elements in turn. The complete algorithm is presented in Figure 3.

An important property of NeighborNet is that it retains the consistency properties of NJ. If the input distance is a circular distance, NeighborNet will return the corresponding collection of circular splits. Consequently, if the input distance is additive, NeighborNet will return the corresponding tree.

4.1   Selection

The selection criterion determines which clusters or nodes are combined at each step. The generic situation is represented in Figure 2. Clusters of size one or two are attached to a central high degree vertex. Each leaf represents a node corresponding to one or more taxa. Initially, all nodes correspond to single taxa and all clusters are of size one.

Let A be the set of nodes, let $N = |A|$ and let d be the distance function defined on pairs of nodes in A. Let $C_1, C_2, \ldots, C_m$ be the clusters of size one (*singletons*) or two (*cherries*). For each pair of clusters C_i, C_j we define

$$d(C_i, C_j) = \frac{\sum_{x \in C_i} \sum_{y \in C_j} d(x, y)}{|C_i||C_j|}$$

There are two steps to the selection process. First we find the pair of clusters C_i, C_j that minimize

$$\hat{Q}(C_i, C_j) = (m - 2)d(C_i, C_j) - \sum_{k \neq i} d(C_i, C_k) - \sum_{k \neq j} d(C_j, C_k).$$

Note that C_i or C_j may contain one or two nodes. We choose $x \in C_i$ and $y \in C_j$ that minimize

$$\hat{Q}(x, y) = (m - 2)d(x, y) - \sum_{k \neq i} d(\{x\}, C_k) - \sum_{k \neq j} d(\{y\}, C_k).$$

If all clusters have size one then this criterion reduces to the standard NJ selection criterion for trees.

The choice of selection criterion was determined by the following four requirements

(P1) *Linearity* The criterion is linear with respect to the input distances.
(P2) *Permutation invariance* The outcome is independent of the choice of labeling
(P3) *Consistency* If the input distance is circular then the criterion selects two taxa that are adjacent in the corresponding circular ordering.

The criterion $\hat{Q}$ satisfies all three properties. The proof of property (P3) is non-trivial, see [10] for details. Furthermore, in some cases $\hat{Q}$ is the *only* selection criterion that satisfies all three properties (up to multiplication or addition by a constant) [9].

4.2   Reduction of the Distance Matrix

Suppose that node y has two neighbors, x and z. In the NeighborNet agglomeration step, we replace x, y, z with two new nodes u, v.

The distances from u and v to the other nodes $a \in A$ are computed using the reduction formulae

$$d(u, a) = \alpha\, d(x, a) + \beta\, d(y, a)$$
$$d(v, a) = \beta\, d(y, a) + \gamma\, d(z, a)$$
$$d(u, v) = \alpha\, d(x, y) + \beta\, d(x, z) + \gamma\, d(y, z)$$

where α, β, γ are positive real numbers with $\alpha + \beta + \gamma = 1$.

If d is a circular distance the reduced matrix is also circular. The reduced distance can be written as a convex combination of three circular distances each with the same ordering and hence lies in the convex cone generated by all circular distance matrices with that ordering.

Gascuel [22] observed that a single degree of freedom can be introduced into the reduction formulae for NJ. In the above formulae we introduce two degrees of freedom, thereby opening the possibility for a variance reduction method in future versions of NeighborNet. In the case study described below we used $a = b = c = \frac{1}{3}$, the equal coefficients being directly analogous to NJ.

4.3   Estimation of the Edge Lengths

At each iteration of the NJ algorithm, every active node corresponds to a subtree of the resulting NJ tree. This subtree is not affected by agglomerations that occur later in the algorithm. We can therefore make estimates of the lengths of the edges in this subtree, and be confident that these edges will appear in the final tree. Thus, edge length estimation is carried on during the agglomeration stage of the NJ algorithm.

The situation is more complicated with circular split systems. The splits in the NeighborNet network are only finalized once all the agglomerations have

Data Structures:
- $n \times n$ distance matrix d, expanded to a total of $3n - 6$ rows and columns.
- A set A of active nodes, initially $\{1, \ldots, n\}$.
- An array of neighbor relations.
- A stack S of five-tuples $[x, y, z, u, v]$ coding agglomerative events.
- A circular ordering θ and weights for the associated splits.
- A collection of splits $\mathcal{NN}$, initially empty.

NEIGHBORNET(d)

[Agglomeration]
1.   **while** $|A| > 3$ **do**
2.   choose odes x, y using the $\hat{Q}$ criterion.
3.   make x and y neighbors.
4.   **while** there exists a vertex y with two neighbors **do**
5.   let x and z denote the neighbors of y.
6.   let u and v be new neighboring nodes.
7.   compute the new entries for d using eqns. 1-1.
8. $A \leftarrow A \cup \{u, v\} - \{x, y, z\}$.
9,   push $[x, y, z, u, v]$ on top of S.
10 .   **end**
11. **end**

[Expansion]
12. let θ be an arbitrary ordering of A.
13. **while** S is non-empty **do**
14.   pop $[x, y, z, u, v]$ off the top of S.
15.   replace u, v in θ by x, y, z.
16 . **end**

[Filtering]
17. denote θ by $(x_1, x_2, \ldots, x_n, x_1)$.
18. Compute split weights by constrained (or unconstrained) least squares and remove splits of zero (or negative) weight.

Fig. 3. The NEIGHBORNET algorithm

taken place. Hence we perform edge length estimation at the end. We borrow our estimation formulae from the theory of circular decompositions, some of which we briefly recall now.

Suppose that d is a *circular distance* on a set X, that is, there is a circular ordering θ of X so that

$$d(x, y) = \sum_{S \in \mathcal{S}(\theta)} \alpha_S \delta_S(x, y) \tag{1}$$

for all $x, y \in X$, where for each $S \in \mathcal{S}(\theta)$, α_S is a non-negative constant and δ_S is the so-called *split metric* that assigns 0 to pairs contained in the same part of S and 1 to all other pairs. A simple formula gives coefficients α_S in terms of d [12,13]: The coefficient for $S := S[x_i, x_j]$, with $1 \leq i \neq j \leq n$ is given by

$$\alpha_S := \tfrac{1}{2} \left(d(x_i x_j) + d(x_{i-1}, x_{j-1}) - d(x_i, x_{j-1}) - d(x_{i-1}, x_j) \right). \tag{2}$$

Note that if d is not circular then some of these coefficients will be negative. As with trees, negative length edges correspond to no real biological process, and their inclusion could lead to distortions in the representation [39]. Nevertheless, they may still contain some signal in the data. With NeighborNet we attempt to minimize this signal loss by using least squares optimization with a non-negativity constraint on the edge lengths. This increases the worst case time complexity of the algorithm.

4.4   Time Complexity of the NeighborNet Method

NeighborNet takes $O(n^3)$ time to construct the split system, the same asymptotic time complexity as NJ. The most time consuming step in the algorithm is the search in line 2 for the pair that minimizes $\hat{Q}(x, y)$. As with NJ, we can perform each of these searches in $O(n^2)$ time by first computing the net divergence indices r_x and ρ_x and then considering each pair of nodes in turn. There are at most $O(n)$ iterations of the loop in lines 1–11, so the total cost of these searches is $O(n^3)$ time.

There is one iteration for the loop in lines 4–10 for each agglomeration. As there are $O(n)$ agglomerations, these lines contribute $O(n^2)$ time in total. The expansion takes $O(n)$ time, and the estimation of all the coefficients α_S takes $O(n^2)$. Finally, it takes $O(n^3)$ time to output the splits and their indices.

The $O(n^3)$ time complexity allows the application of computationally intensive validation techniques such as bootstrapping which can be conducted without edge length estimation. Note that drawing the actual graph can take $O(n^4)$ time [18], and performing constrained least squares can take even longer.

5   NeighborNet Case Studies

We discuss three preliminary case studies, applying the method to two real data sets and one synthetic data set. Our goal is to illustrate the application and potential of the method. A more comprehensive study, involving both synthetic and real data, is currently underway. We are grateful to David Sankoff and Pete Lockhart for background information and interpretation of each data set.

5.1   Early Branching Eukaryotes

The origin and early diversification of the eukaryotes is one of the fundamental problems of evolutionary theory. Molecular phylogenies based on a number of mitochondrial genes have led to a far clearer understanding of the phylogeny of unicellular eukaryotes—the protists and the fungi—than was ever possible based on morphological classifications alone. Nevertheless, this approach is limited by the relatively small number of genes present in all or most mitochondria, and the finite amount of phylogenetic information that can be extracted from the sequence comparison of any of these genes.

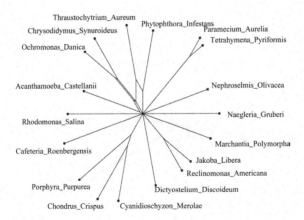

Thraustochytrium_Aureum

Chrysodidymus_Synuroideus   Phytophthora_Infestans

Ochromonas_Danica   Paramecium_Aurelia
                                                     Tetrahymena_Pyriformis

Acanthamoeba_Castellanii   Nephroselmis_Olivacea

Rhodomonas_Salina   Naegleria_Gruberi

Cafeteria_Roenbergensis   Marchantia_Polymorpha
                                                     Jakoba_Libera
                                                     Reclinomonas_Americana

Porphyra_Purpurea   Dictyostelium_Discoideum

Chondrus_Crispus   Cyanidioschyzon_Merolae

Fig. 4. Split decomposition split graph for the normalized breakpoint distances from the 18 mtDNA gene orders. Edge lengths re-estimated using constrained weighted least squares

The approach of [35] is to analyze gene order data of mitochondria using the normalized breakpoint distance. The normalized breakpoint distance is the gene order equivalent of uncorrected p-distances. Biologists typically use uncorrected distances when the model, or parameters of the model, is uncertain. The distance data was analyzed with NJ and FITCH, both of which produced the same tree. The trees contained several of the important groupings (red alga group, stramenopiles, early branching lineages, and the clustering of stramenopiles with the ciliates) but missed others (e.g. red and green alga).

The split graph produced by SPLITSTREE using split decomposition is poorly resolved (fig. 4), only picking out a few small groups. The contrast with the NeighborNet split graph (fig. 5) is marked. Not only did NeighborNet detect all the groupings present in the tree based analysis, but also several well established groups not picked up by any of the tree based methods. We have indicated several of the major groupings in the figure. Also of interest are the pairing of red and green algae, as well as stramenopiles with alveolates. Both of these are consistent with the latest hypotheses. NeighborNet is picking up clear phylogenetic signal.

5.2 Buttercup Sequences

The alpine buttercups (*Ranunculi*) of New Zealand evolved from a common ancestor some five million years ago and are noted for their diverse morphologies and the range of different habitats they occupy. Evidence suggests that the diversification of alpine buttercups resulted from the processes of hybridization and differentiation, both phenomena heavily influenced by the rapid geological and climatic changes in New Zealand over the last five million years. The arrival and departure of ice ages creates new habitats. These can in turn be colonized by hybrid species that may otherwise become extinct [29].

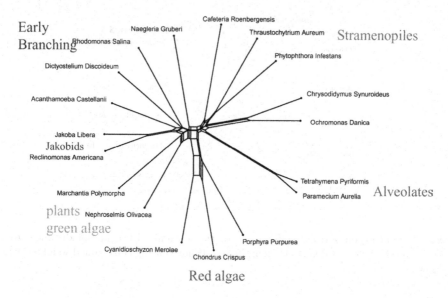

Fig. 5. NeighborNet for the 18 mtDNA gene orders. Major groupings indicated. Edge lengths re-estimated using constrained weighted least squares

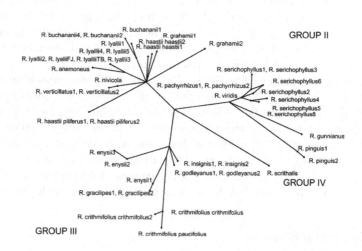

Fig. 6. Splitstree for the observed distances between 46 alpine buttercup nuclear ITS sequences

The recent divergence creates necessitates the use of rapidly evolving genetic markers: Lockhart et al. [29] used sequences from the nuclear ITS region. The evolution of these regions is considerably more complex than what is captured by established stochastic models. Polyploidy, polymorphism, and apparent con-

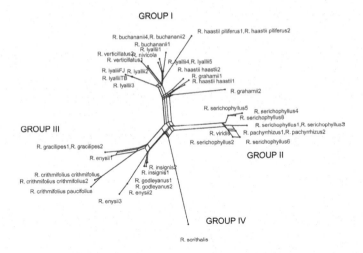

Fig. 7. NeighborNet for the observed distances between 46 alpine buttercup nuclear ITS sequences

certed evolution all confound standard phylogenetic analysis methods. Shortcomings in existing methods forced Lockhart et al. [29] to use a combination of median networks, split decomposition, and quartet puzzling, and to analyze different groups in the data separately.

The graphs produced by split decomposition and NeighborNet are presented in figure 6 and figure 7. Both graphs indicate the four distinct phylogenetic groups described by Lockhart et al. (see also Fisher [21]) and identified on the figures.

If we now turn to the level of resolution within these groups we see that NeighborNet captures substantially more detail than split decomposition. Lockhart et al. [29] were forced to analyze each group separately. Their separate analyses still uncovered less detail than that obtained by NeighborNet applied to the entire set.

One interesting feature of the NeighborNet graph is the indication of geographic centres of phylogenetic diversity. The species *R. lyalli* and *R. sericophyllus* are both paraphylectic. *R. lyalli* splits clearly between a group from the central South Island Alps (*R. lyalli2,3,TB,FJ*) and the southern South Island Alps (*R. lyalli1,4,5*). A similar pattern can be seen among the *R. sericophyllus* taxa.

To summarize, a single analysis with NeighborNet captured a greater level of detail than the multiple analyses carried out by Lockhart et al. It would, of course, be foolish to assume that every split in the NeighborNet corresponded to some biological reality. The key feature is the finer resolution of the representation. The graph is a tool for generating hypotheses that can be tested using specialized methods.

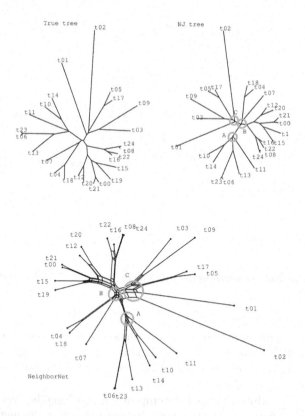

Fig. 8. Model tree on 25 taxa (top left), estimated NJ tree (top right) and NeighborNet. Regions of difference between the NJ tree and the model tree, and corresponding regions of the NeighborNet, are marked (A,B,C)

5.3   Poorly Corrected Data

There are many reasons for data to be non-tree like. The presence of boxes in the split graph could indicate both conflicting evolutionary history (e.g. from reticulations) or 'noise' in the data due to, for example, sampling error or incorrect model based corrections. As an illustration, we generated random sequences on a 25 taxa tree. The tree was generated according to the standard Yules-Harding coalescent model, applying the rates variation method of [25] to simulate varying rates among different lineages. Sequences of 150 nucleotide bases were produced using Seq-Gen under the K2P model (transition-transversion ration of 5.0) and unequal rates across sites ($\Gamma = 0.5$). To imitate incorrect model choice we used uncorrected p-distances to construct the distance matrix.

Both NJ and NeighborNet produced trees (or networks) similar to the true tree. There were three incorrect resolutions in the NJ tree. In the corresponding portions of the NeighborNet, conflicting splits indicated a level of uncertainty or ambiguity. The true tree, NJ tree and NeighborNet are presented in figure 8.

6   Endnote

We have introduced an agglomerative method for constructing planar split graphs. NeighborNet extends the the popular NJ tree construction method to produce networks rather than trees. Preliminary results are encouraging, although further analyses are needed before we can understand the advantages and biases of the method.

There are theoretical aspects that need to be explored. The choice of selection criterion was mostly determined by the three properties (P1)—(P3), though there are still some degrees of freedom. A full characterization of criteria satisfying these four properties could lead to improved selection criteria.

The NJ algorithm was originally invented as a heuristic for the Minimum Evolution (ME) criterion. To evaluate the ME criterion on a tree we first estimate edge lengths using ordinary least squares, and then sum the estimated edge lengths (c.f. [39] for a complete discussion). This raises the question: what is ME for circular splits? The answer – the practical implications of which we have not fully explored – is that minimizing the ME criterion for circular splits is exactly the same as solving the Traveling Salesman Problem (TSP), a classic combinatorial optimization problem. Note that the TSP can be solved efficiently for circular distances [3,13].

The general agglomerative framework can be adapted to many other applications requiring non-hierarchical classifications. We would be interested in tailoring NeighborNet for clustering gene expression data, to name one particularly topical example.

Acknowledgements

We thank Daniel Huson for providing us with the source code of SPLITSTREE and David Sankoff and Pete Lockhart for assistance with interpretation of the eukaryote and buttercup data.

References

1. K. Atteson, The performance of Neighbor-Joining methods of phylogenetic reconstruction, Algorithmica, **25** (1999) 251–278.
2. J.C. Aude, Y. Diaz-Lazcoz Y., J.J. Codani and J.L. Risler, Application of the pyramidal clustering method to biological objects, Comput. Chem. **23**, (1999) 303-315.
3. H.-J. Bandelt, A. Dress, A canonical decomposition theory for metrics on a finite set, Advances in Mathematics, **92** (1992) 47-105.
4. H.-J. Bandelt, P. Forster, B. Sykes, M. Richards, Mitochondrial portraits of human populations, Genetics **141** (1995) 743-753.
5. H.-J. Bandelt, P. Forster, A. Röhl, Median-joining networks for inferring intraspecific phylogenies, Mol. Biol. Evol., **16** (1999) 37-48.
6. A.C. Barbrook, C.J. Howe, N. Blake, P. Robinson, The phylogeny of The Canterbury Tales, Nature, **394** (1998) 839.

7. J. Barthélemy, A. Guenoche, Trees and Proximity Representations, John Wiley & Sons, Chichester New York Brisbane Toronto Singapore, 1991.
8. P. Bertrand, Structural properties of pyramidal clustering, DIMACS, **19** (1995), 35–53.
9. D. Bryant. Canonizing neighbor-joining. *in preparation.*
10. D. Bryant, V. Moulton. The consistency of NeighborNet. *in preparation.*
11. P. Buneman, The recovery of trees from measures of dissimilarity. In F. Hodson et al., Math. in the Archeological and Historical Sci., (pp.387-395), Edinburgh University Press, 1971.
12. V. Chepoi, B. Fichet, A note on circular decomposable metrics. Geom. Dedicata, **69** (1998) 237-240.
13. G. Christopher, M. Farach, M. Trick, The structure of circular decomposable metrics. Algorithms—ESA '96 (Barcelona), 486–500, LNCS 1136, Springer, Berlin, 1996.
14. E. Diday, Une representation des classes empi tantes: les pyramides. Rapport de recherche INRIA **291** (1984).
15. J. Dopazo, A. Dress, A. von Haeseler, Split decomposition: A technique to analyze viral evolution, PNAS, **90** (1993) 10320-10324.
16. A.Dress, M.Hendy, K.Huber, V.Moulton, On the number of vertices and edges of the Buneman graph, Annals Comb., **1** (1997) 329-337.
17. A.Dress, K.Huber, V.Moulton, An exceptional split geometry, Annals Comb., **4** (2000) 1-11.
18. A. Dress, D. Huson, Computing phylogenetic networks from split systems, Mnscrpt, 1998.
19. P.L. Erdös, M. Steel, L.A. Szkely, and T. Warnow, A few logs suffice to build (almost) all trees (Part 2) Theoretical Computer Science 221, (1999) 77-118.
20. M. Farach, Recognizing circular decomposable metrics, J. Comp. Bio., **4** (1997) 157-162.
21. F.J.F. Fisher. The alpine ranunculi of New Zealand. DSIR publishing, New Zealand. 1965.
22. O. Gascuel, BIONJ: an improved version of the NJ algorithm based on a simple model of sequence data, Molecular Biology and Evolution, **14(7)**, (1997) 685-695.
23. O.Gascuel, Concerning the NJ algorithm and its unweighted version, UNJ. in B.Mirkin, F.R.McMorris, F.S.Roberts, A.Rzhetsky, Math. Hierarch. and Biol., AMS, (1997) 149–170.
24. O. Gascuel, Data model and classification by trees: the minimum variance reduction (MVR) method, Journal of Classification, **17** (2000) 67–99.
25. S. Guindon and O. Gascuel Efficient Biased Estimation of Evolutionary Distances When Substitution Rates Vary Across Sites *Mol. Biol. Evol.*, **19**, (2002) 534-543.
26. B.Holland, K.Huber, A.Dress, V.Moulton, Some new techniques in statistical geometry, (in preparation).
27. E.Holmes, M.Worobey, A.Rambaut, Phylogenetic evidence for recombination in dengue virus, Mol. Bio. Evol., **16** (1999) 405-409.
28. D. Huson, SPLITSTREE: a program for analyzing and visualizing evolutionary data, Bioinformatics, **14** (1998) 68-73.
29. P.Lockhart, P.McLenachan, D.Havell, D.Glenny, D.Huson, U.Jensen, Phylogeny, dispersal and radiation of New Zealand alpine buttercups: molecular evidence under split decomposition, Ann. Missouri. Bot. Gard., **88** (2001) 458–477.
30. Maddison, D. R., Swofford, D. L., Maddison, W. P. NEXUS: An extensible file format for systematic information. Systematic Biology **46(4)** 1997, 590-621.

31. V. Makarenkov, T-REX: reconstructing and visualizing phylogenetic trees and reticulation networks, Bioinformatics, **17** (2001) 664-668.

32. P. Legendre, V. Makarenkov, Reconstruction of biogeographic and evolutionary networks using retiulograms. *Syst. Biol.* **51** (2) (2002) 199–216.

33. N. Saitou and M. Nei, The neighbor-joining method: a new method for reconstruction of phylogenetic trees, Mol. Bio. Evol., **4** (1987) 406-425.

34. M.Salemi, M.Leiws, J.Egan, W.Hall, J.Desmyter, A.-M.Vandamme, Different population dynamics of human T cell lymphotropic virus type II in intrevenous drug users compared with endemically infected tribes, PNAS, **96** (1999) 13253-13258.

35. D. Sankoff, D. Bryant, M. Denault, B.F. Lang, and G. Burger, Early eukaryote evolution based on mitochondrial gene order breakpoints. J. of Comp. Biology, **7**(3) (2000) 521–536.

36. S. Sattath, A. Tversky., Additive similarity trees, Psychometrika, **42** (3) 319–345.

37. R.R. Sokal and C.D. Michener. A statistical method for evaluating systematic relationships. Univ. Kansas Science Bull., **38** (1958) 1409–1438.

38. K.Strimmer, C.Wiuf, V.Moulton, Recombination analysis using directed graphical models, Molecular Biology and Evolution, **18** (2001) 97-99.

39. D. Swofford, G. J. Olsen, P. J. Waddell and D. M. Hillis. Phylogenetic Inference, *in Molecular Systematics 2nd Edition*, Hillis, D.M. and Moritz, C. and Mable, B.K. (eds). Sinauer (1996) 407–514.

On the Control of Hybridization Noise in DNA Sequencing-by-Hybridization

H.-W. Leong[1], F.P. Preparata[2,*], W.-K. Sung[1], and H. Willy[1]

[1] School of Computing, National University of Singapore, Singapore
{leonghw,ksung,hugowill}@comp.nus.edu.sg
[2] Computer Science Department, Brown University
115 Waterman Street, Providence, RI 02912-1910, USA
franco@cs.brown.edu

Abstract. DNA sequencing-by-hybridization (SBH) is a powerful potential alternative to current sequencing by electrophoresis. Different SBH methods have been compared under the hypothesis of error-free hybridization. However both false negatives and false positive are likely to occur in practice. Under the assumption of random independent hybridization errors, Doi and Imai [3] recently concluded that the algorithms of [15], which are asymptotically optimal in the error-free case, cannot be successfully adapted to noisy conditions. In this paper we prove that the reported dramatic drop in performance is attributable to algorithmic artifacts, and present instead an algorithm for sequence reconstruction under hybridization noise, which exhibits graceful degradation of performance as the error-rate increases. As a downside, the computational cost of sequence reconstruction rises noticeably under noisy conditions.

1 Introduction

DNA sequencing-by-hybridization (SBH) was proposed over a decade ago [1,7,4,11,12,16] as a potentially powerful alternative to current electrophoresis techniques. As is well known, sequencing by hybridization consists of two fundamental steps. The first, biochemical in nature, is the acquisition, by complementary hybridization with a complete library of probes, of all subsequences (of a selected pattern) of a given unknown target sequence; the set of such subsequences is called the sequence *spectrum*. The second step, combinatorial in nature, is the algorithmic reconstruction of the sequence from its spectrum. Since its original proposal, serious difficulties of biochemical as well as combinatorial nature have prevented SBH from becoming operational. Substantial progress has been recently made in overcoming the combinatorial difficulties [14,15], by adopting probing patterns that use nonspecific (universal) bases, which ideally act as wild cards in the hybridization process. Clearly, the adoption of universal bases further complicates the biochemistry of SBH.

---

* Supported partially by the National Science Foundation under Grant DBI-9983081 and by the Kwan Im Thong Chair at the National University of Singapore.

R. Guigó and D. Gusfield (Eds.): WABI 2002, LNCS 2452, pp. 392–403, 2002.

Comparative evaluations of different approaches (i.e., different probing patterns and reconstruction algorithms) have been normally carried out in the following general setting (*standard model*):

- Target sequences have maximum entropy, i.e., they consist of independent identically distributed DNA symbols (of course, natural DNA sequences do not exactly conform to this model; on the other hand, this model lends itself to meaningful probabilistic analysis);
- The hybridization process is assumed to be error-free, i.e., the spectrum contains *exactly* the subsequences of the target sequence conforming to a chosen pattern.

The serious inadequacy of the latter assumption was early recognized and several suggestions have been made [12,5,8,13] to devote some of the apparent redundancy of the probing scheme to error control. Of course, any approach to error control presupposes an error model, i.e., a formalization of the random process which produces the hybridization errors (hybridization noise), in the form of *false negatives* (errors of the first kind or misses) and of *false positives* (errors of the second kind or false-hits). Unfortunately, knowledge of the hybridization process is currently inadequate for a precise quantification of the hybridization model; judging from available data for natural bases [10,9], it appears likely that a realistic oligonucleotide hybridization model may have to be probe-specific.

In the absence of sufficient data, researchers have directed their attention to the question of a graceful degradation of the efficiency of the reconstruction process in the presence of noise. Such studies are typically based on the following error process (*modified standard model*):

1. Any spectrum probe can be suppressed with a fixed probability (false negatives);
2. Any probe at Hamming distance 1 from a correct spectrum probe can be added to the spectrum with fixed probability (false positives);
3. Hybridization noise is expressed in terms of error rates for false negatives and positives.

In this model, Doi and Imai [3] recently investigated whether the method of [14,15], which provably achieves asymptotic optimality in the standard model, remains viable in the modified standard model. Their study, experimental in nature, reached very negative conclusions, exhibiting a dramatic drop in performance.

It is the purpose of this paper to challenge the conclusions of [3], by showing that proper adaptation of the reconstruction algorithm achieves an acceptably graceful performance degradation. Our analysis in the modified standard model adequately agrees with extensive experimental data; analytically, we also justify in Appendix 1 the poor effectiveness of the approach outlined in [3]. On the negative side, the preservation of acceptable reconstruction effectiveness is achieved at the price of a drop in algorithmic efficiency.

2 Preliminaries

We briefly review the probing scheme and the reconstruction algorithm in the error-free case.

Definition 1. *A probing pattern is a binary string (beginning and ending with a 1), where a 1 denotes the position of a natural base and a 0 that of a universal base.*

Definition 2. *An (s, r) probing scheme, of weight $k = r + s$, has direct pattern $1^s(0^{s-1}1)^r$ and reverse pattern $1(0^{s-1}1)^r1^s$. The spectrum of a given target sequence is the collection of all its subsequences conforming to the chosen probing pattern.*

For notational convenience a probe is viewed as a string of length $(r+1)s = \nu$ over the extended alphabet $\mathcal{A} = \{$ A,C,G,T,*$\}$, where $*$ denotes the "wild-card". A probe occurs at position i of the target if i is the position of its rightmost symbol. Two strings over $\mathcal{A}$ of identical length *agree* if they coincide in the positions where both have symbols different from $*$.

All proposed reconstruction algorithms construct a *putative* sequence symbol-by-symbol. A reconstruction is successful if the completed putative sequence coincides with the original sequence. We ignore here the details, discussed elsewhere [14], of the initiation and termination of the reconstruction process.

Definition 3. *A probe is said to be a* feasible extension *if its $(\nu - 1)$-prefix coincides with the corresponding suffix of the putative sequence.*

Definition 4. *With reference to an ambiguous branching at position i, a probe is said to be* in-phase, *and* out-of-phase *otherwise, if one of its specified symbols occurs at position i.*

For example, if i is the branching position, in-phase probes are those ending at positions $i, i + s, i + 2s, \ldots, i + rs, i + rs + 1, \ldots, i + (r + 1)s - 1$ for a direct pattern, and at position $i, i + 1, \ldots, i + s - 1, i + 2s - 1, \ldots, i + (r + 1)s - 1$ for a reverse pattern.

The standard reconstruction algorithm in the absence of noise proceeds as follows[15]:

- Given the current putative sequence, the spectrum query returns the set of feasible-extension probes (which is necessarily nonempty in the error-free case if the putative sequence is correct). If only one probe is returned, then we have trivial one-symbol extension of the putative sequence viewed as a graph-theoretic path (algorithm in **extension mode**). Otherwise, we have an ambiguous branching and two or more competing paths are spawned; subsequently the algorithm attempts the breadth-first extension of all paths issuing from the branching (and of all other paths spawned in turn by them) on the basis of spectrum probes (algorithm in **branching mode**). The construction of such tree of paths is pursued up to a maximum depth H (a design parameter), unless at some stage of this construction it is found that all surviving paths have a common prefix. In the latter case, this prefix is concatenated to the putative sequence, and the process is iterated.

The occurrence of ambiguous branchings is due to the following construct:

Definition 5. *A fooling probe at position i is a feasible extension for position i which occurs as a subsequence at some position $j \neq i$ in the target sequence.*

Indeed, fooling probes are the cause of reconstruction failures. Obviously, for a given probe weight k, as the length m of the target sequence increases, the spectrum becomes more densely populated, and correspondingly the probability of encountering fooling probes increases. In fact, the probability that a probe, whose $(k-1)$-prefix is specified, belongs to the spectrum is correctly approximated as $1 - e^{-m/4^{k-1}}$, denoted hereafter as α. The rationale for the described extension mechanism is that, whereas the correct path is deterministically extended, the extension of spurious paths rests on the existence in the spectrum of fooling probes, and the parameter H should be chosen large enough to make their joint probability vanishingly small.

3   The Modified Reconstruction Algorithm

Intuitively, a false positive is much less detrimental to the reconstruction than a false negative. Indeed, referring to the above description of the algorithm, a false negative irretrievably interrupts the extension process, whereas a false positive simply adds one probe to the spectrum. Such probe will be accessed only if it may act as a feasible extension (in the same way as a fooling probe). In other words, false positives simply increase the pool of fooling probes, and, provided the false-positive rate remain reasonably small, their effect is truly negligible. This phenomenon was clearly noted by Doi and Imai [3]; in the interest of simplicity and clarity, in this paper we assume that the false-positive rate is zero. Inclusion of false positives should only minutely complicates the analysis.

Essentially, the success of reconstruction depends upon our ability to recover false negatives. In fact, we shall explain analytically that the poor behavior reported in [3] is due to inadequate recovery of false negatives.

We therefore propose, and later analyze, a robust reconstruction algorithm, which interacts with a noisy spectrum. The latter is obtained by eliminating with probability ϵ each probe of the noiseless spectrum. The previous standard algorithm can be viewed as a special case of the following procedure:

1. The spectrum query always returns four scored extensions (for each of the four DNA bases); the score is 0 if the corresponding probe exists in the spectrum and is 1 otherwise;
2. All extant paths have identical length and are extended with additive scoring. A path is terminated when its score exceeds by (an integer) threshold θ the score of the lowest-score extant path.

Clearly, the standard algorithm corresponds to the choice $\theta = 1$. Indeed, we may view the query, described in the previous section, as first returning all four possible extensions and then suppressing those which have no supporting probe in the spectrum. On the other hand, choices of $\theta > 1$ allow recovery of all false negatives, and potential successful reconstruction rests on the fact that lack of fooling probes (for spurious paths) is more likely than the occurrence of closely spaced false negatives (on the correct path). It is also intuitive that higher reliability is achieved for higher values of θ.

However, values of $\theta > 1$ as required to cope with positive error rate ϵ, imply that the algorithm *always* operates in the branching mode, since a score difference of at least 2 must be accumulated before an extension symbol may be decided. This entails the construction of larger extension trees, and, correspondingly, an increase in computational cost. Further analysis and experimentation are needed to ascertain how much of this increase is inherent and how much of it is due to implementation deficiencies.

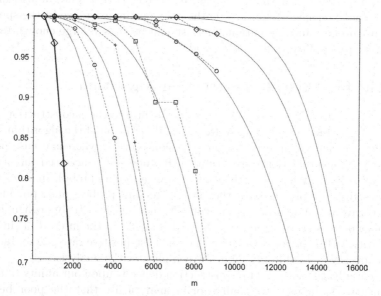

Fig. 1. Analytical curves of the probability of correct reconstruction with direct $(4, 4)$ pattern for different values of the false-negative error rate (from left to right $\epsilon = 0.02, 0.01, 0.005, 0.002, 0.001, 0)$, along with the corresponding frequency data collected in extensive simulations. In the same diagram we display for comparison the experimental performance of the algorithm of [3] for $\epsilon = 0.001$ (thick polygonal line at the far left).

Of course, when evaluating the overall probability of failure, we must also take into consideration the well-known failures attributable to the action of fooling probes, which have been extensively analyzed in prior publications [15,6]. Such failures occur when the algorithm extends up to length H beyond a branching

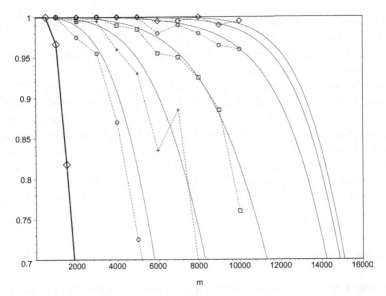

Fig. 2. Analytical curves of the probability of correct reconstruction with reverse $(4, 4)$ pattern for different values of the false-negative error rate (from left to right $\epsilon = 0.02, 0.01, 0.005, 0.002, 0.001, 0$), along with the corresponding frequency data. Again, the Doi-Imai curve is reported at the far left.

position paths that disagree in their initial symbol. As discussed earlier, two distinct modes of failure had been previously identified, for which we simply report here the approximate expressions of the probabilities of occurrence at a fixed sequence position, and refer the reader to [15,6] for details:

$$\eta_1 = (1 - e^{-3m/4^k})\alpha^k \left(1 + \frac{4^{r+1}}{m}\right)^r \left(1 + \frac{4^s}{m}\right)^{s-1} \quad , \quad \eta_2 = \frac{m}{2} \frac{(1 + 3\alpha \sum_{i=0}^{r-1} 4\alpha)^s}{4^{\nu-1}}.$$

Since the three causes of failure are independent, the probability of successful reconstruction (over the entire sequence) is conveniently expressed as

$$e^{-m(\eta + \eta_1 + \eta_2)}$$

where η is either η_D or η_R defined in Appendix 2, i.e., the probability of Mode 3 failures for direct and reverse patterns, respectively.

As it turns out, η_1 and η_2 provide noticeable contributions only for large values of m, or, equivalently, for very small values of the noise ϵ.

Figures 1 and 2, based on Formulae (6) and (7), refer to the case of $(4, 4)$ direct and reverse probes, respectively. The $(4, 4)$ scheme is of particular interest since its microarray implementation falls within the state of the art, and has been frequently considered in previous studies [14,15,3,6]. These figures clearly illustrate the superiority of reverse patterns in protecting against false negatives. Along with the analytical diagrams of the probabilities of successful reconstruction for various values of the parameter ϵ, we also report the corresponding experimental data, which are the result of extensive experimentation with the following protocol: A data point, corresponding to a choice of (s, r, m, ϵ) is obtained on the basis of a large number of runs (typically 200). In each run a random sequence is generated, prepended and postpended with short primers to separate the initiation and termination tasks from the reconstruction task. The spectrum of the sequence is obtained and subsequently modified to reflect the selected error-rate ϵ. Next the reconstruction algorithm is applied to the modified spectrum, adopting the threshold $\theta = 2$ for the termination of extension paths. If no premature failure is reported, the completed putative sequence is compared with the original sequence. Success is reported only when the two sequences coincide. The successes are tallied and their frequency finally reported. The agreement between the two sets of curves appears satisfactory, although not ideal.

4   Conclusions

In this paper we have shown, albeit in a restricted error model, that the gapped-probe method can be adapted to deal with hybridization errors, contradicting previous negative prognoses on this issue. In other words, part of the information-extracting ability of gapped probing can be successfully used to recover false negatives, which are otherwise fatal to the reconstruction process. We have also illustrated, analytically as well as experimentally, a graceful-degradation behavior, in the sense that the loss of reconstruction effectiveness grows smoothly, albeit rapidly, with the error rate.

Despite the above positive conclusions, there are several quantitative issues that deserve further attention. The first is the rather rapid degradation of reconstruction effectiveness as the error rate grows; this phenomenon is certainly partly due to the conservative threshold ($\theta = 2$) adopted in the reconstruction process, as demanded by the necessity to control the computational inefficiency. On a more fundamental level, one may try a multithreshold approach, based on the observation that false negatives, albeit fatal for the noiseless algorithm, are sufficiently rare events. Thus it seems plausible to carry out most of the reconstruction with the very efficient noiseless algorithm ($\theta = 1$) and to switch to larger value of θ upon detection of a reconstruction failure. Such approach raises several implementation issues, and we plan to undertake such project in the future.

References

1. W. Bains and G.C. Smith, A novel method for DNA sequence determination. *Jour. of Theoretical Biology*(1988), 135, 303-307.
2. M.E.Dyer, A.M.Frieze, and S.Suen, The probability of unique solutions of sequencing by hybridization. *Journal of Computational Biology*, 1 (1994) 105-110.
3. K. Doi and H. Imai Sequencing by hybridization in the presence of hybridization errors, *Genome Informatics*, 11 (2000)53-62.

4. R. Drmanac, I. Labat, I. Bruckner, and R. Crkvenjakov, Sequencing of megabase plus DNA by hybridization. *Genomics*,(1989),4, 114-128.

5. R. Drmanac, I. Labat, and R. Crkvenjakov, An algorithm for the DNA sequence generation from k-tuple word contents of the minimal number of random fragments. *Jour. Biomolecular Structure and Dynamics*,(1991) 8, 1085-1102.

6. S.A. Heath and F.P. Preparata, Enhanced sequence reconstruction with DNA microarray application. *COCOON 2001*, 64-74 (2001).

7. Yu.P. Lysov, V.L. Florentiev, A.A. Khorlin, K.R. Khrapko, V.V. Shih, and A.D. Mirzabekov, Sequencing by hybridization via oligonucleotides. A novel method. *Dokl. Acad. Sci. USSR*,(1988) 303, 1508-1511.

8. R.J.Lipshutz, Likelihood DNA sequencing by hybridization. *Jour. Biomolecular Structure and Dynamics* (1993) 11, 637-653.

9. N. Le Novere, MELTING, computing the melting temperature of nucleic acid duplex. *Bioinformatics*, 17, 12, 1226-1227 (2001).

10. J.J. SantaLucia, A unified view of polymer, dumbells, and oligonucleotide DNA nearest-neighbor thermodynamics. *Proc. Natl. Acad. Sci USA*, 95, 1460-1465 (1998).

11. P.A.Pevzner, 1-tuple DNA sequencing: computer analysis. *Journ. Biomolecul. Struct. & Dynamics* (1989) 7, 1, 63-73.

12. P.A.Pevzner, Yu.P. Lysov, K.R. Khrapko, A.V. Belyavsky, V.L. Florentiev, and A.D. Mirzabekov, Improved chips for sequencing by hybridization. *Journ. Biomolecul. Struct. & Dynamics* (1991) 9, 2, 399-410.

13. P.A.Pevzner and R.J. Lipshutz, Towards DNA-sequencing by hybridization. *19th Symp. on Mathem. Found. of Comp. Sci.*, (1994), LNCS-841, 143-258.

14. F.P. Preparata, A.M. Frieze, E. Upfal. On the Power of Universal Bases in Sequencing by Hybridization. *Third Annual International Conference on Computational Molecular Biology*. April 11 - 14, 1999, Lyon, France, pp. 295–301.

15. F.P. Preparata and E. Upfal. Sequencing-by-Hybridization at the information-theory bound: An optimal algorithm. *Journal of Computational Biology*, 7, 3/4, 621-630 (2000).

16. M.S. Waterman, *Introduction to Computational Biology* Chapman and Hall (1995).

5   Appendix 1: Analysis of the Doi-Imai Approach

The discouraging performance of the Doi-Imai approach [3] is fully explainable by analyzing their sequence reconstruction algorithms. We shall just consider their modification of the so-called "basic" algorithm, since the main limitation for this algorithm acts unaltered in their modification of the "advanced" algorithm. In fact, in the rendition of [3], the algorithm enters the "advanced" mode only upon failure of the "basic" mode. We recall here that the basic algorithm [14] fails when the spurious extension is supported by $(r+1)$ fooling probes terminating at in-phase positions (confirming probes). Therefore, the algorithm will fail in the basic mode if the confirmation of the correct extension is prematurely terminated by an FN.

Doi and Imai propose the following algorithmic modification. Here C denotes the set of probes returned as a response to the spectrum query.

1. $|C| = 1$. Extend putative sequence as in the noiseless case (concatenate symbol and proceed).

2. $|C| > 1$. Seek confirming probes as in the noiseless case.
3. $|C| = 0$. Proceed as if $|C| = 4$ and seek confirming probes as in the noiseless case.

In Cases 1 and 2 jointly (i.e., $|C| > 0$) the algorithm fails if there is an FN for the correct response (an unrecovered false negative). The corresponding probability is (neglecting ϵ with respect to 1)

$$\phi_1 = \epsilon(3\alpha(1-\alpha)^2 + 3\alpha^2(1-\alpha) + \alpha^3) = \epsilon\left(1 - (1-\alpha)^3\right)$$

In Case 3 (of probability $\epsilon(1-\alpha)^3$), as well as in Case 2 with no FN for the correct response (of probability $(1 - (1-\alpha)^3)$), the reconstruction *may* fail in the usual probe-confirmation process. Specifically, using the notation of Section 4, failure occurs if
- For any of the subsequent r in-phase positions we have a pair (FN, fp).
The probability of the described event is easily found to be

$$\phi_2 = \left(\epsilon(1-\alpha)^3 + (1 - (1-\alpha)^3)\right)\epsilon\alpha_1\frac{1-\alpha_1^r}{1-\alpha_1}$$

Since the two described failure events are disjoint, an upper bound to the total probability of correct reconstruction is expressed as

$$e^{-m(\phi_1+\phi_2)}$$

This function of m is plotted in Figure 3 for $(4,4)$-gapped probes and $\epsilon = 0.001$, along with the corresponding experimental graph reported in [3], thereby illustrating the validity of our analysis.

6   Appendix 2: Evaluation of the Probability of Mode 3 Failure

We illustrate the analysis approach with reference to threshold $\theta = 2$; generalization to larger threshold values can be carried out using the same ideas.

We begin by considering a direct probing scheme with pattern $1^s(0^{s-1}1)^r$. A failure string is constructed by placing two (FN, fp) pairs at positions to the right of the branching. We may limit our analysis to the position interval $[0, (r+1)s]$ since failure strings terminating beyond position $(r+1)s$ have utterly negligible probability.

The position range $[0, (r+1)s-1]$ is partitioned into subranges $I_1 = [0, rs-1]$ and $I_2 = [rs+1, (r+1)s-1]$.Each placement will be denoted by a pair (L, R) where $L, R \in \{\phi, \bar{\phi}\}$ [1] respectively specify the memberships of the leftmost and rightmost pairs. Recall (from Section 2) that

$$0, s, 2s, \ldots, rs, rs+1, \ldots, (r+1)s - 1$$

---

[1]   The symbols ϕ and $\bar{\phi}$ to respectively denote "in-phase" and "out-of-phase".

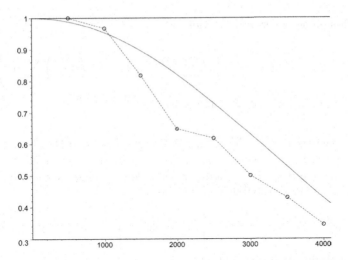

Fig. 3. Analytical diagram of the probability of successful reconstruction for $(4, 4)$-gapped probes and $\epsilon = 0.001$, along with the corresponding experimental result of [3].

are the in-phase positions. Note that, since the branching fooling probe can be selected in three out of four choices in its rightmost symbol, the probability of the branching fooling probe is $(3/4)\alpha$.

1. $(I_1, I_1 \wedge \phi) \equiv$ (left pair in I_1, right pair in I_1 and in-phase). Let us, with $u = 1, \ldots, r$, be the position of the rightmost pair. This implies that there are $u+1$ in-phase fooling probes. The leftmost pair can be chosen in-phase in u ways and out-of-phase in $u(s-1)$ ways. We conclude that the probability of the event is

$$\frac{3}{4}\epsilon^2 \alpha^{u+1}\left(u + u(s-1)\frac{3\alpha}{4}\right),$$

which summed over u yields:

$$\epsilon^2 \alpha\left(1 + (s-1)\frac{3\alpha}{4}\right)\sum_{u=1}^{r} u\alpha^u = \epsilon^2 \alpha A U_1 \tag{1}$$

where we have let $A = (1 + (s-1)(3\alpha/4))$ and $U_1 = \sum_{u=1}^{r} u\alpha^u$.

2. $(I_1, I_1 \wedge \bar{\phi})$. Let $us + v$, $u = 0, \ldots, r-1$ and $v = 1, \ldots, s-1$, be the position of the rightmost pair. Again, there are $u+1$ in-phase fooling probes and the leftmost pair can be chosen in $u + 1$ ways in-phase and in $u(s-1) + v - 1$ ways out-of-phase. The probability of the event is therefore

$$\frac{3}{4}\epsilon^2 \frac{3\alpha}{4}\alpha^{u+1}\left(u + 1 + \frac{3\alpha}{4}(u(s-1) + v - 1)\right)$$

which summed over u and v yields

$$\frac{3}{4}\epsilon^2\alpha\frac{3\alpha}{4}(s-1)\left(A\sum_{u=1}^{r-1}u\alpha^u+\left(1+(s-2)\frac{3\alpha}{8}\right)\sum_{u=0}^{r-1}\alpha^u\right)$$
$$=\frac{3}{4}\epsilon^2\alpha\frac{3\alpha}{4}(s-1)(AU_1'+BU_0') \tag{2}$$

where we have let $U_1'=\sum_{u=1}^{r-1}u\alpha^u$, $U_0'=\sum_{u=0}^{r-1}\alpha^u$, and $B=1+(3\alpha/8)(s-2)$.

3. $(I_1\wedge\phi,I_2)$. Let $rs+v$, with $v=1,\ldots,s-1$ be the position of the rightmost pair. There are $r+1+v$ fooling probes. The leftmost pair can be chosen in $r+1$ ways. The probability of the event is

$$\frac{3}{4}\epsilon^2\alpha^{r+1+v}(r+1)$$

and summing over v we obtain

$$\frac{3}{4}\epsilon^2(r+1)\alpha^{r+1}\sum_{v=1}^{s-1}\alpha^v=\frac{3}{4}\epsilon^2(r+1)\alpha^{r+1}V_0 \tag{3}$$

where we have let $V_0=\sum_{v=1}^{s-1}\alpha^v$.

4. $(I_1\wedge\bar{\phi},I_2)$. Again, $rs+v$ is the position of the rightmost pair and there are $r+1+v$ in-phase fooling probes. The leftmost pair can be chosen in $r(s-1)$ ways (each of probability $3\alpha/4$) so that the event's probability is

$$\frac{3}{4}\epsilon^2\frac{3\alpha}{4}\alpha^{r+1+v}r(s-1)$$

which summed over v yields

$$\frac{3}{4}\epsilon^2\frac{3\alpha}{4}r(s-1)\alpha^{r+1}V_0. \tag{4}$$

5. (I_2,I_2). Either pair is in-phase. If $rs+v$, with $v=2,\ldots,s-1$, is the position of the rightmost pair, there are $r+1+v$ fooling probes and $v-1$ choices for the position of the leftmost pair. The probability of the event is therefore

$$\frac{3}{4}\epsilon^2\alpha^{r+1+v}(v-1)$$

which summed over v yields

$$\frac{3}{4}\epsilon^2\alpha^{r+2}\sum_{v=1}^{s-2}v\alpha^v=\frac{3}{4}\epsilon^2\alpha^{r+2}V_1. \tag{5}$$

where we have let $V_1=\sum_{v=1}^{s-2}v\alpha^u$.

We summarize the preceding analysis with the following theorem:

Theorem 1. *The probability of a Mode 3 failure at a specific step of the reconstruction process with false-negative rate ϵ, for a direct (s,r) pattern, is expressed as*

$$\eta_D = \frac{3}{4}\epsilon^2\alpha\left(AU_1 + \frac{3\alpha}{4}(s-1)(AU_1' + BU_0') + \alpha^r(V_0(1+rA) + \alpha V_1)\right) \quad (6)$$

where $U_1 = \sum_{i=1}^{r}i\alpha^i, U_1' = \sum_{i=1}^{r-1}[\,i\alpha^i, U_0' = \sum_{i=0}^{r-1}\alpha^i, V_1 = \sum_{i=1}^{s-2}i\alpha^i,,$ *and* $V_0 = \sum_{i=1}^{s-1}\alpha^i.$

The same analysis may be carried out for the corresponding reverse probing scheme with pattern $(10^{s-})^r1^s$. For this case we define subintervals $I_1 = [0, s-1]$ and $I_2 = [s, (r+1)s - 1]$. The case analysis closely parallels the preceding one and is therefore omitted. The results are reported below

1. (I_1, I_1) $\frac{3}{4}\epsilon^2\alpha \cdot \sum_{v=1}^{s-2}v\alpha^v$
2. $(I_2, I_2 \wedge \phi)$ $\frac{3}{4}\epsilon^2\alpha \cdot \alpha^{s-1}AU_1$
3. $(I_2, I_2 \wedge \overline{\phi})$ $\frac{3}{4}\epsilon^2\alpha \cdot \alpha^{s-1}(3\alpha/4)(s-1)(AU_1' + BU_0')$
4. $(I_1, I_2 \wedge \phi)$ $\frac{3}{4}\epsilon^2\alpha \cdot \alpha^{s-1}(s-1)\sum_{u=0}^{r}\alpha^u$
5. $(I_1, I_2 \wedge \overline{\phi})$ $\frac{3}{4}\epsilon^2\alpha \cdot \alpha^{s-1}(3\alpha/4)(s-1)^2\sum_{u=0}^{r-1}\alpha^u.$

and are summarized with the following theorem:

Theorem 2. *The probability of a Mode 3 failure at a specific step of the reconstruction process with false-negative rate ϵ, for a reverse (s,r) pattern, is expressed as η_R, which is equal to*

$$\frac{3}{4}\epsilon^2\alpha\left(V_1 + \alpha^{s-1}\left(AU_1 + \frac{3\alpha}{4}(s-1)(AU_1' + BU_0') + (s-1)(U_0'' + \frac{3\alpha}{4}U_0')\right)\right) \quad (7)$$

where $U_0'' = \sum_{i=0}^{r}\alpha^i.$

Restricting SBH Ambiguity
via Restriction Enzymes

Steven Skiena[1] and Sagi Snir[2]

[1] Department of Computer Science, SUNY Stony Brook
Stony Brook, NY 11794-4400
skiena@cs.sunysb.edu
[2] Department of Computer Science, The Technion – Israel Institute of Technology
ssagi@cs.technion.ac.il

Abstract. The expected number of n-base long sequences consistent with a given SBH spectrum grows exponentially with n, which severely limits the potential range of applicability of SBH even in an error-free setting. Restriction enzymes (RE) recognize specific patterns and cut the DNA molecule at all locations of that pattern. The output of a restriction assay is the set of lengths of the resulting fragments. By augmenting the SBH spectrum with the target string's RE spectrum, we can eliminate much of the ambiguity of SBH. In this paper, we build on [20] to enhance the resolving power of restriction enzymes. We give a hardness result for the SBH+RE problem, and supply improved heuristics for the existing backtracking algorithm. We prove a lower bound on the number restriction enzymes required for unique reconstruction, and show experimental results that are not far from this bound.

1 Introduction

Sequencing by hybridization (SBH) [3,5,7,8,11,13] is a proposed approach to DNA sequencing where a set of single-stranded fragments (typically all possible 4^k oligonucleotides of length k) are attached to a substrate, forming a *sequencing chip*. A solution of single-stranded target DNA fragments are exposed to the chip. The resulting hybridization experiment gives the *spectrum* of the target, namely a list of all k-mers occurring at least once in the sequence. Sequencing is successful when only a single sequence is consistent with this spectrum, but *ambiguous* if multiple sequences are.

Unfortunately, the expected number of sequences consistent with a given spectrum increases exponentially with the sequence length. For example, the classical chip $C(8)$, with $4^8 = 65,536$ 8-mers suffices to reconstruct 200 nucleotide long sequences in 94 of 100 cases [12] in error-free experiments. For $n = 900$, however, the expected number of consistent sequences rises to over 35,000. In this paper, we build on previous work [20] to increase the resolving power of SBH by including information from enzymatic digestion assays.

Several alternate approaches for increasing the resolving power of SBH have been proposed in recent years, including positional SBH [4], arrays with universal

R. Guigó and D. Gusfield (Eds.): WABI 2002, LNCS 2452, pp. 404–417, 2002.

bases [16], interactive protocols [9,15], and tiling sequences with multiple arrays [18]. In contrast, the approach of [20] uses a very standard technique in molecular biology that predates oligonucleotide arrays by twenty years. *Restriction enzymes* identify a specific short recognition site in a DNA sequence, and cleave the DNA at all locations of this recognition site. In a complete digestion experiment, the product is a list of DNA fragment lengths, such that no fragment contains the recognition site.

Snir, et.al [20] propose the following procedure. In addition to the hybridization assay, we conduct a small number of complete digestion assays using different restriction enzymes. The computational phase of identifying consistent sequences then combines the hybridization and digestion information. This combination of SBH and digestion assays significantly increases the length of sequences that can be uniquely determined.

In this paper, we study the power of combining SBH with restriction enzymes. Our results include:

- Although the idea of augmenting SBH with restriction enzymes is appealing, the algorithmic question of how to efficiently reconstruct sequences from such data remained open. In [20], a backtracking algorithm was proposed for reconstruction. In Section 3, we show that the reconstruction problem is NP-complete, putting the complexity issue to rest.
- In Section 3, we also propose additional search heuristics which significantly reduce the computation time. Such improvements are important, because for certain sequence lengths and enzyme set sizes the sequence is typically uniquely determined, yet naive backtracking cannot expect to find it within an acceptable amount of time.
- To gain more insight into the power of SBH plus restriction enzymes, in Section 4 we study the case of one digest from a theoretical perspective. This analysis shows when the restriction digest does and does not help uniquely determine the sequence from its SBH-spectrum.
- This analysis also suggests approaches for selecting restriction enzymes in response to the observed SBH-spectrum, so as to maximize the likelihood of unambiguous reconstruction. In Section 5 we give a heuristic to select the most informative restriction enzymes for a given SBH-spectrum, and demonstrate its effectiveness via simulations.
- The resolving power of SBH plus restriction digests increases rapidly with the number of digests. In Section 7 we establish information-theoretic bounds on how many digests are necessary to augment the SBH-spectrum. We use insights from this analysis to select the cutter length and frequency for each digest to provide the optimal design of a sequencing experiment. Our theoretical results compare well to our simulations.

2   Background: SBH and Restriction Digests

As with most SBH algorithms, our work is based on finding postman walks in a subgraph of the de Bruijn digraph [6]. For a given alphabet Σ and length k,

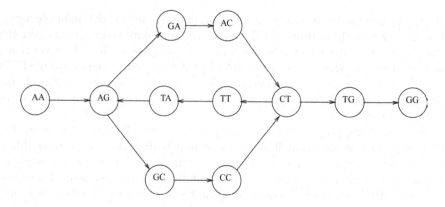

Fig. 1. The de Bruijn graph of 3-mers of the sequence AAGACTTAGCCTGG.

the *de Bruijn digraph* $G_k(\Sigma)$ will contain $|\Sigma|^{k-1}$ vertices, each corresponding to a $(k-1)$-length string on Σ. As shown in Figure 1, there will be an edge from vertex u to v labeled $\sigma \in \Sigma$ if the string associated with v consists of the last $k-2$ characters of u followed by σ. In any walk along the edges of this graph, the label of each vertex will represent the labels of the last $k-1$ edges traversed. Accordingly, each directed edge (u, v) of this graph represents a unique string of length k, defined by the label of u followed by the label of (u, v).

Pevzner's algorithm [14] interprets the results of a sequencing experiment as a subgraph of the de Bruijn graph, such that any Eulerian path corresponds to a possible sequence. As is typical (although regrettable) in SBH papers, here we assume error-free experimental data. Thus the reconstruction is not unique unless the subgraph consists entirely of a directed induced path.

Restriction enzymes recognize and cut DNA molecules at particular patterns. For example, the enzyme *Eco*RI cuts at the pattern *GAATTC*. Enzymes are indigenous to specific bacteria, with the name of each enzyme denoting the order of discovery within the host organism (e.g. *Eco*RI was the first restriction enzyme discovered in E. Coli). Rebase [17] maintains a complete list of all known restriction enzymes, including cutter sequences, literature references, and commercial availability. As of January 1, 2001, 3487 different enzymes were known, defining at least 255 distinct cutter sequences. Cutter sequence lengths range in length from 2 to 15 bases. Although most enzymes cut at specific oligonucleotide base patterns, other enzymes recognize multiple sequences by allowing variants at specific base positions. For example, the cutter AACNNNNNNGTGC matches from the 5' to 3' end any sequence starting AAC, ending GTGC where they are separated any sequence of exactly six bases.

In this paper, we will limit our attention to cutter sequences without wild card bases. As in [20], we assume that all possible cutter sequences of a given length have associated enzymes. While this assumption is not true (only 17 distinct 4-cutters and 85 distinct 6-cutters currently appear in Rebase), we do not believe this materially changes the significance of our results.

3   Complexity of Reconstruction

The algorithmic problem of reconstructing SBH data augmented with restriction digests is not as simple as that of pure SBH, however. Snir, et.al. proposed a backtracking algorithm to enumerate all possible sequences consistent with the data. This algorithm backtracks whenever a prefix sequence violated some property of the data, such as length, oligonucleotide content, or position of restriction sites, and is described in detail in [20].

The question of whether there existed a provably efficient reconstruction algorithm for the *sequence reconstruction from SBH plus restriction digests* problem remained open:

Input: An SBH spectrum S for the array of all k-mers, a set of e partitions of integer n, each corresponding to the results of a restriction digest with a particular cutter sequence.
Output: Does there exist a sequence T which is consistent with both the SBH spectrum S and the set of restriction digest results?

We answer this question in the negative:

Theorem 1. *Sequence reconstruction from SBH plus restriction digests is NP-complete, even with just one restriction digest.*

Proof. We sketch a reduction from Hamiltonian cycle in directed Eulerian graphs [10], by constructing a set of strings returned from an SBH-spectrum, and an associated set of restriction digest lengths such that reconstructing a consistent sequence involves solving the Hamiltonian cycle problem.

Our reduction from input graph $G = (V, E)$ is illustrated in Figure 2. We use a large alphabet Σ, and spectrum length $k = 2$. There will be a distinct letter in Σ for each vertex in V, with each directed edge $(x, y) \in V$ represented by the string xy. We then double all the edges of G, turning it into a multigraph. Note that this has no impact on whether G contains a Hamiltonian cycle. We then augment this graph with the following structures to construct a new graph $G' = (V', E')$:

- Starting and ending paths, S and T, of length l_s and l_t respectively, joining the original graph at an arbitrary vertex w. These define the beginning and end of the sequence,
- Length-$|A|$ loops hanging off all of the original vertices $v \in V$, with $d(v) - 1$ such loops incident on v, where $d(v)$ is the out-degree of $v \in V$ in the doubled graph. All $|E| - |V| + 1$ such loops contain a given string A, which appears nowhere else in the given construction.

In addition, we provide the results of the restriction assay with A as the cutter, yielding the multiset of distances $l_s + |V|$, l_t, and $2|E| - |V|$ fragments of length $|A| + 1$.

We claim that the only sequences consistent with this data describe Hamiltonian cycles in G. To construct a fragment of length $l_s + |V|$, the sequence must

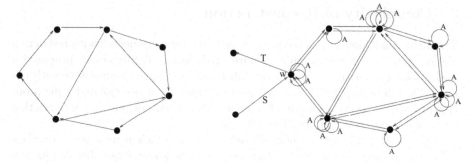

Fig. 2. Reducing Hamiltonian Cycle in G to SBH+RE in G'.

begin by visiting $|V|$ vertices without taking any loop detours. A collection of length-$(|A|+1)$ fragments then follow, picking up all remaining uncovered edges of E', each followed immediately by A. Finally comes the tail fragment of length l_t.

We must now show that the fragment of length $l_s + |V|$ describes a Hamiltonian cycle. Clearly it contains $|V|$ vertices, the length of such a cycle. Further it must include each vertex exactly once, since there are exactly enough detour loop gadgets at each vertex to cover one less than its out-degree. Any repetition of vertices in the length $l_s + |V|$ fragment implies that the detour loops are not positioned so as to generate length $|A|+1$ fragments after each remaining edge.

Finally, we must show that the graph remaining after deleting the Hamiltonian cycle is Eulerian. Clearly all vertices maintain the in-degree equals out-degree condition after deleting the cycle. By doubling each of the directed edges, we ensure the remaining graph remains strongly-connected. Thus it remains Eulerian. Every edge must be visited exactly once due to the length constraints of the digest.   □

Beyond this negative result, we improved the search algorithm of [20] with two new pruning criteria which yield better performance:

1. *Improved Sequence Length Cutoffs* – The results of restriction digests implicitly tell of the length of the target sequence. Thus we are interested in finding all SBH-consistent sequences of a given length. Such sequences correspond to a given postman walk on the appropriate de Bruijn subgraph.

   This gives rise to the following pruning technique: A vertex v with u_i uncovered in-edges and u_o uncovered out-edges is *in-imbalanced* if $u_i > u_o$ and *out-imbalanced* if $u_i < u_o$. Let v be an in-imbalanced vertex and let e_m be v's out-going edge (covered or uncovered) with minimal length l_m. Then we can bound the missing length by at least $(u_i - u_o)(l_m - k + 1)$ and potentially backtrack earlier. This claim is true for all vertices of G_F, however we must take precaution not to double-count the same edge by two adjacent vertices. To prevent this, we separate the extra length implied by in-imbalanced vertices from that implied by out-imbalanced vertices, and add the bigger of these quantities to be the missing length.

2. *Strong Connectivity Cutoffs* – The strongly connected components of a digraph are the maximal subgraphs such that every pair of vertices in a component are mutually reachable. Partitioning any digraph into strongly connected components leaves a directed tree of components such that it is impossible to return to a parent component by a directed path.

This gives rise to the following pruning principle. Let $G_r = (V, E_r)$ be the residual graph of uncovered edges. Let C_1 and C_2 be two different strongly-connected components in G_r, and edge $(u \to v) \in E_r$ link the two components, i.e. $u \in C_1$ and $v \in C_2$. Then we can backtrack on any prefix which traverses edge (u, v) before covering all edges in C_1.

Each of these two techniques reduced search time by roughly 70% on typical instances. When operated in conjunction, the search time was typically reduced by about 80%. All the algorithms were implemented in Java, and run on Pentium 3 PCs.

4   Understanding Single Digests

Restriction digest data usually reduces the ambiguity resulting from a given SBH-spectrum, but not always. We can better understand the potential power of restriction digests by looking at the topology of the given de Bruijn subgraph.

We define the notion of a partially colored graph to integrate the information from an SBH-digest with a given collection of RE-digests. A graph $G(V, E)$ is *partially colored* if a subset of vertices $V' \subset V$ are assigned colors, and the vertices $V - V'$ remain uncolored. Let $G = (V, E)$ be the subgraph of the de Bruijn graph of order $k - 1$ defined by a given SBH-spectrum S, and R be a set of strings $\{r_1, \ldots, r_c\}$. We say that a graph G is *partially colored with respect to R* iff the coloring of a given $v \in V$ implies that there exists a string $r \in R$ where the $|r|$-prefix of the $k - 1$-mer associated with v equals r. We assume that $r \leq k - 1$, as will naturally be the case in reasonable problem instances.

For certain partially colored graphs, the restriction digest data is sufficient to unambiguously reconstruct sequences not completely defined by the SBH-spectrum alone. Figure 3 depicts such a graph. Denote a colored vertex (restriction site) by a filled circle. Without a restriction digest, the postman walk *abcde* will be SBH-consistent with *adbce*. However, since the restriction digests of the two sequences are $\{|ab|, |cde|\}$ and $\{|adb|, |ce|\}$ respectively, such a digest will be sufficient to disambiguate them provided $|ce| \neq |ab|$.

Not all partially colored graphs G have this property. We say that G is *hopeless* with respect to its partial coloring if every postman walk P on G has another postman walk P' on G such that $|P| = |P'|$ and the multisets of distances between successive colored vertices along P and P' will be the same.

Figure 4 depicts three cases of hopeless graphs. The graph in Figure 4(I) is topologically the same as in Figure 3, but now the colored vertex is the junction of two loops. The order of traversal of the two loops cannot be distinguished by the RE-spectrum. The graph in Figure 4(II) is hopeless because the cut at u

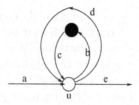

Fig. 3. A partially colored de Bruijn graph which might be disambiguated on the basis of restriction digest data.

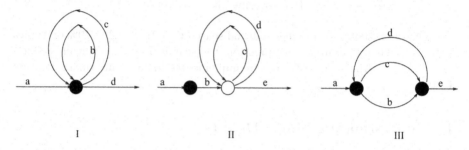

Fig. 4. Three hopeless digraphs. For every postman walk in G, there exists another postman walk with the same length and same set of distances between colored vertices.

can't eliminate the ambiguity from elsewhere in the graph. Finally, the graph in Figure 4(III) is hopeless since every postman walk must traverse paths c and b in some order. Reversing this order (by a tandem repeat) causes no change to either the SBH or RE-spectrums.

We now consider the problem of characterizing sequences which are uniquely defined by SBH plus restriction digests. Conditions under which a sequence is uniquely defined by its SBH-spectrum were established by Pevzner and Ukkonen and formalized in [2]:

Theorem 2. *A word W has another word W' with the same SBH-spectrum iff the sequence of overlapping $k-1$-mers comprising W denoted $\overrightarrow{W}$ has one of the forms:*

- $\alpha a \beta a \gamma a \delta$
- $\alpha a \beta b \gamma a \delta b \epsilon$

where $a, b \in \Sigma^{k-1}$ and $\alpha, \beta, \gamma, \delta, \epsilon \in (\Sigma^{k-1})^*$

As demonstrated by the figures above, a single restriction digest R can resolve ambiguities in the SBH-spectrum S. We make this intuition formal below. We say that a $(k-1)$-mer is *colored* if it defines a colored vertex in the partially colored de Bruijn graph of S plus R.

Theorem 3. *A word W with k-mer representation $\overrightarrow{W}$ which is uniquely defined by its SBH-spectrum and a single restriction digest satisfies all of the following properties.*

1. $\overrightarrow{W}$ *does not contain a colored $k-1$-mer a such that $\overrightarrow{W} = \alpha a \beta a \gamma a \delta$, i.e. a does not occur 3 times in W.*
2. $\overrightarrow{W}$ *does not contain two colored $k-1$-mers a and b such that $\overrightarrow{W} = \alpha a \beta b \gamma a \delta b \epsilon$.*
3. $\overrightarrow{W}$ *does not contain a substring $\overrightarrow{W'}$ consisting entirely of uncolored $k-1$-mers, where $\overrightarrow{W}' = \alpha a \beta a \gamma a \delta$ or $W' = \alpha a \beta b \gamma a \delta b \epsilon$.*

Proof. We analyze each case separately. For case 1, let W_1, W_2, W_3 , W_4 such that $\overrightarrow{W_1} = \alpha$; $\overrightarrow{W_2} = a, \beta$; $\overrightarrow{W_3} = a, \gamma$ and $\overrightarrow{W_4} = a, \delta$. Since a is colored then $\rho(W) = \rho(W_1) \cup \rho(W_2) \cup \rho(W_3) \cup \rho(W_4)$.
Now let W^* such that $\overrightarrow{W^*} = \alpha a \gamma a \beta a \delta$. Then by Theorem 2 W and W^* are SBH-consistent, and $\rho(W^*) = \rho(W_1) \cup \rho(W_3) \cup \rho(W_3) \cup \rho(W_4) = \rho(W)$ - contradiction to the fact that W is uniquely defined.

For case 2, let W_1, W_2, W_3 , W_4 , W_5 such that $\overrightarrow{W_1} = \alpha$; $\overrightarrow{W_2} = a, \beta$; $\overrightarrow{W_3} = a, \gamma$; $\overrightarrow{W_4} = a, \delta$ and $\overrightarrow{W_5} = a, \epsilon$. Since a is colored then $\rho(W) = \rho(W_1) \cup \rho(W_2) \cup \rho(W_3) \cup \rho(W_4 \cup \rho(W_5)$. Now let W^* such that $\overrightarrow{W^*} = \alpha a \delta b \gamma a \beta b \epsilon$. Then by Theorem 2 W and W^* are SBH-consistent, and $\rho(W^*) = \rho(W_1) \cup \rho(W_3) \cup \rho(W_3) \cup \rho(W_4) \cup \rho(W_5) = \rho(W)$ - contradiction to the fact that W is uniquely defined.

Case 3 follows by arguments analogous to those of the previous two cases. □

Theorem 4. *Let $i_1, \ldots, i_d$ represent the position of all interesting positions in a word W, where a position is interesting if (1) it corresponds to a colored vertex, (2) it corresponds to a vertex which appears three or more times, or (3) it corresponds to a vertex of a tandem repeat in W. Then W is uniquely defined by its SBH-spectrum and a single restriction digest if it satisfies all of the above conditions and no two subsets of $\cup_{j=1}^{d+1} i_j - i_{j-1}$ sum to the same value, where $i_0 = 0$ and $i_{d+1} = n$ (the sequence length).*

Proof. Let f_j be the fragment defined between two consecutive interesting points i_j and i_{j+1}. It is clear that every SBH-consistent sequence is a permutation of the fragments between points of type (2) or (3), and such a permutation can not yield a new fragment. Now, by condition 3 of Theorem 3, every triple or tandem repeat contains at least one restriction site. Thus, every shuffling of fragments yields a new set of fragments between restriction sites, and by the assumption, this new set has total length not existing in the original sequence □

5   Selecting Enzymes to Maximize Resolution

Observe that there is nothing in the SBH+RE protocol of [20] which requires that the experiments be done in parallel. This means that we could wait for the SBH-spectrum of the target sequence and use this information to select the

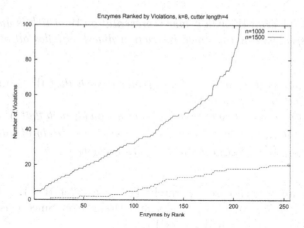

Fig. 5. Number of violations per enzyme in typical random sequences of $n = 1000$ and $n = 1500$.

restriction enzymes which can be expected to most effectively disambiguate the spectrum.

Based on the observations of Theorem 3, we note that digests which color high-degree vertices inherently leave ambiguities. Further, digests which do not include a restriction site breaking regions between tandem repeats or sequence triples cannot resolve the alternate sequences described in Theorem 2.

These observations suggest the following heuristic to select good enzymes. Randomly sample a number of appropriate length sequences consistent with the observed SBH-spectrum. Simulate a restriction digest with each possible cutter sequence on each sampled sequence. For each, count the number of forbidden structures which lead to ambiguous reconstructions of the sampled sequences. Select the enzymes leading to the smallest number of such structures.

The forbidden structures (or *violations*) we seek to avoid in sequence s for enzyme cutter sequence e are:

– Tandem repeat $\alpha a \beta b \gamma a \delta b \epsilon$, where e is the prefix of a and b, or e does not cut $a\beta b\gamma a\delta b$.
– Triple repeat $\alpha a \beta a \gamma a \delta$, where e is the prefix of a or e does not cut $a\beta a\gamma a$.

We define a violation with respect to sequence s and cutter-sequence e for every one of these forbidden structures. We sort the enzymes first according to the total number of violations, breaking ties based on the maximum number of violations. We select the first k enzymes in this sorted list for use in our experiment. The distribution of violations per enzyme for sample sequences of length $n = 1000$ and $n = 1500$ are shown in Figure 5.

Generating random SBH-consistent sequences of the given length is a non-trivial problem. Indeed, [19] proves the problem of generating an SBH-consistent sequence of a given length is NP-complete. Thus we employ a search-based algorithm to find a suitable sequence. Once we have a starting sequence, we can

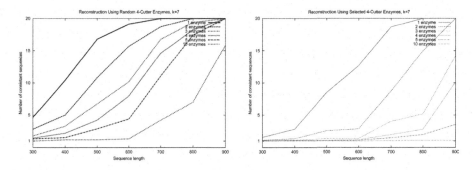

Fig. 6. Reconstruction using random and selected enzymes, for $k = 7$.

interchange fragments at triple subsequences and tandem repeats to generate other sequences by the arguments of Theorem 2.

6   Experimental Results

We used simulations in order to test the effectiveness of our enzyme selection algorithm. All input sequences were randomly chosen strings over $\{a, c, t, g\}$. We restricted ourselves to four-cutter enzymes, and assumed there existed an enzyme to realize every 4-mers.

We sought to test the performance of randomly chosen enzymes versus our method of enzymes selection. Here we held constant the number of consistent sequence samples used to produce the enzymes and varied the target sequence length. For every input sequence s we compared the number of sequences consistent with a random set of enzymes versus the number of sequences consistent with a set of identical size of enzymes, selected by our method.

When we chose random enzymes, we discarded enzymes whose restriction site did not appear at all in the target sequence. This biased the results somewhat in favor of random enzyme selection, but even so, our selection method shows much better results.

We compare the number of sequences consistent with the random set of enzymes with the number of sequences consistent with the "wise" set of enzymes. For practical reasons we interfered in the runs in two different ways:

1. We set a bound of 20 for the number of consistent sequences in every run. So whenever a run reached 20 consistent sequences, it terminated, regardless of how many more sequences are consistent with the input spectra.
2. We set a time counter for each run for cases when the number of consistent sequences is not very large but it still requires too long to find them all. To take these runs in consideration, we assigned them a result of 20 consistent sequences.

Figures 6 and 7 show the performance of enzymes selected by our method versus random enzymes for $k = 7$ and $k = 8$ respectively. The x-axis represents

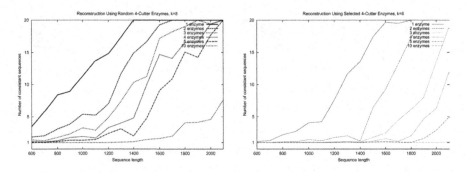

Fig. 7. Reconstruction using random and selected enzymes, for $k = 8$.

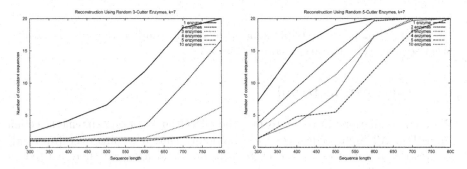

Fig. 8. Reconstruction using random 3-cutters (left) and 5-cutters (right), for $k = 7$.

the input sequence length, and the y-axis the number of consistent sequences. Each curve describes the level of sequence ambiguity for a given set of enzymes, both for random (left) and selected (right) enzyme sets. Our design procedure clearly outperforms randomly selected enzymes.

In fact, our results are even stronger, because we charged unterminated runs the maximum amount of ambiguity. For sequences of length 1700 and $k = 8$, 88% of all runs with 3 "wise" enzymes terminated, 85% of which uniquely returned the target sequence. This result is equivalent to using 10 random enzymes, as reported in [20].

Another advantage of using "smart" enzymes is the reduction in running time. For $n = 1700$ and $k = 8$, 88% of the runs terminated using 3 "smart" enzymes versus *no* completed runs for 3 random enzymes. For four enzymes, 100% of the "smart" runs completed versus 10% for random enzymes.

7   Selecting Cutter Length and Frequency

We use insights from this analysis to study the impact of cutter length and number of digests to provide the best design of an SBH+RE experiment.

Figure 8 shows the impact of cutter length on an enzyme's effectiveness at resolving sequence ambiguity. Comparing these results on 3- and 5-cutters to the 4-cutter results in Figure 6(a), we see that 3-cutters have significantly greater resolving power than 4-cutters, although 4-cutters are much better than 5-cutters.

This preference for short cutters can be understood through the theory of integer partitions [1]. We observe that each complete digest (including multiplicities) returns a partition of the sequence length n. Since, on average, an r-cutter cuts every 4^r bases, the expected number of parts resulting from such a digest is $n/4^r$. We seek to get the maximum amount of information from each restriction digest. The number of partitions of n with p parts peaks around $p = 2\sqrt{n}$, so the ideal cutter-length r will yield this many parts. This occurs at $r = (\log_2 n)/4 - 1$, a function growing slowly enough that it remains less than $r = 2.4$ for $n \leq 10,000$.

We can use similar arguments to obtain a lower bound on the number of digests needed as a function of the length of the sequence:

Theorem 5. *Let S be a random sequence over a four letter alphabet. The expected number of restriction digests needed to disambiguate the k-mer SBH-spectrum of S is at least D, where*

$$D \geq n^{3.5}/(24(\lg e)(\sqrt{2/3})(4^{k-1})^2)$$

Proof. Consider the SBH spectrum associated with all k-mers of a sequence S of length n. The probability $P(k, n)$ that any given k-mer occurs more than once in S may be calculated as

$$P(k, n) \approx \sum_{i=1}^{n} \sum_{j=i+1}^{n} (1/4^k)^2 \approx (1/2)(n/4^k)^2$$

Thus the expected number of vertices of out-degree ≥ 2 in the resulting de Bruijn subgraph is

$$v \approx 4^{k-1} \times P(k-1, n) \approx n^2/(2 \cdot 4^{k-1})$$

Given that we have v vertices of out-degree greater than 2, we can compute a bound on number of postman paths satisfying the spectrum. Whenever a tandem repeat $a, \ldots, b, \ldots, a, \ldots, b$ occurs in S, the two subpaths can be shuffled, creating sequence ambiguity. Thus the number of paths is approximately 2^t, where t is the number of tandem repeats.

The probability that two out-degree 2 vertices create a tandem repeat between them is $1/3$, since there are six possible orderings of the four sites, two of which are tandem. Thus v high degree vertices gives rise to an expected $\approx 2^{v^2/6}$ paths.

The output of each restriction digest is an integer partition of n describing the number and length of the fragments. The number of integer partitions of n is asymptotically:

$$a(n) \approx (1/4n)(1/\sqrt{3})e^{(\pi\sqrt{2n/3})}$$

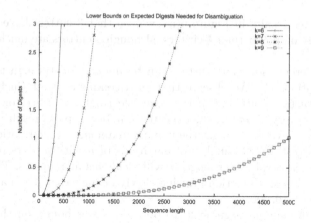

Fig. 9. Lower bounds on expected number of digests required as a function of sequence and k-mer length.

as $n \to \infty$, by Hardy and Ramanujan [1]. Thus the information content of a restriction digest is $\lg(a(n)) \approx (\lg e)(\sqrt{2n/3})$ bits.

We need at least enough bits from the digests to distinguish between the $\approx 2^{v^2/6}$ paths, i.e. the binary logarithm of this number. Since $v \approx n^2/(2 \cdot 4^{k-1})$, we need approximately $n^4/(24 \cdot 4^{2(k-1)})$ bits. Therefore the number of digests D is $D \geq n^{3.5}/(24(\lg e)(\sqrt{2/3})(4^{k-1})^2)$ □

Despite the coarseness of this bound (e.g. ignoring all sequence ambiguities except tandem repeats, and assuming each partition returns a random integer partition instead of one biased by expected number of parts) it does a nice job matching our experimental data. Note that the bound holds for smart enzyme selection as well as random selection. Figure 9 presents this lower bound for $6 \leq k \leq 9$ over a wide range of sequence lengths. In all cases, the expected number of enzymes begins to rise quickly around the lengths where sequence ambiguity starts to grow.

References

1. G. Andrews. *The Theory of Partitions*. Addison-Wesley, Reading, Mass., 1976.
2. R. Arratia, D. Martin, G. Reinert, and M. Waterman. Poisson process approximation for sequence repeats, and sequencing by hybridization. *J. Computational Biology*, 3:425–463, 1996.
3. W. Bains and G. Smith. A novel method for nucleic acid sequence determination. *J. Theor. Biol.*, 135:303–307, 1988.
4. A. Ben-Dor, I. Pe'er, R. Shamir, and R. Sharan. On the complexity of positional sequencing by hybridization. *Lecture Notes in Computer Science*, 1645:88–98, 1999.
5. A. Chetverin and F. Kramer. Oligonucleotide arrays: New concepts and possibilities. *Bio/Technology*, 12:1093–1099, 1994.
6. N.G. de Bruijn. A combinatorial problem. *Proc. Kon. Ned. Akad. Wetensch*, 49:758–764, 1946.

7. R. Dramanac and R. Crkvenjakov. DNA sequencing by hybridization. Yugoslav Patent Application 570, 1987.
8. S. Fodor, J. Read, M. Pirrung, L. Stryer, A. Lu, and D. Solas. Light-directed, spatially addressable parallel chemical synthesis. *Science*, 251:767–773, 1991.
9. A. Frieze and B. Halldorsson. Optimal sequencing by hybridization in rounds. In *Proc. Fifth Conf. on Computational Molecular Biology (RECOMB-01)*, pages 141–148, 2001.
10. M. R. Garey and D. S. Johnson. *Computers and Intractability: A Guide to the theory of NP-completeness*. W. H. Freeman, San Francisco, 1979.
11. Y. Lysov, V. Florentiev, A. Khorlin, K. Khrapko, V. Shik, and A. Mirzabekov. Determination of the nucleotide sequence of dna using hybridization to oligonucleotides. *Dokl. Acad. Sci. USSR*, 303:1508–1511, 1988.
12. P. Pevzner, Y. Lysov, K. Khrapko, A. Belyavski, V. Florentiev, and A. Mizabelkov. Improved chips for sequencing by hybridization. *J. Biomolecular Structure & Dynamics*, 9:399–410, 1991.
13. P. A. Pevzner and R. J. Lipshutz. Towards DNA sequencing chips. *19th Int. Conf. Mathematical Foundations of Computer Science*, 841:143–158, 1994.
14. P.A. Pevzner. *l*-tuple DNA sequencing: Computer analysis. *J. Biomolecular Structure and Dynamics*, 7:63–73, 1989.
15. V. Phan and S. Skiena. Dealing with errors in interactive sequencing by hybridization. *Bioinformatics*, 17:862–870, 2001.
16. F. P. Preparata and E. Upfal. Sequencing-by-hybridization at the information-theory bound: An optimal algorithm. In *Proc. Fourth Conf. Computational Molecular Biology (RECOMB-00)*, pages 245–253, 2000.
17. R. Roberts. Rebase: the restriction enzyme database. http://rebase.neb.com, 2001.
18. R. Shamir and D. Tsur. Large scale sequencing by hybridization. In *Proc. Fifth International Conf. on Computational Molecular Biology (RECOMB-01)*, pages 269–277, 2001.
19. S. Skiena and G. Sundaram. Reconstructing strings from substrings. *J. Computational Biology*, 2:333–353, 1995.
20. S. Snir, E. Yeger-Lotem, B. Chor, and Z. Yakhini. Using restriction enzymes to improve sequencing by hybridization. Technical Report CS-2002-14, Department of Computer Science,The Technion, Haifa, Israel, 2002.

Invited Lecture –
Molecule as Computation: Towards an Abstraction of Biomolecular Systems

Ehud Shapiro*

Weizmann Institute

While the "sequence" and "structure" branches of molecular biology are successfully consolidating their knowledge in a uniform and easily accessible manner, the mountains of knowledge regarding the function, activity and interaction of molecular systems in the cell remain fragmented. The reason is the adoption of good abstractions – "DNA-as-string" and "Protein-as-3D-labeled-graph" – in sequence and structure research, respectively, and the lack of a similarly successful abstraction in molecular systems research. We believe that computer science is the right place to search for this much needed abstraction. In particular, we believe that the "Molecule-as-computation" abstraction could be the basis for consolidating the knowledge of biomolecular systems. With such an abstraction, a system of interacting molecular entities is viewed as, and is described and modeled by, a system of interacting computational processes. Basic molecular entities are described by primitive processes and complex entities are described hierarchically, through process composition. For example, if a and b are abstractions of two molecular domains of a single molecule, then (a parallel b) is an abstraction of the corresponding two-domain molecule. Similarly, if a and b are abstractions of the two possible behaviors of a molecule that can be in one of two conformational states, depending on the ligand it binds, then (a choice b) is an abstraction of the molecule, with the behavioral choice between a and b determined by its interaction with a ligand process. We note that this abstraction is similar to cellular automata in that the behavior of a system emerges from the composition of interacting computational entities, but differs in that the computational entities are abstractions of mobile molecular entities rather than of static portions of space.

Using this abstraction opens up new computational possibilities for understanding molecular systems. Once biological behavior is mapped to computational behavior we can treat a biomolecular system, for example the molecular machinery of a circadian clock, an an implementation and its biological function, for example, a "black-box" abstract oscillator, as its specification. In doing so, ascribing a biological function to a biomolecular system, currently an informal process, becomes an objective question on the semantic equivalence between low-level and high-level computational descriptions.

A key feature of both computer systems and biomolecular systems is that, starting from a relatively small set of elementary components, more complex entities, with ever-increasing sophisticated functions, are constructed, a layer by layer, through the composition of elements from lower layers. A key difference is that biomolecular systems exist independently of our awareness or understanding of them, whereas computer systems exist because we understand, design and build them. We believe that the abstract thinking and communication methods and tools developed for over half a century for grappling with complex computer systems could be used for distilling, consolidating, and communicating the scientific knowledge of biomolecular systems.

---

* Joint work with Aviv Regev and Bill Silverman.

R. Guigó and D. Gusfield (Eds.): WABI 2002, LNCS 2452, p. 418, 2002.
© Springer-Verlag Berlin Heidelberg 2002

Fast Optimal Genome Tiling with Applications to Microarray Design and Homology Search

Piotr Berman[1,*], Paul Bertone[2,**], Bhaskar DasGupta[3,***], Mark Gerstein[4],
Ming-Yang Kao[5,†], and Michael Snyder[6,**]

[1] Department of Computer Science and Engineering, Pennsylvania State University
University Park, PA 16802
berman@cse.psu.edu

[2] Department of Molecular, Cellular, and Developmental Biology, Yale University
New Haven, CT 06520
paul.bertone@yale.edu

[3] Department of Computer Science, University of Illinois at Chicago
Chicago, IL 60607
dasgupta@cs.uic.edu

[4] Department of Molecular Biophysics and Biochemistry and Department of
Computer Science, Yale University, New Haven, CT 06520
mark.gerstein@yale.edu

[5] Department of Computer Science, Northwestern University, Evanston, IL 60201
kao@cs.northwestern.edu

[6] Department of Molecular, Cellular, and Developmental Biology and Department of
Molecular Biophysics and Biochemistry, Yale University, New Haven, CT 06520
michael.snyder@yale.edu

Abstract. In this paper we consider several variations of the following basic tiling problem: given a sequence of *real* numbers with two *size bound* parameters, we want to find a set of tiles such that they satisfy the size bounds and the total weight of the tiles is *maximized*. This solution to this problem is important to a number of computational biology applications, such as selecting genomic DNA fragments for amplicon microarrays, or performing homology searches with long sequence queries. Our goal is to design efficient algorithms with linear or near-linear time and space in the normal range of parameter values for these problems. For this purpose, we discuss the solution of a basic *online interval maximum problem* via a *sliding window* approach and show how to use this solution in a nontrivial manner for many of our tiling problems. We also discuss NP-hardness and approximation algorithms for generalization of our basic tiling problem to higher dimensions.

---

* Supported in part by National Library of Medicine grant LM05110
** Supported in part by NIH grants P50 HG02357 and R01 CA77808
*** Supported in part by NSF grant CCR-0296041 and a UIC startup fund
† Supported in part by NSF grant EIA-0112934

R. Guigó and D. Gusfield (Eds.): WABI 2002, LNCS 2452, pp. 419–433, 2002.

1   Introduction and Motivation

There are currently over 800 complete genome sequences available to the scientific community, representing the three principal domains of life: bacteria, archaea, and eukaryota [19]. The recent sequencing of several large eukaryotic genomes, including the nematode *Caenorhabditis elegans* [26] (100 Mb), the flowering plant *Arabidipsis thaliana* [25] (125 Mb), the fruitfly *Drosophila melanogaster* [1] (180 Mb) and a working draft sequence for *Homo sapiens* [11,27] (3.4 Gb), has enabled advances in biological research on an unprecedented scale.

Genome sequences vary widely in size and composition. In addition to the thousands of sequences that encode functional proteins and genetic regulatory elements, most eukaryotic genomes also possess a large number of non-coding sequences which are replicated in high numbers throughout the genome. These repetitive elements were introduced over evolutionary time, and comprise families of transposable elements that can move from one chromosomal location to another, retroviral sequences integrated into the genome via an RNA intermediate, and simple repeat sequences that can originate de novo at any location. Nearly 50% of human genomic DNA is associated with repetitive elements.

The presence of these repeat sequences can be problematic for both computational and experimental biology research. **Homology searches** [2] of repeat-laden queries against large sequence databases often result in many spurious matches, obscuring significant results and wasting computational resources. Although it is now standard practice to screen query sequences for repetitive elements, doing so subdivides the query into a number of smaller sequences that often produce a less specific match than the original. In an experimental context, when genomic sequence is used to investigate the binding of complementary DNA, repetitive elements can generate false positive signals and mask true positives by providing highly redundant DNA binding sites that compete with the meaningful targets of complementary probe molecules.

Genomic DNA can be screened for repeat sequences using specialized programs such as RepeatMasker [22] which performs local subsequence alignments [24] against a database of known repetitive elements [13]. Repeats are then masked within the query sequence, where a single non-nucleotide character is substituted for the nucleotides of each repeat instance. This global character replacement preserves the content and relative orientation of the remaining subsequences, which are then interspersed with character blocks representing the repetitive elements identified during the screening process.

Although the screening and/or removal of repeats is generally beneficial, additional problems may arise from the resulting genomic sequence fragmentation. Following repeat sequence identification, the remaining high-complexity component (i.e., non-repetitive DNA) exists as a population of fragments ranging in size from a few nucleotides to several kilobases. For organisms such as *Homo sapiens*, where the genome contains many thousands of repeat elements, the vast majority of these high-complexity sequence fragments are below 1 Kb in size. This situation presents a significant impediment to both computational and experimental research. Bioinformatics analyses often benefit from the availability of larger contiguous sequences, typically 1 Kb and larger, for homology searches and gene predictions. Similarly, extremely small sequences ($< 200bp$) are of limited use in many high-throughput experimental applications. These constraints provide the basis of the tiling problems formalized in the subsequent sections of this paper.

Another motivation for looking at the tiling problems considered in this paper is their application to **DNA microarray designs** for efficient genome analysis. A number of large-scale techniques have recently been developed for genome-wide data acquisition. Of these, a variety of microarray technologies have become ubiquitous in genome analysis centers, due to the highly parallel nature of these experiments. DNA microarray experiments rely on the property of *complementarity*, whereby each of the four nucleotide bases can preferentially couple with another base: adenine (A) couples with thymine (T) and vice versa, and cytosine (C) couples with guanine (G) and vice versa. When nucleic acid molecules recognize and anneal to molecules having complementary sequences, they are said to *hybridize*. DNA microarrays exploit this phenomenon by measuring the degree of hybridization that takes place when a battery of different sequences is exposed to a test molecule or *probe*. Each array element consists of immobilized DNA sequences which are exposed to a fluorescence-labeled probe. The signals emitted by the bound probe molecules are detected and quantified with a laser scanner. Using microarray systems, researchers can simultaneously interrogate thousands of individual molecules in a single experiment. Designing microarrays that contain sequences representing large spans of genomic DNA provides a vehicle for the global analysis of gene expression or other types of molecular interactions.

Various types of microarrays have been developed, each designed to capture a different kind of information. The most include cDNA microarrays [21], high-density oligonucleotide systems [16], and microarrays whose elements are composed of amplified genomic DNA [12,10]. The principal differences between these systems lie in the substrate DNA molecules that are used. cDNA microarrays are constructed using gene sequences which have been previously been observed to be expressed within a cell. High-density oligonucleotide systems employ an *in situ* DNA synthesis method to deposit short (25-70 nucleotide) DNA fragments using a modified inkjet technology. These typically represent sequences internal to known genes, providing an anchor to which complementary probe molecules may hybridize.

In contrast, the microarrays we consider here are composed of larger (typically 500bp-1.5 Kb) sequences of genomic DNA that are acquired via the *polymerase chain reaction* (PCR) [17], in which segments of DNA may be selectively amplified using a chemical system that recreates DNA replication *in vitro*. Although the size resolution of these array elements is not as fine as that of high-density oligonucleotide systems, PCR-based (or *amplicon*) microarrays provide experimental access to much larger regions of genomic sequence.

A maximal-coverage amplicon microarray can be designed by deriving a tile path through a target genomic sequence such that the best set of tiles is selected for PCR amplification. We are therefore concerned with finding the maximum number of high-complexity subsequence fragments (tiles) given a genomic DNA sequence whose repetitive elements have been masked. Deriving this tile set allows one to achieve optimal coverage of high-complexity DNA across the target sequence, while simultaneously maximizing the number of potential subsequences of sufficient size to facilitate large-scale biological research.

2   Definitions, Notations and Problem Statements

Based on the applications discussed in the previous section, we formalize a family of tiling problems as discussed below. These problems build upon a basic genome tiling problem in [6], which we call the *GTile problem* and describe as follows. For notational convenience, let $[i, j)$ denote integer set $\{i, i+1, \ldots, j-1\}$, $[i, j]$ denote $[i, j+1)$, and $f[i, j)$ denote the elements of an array f with indices in $[i, j)$. Our inputs consist of an array $c[0, n)$ of *real* numbers and two size parameters ℓ and u. A subarray $B = c[i, j)$ is called a *block*, of *length* $j - i$ and *weight* $w(B) = \sum_{k=i}^{j-1} c_k$, the weight of a set of blocks is the sum of their weights and a block is called a *tile* if its length belongs to $[l, u]$ Our goal is to find a set of *pairwise disjoint* tiles with the maximum possible weight. The tiling problems of interest in this paper are variations, restrictions, and generalizations of the GTile problem specified by a certain combinations of the following items:

Compressed versus uncompressed input data: In the important special case of the GTile problem when all entries of $c[0, n)$ belong to $\{-x, x\}$ for some fixed x, the input sequence can be more efficiently represented by specifying blocks of identical values. In other words, we can compress the input $c[0, n]$ to an integer sequence $d[0, m]$ such that
- $S_0 = 0$, $S_{i+1} = S_i + d_i$, $S_m = n + 1$, where $d_0 = 0$ is allowed;
- each element of $c[S_{2j}, S_{2j+1})$ is x for all j;
- each element of $c[S_{2j-1}, S_{2j})$ is $-x$ for all j.

Notice that the input size of such a compressed input data is m, which is typically much smaller than n (see Section 2.2).

Unbounded versus bounded number of tiles: Another important item of interest is when the number of tiles that may be used is at most a given value t, which could be considerably smaller than the number of tiles used by a tiling with no restrictions on the number of tiles.

No overlap versus overlap: It may be useful to relax this condition by allowing two tiles to share at most p elements, for some given (usually small) $p > 0$. However, we will penalize this overlap by subtracting the weight of an overlapped region from the sum of weights of all tiles, where the *weight* of each overlapped region is the sum of the elements in it. In other words, if $\mathcal{T}$ is the set of tiles and $\mathcal{R}$ is the set of elements of $\mathbf{C}$ that belong to more than one tile in $\mathcal{T}$, then the weight is $w(\mathcal{T}) - \sum_{c_i \in \mathcal{R}} c_i$.

1-dimensional versus d-dimensional: Generalization of the GTile problem in d dimensions has potential applications in database designs. In this case, we are given a d-dimensional array $\mathbf{C}$ of size $n_1 \times n_2 \times \cdot \times n_d$ and $2d$ size parameters $\ell_1, \ell_2, \ldots, \ell_d, u_1, u_2, \ldots, u_d$. A tile is a rectangular subarray of $\mathbf{C}$ of size $p_1 \times p_2 \times \cdots \times p_d$ satisfying $\ell_i \leq p_i \leq u_i$ for all i. The weight of a tile is again the sum of all the elements in the tile and our goal is again to find a set of tiles such that the sum of weights of the tiles is maximized.

We examine only those combinations of the above four items which are of importance in our applications as discussed before. To simplify exposition, unless otherwise stated explicitly, the GTile problem we consider is 1-*dimensional*, has an *uncompressed* input, an *unbounded* number of tiles, and *no overlaps*. In addition to the previously defined notations, unless otherwise stated, we use the following notations and variables with their designated meanings throughout the rest of the paper: $n + 1$ is the number of elements of the (uncompressed) 1-dimensional input array $c[i, j]$, $n_1 \leq n_2 \leq \cdots \leq n_d$ are the sizes of each dimension for the d-dimensional input array, $d[0, m]$ is the sequence of $m + 1$ integers for the compressed input data in 1-dimension, $w(\mathcal{T})$ is the weight for a *set* of tiles $\mathcal{T}$, t is the given number of tiles when the number of tiles is bounded and p is the maximum overlap between two tiles in 1-dimension. Finally, all logarithms are in base 2 unless stated otherwise explicitly. Many proofs are omitted due to page limitations; they can be found in the full version of the paper.

2.1   Related Work

Tiling an array of numbers in one or more dimensions under various constraints is a very active research area (for example, see [4,3,5,14,15,18,23]) and has applications in several areas including database decision support, two-dimensional histogram computation and resource scheduling. Several techniques, such as the slice-and-dice approach [4], the shifting technique [9, Chapter 9] and dynamic programming methods based on binary space partitions [3,14,18] have proven useful for these problems. However, to the best of our knowledge, the particular tiling problems that we investigate in this paper have not been looked at before. Our problems are different from the tiling problems in [4,3,14,15,18,23]; in particular, we do not require partitioning of the entire array, the array entries may be negative and there are lower and upper bounds on the size of a tile. The papers which most closely relate to our work are the references [20] and [28]. The authors in [20] provide an $O(n)$ time algorithm to find all *maximal* scoring subsequences of a sequence of length n. In [28] the authors investigate computing maximal scoring subsequences which contain no subsequences with weights below a particular threshold.

2.2   Typical Parameter Values for Microarray Design and Homology Search Applications

$n + 1$ (**the DNA sequence length**): Although the sizes of sequenced eukaryotic genomes range from 12 Mb (for the budding yeast *Saccharomyces cerevisiae*) to 3.4 Gb (*H. sapiens*), these exist as separate chromosomes that are

treated as individual sequence databases by our tiling algorithms. Eukaryotic chromosomes range in size from approximately 230 Kb (*S. cerevisiae*
chromosome I) to 256 Mb (human chromosome 1), with the average human
chromosome being 150 Mb in size.

ℓ and u **(lower and upper bounds for tile sizes):** In computing an optimal tile path for microarray design, tile sizes can range from 200 bp to
1.5 Kb. Sequence fragments below 200 bp become difficult to recover when
amplified in a high-throughput setting. An upper bound of 1.5 Kb balances
two factors: (1) obtaining maximal sequence coverage with a limited number
of tiles, and (2) producing a set of tiles which are small enough to achieve
sufficient array resolution. In practice the average tile size is 800 when ℓ and
u are set to 300 and 1500, respectively. For the homology search problem it
is desirable to extend the upper bound from 1.5 Kb to 2 Kb, representing
the average size of mammalian messenger RNA transcripts.

p **(maximum overlap between two tiles):** For microarray applications,
tiles are disjoint; that is, the overlap parameter p is 0, and no sequence
elements are shared between any two tiles. However, searching sequence
databases for homology matches can be enhanced by introducing a maximum overlap of $p \leq 100$ nucleotides. This condition provides for the case
when potential matches can be made at tile boundaries.

t **(maximum number of tiles, when the number of tiles is bounded):**
In selecting tiles for microarray applications, t can be specified to limit
the number of sequence fragments considered for PCR amplification. For
mammalian DNA where repeat content (and subsequent sequence fragmentation) is high, we can expect the high-complexity sequence nucleotides to
cover $n/2$ sequence elements; the desired number of tiles to be computed
will thus be $\frac{n}{2}$ divided by $\frac{u+\ell}{2}$ (the average of u and ℓ). For homology search
problems t is unbounded.

m **(size of compressed input):** It is difficult to give an accurate estimate of
the number of high-complexity sequence fragments in the target sequence
following repeat screening since it varies greatly with the organism. In our
experience, human chromosomes end up having between 2 to 3 times as
many high-complexity sequence fragments (before processing) as there are
final tiles (after processing), that is, m is roughly between $2t$ and $3t$. In other
words, in practice m may be smaller than n by a factor of at least 600 or
more.

2.3   Synopsis of Results

All our results are summarized in Table 1, except the result that GTile problem
is NP-hard in two or more dimensions. Most of our algorithms use simple data
structures (such as a double-ended queue) and are easy to implement. The techniques used for many of these tiling problems in one dimension use the solution
of an Online Interval Maximum (OLIM) problem. In Section 3, we discuss the
OLIM problem together with a solution for it using a windowing scheme reminis-
cent of that in [8]. However, the primary consideration in the applications in [8]

Table 1. A summary of the results in this paper. The parameter $\varepsilon > 1$ is any arbitrary *constant*. For the d-dimensional case, $M = \Pi_{i=1}^{d} n_i(u_i - \ell_i + 1)$, $N = \max_{1 \leq i \leq d} n_i$ and $\frac{u}{\ell} = \max_i \frac{u_i}{\ell_i}$. For our biology applications $p \leq 100 < \frac{\ell}{2} \ll n$, $t \simeq \frac{n}{u+\ell}$, m is much smaller than n and $\frac{\ell}{u-\ell} < 6$. The column of Approximation Ratio indicates whether the algorithm computes the optimal solution exactly or, for an approximation algorithm, the ratio of our total weight to that of the optimum.

Version of GTile	Time	Space	Approximation Ratio	Theorem
basic	n	n	exact	2
overlap is 0 or $p < \ell/2$	n	n	exact	3
compressed input	$\frac{\ell}{u-\ell}(m + \log \frac{\ell}{u-\ell})$	m	exact	4
number of tiles given	$\min\{n \log \frac{n}{\ell}, tn\}$	n	exact	5
d-dimensional	$\left(\left(\frac{u}{\ell}\right)\varepsilon\right)^{4\left(\frac{u}{\ell}\right)^2\varepsilon^2} M\varepsilon^2$	M	$\left(1 - \frac{1}{\varepsilon}\right)^d$	6
d-dimensional number of tiles given	$tM + dM\log^\varepsilon M$ $+dN\frac{\log N}{\log\log N}$ $M^{(2^\varepsilon-1)^{d-1}+1}\, dt$	M $M^{(2^\varepsilon-1)^{d-1}+1}\, dt$	$\left(\Pi_{i=1}^{d-1}(\lfloor 1 + \log n_i \rfloor)\right)^{-1}$ $\left(\Pi_{i=1}^{d-1}\left(\lfloor 1 + \frac{\log n_i}{\varepsilon}\rfloor\right)\right)^{-1}$	6 6

was reduction of space because of the online nature of the problems, whereas we are more concerned with time-complexity issues since our tiling problems are off-line in nature (and hence space for storing the entire input is always used). Moreover, our windowing scheme needed is somewhat different from that in [8] since we need to maintain multiple windows of different sizes and data may not arrive at evenly spaced time intervals.

3   Solving an Online Interval Maximum Problem via Sliding Window Approach

In this section, we discuss the Online Interval Maximum (OLIM for short) problem, which is used to design many of our algorithms. The authors in [8] mentions a restricted version of the OLIM problem in the context of maintaining stream statistics in the *sliding window* model and briefly mentions a solution for this problem. We expect the OLIM problem to be useful in other contexts as well. The problem in its most general form can be stated as follows.

Input: (1) a sequence $a[0, n)$ of real values in increasing order where each value a_i is an *argument* or a *test* (possibly both); (2) 2α real numbers $\ell_1, u_1, \ell_2, u_2, \ldots, \ell_\alpha, u_\alpha$ with $0 \leq \ell_1 < u_1 < \ell_2 < u_2 < \cdots < \ell_\alpha < u_\alpha$ and (3) a real function g defined on the *arguments*.

Output: for every test number a_k compute the maximum b_k of the α quantities $b_{k,1}, b_{k,2}, \ldots, b_{k,\alpha}$, where $b_{k,i}$ is given by

$$b_{k,i} = \max\{g(a_j) : a_k - u_i \leq a_j < a_k - \ell_i \text{ and } a_j \text{ is an argument}\}.$$

On-line limitation: we read the elements of the sequence $a[0, n)$ one at a time from left to right, and we need to compute b_k (if a_k is a test) before computing $g(a_k)$.

Theorem 1. *Let n_1 and n_2 be the numbers of arguments and tests in the input and β be the maximum time to compute $g(x)$ for any given x. Then, the OLIM problem can be solved in $O(n_1\beta + n\alpha)$ time using $O(n_1 + \alpha)$ space.*

Proof. In our algorithm we will maintain a queue Q_i for each (ℓ_i, u_i). We will compute the pair $(a, g(a))$ for each argument a; these pairs will be also stored in the abovementioned queues with the following property: Q_i stores a minimal set of argument-value pairs such that it is possible for some test a_m that as yet has not been read we have $b_{m,i} = g(x)$ for some $(x, g(x))$ in Q_i. Q_i can be maintained using the following two rules:

Rule 1. After reading a_k, remove from Q_i every $(x, g(x))$ such that $x < a_k - u_i$.

Rule 2. Let p be the smallest index of an argument such that $a_k - u_i \le a_p < a_k - \ell_i$ and $(a_p, g(a_p)) \notin Q_i$. After reading a_k, remove from Q_i every $(x, g(x))$ such that $g(x) \le g(a_p)$. Then insert $(a_p, g(a_p))$ in Q_i.

Rule 2 is valid because for $m \ge k$ if we compute $b_{m,i}$ as the maximum value of a set that contains a removed $(x, g(x))$, then this set must also contain $(a_p, g(a_p))$. This is true because $x < a_p$ and therefore rule 1 would remove $(x, g(x))$ earlier.

If we perform all needed insertions to Q_i using Rule 2 then the following holds: if $j < m$ and $(a_j, g(a_j))$ and $(a_m, g(a_m))$ are simultaneously present in Q_i, then $g(a_j) > g(a_m)$. Consequently, the maximum of the g-values in Q_i is contained in the oldest pair in Q_i. These observations show that we need to maintain Q_i as a double-ended queue where $\texttt{front}(Q_i)$ stores the maximum of all the g-values of the elements in Q_i (needed to compute $b_{m,i}$, and to perform Rule 1), while $\texttt{tail}(Q_i)$ has the minimum g-value (needed to perform Rule 2).

For each queue Q_i of the α queues and each a_k we need to check if the queue has to be updated using either of the two rules. This takes $O(n\alpha)$ time. Because each argument is inserted and deleted to a queue exactly once, the aggregate of these updates takes $O(n_1\alpha)$ time. For every test we compute the maximum of the maxima of the queues, this takes $O(n_2\alpha)$ time.

4   Basic GTile

Theorem 2. *The GTile problem can be solved in $O(n)$ time using $O(n)$ space.*

Proof. We use dynamic programming to reduce the GTile problem to OLIM. Subproblem k has as input $c[0, k]$. Let m_k be the sum of weights of the tiles and d_k and e_k be the left (beginning) and right (end) index of the last tile in an optimum solution of subproblem k. If $m_k = m_{k-1}$, then subproblem k has the same solution as subproblem $k - 1$, otherwise this solution consists of tile $c[d_k, k)$ and the tiles in the solution of subproblem d_k. Let $s_k = w(c[0, k))$, hence

$w(c[i,j)) = s_j - s_i$. It is trivial to compute $s[0,n)$ in $O(n)$ time and space. Obviously, $m_k = 0$ for $0 \leq k < \ell$. For $k \geq \ell$ we can compute m_k and d_k recursively:

> let $i \in [k - u, k - \ell] \cap [0, \infty)$ be an index that maximizes
> $$v_i = m_i + s_k - s_i$$
> if $v_i > m_{k-1}$, then $m_k = v_i$, $d_k = i$ and $e_k = k$
> $\quad\quad$ else $m_k = m_{k-1}$, $d_k = d_{k-1}$ and $e_k = e_{k-1}$

To prove our claim, it suffices to show how to compute such an index i for every $k \in [\ell, n]$ in a total of $O(n)$ time and space. Equivalently, for each k we can search for $i \in [k - u, k - \ell] \cap [0, \infty)$ that maximizes $y_i = m_i - s_i$; then we know $v_i = y_i + s_k$. This is the OLIM problem with input array $a[0, n)$, $a_i = i$, each a_i is both an argument and a test, $\alpha = 1$, $\ell_1 = \ell$, $u_1 = u$, and, for $0 \leq i < n$, $g(i) = m_i - s_i$.

It is also easy to recover an optimal tiling via the d_i and e_i values.

5   GTile with Restricted Overlap

In this section, we consider the GTile problem when either two tiles are disjoint or they have exactly p elements in common, for some fixed $p < \ell/2$. The constraint $p < \ell/2$ is true for our biology applications since typically $p \leq 100$ and $\ell \simeq 300$.

Theorem 3. *The GTile problem with restricted overlap can be solved in $O(n)$ time using $O(n)$ space.*

Remark 1. Consider the GTile problem with overlap in which the overlap can be any integer from $[1, p]$ with $p < \ell/2$. An approach similar to above shows how to solve this problem in $O(pn)$ time using $O(pn)$ space.

6   GTile with Compressed Input

Theorem 4. *The GTile problem with compressed input data can be solved in $O(\alpha(m + \log \alpha))$ time using $O(m)$ space where $\alpha = \lceil \ell/(u - \ell) \rceil$.*

7   GTile with Bounded Number of Tiles

In this section, we consider the case when the maximum number of tiles t is given.

Theorem 5. *The GTile problem with bounded number of tiles can be solved in $O(\min\{n \log n, nt\})$ time using $O(n)$ space.*

It is not too difficult to provide an algorithm with $O(nt)$ time and space using the approach of Theorem 2 and maintaining queues for each possible value of number of tiles. In the rest of this section, we will use a different approach to reduce the space to $O(n)$ which is significant since t could be large. We will also provide another algorithm that runs in $O(n \log n)$ time using $O(n)$ space which is also significant since $\log n$ is usually much smaller than t.

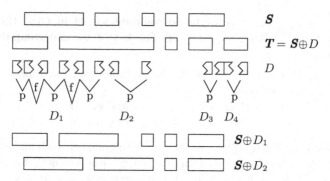

Fig. 1. Alterations, partners and friends. A set of tiles is altered with a set of tile ends. Partners are indicated with p and friends with f. D_i's are atomic alterations.

7.1   Sets and Sequences of Block Ends

Recall that a *block* is contiguous subsequence $c[p, q)$ of the given input sequence, a block of length at least ℓ and at most u is a tile and our solution consists of a set of disjoint tiles. A set of blocks $\boldsymbol{S}$ can be uniquely characterized by the set of endpoints of its blocks by using the following two quantities (where the first component of an ordered pair is λ of μ depending on whether the endpoint is the left or the right endpoint of the block, respectively):

$$ends(c[a, b)) = \{(\lambda, a), (\rho, b)\}, \quad ends(\boldsymbol{S}) = \bigcup_{T \in \boldsymbol{S}} ends(T)$$

A *block end* $e = (\mu, m)$ has *side* $side(e) = \mu$ and *position* $pos(e) = m$. A set of ends E is *consistent* if $E = ends(\boldsymbol{S})$ for some set of non-empty blocks. We introduce a partial order $\prec$ among block ends as follows: $e \prec f$ if $pos(e) < pos(f)$ or if $pos(e) = pos(f)$, $side(e) = \rho$ and $side(f) = \lambda$. A set of ends E ordered according to $\prec$ is the sequence $\vec{E} = (e_0, e_1, \ldots, e_m)$.

The test for consistency of E is obvious: the number of endpoints m in E has to be even and the sequence $side(e_0)$, $side(e_1)$, $\ldots$, $side(e_m)$ has to be $(\lambda, \rho, \ldots, \lambda, \rho)$. Our insistence that $(\rho, k) \prec (\lambda, k)$ reflects the fact that we do not allow empty blocks, *i.e.* blocks of the form $c[k, k)$.

In this subsection we will assume that $\boldsymbol{S}$ and $\boldsymbol{T}$ are sets of blocks with $A = ends(\boldsymbol{S})$ and $A' = ends(\boldsymbol{T})$; hence both A and A' are consistent. We also assume that $B = A \oplus A' = (A - A') \cup (A' - A)$, $C = A \cup A'$ and $\vec{B} = (b_0, \ldots, b_{2k-1})$.

If $A \oplus D$ is consistent, we will say that D is an *alteration* of $\boldsymbol{S}$, and $\boldsymbol{S} \oplus D$ is the set of blocks $\boldsymbol{U}$ such that $ends(\boldsymbol{U}) = A \oplus D$. Obviously, B is an alteration of $\boldsymbol{S}$. We want to characterize the subsets of B that are alterations as well. For every $i \in [0, k)$ we say that b_{2i} and b_{2i+1} are *partners* in B. See Figure 1 for an illustration.

Lemma 1 *Partners in B are adjacent in $\vec{C}$.*

Proof. For the sake of contradiction, suppose that in $\vec{C}$ entries b_{2i} and b_{2i+1} are separated by another entry, say a, not in B. Then a is preceded by an odd number of elements of B. Consequently, if a is preceded by an odd (respectively, even) number of elements of A, then it is preceded by an even (respectively, odd) number of elements of $A \oplus B$. Thus if the consistency of A dictates that $side(a) = \lambda$ (respectively, $side(a) = \rho$) then the consistency of $A \oplus B$ dictates that $side(a) = \rho$ (respectively, $side(a) = \lambda$), a contradiction.

Lemma 2 *Assume that $D \subset B$ does not separate any pair of partners of B, i.e. for each pair of partners in B, D either has either both or none of them. Then $A \oplus D$ is consistent.*

7.2 Modifying a Set of Tiles

We now slightly modify the assumptions of the previous subsection. Now we assume that $\boldsymbol{S}$ and $\boldsymbol{T}$ are two sets of *tiles* (*i.e.* they satisfy the size bounds), and we redefine the notion of alteration as follows: D is an *alteration* of $\boldsymbol{S}$ if $\boldsymbol{S} \oplus D$ is a set of *tiles*. Again, we want to characterize the alterations of $\boldsymbol{S}$ that are subsets of B.

If $g < h < i < j$ and $c[g, i]$, $c[h, j) \in \boldsymbol{S} \cup \boldsymbol{T}$ we say that (λ, h) and (ρ, i) are friends; see Figure 1 for an illustration. We need the following lemma.

Lemma 3 *Two friends must be adjacent in $\overrightarrow{C}$, they must both belong to B and they are not partners.*

Note that a pair of friends is easy to recognize: it must be a pair of the form $\{b_{2i-1}, b_{2i}\} = \{(\lambda, g), (\rho, h)\}$ where either $b_{2i-1} \in A - A'$ and $e_{2i} \in A' - A$, or $b_{2i-1} \in A' - A$ and $b_{2i} \in A - A'$, Let G_B be the graph with the vertex set as the set of block ends B and with two kinds of edges: between pairs of partners and between pairs of friends; see Figure 1 for an illustration. By Lemmas 1 and 3 these sets of edges form two disjoint matchings of G_B. Now we have our crucial lemma.

Lemma 4 *If $D \subseteq B$ is the set of vertices in a connected component of G_B, then D is an alteration of $\boldsymbol{S}$.*

Alterations that are vertices in a connected component of G_B will be called *atomic*; see Figure 1 for an illustration. Obviously any alteration can be expressed as a union of one or more disjoint atomic alterations and two disjoint atomic alterations can be applied in any order on a given set of tiles to obtain the same set of tiles. We will say that an atomic alteration is *increasing, neutral* or *decreasing* if $|\boldsymbol{S} \oplus D| - |\boldsymbol{S}|$ equals 1, 0 or -1, respectively. See Figure 2 for an illustration.

Lemma 5 *If D is an atomic alteration of $\boldsymbol{S}$, then $-1 \le |\boldsymbol{S} \oplus D| - |\boldsymbol{S}| \le 1$.*

7.3 Computing S_t in $O(nt)$ Time Using $O(n)$ Space

Let $\boldsymbol{S}_0$ be the empty set of tiles, and let $\boldsymbol{S}_{t+1}$ be a set of $t + 1$ tiles of maximum weight, and, under this limitation, one that is obtained by applying a *minimal*

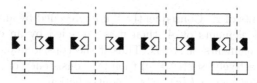

Fig. 2. An increasing atomic alteration. White tile ends are old, and black tile ends are new. There can be also an increasing atomic alteration that does not have the first two tile ends, or the last two: this is the case when ends of new and old tiles coincide.

alteration (*i.e.* an alteration that is not properly contained in another alteration) to S_t.

Lemma 6 *If $S_{t+1} = S_t \oplus D$ then D is an atomic alteration.*

Our algorithm is straightforward. We start with $S_0 = \emptyset$ and then, for $0 < p \le t$ we compute S_p from S_{p-1} if the total weight of S_p if greater than that of S_p (otherwise we stop). By Lemma 6 we need to find an increasing atomic alteration that results in the maximum possible weight gain. Our claim on the time and space complexity of the algorithm follows from the following lemma.

Lemma 7 *Given S_{p-1}, we can find in $O(n)$ time and space an atomic alteration D such that $S_p = S_{p-1} \oplus D$.*

Remark 2. The claim of Lemma 7 holds even if D is a neutral or decreasing alteration by using a very similar algorithm. Moreover, we actually compute, for each possible tile end e the optimum alteration of the prescribed type in which all elements precede e.

7.4   Computing S_t in $O(n \log \frac{n}{t})$ Time Using $O(n)$ Space

The idea of this algorithm is the following. We will proceed in phases. Before a phase, we computed a set of t disjoint tiles, say S, that has the *largest* weight under the constraint that each tile is contained in one of the blocks $c(a_0, a_1), c(a_1, a_2), \ldots, c(a_{k-1}, a_k)$. For this phase, we will select some a_i such that, after the phase, we replace S with some $S \oplus B$ that maximizes the sum of weights under the constraint that each tile in $S \oplus B$ is contained in $c(a_0, a_1)$, $c(a_1, a_2)$, $c(a_{i-2}, a_{i-1})$, $c(a_{i-1}, a_{i+1})$, $c(a_{i+1}, a_{i+2}), \ldots, c(a_{k-1}, a_k)$ (*i.e.* the new set of blocks is obtained from the old set of blocks by coalescing the two blocks $c(a_{i-1}, a_i)$ and $c(a_i, a_{i+1})$ into one block $c(a_{i-1}, a_{i+1})$).

We can start with $a_i = i\ell$; each block is a tile of minimum length, thus we can compute the weight of each tile and select the tiles with t largest weights. Using any linear time algorithm for order statistics [7], we can complete the first phase in $O(n)$ time and space. In Lemma 8 below we show that we can perform a single phase in $O(M + \log n)$ time and $O(n)$ space, where $M = \max_{i=1}^{k}\{a_i - a_{i-1}\}$. This will be sufficient for our claim on the time and space complexity of our complete algorithm by the following analysis[1]. We first coalesce adjacent pairs of blocks of length ℓ into blocks of length 2ℓ (unless there is only one block of length ℓ left). This requires at most n/ℓ phases and each of them takes $O(\ell)$ time, because during these phases the longest block has length 2ℓ. Hence the total time and space complexity for these n/ℓ phases is $O(n)$. Repeating the same procedure for blocks of length $2\ell, 4\ell, \ldots$, it follows that that if all blocks but one have the maximum length then we can half the number of blocks in $O(n)$ time and space, and again, all blocks but one will have the maximum length. Obviously, we are done when we have one block only. Since each phase can be carried out independently of any other phase, the space complexity is $O(n)$, and the total time complexity is $O\left(\sum_{i=1}^{\log(n/\ell)}\left((2^{i+1}\ell + \log n)\,\frac{n}{2^i\ell}\right)\right) = O\left(n\log\frac{n}{\ell}\right)$. Hence it suffices to prove the following lemma to prove our claim.

Lemma 8. *We can perform a single phase in $O(M + \log n)$ time and $O(n)$ space.*

8   GTile in d-Dimensions

It is not difficult to see from previously known results that the GTile problem is NP-hard even for $d = 2$. The following theorem summarizes approximability issues for the higher dimensional cases.

Theorem 6. *Let $M = \Pi_{i=1}^d n_i(u_i - \ell_i + 1)$, $N = \max_{1\le i\le d} n_i$, $\frac{u}{\ell} = \max_i \frac{u_i}{\ell_i}$, and $\varepsilon > 0$ and $c \ge 1$ be any two arbitrary given constants. Then, it is possible to design the following approximation algorithms for the GTile problem in d-dimension:*

(1) *if the number of tiles is unbounded, then it is possible to design an*
$O\left(\left(\left(\frac{u}{\ell}\right)\varepsilon\right)^{4\left(\frac{u}{\ell}\right)^2\varepsilon^2} M\varepsilon^2\right)$ *time algorithm using $O(M)$ space with an approximation ratio of $\left(1 - \frac{1}{\varepsilon}\right)^d$.*

(2) *if the number of tiles is bounded, then approximation algorithms with the following bounds are possible:*
  - *an $O(tM + dM\log^\varepsilon M + dN\frac{\log N}{\log\log N})$ time algorithm using $O(M)$ space with an approximation ratio of $1/\left(\Pi_{i=1}^{d-1}\left(\lfloor 1 + \log n_i\rfloor\right)\right)$.*
  - *an $O(M^{(2^\varepsilon-1)^{d-1}+1}\,dt)$ time algorithm using $O(M^{(2^\varepsilon-1)^{d-1}+1}\,dt)$ space with an approximation ratio of $1/\left(\Pi_{i=1}^{d-1}\left(\lfloor 1 + \frac{\log n_i}{\varepsilon}\rfloor\right)\right)$.*

Proof. The GTile problem in d dimension can be easily reduced to the d-RPACK problem in [5] in which the number of rectangles is M and the coordinates of the ith dimension has coordinates from $\{1, 2, \ldots, n_i\}$. This gives us the results in **(2)**. Since the ith dimension of any tile has a length of at least ℓ_i and at most u_i, the aspect ratio of any tile is at most $\frac{u}{\ell}$ and hence we can use the shifting strategy of [9, Chapter 9] to get the result in **(1)**.

---

[1] This is similar to the analysis of mergesort in which two blocks of sorted numbers are merged during one step [7].

References

1. M. D. Adams et al. The genome sequence of Drosophila melanogaster. *Science*, 287:2185–2195, 2000.
2. S. F. Altschul, W. Gish, W. Miller, E. W. Myers, and D. J. Lipman. A basic local alignment search tool. *Journal of Molecular Biology*, 215:403–410, 1990.
3. P. Berman, B. DasGupta, and S. Muthukrishnan. On the exact size of the binary space partitioning of sets of isothetic rectangles with applications. *SIAM Journal of Discrete Mathematics*, 15 (2): 252-267, 2002.
4. P. Berman, B. DasGupta, and S. Muthukrishnan. Slice and dice: A simple, improved approximate tiling recipe. In *Proceedings of the 13th Annual ACM-SIAM Symposium on Discrete Algorithms*, 455–464, January 2002.
5. P. Berman, B. DasGupta, S. Muthukrishnan, and S. Ramaswami. Improved approximation algorithms for tiling and packing with rectangles. In *Proceedings of the 12th Annual ACM-SIAM Symposium on Discrete Algorithms*, 427–436, January 2001.
6. P. Bertone, M. Y. Kao, M. Snyder, and M. Gerstein. The maximum sequence tiling problem with applications to DNA microarray design, submitted for journal publication.
7. T. H. Cormen, C. L. Leiserson and R. L. Rivest, *Introduction to Algorithms*, MIT Press, Cambridge, MA, 1990.
8. M. Datar, A. Gionis, P. Indyk, and R. Motwani. Maintaining stream statistics over sliding windows. In *Proceedings of the 13th Annual ACM-SIAM Symposium on Discrete Algorithms*, 635–644, January 2002.
9. D. S. Hochbaum. *Approximation Algorithms for NP-Hard Problems.* PWS Publishing, Boston, MA, 1997.
10. C. E. Horak, M. C. Mahajan, N. M. Luscombe, M. Gerstein, S. M. Weissman, and M. Snyder. GATA-1 binding sites mapped in the beta -globin locus by using mammalian chip-chip analysis. *Proceedings of the National Academy of Sciences of the U.S.A.*, 995:2924–2929, 2002.
11. International Human Genome Sequencing Consortium. Initial sequencing and analysis of the human genome. *Nature*, 15:860–921, 2001.
12. V. R. Iyer, C. E. Horak, C. S. Scafe, D. Botstein, M. Snyder, and P. O. Brown. Genomic binding sites of the yeast cell-cycle transcription factors SBF and MBF. *Nature*, 409:33–538, 2001.
13. J. Jurka. Repbase Update: a database and an electronic journal of repetitive elements. *Trends in Genetics*, 9:418–420, 2000.
14. S. Khanna, S. Muthukrishnan, and M. Paterson. On approximating rectangle tiling and packing. In *Proceedings of the 9th Annual ACM-SIAM Symposium on Discrete Algorithms*, 384–393, 1998.
15. S. Khanna, S. Muthukrishnan, and S. Skiena. Efficient array partitioning. In G. Goos, J. Hartmanis, and J. van Leeuwen, editors, *Lecture Notes in Computer Science 1256: Proceedings of the 24th International Colloquium on Automata, Languages, and Programming*, 616–626. Springer-Verlag, New York, NY, 1997.
16. D. J. Lockhart, H. Dong, M. C. Byrne, M. T. Follettie, M. V. Gallo, M. S. Chee, M. Mittmann, C. Wang, M. Kobayashi, and H. Horton et al. Expression monitoring by hybridization to high-density oligonucleotide arrays. *Nature Biotechnology*, 14:1675–1680, 1996.
17. K. Mullis, F. Faloona, S. Scharf, R. Saiki, G. Horn, and H. Erlich. Specific enzymatic amplification of DNA in vitro: the polymerase chain reaction. *Cold Spring Harbor Symposium in Quantitative Biology*, 51:263–273, 1986.

18. S. Muthukrishnan, V. Poosala, and T. Suel. On rectangular partitions in two dimensions: Algorithms, complexity and applications. In *Proceedings of the 7th International Conference on Database Theory*, 236–256, 1999.

19. National Center for Biotechnology Information (NCBI). www.ncbi.nlm.nih.gov, 2002.

20. W. L. Ruzzo and M. Tompa. Linear time algorithm for finding all maximal scoring subsequences. In *Proceedings of the 7th International Conference on Intelligent Systems for Molecular Biology*, 234–241, 1999.

21. D. D. Shalon and P. O. B. S. J. Smith. A DNA microarray system for analyzing complex DNA samples using two-color fluorescent probe hybridization. *Genome Research*, 6(7):639–645, July 1996.

22. A. F. A. Smit and P. Green. RepeatMasker, repeatmasker.genome.washington.edu, 2002.

23. A. Smith and S. Suri. Rectangular tiling in multi-dimensional arrays. In *Proceedings of the 10th Annual ACM-SIAM Symposium on Discrete Algorithms*, 786–794, 1999.

24. T. F. Smith and M. S. Waterman. Identification of common molecular subsequences. *Journal of Molecular Biology*, 147:195–197, 1981.

25. The Arabidipsis Genome Initiative. Analysis of the genome sequence of the flowering plant arabidopsis thaliana. *Nature*, 408:796–815, 2000.

26. The C. elegans Sequencing Consortium. Genome sequence of the nematode c. elegans: a platform for investigating biology. *Science*, 282:2012–2018, 1998.

27. J. C. Venter et al. The sequence of the human genome. *Science*, 291:1304–1351, 2001.

28. Z. Zhang, P. Berman, and W. Miller. Alignments without low-scoring regions. *Journal of Computational Biology*, 5(2):197–210, 1998.

Rapid Large-Scale Oligonucleotide Selection for Microarrays

Sven Rahmann[1,2]

[1] Max-Planck-Institute for Molecular Genetics
Dept. of Computational Molecular Biology, Ihnestraße 63–73, D-14195 Berlin
Sven.Rahmann@molgen.mpg.de
[2] Free University of Berlin, Dept. of Mathematics and Computer Science

Abstract. We present the first program that selects short oligonucleotide probes (e.g. 25-mers) for microarray experiments on a large scale. Our approach is up to two orders of magnitude faster than previous approaches (e.g. [2], [3]) and is the first one that allows handling truly large-scale datasets. For example, oligos for human genes can be found within 50 hours. This becomes possible by using the longest common substring as a specificity measure for candidate oligos. We present an algorithm based on a suffix array [1] with additional information that is efficient both in terms of memory usage and running time to rank all candidate oligos according to their specificity. We also introduce the concept of master sequences to describe the sequences from which oligos are to be selected. Constraints such as oligo length, melting temperature, and self-complementarity are incorporated in the master sequence at a preprocessing stage and thus kept separate from the main selection problem. As a result, custom oligos can be designed for any sequenced genome, just as the technology for on-site chip synthesis is becoming increasingly mature. Details will be given in the presentation and can be found in [4].

Keywords: microarray, oligo selection, probe design, suffix array, matching statistics, longest common substring.

References

1. Manber, U., Myers, G.W.: Suffix arrays: A new method for on-line string searches. In: Proceedings of the first annual ACM-SIAM Symposium on Discrete Algorithms, SIAM (1990) 319–327
2. Kaderali, L., Schliep, A.: Selecting signature oligonucleotides to identify organisms using dna arrays. Bioinformatics (2002) in press
3. Li, F., Stormo, G.: Selection of optimal dna oligos for gene expression analysis. Bioinformatics **17** (2001) 1067–1076
4. Rahmann, S.: Rapid large-scale oligonucleotide selection for microarrays. In: Proceedings of the IEEE Computer Society Bioinformatics Conference (CSB), IEEE (2002)

R. Guigó and D. Gusfield (Eds.): WABI 2002, LNCS 2452, p. 434, 2002.
© Springer-Verlag Berlin Heidelberg 2002

Border Length Minimization
in DNA Array Design*

A.B. Kahng[1], I.I. Măndoiu[1], P.A. Pevzner[1], S. Reda[1], and A.Z. Zelikovsky[2]

[1] Department of Computer Science & Engineering
University of California at San Diego, La Jolla, CA 92093
{abk,mandoiu,ppevzner,sreda}@cs.ucsd.edu
[2] Department of Computer Science
Georgia State University, Atlanta, GA 30303
alexz@cs.gsu.edu

Abstract. Optimal design of DNA arrays for very large-scale immobilized polymer synthesis (VLSIPS) [3] seeks to minimize effects of unintended illumination during mask exposure steps. Hannenhalli et al. [6] formulate this requirement as the Border Minimization Problem and give an algorithm for placement of probes at array sites under the assumption that the array synthesis is synchronous, i.e., nucleotides are synthesized in a periodic sequence $(ACGT)^k$ and every probe grows by exactly one nucleotide with every group of four masks. Their method reduces the number of conflicts, i.e., total mask border length, by 20-30% versus the previous standard method for array design. In this paper, we propose a probe placement algorithm for synchronous array synthesis which reduces the number of conflicts by up to 10% versus the method of Hannenhalli et al [6]. We also consider the case of *asynchronous* array synthesis, and present new heuristics that reduce the number of conflicts by up to a further 15.5-21.8%. Finally, we give lower bounds that offer insights into the amount of available further improvements. The paper concludes with several directions for future research.

1 Introduction

DNA *probe arrays* are used in a wide range of genomic analyses. As described in [3], during very large-scale immobilized polymer synthesis (VLSIPS) the *sites* of a DNA probe array are selectively exposed to light in order to activate oligonucleotides for further synthesis. The selective exposure is achieved by a sequence $M_1, M_2, \ldots, M_K$ of *masks*, with each mask M_i consisting of nontransparent and

* The work of ABK, IIM, and SR was partially supported by Cadence Design Systems, Inc. and the MARCO Gigascale Silicon Research Center, the work of PAP was supported by Ronald R. Taylor chair fund and the work of AZZ was partially supported by NSF Grant CCR-9988331, Award No. MM2-3018 of the Moldovan Research and Development Association (MRDA) and the U.S. Civilian Research & Development Foundation for the Independent States of the Former Soviet Union (CRDF), and by the State of Georgia's Yamacraw Initiative.

R. Guigó and D. Gusfield (Eds.): WABI 2002, LNCS 2452, pp. 435–448, 2002.

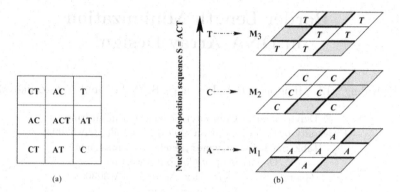

Fig. 1. (a) 2-dimensional probe placement. (b) 3-dimensional probe embedding: the nucleotide deposition sequence $S = (ACT)$ corresponds to the sequence of three masks M_1, M_2 and M_3. In each mask the masked sites are shaded and the borders between exposed and masked sites are thickened.

transparent areas corresponding to the masked and exposed array sites. Each mask induces deposition of a particular nucleotide $s_i \in \{A, C, T, G\}$) at its exposed array sites. The *nucleotide deposition sequence* $S = s_1 s_2 \ldots s_K$ corresponding to the sequence of masks is therefore a supersequence of all probe sequences in the array (see Figure 1). Typically, S is assumed to be periodic, e.g., $S = (ACGT)^k$, where $(ACGT)$ is a *period* and k is the (uniform) length of all probes in the array. The design of DNA arrays raises a number of combinatorial problems that have been discussed by Hubbell and Pevzner [5], Alon et al. [2], Heath and Preparata [4], Li and Stormo [8], Sengupta and Tompa [10], etc. In this paper, we study the *Border Minimization Problem* that was recently introduced by Hannenhalli et al. [6].

Optical effects (diffraction, reflections, etc.) can cause unwanted illumination at masked sites that are adjacent to the sites intentionally exposed to light - i.e., at the *border* sites of clear regions in the mask. This results in synthesis of unforeseen sequences in masked sites and compromises interpretation of experimental data. To reduce such uncertainty, one can exploit freedom in how probes are assigned to array sites. The *Border Minimization Problem (BMP)* [6] seeks a placement of probes that minimizes the sum of border lengths in all masks.

Observe that in general, a given probe can be *embedded* within the nucleotide deposition sequence S in several different ways. We may view the array design as a *three-dimensional placement problem* (see Figure 1): two dimensions are represented by the site array, and the third dimension is represented by the sequence S. Each layer in the third dimension corresponds to a mask inducing deposition of a particular nucleotide (A, C, G, or T), and a probe is a "column" within this three-dimensional placement representation. Border length of a given mask is computed as the number of *conflicts*, i.e., pairs of adjacent exposed and masked sites in the mask. Given two adjacent embedded probes p and p', the *conflict distance* $d(p, p')$ is the number of conflicts between the corresponding

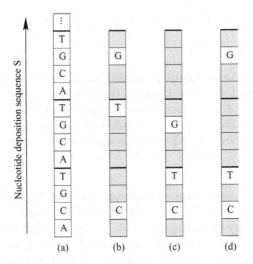

Fig. 2. (a) Periodic nucleotide deposition sequence S. (b) Synchronous embedding of probe CTG into S; the shaded sites denote the masked sites in the corresponding masks. (c-d) Two different asynchronous embeddings of the same probe.

columns. The border length of the embedding is the sum of conflict distances between adjacent probes.

We distinguish two types of DNA array synthesis. In *synchronous* synthesis, the ith period ($ACGT$) of the periodic nucleotide deposition sequence S synthesizes a single (the ith) nucleotide in each probe. This corresponds to a unique (and trivially computed) embedding of each probe p in the sequence S; see Figure 2(b). On the other hand, *asynchronous* array synthesis permits arbitrary embeddings, as shown in Figure 2(c-d).

In the remainder of this paper, we make the following contributions.

Improved Assignment of Probes to Sites for Synchronous Array Synthesis. Previous work on DNA array synthesis has considered only the synchronous context, when the conflict distance between two probes is $d(p, p') = 2h(p, p')$, with $h(p, p')$ denoting the Hamming distance between p and p' (i.e., the number of positions in which p and p' differ). As recounted in [6], the first array design at Affymetrix used a traveling salesman problem (TSP) heuristic to arrange all probes in a tour that heuristically minimized Hamming distance between neighboring probes in the tour. The tour was then *threaded* into the two-dimensional array of sites. [6] enhanced this threading approach to achieve up to 20% border length reduction for large chips. In Section 3, we suggest an *epitaxial* placement heuristic which places a random probe in the center of the array and then continues to insert probes in sites adjacent to already-placed probes, so as to greedily minimize the number of induced conflicts[1]. Our epitax-

---

[1] Recently a similar heuristic has been explored by Abdueva and Skvortsov [1].

ial placement method improves over the previous TSP-based approach by up to 10%.

Locally Optimal Asynchronous Embedding. Note that in the asynchronous context, the conflict distance between two adjacent probes depends on their embedding. Section 4 proposes dynamic programming algorithms that embed a given probe optimally with respect to fixed embeddings of the probe's neighbors. This dynamic programming algorithm is useful whenever we need to improve a given embedding.

Post-placement Improvement of Asynchronous Embeddings. In Section 5 we suggest two ways to improve the embedding of probes that have already been placed (e.g., by threading as in [6], or by epitaxial methods) onto array sites. We describe the *chessboard* and *batched greedy* methods for optimizing probe embeddings *after* assignment of probes to their positions in the array. The *chessboard* method alternates between optimal embedding adjustment of "black" and "white" sites with respect to their neighbors (which always have different color). The *batched greedy* method implements non-conflicting optimal embeddings in a greedy fashion. Our experiments show that the chessboard method (winning slightly over the batched greedy method) decreases by 15.5-21.8% the number of conflicts in the original synchronous placement.

Lower Bounds for the Border Minimization Problem. In Section 2 we give *a priori* lower bounds on the total border length of the optimum synchronous solution based on Hamming distance, and of the optimum asynchronous solution based on the length of the Longest Common Subsequence (LCS). These afford an estimate of potential future progress due to improved probe placement algorithms for synchronous embedding. In Section 5 a different LCS-distance based lower bound is applied to the probe embedding, yielding bounds on possible improvement from exploiting this degree of freedom.

2   Array Design Problem Formulations and Lower Bounds

In this section we give a graph-theoretical definition of probe placement and then formulate the synchronous and asynchronous array design problems. For both formulations we give lower bounds on the cost of optimum solutions.

Following the development in [6], let $G_1(V_1, E_1, w_1)$ and $G_2(V_2, E_2, w_2)$ be two edge-weighted graphs with weight functions w_1 and w_2. (In the following, any edge not explicitly defined is assumed to be present in the graph with weight zero.) A bijective function $\phi : V_2 \to V_1$ is called a *placement* of G_2 on G_1. The cost of the placement is defined as

$$cost(\phi) = \sum_{x,y \in V_2} w_2(x,y) w_1(\phi(x), \phi(y)).$$

The *optimal placement problem* is to find a minimum cost placement of G_2 on G_1.

In synchronous array design, the Border Minimization Problem (BMP) corresponds to the optimal placement problem for appropriately defined graphs. Let $G2$ be a *2-dimensional grid graph* corresponding to the arrangement of sites in the DNA array. The vertices of $G2$ correspond to $N \times N$ array sites, and edges of $G2$ have weight 1 if endpoints correspond to adjacent sites and 0, otherwise. Also, let H be the *Hamming graph* defined by the set of probes, i.e., the complete graph with probes as vertices and each edge weight equal to twice the Hamming distance between corresponding probes.

Synchronous Array Design Problem (SADP). Find a minimum-cost placement of the Hamming graph H on the 2-dimensional grid graph $G2$.

Let L be the directed graph over the set of probes obtained by including arcs from each probe to the 4 closest probes with respect to Hamming distance, and then deleting the heaviest $4N$ arcs. Since the graph L has weight no more than the conflict cost of any placement of the set of probes on the grid graph $G2$ we obtain the following

Theorem 1. *The total arc weight of L is a lower bound on the cost of the optimum SADP solution.*

For asynchronous array design, the BMP is more involved. The asynchronous design consists of two steps: (i) the embedding e of each probe p into the nucleotide deposition sequence S, i.e., $e : p \to S$ such that for each nucleotide $x \in p$, $e(x) = x$ and (ii) the placement of embedded probes into the $N \times N$ array of sites. An embedded probe p may be viewed as a sequence obtained from S by replacing $|S| - |p|$ nucleotides with blanks. Let the *Hamming graph H'* be the graph with vertices corresponding to embedded probes and with edge weights equal to the Hamming distance (we add 1 to the distance between two embedded probes for each nucleotide in one that corresponds to a blank in the other).

Asynchronous Array Design Problem (AADP). Find an embedding of each given probe, and a placement, which minimizes the cost of the placement of the Hamming graph H' on the 2-dimensional grid graph $G2$.

To estimate the cost of the optimum AADP solution, it is necessary to estimate the distance between two probes independent of their embedding into S. Observe that the number of nucleotides (mask steps) common to two embedded probes cannot exceed the length of the *longest common subsequence* (LCS) of these two probes. Define the LCS distance between probes p and p' as $lcsd(p, p') = k - |LCS(p, p')|$, where $k = |p| = |p'|$. Let L' be the directed graph over the set of probes obtained by including arcs from each probe to the 4 closest probes with respect to LCS distance, and then deleting the heaviest $4N$ arcs. Similar to Theorem 1 we get:

440   A.B. Kahng et al.

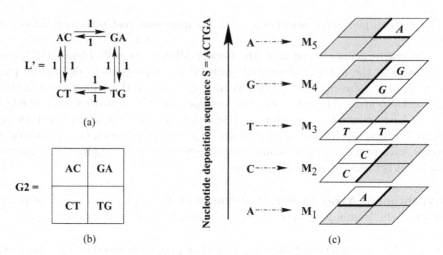

(a)

G2 =

AC	GA
CT	TG

(b)

(c)

Fig. 3. (a) Lower-bound digraph L' for four probes AC, GA, CT and TG. The lower bound on the number of conflicts is 8 which is the arc weight of L'. (b) Optimum 2-dimensional placement of the probes. (c) Optimum embedding of the probes into the nucleotide deposition supersequence $S = ACTGA$. The optimum embedding has 10 conflicts, exceeding by 2 the lower bound.

Theorem 2. *The total arc weight of L' is a lower bound on the cost of the optimum AADP solution.*

Example. Note that weight of the graph L' may be much smaller than the optimum cost, since in addition to aligning with each other the probes should also be aligned with the nucleotide deposition sequence S. Figure 3 illustrates a case of four 2-nucleotide sequences CT, GC, AG, and TA which should be placed on a 2×2 grid. The lower bound on the number of conflicts is 8 while the optimum number of conflicts is 10.

3   Epitaxial Algorithm for SADP

In this section, we describe the so-called *epitaxial placement* approach to SADP and discuss some efficient implementation details. Epitaxial, or "seeded crystal growth", placement is a technique that has been well-explored in the VLSI circuit placement literature [9,11]. The technique essentially grows a 2-dimensional placement around a single starting "seed".

The intuition behind our use of epitaxial placement is that if more information (i.e., about placed neighboring probes) is available when a site is about to be filled, we can make a better choice of the probe to be placed. For example, if we already know all four neighboring probes for a given site, then our choice of a probe to be placed in this site will be the "best possible", i.e., it will guarantee minimum sum of conflict distances to the four neighbors. The TSP method

---

Input: Set P of N^2 probes, scaling coefficients k_i, $i = 1, \ldots, 3$
Output: Assignment of the probes to the sites of an $N \times N$ grid

---

1. Mark all grid sites as empty
2. Assign a randomly chosen probe to the center site and mark this site as full
3. While there are empty sites, do
       If there exists an empty site c with all 4 neighbors full, then
             Find probe $p(c) \in P$ with minimum sum of Hamming distances to the
                neighboring probes
             Assign probe $p(c)$ to site c and mark c as full
       Else
             For each empty site c with $i > 0$ adjacent full sites, find probe
                $p(c) \in P$ with minimum sum S of Hamming distances to the
                probes in full neighbors, and let $norm\_cost(c) = k_i S / i$.
             Let c^* be the site with minimum $norm\_cost$
             Assign probe $p(c^*)$ to site c^* and mark c^* as full

---

Fig. 4. The Epitaxial Algorithm.

that has been previously used for DNA array synthesis[2] essentially chooses each probe in the tour based on knowledge about a single neighbor. In contrast, in epitaxial placement an average of 2 neighbors are already known when choosing the probe to be assigned to a site.

The epitaxial algorithm (see Figure 4) places a random probe in the center of the array, and then iteratively places probes in sites adjacent to already-placed probes so as to greedily minimize the *average* number of conflicts induced between the newly created pairs of neighbors. We have found that sites with more filled neighbors should have higher priority to be filled. In particular, we give highest priority to sites with 4 known neighbors. In the remaining cases we apply scaling coefficients to prioritize candidate probe-site pairs. In our implementation, if a probe is to be placed at a site with $i < 4$ placed neighbors, then the average number of conflicts caused by this placement is multiplied by a coefficient $0 < k_i \leq 1$. Based on our experiments, we set $k_1 = 1$, $k_2 = 0.8$, and $k_3 = 0.6$. In our implementation we avoid repeated distance computations by keeping with each border site a list of probes sorted by normalized cost. For each site this list is computed at most four (and on the average two) times, i.e., when one of the neighboring sites is being filled while the site is still empty.

4 Optimum Probe Alignment Algorithms

In this section we first give an efficient dynamic programming algorithm for computing the optimal embedding of a single probe whose neighbors are already embedded. We then generalize this algorithm for optimal simultaneous alignment several neighboring probes with respect to already embedded neighbors.

---

[2] Recall that the method of Hannenhalli et al. [6] first finds a TSP-tour and then threads it into the array. See Section 6 for more details.

Input: Nucleotide deposition sequence $S = s_1 s_2 \ldots s_K$, $s_i \in \{A, C, G, T\}$; set X of probes already embedded into S; and unembedded probe $p = p_1 p_2 \ldots p_k$, $p_i \in \{A, C, G, T\}$

Output: The minimum number of conflicts between an embedding of p and probes in X, along with a minimum-conflict embedding

---

1. For each $j = 1, \ldots, K$, let x_j be the number of probes in X which have a blank in j^{th} position.
2. $cost(0,0) = 0$; For $i = 1, \ldots, k$, $cost(i, 0) = \infty$
3. For $j = 1, \ldots, K$ do
   $$cost(0, j) = cost(0, j - 1) + |X| - x_j$$
   For $i = 1, \ldots, k$ do
   If $p_i = s_j$ then
   $$cost(i, j) = \min\{cost(i, j - 1) + x_j, \ cost(i - 1, j - 1) + |X| - x_j\}$$
   Else $cost(i, j) = cost(i, j - 1) + x_j$
4. Return $cost(k, K)$ and the corresponding embedding of s

Fig. 5. The Single Probe Alignment Algorithm.

4.1   Optimum Single Probe Alignment Algorithm

The basic operation that we use in the post-placement improvement algorithms (Section 5) finds an optimum embedding of a probe when some (or all) adjacent sites contain already embedded probes. In other words, our goal is to simultaneously align the given probe s to its embedded neighboring probes, while making sure this alignment gives a feasible embedding of s in the nucleotide deposition sequence S. In this subsection we give an efficient dynamic programming algorithm for computing an optimum alignment.

The Single Probe Alignment algorithm (see Figure 5) essentially computes a shortest path in a specific directed acyclic graph $G = (V, E)$. Let p be the probe to be aligned, and let X be the set of already embedded probes adjacent to p. Each embedded probe $q \in X$ is a sequence of length $K = |S|$ over the alphabet $\{A, C, G, T, b\}$, with the jth letter of q being either a blank or s_j, the jth letter of the nucleotide deposition sequence S. The graph G (see Figure 6) is a directed "tilted mesh" with vertex set $V = \{0, \ldots, k\} \times \{0, \ldots, K\}$ (where k is the length of p) and edge set $E = E_{horiz} \cup E_{diag}$ where

$$E_{horiz} = \{(i, j - 1) \rightarrow (i, j) \mid 0 \leq i \leq k, 0 < j \leq K\}$$

and

$$E_{diag} = \{(i - 1, j - 1) \rightarrow (i, j) \mid 0 < i \leq k, 0 < j \leq K\}.$$

The cost of a "horizontal" edge $(i, j - 1) \rightarrow (i, j)$ is defined as the number of embedded probes in X which have a non-blank letter on j^{th} position, while the cost of a "diagonal" edge $(i - 1, j - 1) \rightarrow (i, j)$ is infinity if the ith letter of p differs from the jth letter of S, and is equal to the number of embedded probes of X with a blank on the jth position otherwise. The Single Probe Alignment algorithm computes the shortest path from the source node $(0, 0)$ to the sink

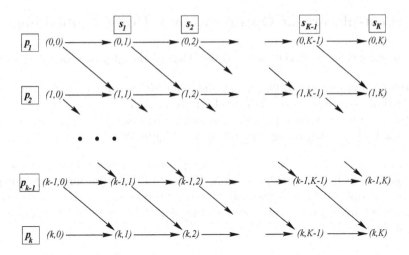

Fig. 6. Directed acyclic graph representing all possible embeddings of probe $p = p_1 p_2 \ldots p_k$ into mask sequence $S = s_1 s_2 \ldots s_K$.

node (k, K) using a topological traversal of G (the graph G is not explicitly constructed).

Theorem 3. *The algorithm in Figure 5 returns, in $O(kK)$ time, the minimum number of conflicts between an embedding of s and the adjacent embedded probes X (along with a minimum-conflict embedding of s).*

Proof. Each directed path from $(0,0)$ to (k, K) in G consists of K edges, k of which must be diagonal. Each such path P corresponds to an embedding of p into S as follows. If the jth arc of P is horizontal, the embedding has a blank in j^{th} position. Otherwise, the jth arc must be of the form $(i-1, j-1) \rightarrow (i, j)$ for some $1 \leq i \leq k$, and the embedding of p corresponding to P has $p_i = s_j$ in the jth position. It is easy to verify that the edge costs defined above ensure that the total cost of P gives the number of conflicts between the embedding of p corresponding to P and the set X of embedded neighbors.

4.2 Optimum Multiple Probe Alignment Algorithm

The dynamic programming algorithm for the optimal simultaneous embedding of $n > 1$ probes (see [7]) is a straightforward generalization of the algorithm above. The corresponding directed acyclic graph G consist of $k^n K$ nodes $(i_1, \ldots, i_n, j)$, where $0 \leq i_l \leq k$, $1 \leq j \leq K$. All arcs into $(i_1, \ldots, i_n, j)$ come from nodes $(i'_1, \ldots, i'_n, j - 1)$, where $i'_l \in \{i_l, i_l - 1\}$. Therefore the indegree of each node is at most 2^n. The weight of each edge is defined as above such that each finite-weight path defines embeddings for all n probes and the weight equals the number of conflicts. Finally, computing the shortest path between $(0, \ldots, 0)$ and $(k, \ldots, k, K)$ can be done in $O(2^n k^n K)$ time.

5   Post-placement Optimization of Probe Embeddings

We now consider an AADP solution flow that is split into two consecutive steps:

Step (i). Find an initial placement and embedding (using e.g., Epitaxial Algorithm (see Figure 4) or TSP+1-Threading), and

Step (ii). Optimize each probe's embedding with respect to its neighbors using the Optimal Alignment Algorithm (see Figure 5).

Theorems 1 and 2 above provide pre-placement lower bounds on the total border (conflict) cost. In this section, we first estimate the limitations of *post-placement* optimization in Step (ii). We then propose two methods of reoptimizing probe embeddings, which we call the *Batched Greedy* and the *Chessboard* algorithms.

5.1   Post-placement Lower Bounds

Once a 2-dimensional placement is fixed, we do not have any freedom left if probes must be embedded synchronously. In the asynchronous array design we can still reembed all probes so as to substantially improve the border length. To get an estimate on the how much improvement is possible, let $LG2$ be a grid graph $G2$ with weights on edges equal to LCS distances between endpoint probes. The following lower bound is obvious.

Theorem 4. *The total edge weight of the graph $LG2$ is a lower bound on the optimum AADP solution cost with a given placement.*

Note. A more accurate lower bound can be obtained by replacing LCS distance with *embedded LCS* distance, $elcsd(p, p')$, which is the minimum number of conflicts over all possible pairs of embeddings of the probes p and p'. The embedded LCS distance can be computed using an $O(|p| \cdot |p'| \cdot |S|)$ dynamic programming algorithm (see [7]). Unfortunately, neither of these lower bounds is tight, as can be seen from the example given in Section 2.

5.2   Batched Greedy Optimization

We have implemented a natural greedy algorithm (GA) for optimizing probe embeddings. The GA finds a probe that offers largest cost gain from optimum reembedding with respect to the (fixed) embeddings of its neighbors; the algorithm then implements this reembedding, updates gains, and repeats. A faster *batched* version of GA (see Figure 7) partially sacrifices its greedy nature in favor of runtime, via the mechanism of less-frequent gain updates. In other words, during a single *batched* phase we reembed probes in greedy order according to the cost gains from reembedding, but we do not update any gain while there are still probes with positive unchanged gains.

Input: Feasible AADP solution, i.e., placement in $G2$ of probes embedded in S
Output: A heuristic low-cost feasible AADP solution

While there exist probes which can be reembedded with gain in cost do
    Compute gain of the optimum reembedding of each probe.
    Unmark all probes
    For each unmarked probe p, in descending order of gain, do
        Reembed p optimally with respect to its four neighbors
        Mark p and all probes in adjacent sites

Fig. 7. The Batched Greedy Algorithm.

Input: Feasible AADP solution, i.e., placement in $G2$ of probes embedded in S
Output: A heuristic low-cost feasible AADP solution

Repeat until there is no gain in cost
    For each site (i, j), $1 \leq i, j \leq N$ with $i + j$ even, reembed probe optimally
        with respect to its four neighbors
    For each site (i, j), $1 \leq i, j \leq N$ with $i + j$ odd, reembed probe optimally
        with respect to its four neighbors

Fig. 8. The Chessboard Algorithm.

5.3 Chessboard Optimization

The main idea behind our so-called "Chessboard" algorithm is to maximize the number of *independent* reembeddings, where two probes are independent if changing the embedding of one does not affect the optimum embedding of the other. It is easy to see that if we bicolor our grid as we would a chessboard, then all white (resp. black) sites will be independent and can therefore be simultaneously, and optimally, reembedded. The Chessboard Algorithm (see Figure 8) alternates reembeddings of black and white sites until no improvement is obtained.

A 2×1 version of the Chessboard algorithm partitions the array into iso-oriented 2×1 tiles and bicolors them. Then using the multiprobe alignment algorithm (see Section 4.2) with $n = 2$ it alternatively optimizes the black and white 2×1 tiles.

6 Experimental Results

We have implemented two placement algorithms: the multi-fragment Greedy TSP algorithm followed by 1-Threading (see Figure 9) as described in [6] and the Epitaxial algorithm. We have also implemented three post-placement optimization algorithms: Batched Greedy, Chessboard and 2×1 Chessboard. To improve the runtime in our implementations of post-placement algorithms we stop as soon as the iteration improvement drops below 0.1% of the total num-

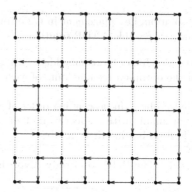

Fig. 9. 1-Threading.

Table 1. Placement heuristics and lower bounds.

Chip Size	Sync LB Cost	TSP+1-Threading			Epitaxial			Async LB	
		Cost	%Gap	CPU	Cost	%Gap	CPU	Cost	%Gap
20	19242	23401	21.6	0	22059	14.6	0	10236	-46.8
40	72240	92267	27.7	3	85371	18.2	6	38414	-46.8
60	156149	204388	30.9	15	187597	20.1	32	83132	-46.8
80	268525	358945	33.7	46	327924	22.1	104	144442	-46.2
100	410019	554849	35.3	113	505442	23.3	274	220497	-46.2
200	1512014	2140903	41.6	1901	1924085	27.3	4441	798708	-47.2
300	3233861	4667882	44.3	12028	—	—	—	—	—
500	8459958	12702474	50.1	109648	—	—	—	—	—

ber of conflicts[3]. For comparison, we include the synchronous, resp. asynchronous pre-placement lower-bounds given by Theorems 1 and 2, and the post-placement lower-bound given by Theorem 4.

All algorithms were coded in C and compiled using g++ version 2.95.3 with -O3 optimization. All reported times are in CPU seconds. Placement algorithms were run on an SGI Origin 2000 with 16 195MHz MIPS R10000 processors (only one of which is actually used by the sequential implementations included in our comparison) and 4 GB of internal memory, running under IRIX 6.4 IP27. Post-placement algorithms were run on a dual-processor 1.4GHz Intel Xeon server with 512MB RAM.

All experiments were performed on DNA arrays of size varying between 20×20 to 500×500. All results reported in this section are averages over 10 sets of probes (of length 25, generated uniformly at random).

Table 1 compares the cost (number of conflicts) and runtime of the two placement algorithms. Missing entries were not computed due to prohibitively large running time or memory requirements. For comparison we include the

---

[3] In our experiments imposing the threshold of 0.1% leads to a total loss in solution quality of at most 1%. On the other hand, the number of iterations and hence the runtime decreases by more than one order of magnitude.

Table 2. Optimization of probe embeddings after to the TSP+1-Threading placement.

Chip Size	LCS LB cost	Initial %Gap	Batched Greedy			Chessboard			2×1 Chessboard		
			%Gap	#Iter	CPU	%Gap	#Iter	CPU	%Gap	#Iter	CPU
20	14859	57.5	24.3	8.8	2	19.1	13.3	2	18.0	11.1	18
40	59500	55.1	25.0	8.9	7	20.0	12.7	9	18.8	11.8	77
60	133165	53.5	25.3	8.5	15	20.2	13.2	20	19.0	11.9	175
80	234800	52.9	25.7	8.7	27	20.5	13.4	35	19.3	12.0	315
100	364953	52.0	25.7	8.8	40	20.5	13.6	54	19.4	11.7	480
200	1425784	50.2	26.3	8.3	154	20.9	13.9	221	19.7	11.8	1915
300	3130158	49.1	26.7	8.0	357	21.5	13.6	522	21.6	11.0	4349
500	8590793	47.9	27.1	8.0	943	21.4	14.0	1423	20.2	12.0	15990

Table 3. Optimization of probe embeddings after the Epitaxial placement.

Chip Size	LCS LB cost	Initial %Gap	Batched Greedy			Chessboard			2×1 Chessboard		
			%Gap	#Iter	CPU	%Gap	#Iter	CPU	%Gap	#Iter	CPU
20	14636	50.7	22.7	8.4	1	17.9	12.3	2	17.0	10.5	17
40	58494	45.9	22.4	8.1	6	17.7	12.9	8	16.8	11.4	74
60	130636	43.6	22.0	8.0	14	17.5	12.3	18	16.6	11.2	164
80	230586	42.2	21.7	8.0	24	17.2	12.6	33	16.3	11.1	289
100	357800	41.3	21.6	8.0	37	17.1	12.1	48	16.3	11.3	461
200	1395969	37.8	20.8	7.1	130	16.4	12.0	190	15.6	10.7	1779

synchronous (resp. asynchronous) placement lower-bounds given by Theorems 1 and 2. The column "%Gap" shows by how much respective heuristic or lower bound exceeds the lower bound for synchronous placement given by Theorem 1. For 200×200 DNA arrays, the Epitaxial algorithm improves by more than 10% over the previously best TSP+1-Threading algorithm of [6], coming within 14-27% of the lower-bound given by Theorem 1.

Tables 2–3 give the results of post-placement probe-embedding heuristics when applied to the TSP+1-Threading and Epitaxial placements, respectively. The column "Gap" shows by how much the initial placement and respective heuristics exceed the post-placement LCS lower bound given by Theorem 4. The results show that the Chessboard algorithm is better than Batched Greedy with comparable running time. 2×1 Chessboard is giving a further improvement of .8-1.2%, coming within 18-20% and 15-17% of the LCS lower-bound for TSPthreading and Epitaxial placements, respectively. Overall, the best conflict reduction is obtained by applying 2×1 Chessboard optimization to the Epitaxial placement improving [6] by, e.g., 25% for 100x100 chip.

7   Conclusions

For synchronous DNA array design, we have suggested a new epitaxial heuristic which substantially decreases unintended illumination as measured by total border length, during mask exposure steps. We have explored advantages of asynchronous DNA array design, giving several new heuristics to reduce the mask border length. Further directions that we address in ongoing research include:

- developing techniques for simultaneous placement and asynchronous embedding;
- improving solution quality and runtime by incorporating methods such as hierarchical placement (this and several other promising approaches have been previously proposed in the VLSI CAD context); and
- working with industry to assess the quality of our methods on real VLSIPS test cases.

References

1. Abdueva and Skvortsov, *Personal Communication.*
2. N. Alon, C. J. Colbourn, A. C. H. Lingi and M. Tompa, "Equireplicate Balanced Binary Codes for Oligo Arrays", *SIAM Journal on Discrete Mathematics* 14(4) (2001), pp. 481-497.
3. S. Fodor, J. L. Read, M. C. Pirrung, L. Stryer, L. A. Tsai and D. Solas, "Light-Directed, Spatially Addressable Parallel Chemical Synthesis", *Science* 251 (1991), pp. 767-773.
4. S. A. Heath and F. P. Preparata, "Enhanced Sequence Reconstruction With DNA Microarray Application", *Proc. 7th Annual International Conf. on Computing and Combinatoris (COCOON)*, Springer-Verlag, Lecture Notes in Computer Science vol. 2108, August 2001, pp. 64-74.
5. E. Hubbell and P. A. Pevzner, "Fidelity Probes for DNA Arrays", *Proc. Seventh International Conference on Intelligent Systems for Molecular Biology*, 1999, pp. 113-117.
6. S. Hannenhalli, E. Hubbell, R. Lipshutz and P. A. Pevzner, "Combinatorial Algorithms for Design of DNA Arrays", in *Chip Technology* (ed. J. Hoheisel), Springer-Verlag, 2002.
7. A.B. Kahng, I.I. Măndoiu, P. Pevzner, S. Reda, and A. Zelikovsky, "Border Length Minimization in DNA Array Design", Technical Report CS2002-0713, Department of Computer Science and Engineering, University of California at San Diego, La Jolla, CA, 2002.
8. F. Li and G.D. Stormo, "Selecting Optimum DNA Oligos for Microarrays", *Proc. IEEE International Symposium on Bio-Informatics and Biomedical Engineering*, 2000, pp. 200-207.
9. B. T. Preas and M. J. Lorenzetti, eds., *Physical Design Automation of VLSI Systems*, Benjamin-Cummings, 1988.
10. R. Sengupta and M. Tompa, "Quality Control in Manufacturing Oligo Arrays: a Combinatorial Design Approach", *Journal of Computational Biology* 9 (2002), pp. 1–22.
11. K. Shahookar and P. Mazumder, "VLSI Cell Placement Techniques", *Computing Surveys* 23(2) (1991), pp. 143-220.

The Enhanced Suffix Array and Its Applications to Genome Analysis

Mohamed Ibrahim Abouelhoda, Stefan Kurtz, and Enno Ohlebusch

Faculty of Technology, University of Bielefeld
P.O. Box 10 01 31, 33501 Bielefeld, Germany
{mibrahim,enno,kurtz}@TechFak.Uni-Bielefeld.DE

Abstract. In large scale applications as computational genome analysis, the space requirement of the suffix tree is a severe drawback. In this paper, we present a uniform framework that enables us to systematically replace every string processing algorithm that is based on a bottom-up traversal of a suffix tree by a corresponding algorithm based on an enhanced suffix array (a suffix array enhanced with the lcp-table). In this framework, we will show how maximal, supermaximal, and tandem repeats, as well as maximal unique matches can be efficiently computed. Because enhanced suffix arrays require much less space than suffix trees, very large genomes can now be indexed and analyzed, a task which was not feasible before. Experimental results demonstrate that our programs require not only less space but also much less time than other programs developed for the same tasks.

1 Introduction

Repeat analysis plays a key role in the study, analysis, and comparison of complete genomes. In the analysis of a single genome, a basic task is to characterize and locate the repetitive elements of the genome.

The repetitive elements can be generally classified into two large groups: dispersed repetitive DNA and tandemly repeated DNA. Dispersed repetitions vary in size and content and fall into two basic categories: transposable elements and segmental duplications [14]. Transposable elements belong to one of the following four classes: SINEs (short interspersed nuclear elements), LINEs (long interspersed nuclear elements), LTR (long terminal repeats), and transposons. Segmental duplications, which might contain complete genes, have been divided into two classes: chromosome-specific and trans-chromosome duplications [17]. Tandemly repeated DNA can also be classified into two categories: simple sequence repetitions (relatively short k-mers such as micro- and minisatellites) and larger ones, which are called blocks of tandemly repeated segments.

The repeat contents of a genome can be very large. For example, 50% of the 3 billion bp of the human genome consist of repeats. Repeats also comprise 11% of the mustard weed genome, 7% of the worm genome and 3% of the fly genome [14]. Clearly, one needs extensive algorithmic support for a systematic study of repetitive DNA on a genomic scale. The software tool *REPuter* has

R. Guigó and D. Gusfield (Eds.): WABI 2002, LNCS 2452, pp. 449–463, 2002.

originally been developed for this task [13]. The core algorithm of *REPuter* locates maximal exact repeats of a sequence in linear space and time. Although it is based on an efficient and compact implementation of suffix trees [12], the space consumption is still quite large: it requires 12.5 bytes per input character in practice. Moreover, in large scale applications, the suffix tree suffers from a poor locality of memory reference, which causes a significant loss of efficiency on cached processor architectures.

The same problem has been identified in the context of genome comparison. Nowadays, the DNA sequences of entire genomes are being determined at a rapid rate. For example, the genomes of several strains of *E. coli* and *S. aureus* have already been completely sequenced. When the genomic DNA sequences of closely related organisms become available, one of the first questions researchers ask is how the genomes align. This alignment may help, for example, in understanding why a strain of a bacterium is pathogenic or resistant to antibiotics while another is not. The software tool *MUMmer* [5] has been developed to efficiently align two sufficiently similar genomic DNA sequences. In the first phase of its underlying algorithm, a maximal unique match (*MUM*) decomposition of two genomes S_1 and S_2 is computed. A *MUM* is a sequence that occurs exactly once in genome S_1 and once in genome S_2, and is not contained in any longer such sequence. (We will show in this paper that *MUM*s are special supermaximal repeats.) Using the suffix tree of $S_1 \# S_2$, *MUM*s can be computed in $O(n)$ time and space, where $n = |S_1 \# S_2|$ and $\#$ is a symbol neither occurring in S_1 nor in S_2. However, the space consumption of the suffix tree is a major problem when comparing large genomes; see [5].

To sum up, although the suffix tree is undoubtedly one of the most important data structures in sequence analysis [2, 7], its space consumption is the bottleneck in genome analysis tasks, whenever the sequences under consideration are large genomes.

More space efficient data structures than the suffix tree exist. The most prominent one is the *suffix array* [16] which requires only $4n$ bytes (4 bytes per input character) in its basic form. In most practical applications, it can be stored in secondary memory because no random access to it is necessary. The direct construction of the suffix array takes $O(n \cdot \log n)$ time in the worst case, but in practice it can be constructed as fast as a suffix tree [15]. Furthermore, it is known that pattern matching based on suffix arrays can compete with pattern matching based on suffix trees; see [1, 16] for details. However, efficient algorithms on suffix arrays that solve typical string processing problems like searching for all repeats in a string have not yet been devised.

In this paper, we present a uniform framework that enables us to systematically replace a string processing algorithm that is based on a bottom-up traversal of a suffix tree by a corresponding algorithm that is based on an enhanced suffix array (a suffix array enhanced with the lcp-table)[1]. This approach has several advantages:

---

[1] It is also possible to replace every string processing algorithm based on a top-down traversal of a suffix tree by an algorithm on an enhanced suffix array; see [1].

1. The algorithms are more space efficient than the corresponding ones based on the suffix tree because our implementation of the enhanced suffix array requires only $5n$ bytes.
2. Experiments show that the running times of the algorithms are much better than those based on the suffix tree.
3. The algorithms are easier to implement on enhanced suffix arrays than on suffix trees.

First, we introduce the unified framework: the lcp-interval tree of a suffix array. Based on this framework, we then concentrate on algorithms that efficiently compute various kinds of repeats. To be precise, we will show how to efficiently compute maximal, supermaximal, and tandem repeats of a string S using a bottom-up traversal of the lcp-interval tree of S. We stress that this tree is only conceptual, in the sense that it is not really built during the bottom-up traversal (i.e., at any stage, only a representation of a small part of the lcp-interval tree resides in main memory). We implemented the algorithms and applied them to several genomes. Experiments show that our programs require less time and space than other programs for the same task. We would like to point out that our framework is not confined to the above-mentioned applications. For example, the off-line computation of the Lempel-Ziv decomposition of a string (which is important in data compression) can also be efficiently implemented using this framework.

2 Basic Notions

Let S be a string of length $|S| = n$ over an ordered alphabet Σ. To simplify analysis, we suppose that the size of the alphabet is a constant, and that $n < 2^{32}$. The latter implies that an integer in the range $[0, n]$ can be stored in 4 bytes. We assume that the special symbol $ is an element of Σ (which is larger then all other elements) but does not occur in S. $S[i]$ denotes the character at position i in S, for $0 \leq i < n$. For $i \leq j$, $S[i..j]$ denotes the substring of S starting with the character at position i and ending with the character at position j. The substring $S[i..j]$ is also denoted by the *pair of positions* (i, j).

A pair of substrings $R = ((i_1, j_1), (i_2, j_2))$ is a *repeated pair* if and only if $(i_1, j_1) \neq (i_2, j_2)$ and $S[i_1..j_1] = S[i_2..j_2]$. The length of R is $j_1 - i_1 + 1$. A repeated pair $((i_1, j_1), (i_2, j_2))$ is called *left maximal* if $S[i_1 - 1] \neq S[i_2 - 1]$ [2] and *right maximal* if $S[j_1 + 1] \neq S[j_2 + 1]$. A repeated pair is called *maximal* if it is left and right maximal. A substring ω of S is a (*maximal*) *repeat* if there is a (maximal) repeated pair $((i_1, j_1), (i_2, j_2))$ such that $\omega = S[i_1..j_1]$. A *supermaximal repeat* is a maximal repeat that never occurs as a substring of any other maximal repeat. A string is a *tandem repeat* if it can be written as $\omega\omega$ for some nonempty string ω. An *occurrence* of a tandem repeat $\omega\omega = S[p..p + |2\omega| - 1]$ is represented by the pair $(|\omega|, p)$. Such an occurrence $(|\omega|, p)$ is *branching* if $S[p + |\omega|] \neq S[p + 2|\omega|]$.

---

[2] This definition has to be extended to the cases $i_1 = 0$ or $i_2 = 0$, but throughout the paper we do not explicitly state boundary cases like these.

i	suftab	lcptab	bwtab	suftab^{-1}	$S_{\mathsf{suftab}[i]}$
0	2	0	c	2	aaacatat\$
1	3	2	a	6	aacatat\$
2	0	1		0	acaaacatat\$
3	4	3	a	1	acatat\$
4	6	1	c	3	atat\$
5	8	2	t	7	at\$
6	1	0	a	4	caaacatat\$
7	5	2	a	8	catat\$
8	7	0	a	5	tat\$
9	9	1	a	9	t\$
10	10	0	t	10	\$

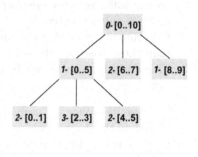

Fig. 1. Enhanced suffix array of the string $S =$ acaaacatat\$ and its lcp-interval tree. Table suftab^{-1} is introduced in Section 6.2.

The *suffix array* suftab is an array of integers in the range 0 to n, specifying the lexicographic ordering of the $n + 1$ suffixes of the string $S\$$. That is, $S_{\mathsf{suftab}[0]}, S_{\mathsf{suftab}[1]}, \ldots, S_{\mathsf{suftab}[n]}$ is the sequence of suffixes of $S\$$ in ascending lexicographic order, where $S_i = S[i..n-1]\$$ denotes the ith nonempty suffix of the string $S\$$, $0 \le i \le n$. The suffix array requires $4n$ bytes. The direct construction of the suffix array takes $O(n \cdot \log n)$ time [16], but it can be build in $O(n)$ time via the construction of the suffix tree; see, e.g., [7].

bwtab is a table of size $n + 1$ such that for every $i, 0 \le i \le n$, bwtab$[i] = S[\mathsf{suftab}[i] - 1]$ if suftab$[i] \ne 0$. bwtab$[i]$ is undefined if suftab$[i] = 0$. bwtab is the *Burrows and Wheeler* transformation [4] known from data compression. It is stored in n bytes and constructed in one scan over suftab in $O(n)$ time.

The lcp-table lcptab is an array of integers in the range 0 to n. We define lcptab$[0] = 0$ and lcptab$[i]$ is the length of the longest common prefix of $S_{\mathsf{suftab}[i-1]}$ and $S_{\mathsf{suftab}[i]}$, for $1 \le i \le n$. Since $S_{\mathsf{suftab}[n]} = \$$, we always have lcptab$[n] = 0$; see Fig. 1. The lcp-table can be computed as a by-product during the construction of the suffix array, or alternatively, in linear time from the suffix array [9]. The lcp-table requires $4n$ bytes in the worst case. However, in practice it can be implemented in little more than n bytes. More precisely, we store most of the values of table lcptab in a table lcptab$_1$ using n bytes. That is, for any $i \in [1, n]$, lcptab$_1[i] = \max\{255, \mathsf{lcptab}[i]\}$. There are usually only few entries in lcptab that are larger than or equal to ≥ 255; see Section 7. To access these efficiently, we store them in an extra table llvtab. This contains all pairs $(i, \mathsf{lcptab}[i])$ such that lcptab$[i] \ge 255$, ordered by the first component. At index i of table lcptab$_1$ we store 255 whenever lcptab$[i] \ge 255$. This tells us that the correct value of lcptab is found in llvtab. If we scan the values in lcptab$_1$ in consecutive order and find a value 255, then we access the correct value in lcptab in the next entry of table llvtab. If we access the values in lcptab$_1$ in arbitrary order and find a value 255 at index i, then we perform a binary search in llvtab using i as the key. This delivers lcptab$[i]$ in $O(\log_2 |\mathsf{llvtab}|)$ time.

3   The lcp-Interval Tree of a Suffix Array

Definition 1. *Interval* $[i..j]$, $0 \leq i < j \leq n$, *is an* lcp-interval *of* lcp-value ℓ *if*

1. $\mathsf{lcptab}[i] < \ell$,
2. $\mathsf{lcptab}[k] \geq \ell$ *for all* k *with* $i+1 \leq k \leq j$,
3. $\mathsf{lcptab}[k] = \ell$ *for at least one* k *with* $i+1 \leq k \leq j$,
4. $\mathsf{lcptab}[j+1] < \ell$.

We will also use the shorthand ℓ-interval (or even ℓ-$[i..j]$) for an lcp-interval $[i..j]$ of lcp-value ℓ. Every index k, $i+1 \leq k \leq j$, with $\mathsf{lcptab}[k] = \ell$ is called ℓ-index. The set of all ℓ-indices of an ℓ-interval $[i..j]$ will be denoted by $\ell Indices(i,j)$. If $[i..j]$ is an ℓ-interval such that $\omega = S[\mathsf{suftab}[i]..\mathsf{suftab}[i] + \ell - 1]$ is the longest common prefix of the suffixes $S_{\mathsf{suftab}[i]}, S_{\mathsf{suftab}[i+1]}, \ldots, S_{\mathsf{suftab}[j]}$, then $[i..j]$ is also called ω-interval.

As an example, consider the table in Fig. 1. $[0..5]$ is a 1-interval because $\mathsf{lcptab}[0] = 0 < 1$, $\mathsf{lcptab}[5+1] = 0 < 1$, $\mathsf{lcptab}[k] \geq 1$ for all k with $1 \leq k \leq 5$, and $\mathsf{lcptab}[2] = 1$. Furthermore, $\ell Indices(0,5) = \{2,4\}$.

Kasai et al. [9] presented a linear time algorithm to simulate the bottom-up traversal of a suffix tree with a suffix array combined with the lcp-information. The following algorithm is a slight modification of their algorithm TraverseWith-Array. It computes all lcp-intervals of the lcp-table with the help of a stack. The elements on the stack are lcp-intervals represented by tuples $\langle lcp, lb, rb \rangle$, where lcp is the lcp-value of the interval, lb is its left boundary, and rb is its right boundary. In Algorithm 1, *push* (pushes an element onto the stack) and *pop* (pops an element from the stack and returns that element) are the usual stack operations, while *top* provides a pointer to the topmost element of the stack.

Algorithm 1 *Computation of lcp-intervals (adapted from Kasai et al. [9]).*

$push(\langle 0, 0, \perp \rangle)$
for $i := 1$ **to** n **do**
   $lb := i - 1$
   while $\mathsf{lcptab}[i] < top.lcp$
      $top.rb := i - 1$
      $interval := pop$
      $report(interval)$
      $lb := interval.lb$
   if $\mathsf{lcptab}[i] > top.lcp$ **then**
      $push(\langle \mathsf{lcptab}[i], lb, \perp \rangle)$

Here, we will take the approach of Kasai et al. [9] one step further and introduce the concept of an lcp-interval tree.

Definition 2. *An m-interval $[l..r]$ is said to be* embedded *in an ℓ-interval $[i..j]$ if it is a subinterval of $[i..j]$ (i.e., $i \leq l < r \leq j$) and $m > \ell$ [3]. The ℓ-interval*

---
[3] Note that we cannot have both $i = l$ and $r = j$ because $m > \ell$.

$[i..j]$ *is then called the interval* enclosing $[l..r]$. *If* $[i..j]$ *encloses* $[l..r]$ *and there is no interval embedded in* $[i..j]$ *that also encloses* $[l..r]$, *then* $[l..r]$ *is called a* child interval *of* $[i..j]$.

This parent-child relationship constitutes a conceptual (or virtual) tree which we call the lcp-interval tree of the suffix array. The root of this tree is the 0-interval $[0..n]$; see Fig. 1. The lcp-interval tree is basically the suffix tree without leaves (note, however, that it is not our intention to build this tree). These leaves are left implicit in our framework, but every leaf in the suffix tree, which corresponds to the suffix $S_{\mathsf{suftab}[l]}$, can be represented by a *singleton interval* $[l..l]$. The parent interval of such a singleton interval is the smallest lcp-interval $[i..j]$ with $l \in [i..j]$. For instance, continuing the example of Fig. 1, the child intervals of $[0..5]$ are $[0..1]$, $[2..3]$, and $[4..5]$. The next theorem shows how the parent-child relationship of the lcp-intervals can be determined from the stack operations in Algorithm 1.

Theorem 2. *Consider the* **for**-*loop of Algorithm 1 for some index* i. *Let top be the topmost interval on the stack and* $(top - 1)$ *be the interval next to it (note that* $(top - 1).lcp < top.lcp)$.

1. *If* $\mathsf{lcptab}[i] \leq (top - 1).lcp$, *then top is the child interval of* $(top - 1)$.
2. *If* $(top - 1).lcp < \mathsf{lcptab}[i] < top.lcp$, *then top is the child interval of those* $\mathsf{lcptab}[i]$-*interval which contains* i.

An important consequence of Theorem 2 is the correctness of Algorithm 3. There, the elements on the stack are lcp-intervals represented by quadruples $\langle lcp, lb, rb, childList \rangle$, where lcp is the lcp-value of the interval, lb is its left boundary, rb is its right boundary, and $childList$ is a list of its child intervals. Furthermore, $add([c_1, \ldots, c_k], c)$ appends the element c to the list $[c_1, \ldots, c_k]$.

Algorithm 3 *Traverse and process the lcp-interval tree*

```
lastInterval := ⊥
push(⟨0, 0, ⊥, [ ]⟩)
for i := 1 to n do
    lb := i − 1
    while lcptab[i] < top.lcp
        top.rb := i − 1
        lastInterval := pop
        process(lastInterval)
        lb := lastInterval.lb
        if lcptab[i] ≤ top.lcp then
            top.childList := add(top.childList, lastInterval)
            lastInterval := ⊥
    if lcptab[i] > top.lcp then
        if lastInterval ≠ ⊥ then
            push(⟨lcptab[i], lb, ⊥, [lastInterval]⟩)
            lastInterval := ⊥
        else push(⟨lcptab[i], lb, ⊥, [ ]⟩)
```

In Algorithm 3, the lcp-interval tree is traversed in a bottom-up fashion by a linear scan of the lcptab, while needed information is stored on a stack. We stress that the lcp-interval tree is not really build: whenever an ℓ-interval is processed by the generic function *process*, only its child intervals have to be known. These are determined solely from the lcp-information, i.e., there are no explicit parent-child pointers in our framework. In contrast to Algorithm 1, Algorithm 3 computes all lcp-intervals of the lcp-table *with* the child information. In the rest of the paper, we will show how to solve several problems merely by specifying the function *process* called on line 8 of Algorithm 3.

4 An Efficient Implementation of an Optimal Algorithm for Finding Maximal Repeated Pairs

The algorithm of Gusfield [7, page 147] computes maximal repeated pairs in a sequence S. It runs in $O(kn + z)$ time where $k = |\Sigma|$ and z is the number of maximal repeated pairs. This running time is optimal. To the best of our knowledge, Gusfield's algorithm was first implemented in the *REPuter*-program [13], based on space efficient suffix trees as described in [12]. In this section, we show how to implement the algorithm using enhanced suffix arrays. This considerably reduces the space requirements, thus removing a bottle neck in the algorithm. As a consequence, much larger genomes can be searched for repetitive elements. The implementation requires tables suftab, lcptab, bwtab, but does not access the input sequence. The accesses to the three tables are in sequential order, thus leading to an improved cache coherence and in turn considerably reduced running time. This is verified in Section 7.

We begin by introducing some notation: Let $\perp$ stand for the undefined character. We assume that it is different from all characters in Σ. Let $[i..j]$ be an ℓ-interval and $u = S[\text{suftab}[i]..\text{suftab}[i] + \ell - 1]$. Define $\mathcal{P}_{[i..j]}$ to be the set of positions p such that u is a prefix of S_p, i.e., $\mathcal{P}_{[i..j]} = \{\text{suftab}[r] \mid i \leq r \leq j\}$. We divide $\mathcal{P}_{[i..j]}$ into disjoint and possibly empty sets according to the characters to the left of each position: For any $a \in \Sigma \cup \{\perp\}$ define

$$\mathcal{P}_{[i..j]}(a) = \begin{cases} \{0 \mid 0 \in \mathcal{P}_{[i..j]}\} & \text{if } a = \perp \\ \{p \in \mathcal{P}_{[i..j]} \mid p > 0 \text{ and } S[p-1] = a\} & \text{otherwise} \end{cases}$$

The algorithm computes position sets in a bottom-up strategy. In terms of an lcp-interval tree, this means that the lcp-interval $[i..j]$ is processed only after all child intervals of $[i..j]$ have been processed.

Suppose $[i..j]$ is a singleton interval, i.e., $i = j$. Let $p = \text{suftab}[i]$. Then $\mathcal{P}_{[i..j]} = \{p\}$ and

$$\mathcal{P}_{[i..j]}(a) = \begin{cases} \{p\} & \text{if } p > 0 \text{ and } S[p-1] = a \text{ or } p = 0 \text{ and } a = \perp \\ \emptyset & \text{otherwise} \end{cases}$$

Now suppose that $i < j$. For each $a \in \Sigma \cup \{\perp\}$, $\mathcal{P}_{[i..j]}(a)$ is computed step by step while processing the child intervals of $[i..j]$. These are processed from left to right. Suppose that they are numbered, and that we have already processed q child intervals of $[i..j]$. By $\mathcal{P}^q_{[i..j]}(a)$ we denote the subset of $\mathcal{P}_{[i..j]}(a)$ obtained after processing the qth child interval of $[i..j]$. Let $[i'..j']$ be the $(q + 1)$th child interval of $[i..j]$. Due to the bottom-up strategy, $[i'..j']$ has been processed and hence the position sets $\mathcal{P}_{[i'..j']}(b)$ are available for any $b \in \Sigma \cup \{\bot\}$.

$[i'..j']$ is processed in the following way: First, maximal repeated pairs are output by combining the position set $\mathcal{P}^q_{[i..j]}(a)$, $a \in \Sigma \cup \{\bot\}$, with position sets $\mathcal{P}_{[i'..j']}(b)$, $b \in \Sigma \cup \{\bot\}$. In particular, $((p, p+\ell-1), (p', p'+\ell-1))$, $p < p'$, are output for all $p \in \mathcal{P}^q_{[i..j]}(a)$ and $p' \in \mathcal{P}_{[i'..j']}(b)$, $a, b \in \Sigma \cup \{\bot\}$ and $a \neq b$.

It is clear that u occurs at position p and p'. Hence $((p, p+\ell-1), (p', p'+\ell-1))$ is a maximal repeated pair. By construction, only those positions p and p' are combined for which the characters immediately to the left, i.e., at positions $p-1$ and $p'-1$ (if they exist), are different. This guarantees left-maximality of the output repeated pairs.

The position sets $\mathcal{P}^q_{[i..j]}(a)$ were inherited from child intervals of $[i..j]$ that are different from $[i'..j']$. Hence the characters immediately to the right of u at positions $p + \ell$ and $p' + \ell$ (if these exist) are different. As a consequence, the output repeated pairs are maximal.

Once the maximal repeated pairs for the current child interval $[i'..j']$ are output, we compute the union $\mathcal{P}^{q+1}_{[i..j]}(e) := \mathcal{P}^q_{[i..j]}(e) \cup \mathcal{P}_{[i'..j']}(e)$ for all $e \in \Sigma \cup \{\bot\}$. That is, the position sets are inherited from $[i'..j']$ to $[i..j]$.

In Algorithm 3 if the function *process* is applied to an lcp-interval, then all its child intervals are available. Hence the maximal repeated pair algorithm can be implemented by a bottom-up traversal of the lcp-interval tree. To this end, the function *process* in Algorithm 3 outputs maximal repeated pairs and further maintains position sets on the stack (which are added as a fifth component to the quadruples). The bottom-up traversal requires $O(n)$ time.

Algorithm 3 accesses the lcp-table in sequential order. Additionally, the maximal repeated pair algorithm does so with table suftab. Now consider the accesses to the input string S: Whenever suftab$[i]$ is processed, the input character $S[\text{suftab}[i] - 1]$ is accessed. Since bwtab$[i] = S[\text{suftab}[i] - 1]$, whenever suftab$[i] > 0$, the access to S can be replaced by an access to bwtab. Since suftab is accessed in sequential order, this also holds for bwtab. In other words, we do not need the input string when computing maximal repeated pairs. Instead we use table bwtab without increasing the total space requirement. The same technique is applied when computing supermaximal repeats; see Section 5.

There are two operations performed when processing an lcp-interval $[i..j]$. Output of maximal repeated pairs by combining position sets and union of position sets. Each combination of position sets means to compute their Cartesian product. This delivers a list of position pairs, i.e., maximal repeated pairs. Each repeated pair is computed in constant time from the position lists. Altogether, the combinations can be computed in $O(z)$ time, where z is the number of repeats. The union operation for the position sets can be implemented in constant time, if we use linked lists. For each lcp-interval, we have $O(k)$ union operations, where $k = |\Sigma|$. Since $O(n)$ lcp-intervals have to be processed, the union and add operations require $O(kn)$ time. Altogether, the algorithm runs in $O(kn + z)$ time.

Next, we analyze the space consumption of the algorithm. A position set $\mathcal{P}_{[i..j]}(a)$ is the union of position sets of the child intervals of $[i..j]$. If the child intervals of $[i..j]$ have been processed, the corresponding position sets are obsolete. Hence it is not required to copy position sets. Moreover, we only have to store the position sets for those lcp-intervals which are on the stack, which is used for the bottom-up traversal of the lcp-interval tree. So it is natural to store references to the position sets on the stack together with other information about the lcp-interval. Thus the space required for the position sets is determined by the maximal size of the stack. Since this is $O(n)$, the space requirement is $O(|\Sigma|n)$. In practice, however, the stack size is much smaller. Altogether the algorithm is optimal, since its space and time requirement is linear in the size of the input plus the output.

5   A New Algorithm for Finding Supermaximal Repeats

An ℓ-interval $[i..j]$ is called a *local maximum* in the lcp-table if $\mathsf{lcptab}[k] = \ell$ for all $i+1 \leq k \leq j$. It is not difficult to see that there is a one-to-one correspondence between the local maxima in the lcp-table and the leaves of the lcp-interval tree.

Lemma 1. *A string ω is a supermaximal repeat if and only if there is an ℓ-interval $[i..j]$ such that*

– *$[i..j]$ is a local maximum in the lcp-table and $[i..j]$ is the ω-interval.*
– *the characters $\mathsf{bwtab}[i], \mathsf{bwtab}[i + 1], \ldots, \mathsf{bwtab}[j]$ are pairwise distinct.*

The preceding lemma does not only imply that the number of supermaximal repeats is smaller than n, but it also suggests a simple linear time algorithm to compute all supermaximal repeats of a string S: Find all local maxima in the lcp-table of S. For every local maximum $[i..j]$ check whether $\mathsf{bwtab}[i], \mathsf{bwtab}[i + 1], \ldots, \mathsf{bwtab}[j]$ are pairwise distinct characters. If so, report $S[\mathsf{suftab}[i]..\mathsf{suftab}[i] + \mathsf{lcptab}[i] - 1]$ as supermaximal repeat. The reader is invited to compare our simple algorithm with the one described in [7, page 146]. An experimental comparison of the two algorithms can be found in Section 7.

As an application of Lemma 1, we will show how to efficiently compute maximal unique matches. This is an important subtask in the software tool *MUMmer* [5], which aligns DNA sequences of whole genomes, and in the identification of X-patterns, which appear in the comparison of two bacterial genomes [6].

Definition 3. *Given two sequences S_1 and S_2, a MUM is a sequence that occurs exactly once in S_1 and once in S_2, and is not contained in any longer such sequence.*

Lemma 2. *Let # be a unique separator symbol not occurring in S_1 and S_2 and let $S = S_1 \# S_2$. u is a MUM of S_1 and S_2 if and only if u is a supermaximal repeat in S such that*

1. *there is only one maximal repeated pair* $((i_1, j_1), (i_2, j_2))$ *with* $u = S[i_1..j_1] = S[i_2..j_2]$,
2. $j_1 < p < i_2$, *where* $p = |S_1|$.

In *MUMmer*, *MUM*s are computed in $O(|S|)$ time and space with the help of the suffix tree of $S = S_1 \# S_2$. Using an enhanced suffix array, this task can be done more time and space economically as follows: Find all local maxima in the lcp-table of $S = S_1 \# S_2$. For every local maximum $[i..j]$ check whether $i + 1 = j$ and bwtab$[i] \neq$ bwtab$[j]$. If so, report $S[$suftab$[i]..$suftab$[i] +$ lcptab$[i] - 1]$ as *MUM*. In Section 7, we also compare the performance of *MUMmer* with the implementation of the preceding algorithms.

6   Efficient Detection of Branching Tandem Repeats

This section is devoted to the computation of tandem repeats via enhanced suffix arrays. Stoye and Gusfield [18] described how all tandem repeats can be derived from branching tandem repeats by successive left-rotations. For this reason, we restrict ourselves to the computation of all branching tandem repeats.

6.1   A Brute Force Method

The simplest method to find branching tandem repeats is to process all lcp-intervals. For a given ω-interval $[i..j]$ one checks whether there is a child interval $[l..r]$ (which may be a singleton interval) of $[i..j]$ such that $\omega\omega$ is a prefix of $S_{\text{suftab}[q]}$ for each $q \in [l..r]$. Such a child interval can be detected in $O(|\omega| \log(j - i))$ time (if it exists), using the algorithm of [16] that searches for ω in $[i..j]$. It turns out that the running time of this algorithm is $O(n^2)$ (take, e.g., $S = a^n$). However, in practice this method is faster and more space efficient than other methods; see Section 7.

6.2   The Optimized Basic Algorithm

The *optimized basic algorithm* of [18] computes all branching tandem repeats in $O(n \log n)$ time. It is based on a traversal of the suffix tree, in which each branching node α is annotated by its leaf list, i.e., by the set of leaves in the subtree below α. The leaf list of a branching node α corresponds to an lcp-interval in the lcp-interval tree. As a consequence, it is straightforward to implement the optimized basic algorithm via a traversal of the lcp-interval tree.

Algorithm 4 *For each ℓ-interval $[i..j]$ determine the child interval $[i_{\max}..j_{\max}]$ of maximal width among all child intervals of $[i..j]$. Then, for each q such that $i \leq q \leq i_{\max} - 1$ or $j_{\max} + 1 \leq q \leq j$, let $p =$ suftab$[q]$ and verify the following:*

(1) $p + 2\ell = n$ *or* $S[p + \ell] \neq S[p + 2\ell]$ (*character-check*)
(2) $p + \ell =$ suftab$[h]$ *for some* $h, i \leq h \leq j$ (*forward-check*)
(3) $p - \ell =$ suftab$[h]$ *for some* $h, i \leq h \leq j$ (*backward-check*)

If (1) *and either* (2) *or* (3) *is satisfied, then output* (ℓ, p).

Algorithm 4 computes all branching tandem repeats. To analyze the efficiency, we determine the number of character-, backward-, and forward-checks. Since the largest child interval is always excluded, Algorithm 4 only performs verification steps for $O(n \log n)$ suffixes; see [18]. A character-check can easily be performed in constant time. For the forward- and backward-check the inverse suffix table suftab^{-1} is used. This is a table of size $n + 1$, such that for any $q, 0 \leq q \leq n$, suftab^{-1}[suftab$[q]$] $= q$. Obviously, $p + \ell =$ suftab$[h]$, for some $h, i \leq h \leq j$, if and only if $i \leq$ suftab$^{-1}[p + \ell] \leq j$. Similarly, $p - \ell =$ suftab$[h]$ for some $h, i \leq h \leq j$, if and only if $i \leq$ suftab$^{-1}[p - \ell] \leq j$. Hence, every forward-check and backward-check requires constant time. The lcp-interval tree is processed in $O(n)$ time. For each interval, a child interval of maximal width can be determined in constant extra time. Therefore, Algorithm 4 runs in $O(n \log n)$ time. Processing the lcp-interval tree requires tables lcptab and suftab plus some space for the stack used during the bottom-up traversal. To perform the forward- and backward-checks in constant time, one also needs table suftab^{-1}, which requires $4n$ bytes. Hence our implementation of the optimized basic algorithm requires $9n$ bytes.

6.3 The Improved $O(n \log n)$-Algorithm

We improve Algorithm 4 by getting rid of the character-checks and reducing the number of forward- and backward checks. That is, we exploit the fact that for an occurrence $(|\omega|, p)$ of a branching tandem repeat $\omega\omega$, we have $S[p] = S[p + |\omega|]$. As a consequence, if $p + |\omega| =$ suftab$[q]$ for some q in the ω-interval $[i..j]$, p must occur in the child interval $[l_a..r_a]$ storing the suffixes of S that have ωa as a prefix, where $a = S[$suftab$[i]] = S[p]$. This is formally stated in the following lemma.

Lemma 3. *The following statements are equivalent:*

(1) $(|\omega|, p)$ *is an occurrence of a branching tandem repeat* $\omega\omega$.
(2) $p + |\omega| =$ suftab$[q]$ *for some* q *in the* ω-*interval* $[i..j]$, *and* $p =$ suftab$[q_a]$ *for some* q_a *in the child interval* $[i_a..j_a]$ *representing the suffixes of* S *that have* ωa *as a prefix, where* $a = S[p]$.

This lemma suggests the following algorithm:

Algorithm 5 *For each* ℓ-*interval* $[i..j]$, *let* $a = S[$suftab$[i]]$ *and determine the child interval* $[i_a..j_a]$ *of* $[i..j]$ *such that* $a = S[$suftab$[i_a] + \ell]$. *Proceed according to the following cases:*

(1) *If* $j_a - i_a + 1 \leq i - j + 1 - (j_a - i_a + 1)$, *then for each* q, $i_a \leq q \leq j_a$, *let* $p =$ suftab$[q]$. *If* $p + \ell =$ suftab$[r]$ *for some* r, $i \leq r \leq i_a - 1$ *or* $j_a + 1 \leq r \leq j$, *then output* (ℓ, p).
(2) *If* $j_a - i_a + 1 > i - j + 1 - (j_a - i_a + 1)$, *then for each* q, $i \leq q \leq i_a - 1$ *or* $j_a + 1 \leq q \leq j$, *let* $p =$ suftab$[q]$. *If* $p - \ell =$ suftab$[r]$ *for some* r, $i_a \leq r \leq j_a$, *then output* (ℓ, p).

One easily verifies for each interval $[l..r]$ that $l - r + 1 - (r_{max} - l_{max} + 1) \geq \min\{r_a - l + 1, l - r + 1 - (r_a - l_a + 1)\}$. Hence Algorithm 5 checks not more suffixes than Algorithm 4. Thus the number of suffixes checked is $O(n \log n)$. For each suffix either a forward-check or backward-check is necessary. This takes constant time. The lcp-interval tree is processed in $O(n)$ time. Additionally, for each interval the algorithm determines the child interval $[l_a..r_a]$. This takes constant extra time. Hence the running time of Algorithm 5 is $O(n \log n)$.

7   Experiments

7.1   Programs and Data

In our experiments we applied the following programs:

- REPuter and esarep implement the algorithm of Gusfield (see Section 4) to compute maximal repeated pairs. While REPuter is based on suffix trees, esarep uses enhanced suffix arrays, as described in Section 4.
- supermax computes supermaximal repeats. It implements the algorithm described in Section 5.
- unique-match and esamum compute MUMs. unique-match is part of the original distribution of MUMmer (version 1.0) [5]. It is based on suffix trees. esamum is based on enhanced suffix arrays and uses the algorithm described at the end of Section 5.
- BFA, OBA, and iOBA compute branching tandem repeats. BFA implements the brute force algorithm described in Section 6. It uses the binary search algorithm of [16] to check for the appropriate child interval. OBA is an implementation of the optimized basic algorithm, and iOBA incorporates the improvements described in Section 6.

We applied the six programs for the detection of repeats to the genome of E. coli (4,639,221 bp) and the genome of yeast (12,156,300 bp). Additionally, we applied unique-match and esamum to the following pairs of genomes:

Mycoplasma 2: The complete genomes of Mycoplasma pneumoniae (816,394 bp) and of Mycoplasma genitalium (580,074 bp).
Streptococcus 2: The complete genomes of two strains of Streptococcus pneumoniae (2,160,837 bp and 2,038,615 bp).
E. coli 2: The complete genomes of two strains of E. coli (4,639,221 bp and 5,528,445 bp).

For E. coli and yeast and for all pairs of genomes we constructed the corresponding enhanced suffix array (tables suftab, lcptab, bwtab, suftab^{-1}) once and stored each table on a separate file. The construction was done by a program that is based on the suffix sorting algorithm of [3]. REPuter and unique-match construct the suffix tree in main memory (using $O(n)$ time) before they search for maximal repeated pairs and MUMs, respectively. All other programs use memory mapping to access the enhanced suffix array from the different files. Of

Table 1. Running times (in sec.) for computing maximal repeated pairs and super-maximal repeats. The column titled #reps gives the number of repeats of length $\geq \ell$.

ℓ	E. coli ($n = 4{,}639{,}221$)				yeast ($n = 12{,}156{,}300$)			
	maximal repeated pairs			supermax	maximal repeated pairs			supermax
	#reps	REPuter	esarep	#reps	#reps	REPuter	esarep	#reps
18	11884	9.68	0.83	1676 0.15	306931	27.81	5.85	12939 0.42
20	7800	9.65	0.77	890 0.15	175456	27.70	4.07	6372 0.41
23	5207	9.68	0.74	635 0.14	84116	27.62	2.91	4041 0.40
27	3570	9.66	0.72	493 0.14	41401	27.62	2.39	2804 0.40
30	2731	9.66	0.71	449 0.14	32200	27.64	2.27	2367 0.40
40	841	9.67	0.69	280 0.14	20768	27.69	2.12	1669 0.40
50	608	9.66	0.68	195 0.14	16210	27.64	2.05	1349 0.40

Table 2. Running times (in sec.) for computing branching tandem repeats. The column titled #reps gives the number of branching tandem repeats of length $\geq \ell$.

ℓ	E. coli ($n = 4{,}639{,}221$)				yeast ($n = 12{,}156{,}300$)			
	#reps	BFA	OBA	iOBA	#reps	BFA	OBA	iOBA
2	298853	1.32	7.83	1.60	932971	3.59	22.77	4.38
5	5996	1.32	4.73	1.47	32034	3.56	14.80	4.07
8	136	1.30	2.87	1.38	4107	3.54	8.72	3.87
11	20	0.84	1.20	1.13	1326	2.83	4.74	3.47
14	17	0.38	0.46	0.52	576	1.20	1.59	1.64

course, a file is mapped into main memory only if the table it stores is required for the particular algorithm.

Our method to construct the enhanced suffix array uses about 30% of the space required by *REPuter*, and 15% of the space required by *unique-match*. The construction time is about the same as the time to construct the suffix tree in *REPuter* and in *unique-match*.

7.2 Main Experimental Results

The results of applying the different programs to the different data sets are shown in Tables 1–3. The running times reported are for an Intel Pentium III computer with a 933 MHz CPU and 500 MB RAM running Linux. For a fair comparison, we report the running time of *REPuter* and of *unique-match* without suffix tree construction.

The running time of *supermax* is almost independent of the minimal length of the supermaximal repeats computed. Since the algorithm is so simple, the main part of the running time is the input and output. The strmat-package of [10] implements a more complicated algorithm than ours for the same task. For example, when applied to *E. coli*, it requires 19 sec. (without suffix tree construction) to compute all 944,546 supermaximal repeats of length at least 2. For this task *supermax* requires 1.3 sec due to the large size of the output.

Table 3. Running times (in sec.) and space consumption (in megabytes) for computing *MUMs*. The column titled #*MUMs* gives the number of *MUMs* of length $\geq \ell$. The time given for *unique-match* does not include suffix tree construction. *esamum* reads the enhanced suffix array from different files via memory mapping.

			unique-match		esamum	
genome pair	ℓ	#*MUMs*	time	space	time	space
Mycoplasma 2	20	10	1.85	65.26	0.03	7.99
Streptococcus 2	50	6613	6.76	196.24	0.29	29.71
E. coli 2	100	10817	17.67	472.47	0.46	62.59

The comparison of *esarep* and *REPuter* underline the advantages of the enhanced suffix array over the suffix tree. *esarep* used about halve of the space of *REPuter* and it is 5 to 10 times faster. The performance gain is due to the improved cache behavior achieved by the linear scanning of the tables suftab, lcptab, and bwtab.

The most surprising result of our experiments is the superior running time of the brute force algorithm to compute branching tandem repeats. *BFA* is always faster than the other two algorithms. *iOBA* is faster than *OBA* if $\ell \leq 11$, and slightly slower if $\ell = 14$. This is due to the fact that the additional access to the text, which is necessary to find the appropriate child interval $[i_a..j_a]$, outweighs the efficiency gain due to the reduced number of suffix checks.

We have also measured the running time of a program implementing the linear time algorithm of [8] to compute all tandem repeats. It is part of the strmat-package. For *E. coli* the program needs about 21 sec. (without suffix tree construction) and 490 MB main memory to deliver all tandem repeats of length at least 2. The linear time algorithm of [11] takes 4.7 sec. using 63 MB of main memory. For the same task our fastest program *BFA* (with an additional post processing step to compute non-branching tandem repeats from branching tandem repeats) requires 1.7 sec and 27 MB of main memory.

The running times and space results shown in Table 3 reveal that *esamum* is at least 20 times faster than *unique-match*, using only 15% of the space.

All in all, the experiments show that our programs based on enhanced suffix arrays define the state-of-the-art in computing different kinds of repeats and maximal matches.

References

1. M.I. Abouelhoda, E. Ohlebusch, and S. Kurtz. Optimal Exact String Matching Based on Suffix Arrays. In *Proceedings of the Ninth International Symposium on String Processing and Information Retrieval*. Springer-Verlag, Lecture Notes in Computer Science, 2002.
2. A. Apostolico. The Myriad Virtues of Subword Trees. In *Combinatorial Algorithms on Words*, Springer-Verlag, pages 85–96, 1985.
3. J. Bentley and R. Sedgewick. Fast Algorithms for Sorting and Searching Strings. In *Proceedings of the ACM-SIAM Symposium on Discrete Algorithms*, pages 360–369, 1997.

4. M. Burrows and D.J. Wheeler. A Block-Sorting Lossless Data Compression Algorithm. Research Report 124, Digital Systems Research Center, 1994.
5. A.L. Delcher, S. Kasif, R.D. Fleischmann, J. Peterson, O. White, and S.L. Salzberg. Alignment of Whole Genomes. *Nucleic Acids Res.*, 27:2369–2376, 1999.
6. J. A. Eisen, J. F. Heidelberg, O. White, and S.L. Salzberg. Evidence for Symmetric Chromosomal Inversions Around the Replication Origin in Bacteria. *Genome Biology*, 1(6):1–9, 2000.
7. D. Gusfield. *Algorithms on Strings, Trees, and Sequences*. Cambridge University Press, New York, 1997.
8. D. Gusfield and J. Stoye. Linear Time Algorithms for Finding and Representing all the Tandem Repeats in a String. Report CSE-98-4, Computer Science Division, University of California, Davis, 1998.
9. T. Kasai, G. Lee, H. Arimura, S. Arikawa, and K. Park. Linear-Time Longest-Common-Prefix Computation in Suffix Arrays and its Applications. In *Proceedings of the 12th Annual Symposium on Combinatorial Pattern Matching*, pages 181–192. Lecture Notes in Computer Science 2089, Springer-Verlag, 2001.
10. J. Knight, D. Gusfield, and J. Stoye. The Strmat Software-Package, 1998. http://www.cs.ucdavis.edu/ gusfield/strmat.tar.gz.
11. R. Kolpakov and G. Kucherov. Finding Maximal Repetitions in a Word in Linear Time. In *Symposium on Foundations of Computer Science*, pages 596–604. IEEE Computer Society, 1999.
12. S. Kurtz. Reducing the Space Requirement of Suffix Trees. *Software—Practice and Experience*, 29(13):1149–1171, 1999.
13. S. Kurtz, J.V. Choudhuri, E. Ohlebusch, C. Schleiermacher, J. Stoye, and R. Giegerich. REPuter: The Manifold Applications of Repeat Analysis on a Genomic Scale. *Nucleic Acids Res.*, 29(22):4633–4642, 2001.
14. E.S. Lander, L.M. Linton, B. Birren, C. Nusbaum, M.C. Zody, J. Baldwin, K. Devon, and K. Dewar, et. al. Initial Sequencing and Analysis of the Human Genome. *Nature*, 409:860–921, 2001.
15. N.J. Larsson and K. Sadakane. Faster Suffix Sorting. Technical Report LU-CS-TR:99-214, Dept. of Computer Science, Lund University, 1999.
16. U. Manber and E.W. Myers. Suffix Arrays: A New Method for On-Line String Searches. *SIAM Journal on Computing*, 22(5):935–948, 1993.
17. C. O'Keefe and E. Eichler. The Pathological Consequences and Evolutionary Implications of Recent Human Genomic Duplications. In *Comparative Genomics*, pages 29–46. Kluwer Press, 2000.
18. J. Stoye and D. Gusfield. Simple and Flexible Detection of Contiguous Repeats Using a Suffix Tree. *Theoretical Computer Science*, 270(1-2):843–856, 2002.

The Algorithmic of Gene Teams[*]

Anne Bergeron[1], Sylvie Corteel[2], and Mathieu Raffinot[3]

[1] LaCIM, Université du Québec à Montréal, Canada, and
Institut Gaspard-Monge, Université Marne-la-Vallée, France
anne@lacim.uqam.ca
[2] CNRS - Laboratoire PRiSM, Université de Versailles
45 Avenue des Etats-Unis, 78035 Versailles cedex, France
syl@prism.uvsq.fr
[3] CNRS - Laboratoire Génome et Informatique, Tour Evry 2, 523
Place des Terrasses de l'Agora, 91034 Evry, France
raffinot@genopole.cnrs.fr

Abstract. Comparative genomics is a growing field in computational biology, and one of its typical problem is the identification of sets of orthologous genes that have virtually the same function in several genomes. Many different bioinformatics approaches have been proposed to define these groups, often based on the detection of sets of genes that are "not too far" in all genomes. In this paper, we propose a unifying concept, called *gene teams*, which can be adapted to various notions of distance. We present two algorithms for identifying gene teams formed by n genes placed on m linear chromosomes. The first one runs in $O(m^2n^2)$ time, and follows a direct and simple approach. The second one is more tricky, but its running time is $O(mn\log^2(n))$. Both algorithms require linear space. We also discuss extensions to circular chromosomes that achieve the same complexity.

1 Introduction

In the last few years, research in genomics science rapidly evolved. More and more complete genomes are now available due to the development of semi-automatic sequencer machines. Many of these sequences – particularly prokaryotic ones – are well annotated: the position of their genes are known, and sometimes parts of their regulation or metabolic pathways.

A new computational challenge is to extract gene or protein knowledge from high level comparison of genomes. For example, the knowledge of sets of orthologous or paralogous genes on different genomes helps to infer putative functions from one genome to the other. Many researchers have explored this avenue, trying to identify groups or clusters of orthologous genes that have virtually the same function in several genomes [1,4,2,5,6,7,8,9,10,12]. These researches are often based on a simple, but biologically verified fact, that proteins that interact are usually coded by genes closely placed in the genomes of different species.

---

[*] Extended abstract.

R. Guigó and D. Gusfield (Eds.): WABI 2002, LNCS 2452, pp. 464–476, 2002.

With the knowledge of the positions of genes, it becomes possible to automate the identification of group of closely placed genes in several genomes.

From an algorithmic and combinatorial point of view, the formalizations of the concept of *closely placed genes* are still fragmentary, and sometimes confusing. The distance between genes is variously defined as differences between physical locations on a chromosome, distance from a specified target, or as a discrete count of intervening actual or predicted genes. The algorithms often lack the necessary grounds to prove their correctness, or assess their complexity. This paper contributes to a research movement of clarification of these notions. We aim to formalize, in the simplest and comprehensive ways, the concepts underlying the notion of distance-based clusters of genes. We can then use these concepts, and their formal properties, to design sound and efficient algorithms.

A first step in that direction has been done in [6,11] with the concept of *common intervals*. A common interval is a set of orthologous genes that appear consecutively, possibly in different orders, on a chromosome of two or more species. This concept covers the simplest cases of sets closely placed genes, but do not take in account the nature of the gaps between genes. In this paper, we extend this notion by relaxing the "consecutive" constraint: we allow genes to be separated by gaps that do not exceed a fixed threshold. We develop a simple formal setting for these concepts, and give two polynomial algorithms that detect maximal sets of closely placed genes, called *gene teams*, in m chromosomes.

The first algorithm is based on a direct approach and its time complexity is $O(n^2)$ time for two chromosomes, where n is the number of genes under study, and $O(m^2 n^2)$ time for m chromosomes. The second algorithm is more involved, but runs in $O(n log^2(n))$ time for two chromosomes, and $O(mn \log^2(n))$ time for m chromosomes, which is a strong improvement over the first one. Both algorithms are linear in space requirements.

This extended abstract is organized as follow. In Section 2, we formalize the concept of *gene teams* that unifies most of the current approaches, and discuss their basic properties. In Section 3 we present two algorithms that identify the gene teams of two chromosomes. The first one uses a straightforward approach, and the second one uses less elementary techniques to reduce the complexity of the search to $\mathcal{O}(n \log^2 n)$. Finally, in Section 4, we extend our algorithms to m chromosomes, and to circular chromosomes.

2   Gene Teams and Their Properties

Much of the following definitions refer to sets of *genes* and *chromosomes*. These are biological concepts whose definitions are outside the scope of this paper. However, we will assume some elementary formal properties relating genes and chromosomes: a chromosome is an ordering device for genes that belong to it, and a gene can belong to several chromosome. If a gene belongs to a chromosome, we assume that its position is known, and unique.

2.1   Definitions and Examples

Let Σ be a set of n genes that belong to a chromosome C, and let P_C be a function:

$$\Sigma \xrightarrow{P_C} R$$

that associates to each gene g in Σ a real number $P_C(g)$, called its *position*.

Function of this type are quite general, and cover a wide variety of applications. The position can be, as in [9,10,7], the physical location of an actual sequence of nucleotides on a chromosome. In more qualitative studies, such as [1,8], the positions are positive integers reflecting the relative ordering of genes in a given set. In other studies [2,4], positions are both negative and positive numbers computed in relation to a target sequence.

The function P_C induces a permutation on any subset S of Σ, ordering the genes of S from the gene of lowest position to the gene of highest position. We will denote the permutation corresponding to the whole set Σ by π_C. If g and g' are two genes in Σ, their *distance* $\Delta_C(g, g')$ in chromosome C is given by $|P_C(g') - P_C(g)|$.

For example, if $\Sigma = \{a, b, c, d, e\}$, consider the following chromosome X, in which genes not in Σ are identified by the star symbol:

$$X = c * * e\,d\,a * b.$$

Define $P_X(g)$ as the number of of genes appearing to the left of g. Then $\Delta_X(c, d) = 4$, $\pi_X = (c\,e\,d\,a\,b)$, and the permutation induced on the subset $\{a, c, e\}$ is $(c\,e\,a)$.

Definition 1. *Let S be a subset of Σ, and $(g_1 \ldots g_k)$ be the permutation induced on S on a given chromosome C. For $\delta > 0$, the set S is called a δ-chain of chromosome C if $\Delta_C(g_j, g_{j+1}) \leq \delta$, for $1 \leq j < k$.*

For example, if $\delta = 3$, then $\{a, c, e\}$ is a δ-chain of X, since each pair of consecutive elements in the permutation $(c\,e\,a)$ is distant by less than δ.

We will also refer to *maximal δ-chains* with respect to the partial order induced on the subsets by the inclusion relation. For example, with $\delta = 2$, the maximal δ-chains of X are $\{c\}$ and $\{a, b, d, e\}$. Note that singletons are always δ-chains, regardless of the value of δ.

Definition 2. *A subset S of Σ is a δ-set of chromosomes C and D if S is a δ-chain both in C and D. A δ-team of the chromosomes C and D is a maximal δ-set with respect to inclusion. A δ-team with only one element is called a* lonely gene.

Consider, for example, the two chromosomes:

$$X = c * * e\,d\,a * b$$
$$Y = a\,b * * * c * d\,e.$$

For $\delta = 3$ then $\{d, e\}$ and $\{c, d, e\}$ are δ-sets, but not $\{c, d\}$ since the latter is not a δ-chain in X. The δ-teams of X and Y, for values of δ from 1 to 4 are given in the following table.

δ	δ-teams	Lonely Genes
1	$\{d, e\}$	$\{a\}, \{b\}, \{c\}$
2	$\{a, b\}, \{d, e\}$	$\{c\}$
3	$\{a, b\}, \{c, d, e\}$	
4	$\{a, b, c, d, e\}$	

Our goal is to develop algorithms for the efficient identification of gene teams. The main pitfalls are illustrated in the next two examples.

The Intersection of δ-Chains Is not Always a δ-Set. A naive approach to construct δ-sets is to identify maximal δ-chains in each sequence, and intersect them. Although this works on some examples, the approach does not hold in the general case. For example, in the chromosomes:

$$X = a\,b\,c$$
$$Y = a\,c * * b,$$

with $\delta = 1$, the maximal δ-chain of X is $\{a, b, c\}$, and the maximal δ-chains of Y are $\{a, c\}$ and $\{b\}$. But $\{a, c\}$ is not a δ-team.

Gene Teams Cannot Be Grown from Smaller δ-Sets. A typical approach for constructing maximal objects is to start with initial objects that have the desired property, and cluster them with a suitable operation. For gene teams, the singletons are perfect initial objects, but there is no obvious operation that, applied to two small δ-sets, produces a bigger δ-set. Consider the following chromosomes:

$$X = a\,b\,c\,d$$
$$Y = c\,a\,d\,b\,.$$

For $\delta = 1$, the only δ-sets are the sets $\{a\}$, $\{b\}$, $\{c\}$ and $\{d\}$, and the set $\{a, b, c, d\}$. In general, it is possible to construct pairs of chromosomes with an arbitrary number of genes, such that the only δ-sets are the singletons and the whole set.

Instead of growing teams from smaller δ-sets, we will extract them from larger sets that contain only teams. This leads to the following definition:

Definition 3. *A* league *of chromosomes C and D is a union of teams of the chromosomes C and D.*

2.2   Properties of δ-Sets and Teams

The first crucial property of δ-teams is that they form a partition of the set of genes Σ. It is a consequence of the following lemma:

Lemma 1. *If S and T are two δ-chains of chromosome C, and $S \cap T \neq \emptyset$, then $S \cup T$ is also a δ-chain.*

Proof: Consider the permutation induced on the set $S \cup T$, and let g and g' be two consecutive elements in the permutation. If g and g' both belong to S (or to T), then they are consecutive in the permutation induced by S (or by T), and $\Delta(g, g') \leq \delta$. If g is in S but not in T, and g' is in T but not in S, then either g is between two consecutive elements of T, or g' is between two consecutive elements of S. Otherwise, the two sets S and T would have an empty intersection. If g is between two consecutive elements of T, for example, then one of them is g', implying $\Delta(g, g') \leq \delta$. ∎

We now have easily:

Proposition 1. *For a given set of genes Σ, the δ-teams of chromosomes C and D form a partition of the set Σ.*

Proposition 1 has the following corollary:

Corollary 1. *If a set S is both a league, and a δ-set, of chromosomes C and D, then S is a δ-team.*

The algorithms described in the next section work on leagues, splitting them while ensuring that a league is split in smaller leagues. The process stops when each league is a δ-set. Corollary 1 provides a simple proof that such an algorithm correctly identifies the teams. The next proposition gives the "initial" leagues for the first algorithm.

Proposition 2. *Any maximal δ-chain of C or D is a league.*

3   Algorithms to Find Gene Teams

It is quite straightforward to develop polynomial algorithms that find gene teams in two chromosomes. In the following section, we present a first $\mathcal{O}(n^2)$ algorithm based on the successive refinement of the partitions induced by the initial maximal δ-chains. However, since the ultimate goal is to be able to upgrade the definitions and algorithms to more than two chromosomes, such a threshold is too high. In Section 3.2, we take a radically different approach to develop an $\mathcal{O}(n \log^2 n)$ algorithm.

3.1   A First Polynomial Algorithm

Assume that we are given two permutations on Σ, π_C and π_D, each already partitioned into maximal δ-chains of chromosomes C and D:

$$\pi_C = (c_1 \ldots c_{k_1})(c_{k_1+1} \ldots c_{k_2}) \ldots (c_{k_s+1} \ldots c_n)$$
$$\pi_D = (d_1 \ldots d_{l_1})(d_{l_1+1} \ldots d_{l_2}) \ldots (d_{l_t+1} \ldots d_n).$$

Let $(c_i \ldots c_j)$ be one of the class of the partition of π_C. By Proposition 2 $(c_i \ldots c_j)$ is a league. Our goal is to split this class in m subclasses $S_1, \ldots, S_v$

such that: a) each subclass is a league; b) each subclass a δ-chain in C; and c) each subclass is contained in one of the class of π_D.

Initially, we create one subclass $S_1 = (c_i)$. We then successively read the genes c_k, for k in $[i + 1..j]$. When reading c_k, assume that the subclasses S_1 to S_u have already been created, that they are all δ-chains, and each of them is contained in one of the class of π_D, then either:

1. The gene c_k can be added as the last element of – exactly – one of the subclasses already created, or
2. The gene c_k is the beginning of a new subclass $S_{u+1} = (c_k)$.

The algorithm repeats this process of splitting classes, alternatively on classes of π_C and π_D, until the classes are equal in the two permutations.

The gene c_k can be added to an already created subclass whose last element is c if both c and c_k belongs to the same class in π_D, and if $\Delta_C(c, c_k) \leq \delta$. Note that only one element c can satisfy these two conditions, since if there were another, it would be in the same subclass, and a subclass has only one last element.

To show that the algorithm is correct, first remark that the way subclasses are constructed ensures that they remain δ-chains of their respective chromosomes, and the stopping condition will make each class a δ-set. In order that these δ-sets be teams, we must show that each subclass is a league. By Proposition 2, each class of two initial partitions is a league, and if T is a team in $(c_i \ldots c_j)$, each successive member of the team T that will be considered will be added to the same subclass.

The fact that the algorithm runs in $\mathcal{O}(n^2)$ comes from the observation that, except for the last one, each loop is guaranteed to split at least one class. Indeed, if the classes are not equal in the two permutations, there is at least one class of either π_C or π_D that is not properly contained in a single class of the other permutation. Since there is a maximal number of n class, the algorithm makes at most n iterations. The cost of each iteration is $\mathcal{O}(n)$ since the number of subclasses that need to be tested in step 1 is at most δ – for a fixed c_k, $\delta_C(c_k, x) \leq \delta$ can be true for at most δ elements preceding c_k.

3.2   A Divide-and-Conquer Algorithm

The following algorithm to identify teams is a divide-and-conquer algorithm that work by extracting small leagues from larger ones. Its basic principle is described in the following paragraph.

Assume that S is a league of chromosomes C and D, and that the genes of S are respectively ordered in C and D as:

$$(c_1 \ldots c_n), \text{ and } (d_1 \ldots d_n).$$

By Proposition 1, if S is a δ-set, then S is a δ-team. If S is not a δ-set, there are at least two consecutive elements, say c_i and c_{i+1}, that are distant by more than δ. Therefore, both $(c_1 \ldots c_i)$ and $(c_{i+1} \ldots c_n)$ are leagues, splitting the initial problem in two sub-problems. The following two lemmas explain how to split a problem efficiently.

Lemma 2. *If S is a league, but not a team, of chromosomes C and D, then there exists a sub-league of S with at most $|S|/2$ genes.*

Proof: Let $|S| = n$, if all sub-leagues of S have more than $n/2$ genes, it follows that each team included in S have more than $\lfloor n/2 \rfloor$ genes, and the intersection of two such team cannot be empty. ∎

The above lemma implies that if S is a league, but not a team, and if the sequences $(c_1 \ldots c_n)$ and $(d_1 \ldots d_n)$ are the corresponding permutations in chromosomes C and D, then there exist a value $p \leq n/2$ such that at least one of the following sequences is a league:

$$(c_1 \ldots c_p),$$
$$(c_{n-p+1} \ldots c_n),$$
$$(d_1 \ldots d_p),$$
$$(d_{n-p+1} \ldots d_n).$$

For example, if

$$X = a\ b\ c * d\ e\ f\ g$$
$$Y = c\ a\ e\ d\ b\ g\ f$$

and $\delta = 1$, then $(a\ b\ c)$ is easily identified as a league, since the distance between c and d is greater than 1 in chromosome X. The next problem is to extract the corresponding permutation in chromosome Y. This is taken care by the following lemma that describes the behavior of the function "Extract$((c_1 \ldots c_p), D)$":

Lemma 3. *Assume that π_C and π_D, and their inverse, are known. If $(c_1 \ldots c_p)$ is a set of genes ordered in increasing position in chromosome C, then the corresponding permutation $(d'_1 \ldots d'_p)$ on chromosome D can be obtained in time $\mathcal{O}(p \log p)$.*

Proof: Given $(c_1 \ldots c_p)$, we first construct the array $A = (\pi_D^{-1}(c_1), \ldots, \pi_D^{-1}(c_p))$. Sorting A requires $\mathcal{O}(p \log p)$ operations, yielding the array A'. The sequence $(d'_1 \ldots d'_p)$ is given by $(\pi_D(A'_1) \ldots \pi_D(A'_p))$. ∎

The last operation needed to split a league is to construct the ordered complement of an ordered league. For example, for the league $\pi_Y = (c\ a\ e\ d\ b\ g\ f)$, the complement of the league $(c\ a\ b)$ is the league $(e\ d\ g\ f)$.

More formally, if $(d'_1 \ldots d'_p)$ is a subsequence of $(d_1 \ldots d_n)$, we will denote by

$$(d_1 \ldots d_n) \setminus (d'_1 \ldots d'_p)$$

the subsequence of $(d_1 \ldots d_n)$ obtained by deleting the elements of $(d'_1 \ldots d'_p)$.

In our particular context, this operation can be done in $\mathcal{O}(p)$ steps. Indeed, once a problem is split in two sub-problems, there is no need to backtrack in the

```
FastFindTeams((c₁ ... cₙ), (d₁ ... dₙ))
1.      SubLeagueFound ← False
2.      p ← 1
3.      While (NOT SubLeagueFound) AND p ≤ ⌊n/2⌋ Do
4.          If Δ_C(c_p, c_{p+1}) > δ OR Δ_C(c_{n-p}, c_{n-p+1}) > δ OR
5.              Δ_C(d_p, d_{p+1}) > δ OR Δ_C(d_{n-p}, d_{n-p+1}) > δ        Then
6.                  SubLeagueFound ← True
7.              Else p ← p + 1
8.          End of if
9.      End of while
10.     If SubLeagueFound Then
11.         If Δ_C(c_p, c_{p+1}) > δ Then
12.             (d'₁ ... d'_p) ← Extract((c₁ ... c_p), D))
13.             FastFindTeams((c₁ ... c_p), (d'₁ ... d'_p))
14.             FastFindTeams((c_{p+1} ... cₙ), (d₁ ... dₙ) \ (d'₁ ... d'_p))
15.         Else If ...
16.                     /* The three other cases are similar */
17.         End of if
18.     Else  (c₁ ... cₙ) is a Team
19.     End of if
```

Fig. 1. Fast recursive algorithm for gene teams identification.

former problems. Therefore, at any point in the algorithm, each gene belongs to exactly two ordered leagues, one in each chromosome. If the gene data structure contains pointers to the previous and the following gene – if any – in both leagues, the structure can be updated in constant time as soon as an extracted gene is identified. Since p genes are extracted, the operation can be done in $\mathcal{O}(p)$ steps.

Fig. 1 contains the formal description of the algorithm FastFindTeams. The three cases that are not shown correspond to the tests $\Delta_C(c_{n-p}, c_{n-p+1}) > \delta$, $\Delta_C(d_p, d_{p+1}) > \delta$ and $\Delta_C(d_{n-p}, d_{n-p+1}) > \delta$, and are duplications of the first case, up to indices.

Theorem 1. *On input π_C and π_D, algorithm FastFindTeams correctly identifies the δ-teams of chromosomes C and D.*

The space needed to execute Algorithm FastFindTeams is easily seen to be $\mathcal{O}(n)$ since it needs the fours arrays containing π_C, π_D, π_C^{-1}, π_D^{-1}, and the n genes, each with four pointers coding implicitly for the ordered leagues.

3.3   Time Complexity of Algorithm FastFindTeams

In the last section, we saw that algorithm FastFindTeams splits a problem of size n in two similar problems of size p and $n - p$, with $p \leq n/2$. The number of operations needed to split the problem is $\mathcal{O}(p \log p)$, but the value of p is not fixed from one iteration to the other. In order to keep the formalism manageable, we will "... *neglect certain technical details when we state and solve recurrences. A good example of a detail that is glossed over is the assumption of integer arguments to functions.*", [3] p. 53.

Assume that the number of operation needed to split the problem is bounded by $\alpha p \log p$, and let $F(n)$ denote the number of operations needed to solve a problem of size n. Then $F(n)$ is bounded by the function $T(n)$ described by the following equation:

$$T(n) = \max_{1 \le p \le \lfloor n/2 \rfloor} \{\alpha p \log p + T(p) + T(n - p)\}. \tag{1}$$

with $T(1) = 1$.

Surprisingly, the worst case scenario of the above equation is when the input is always split in half. Indeed, we will show that $T(n)$ is equal to the function:

$$T_2(n) = (\alpha n/2) \log(n/2) + 2T_2(n/2), \tag{2}$$

with $T_2(1) = 1$. One direction is easy since $T(n) \ge T_2(n)$.

In order to show the converse, we first obtain a closed form for $T_2(n)$, which can be verified by direct substitution.

Lemma 4. $T_2(n) = n - (\alpha n/4) \log n + (\alpha n/4) \log^2 n$.

We use this relation to show the following remarkable property of $T_2(n)$. It says that when a problem is split in two, the more unequal the parts, the better.

Proposition 3. If $x < y$ then $T_2(x) + T_2(y) + \alpha x \log x < T_2(x + y)$.

Sketch of Proof: We will show that $[T_2(x + y) - T_2(x) - T_2(y)]/(\alpha x) > \log x$. Using the closed form for T_2, we have:

$$\begin{aligned}
&[T_2(x + y) - T_2(x) - T_2(y)]/(\alpha x) \\
&= (1/4)[\log^2(x + y) - \log^2 x] + (y/4x)[\log^2(x + y) - \log^2 y] \\
&\quad -(1/4)[\log(x + y) - \log x] - (y/4x)[\log(x + y) - \log y].
\end{aligned}$$

Consider the variable $z = y/x$. Define $H(z) = \log(1 + z) + z \log(1 + 1/z)$. Its value for $z = 1$ is 2, and its derivative is $\log(1 + 1/z)$, implying that the $H(z)$ is strictly increasing. Substituting y/x by z, the last expression becomes:

$$\begin{aligned}
&(H(z)/4)[2 \log x + \log(1 + z) - 1] + (1/4)z \log z \log(1 + 1/z) \\
&\ge (H(z)/2) \log x \\
&> \log x, \text{ since } H(z) > 2, \text{ when } z > 1.
\end{aligned}$$

∎

Using Proposition 3, we get:

Proposition 4. $T(n) \leq T_2(n)$.

Proof: Suppose that $T(i) \leq T_2(i)$ for all $i < n$, then

$$
\begin{aligned}
T(n) &= \max_{1 \leq p \leq \lfloor n/2 \rfloor} \{\alpha p \log p + T(p) + T(n - p)\} \\
&\leq \max_{1 \leq p \leq \lfloor n/2 \rfloor} \{\alpha p \log p + T_2(p) + T_2(n - p)\} \\
&\leq \max_{1 \leq p \leq \lfloor n/2 \rfloor} \{T_2(p + n - p)\} \\
&\leq \max_{1 \leq p \leq \lfloor n/2 \rfloor} \{T_2(n)\} \\
&= T_2(n).
\end{aligned}
$$

■

We thus have:

Theorem 2. *The time complexity of algorithm FastFindTeams is* $\mathcal{O}(n \log^2 n)$.

Theorem 2 is truly a worst case behavior. It is easy to construct examples in which its behavior will be linear, taking, for example, an input in which one chromosome has only singletons as maximal δ-chains.

4   Extensions

4.1   Multiple Chromosomes

The most natural extension of the definition of δ-teams to a set of $\{C_1, \ldots, C_m\}$ of chromosomes, is to define a δ-set S as a δ-chain in each chromosome C_1 to C_m, and consider maximal δ-sets as in Definition 2. For example, with $\delta = 2$, the only δ-team of chromosomes:

$$
\begin{aligned}
X &= c * * e \, d \, a * b \\
Y &= a \, b * * c * d \, e \\
Z &= b \, a \, e * * c * d,
\end{aligned}
$$

that is not a lonely gene is the set $\{a, b\}$.

All the definitions and results of Section 2 apply directly to this new context, replacing C and D by the m chromosomes. Each iteration of the first algorithm, who worked by successive refinements of partitions on the two chromosomes, must consider the m chromosomes, and assignment of a gene to a class requires $m - 1$ operations, yielding a time complexity of $\mathcal{O}(m^2 n^2)$.

The second algorithm upgrades more elegantly. The core of the algorithm is still finding and extracting small leagues: identifying the league to extract is done in $\mathcal{O}(mp)$ operations, where $p \leq n/2$, and the problem is split in two subproblems by extracting $m - 1$ sub-leagues of size p. The analysis of Section 3.3 yields directly an $\mathcal{O}(mn \log^2 n)$ time bound for this algorithm, since the parameter α in Equation 1 was arbitrary.

4.2   Extension to Circular Chromosomes

In the case of circular chromosomes, we first modify slightly the assumptions and definitions. The positions of genes are given here as values on a finite interval:

$$\Sigma \xrightarrow{P_C} [0..L],$$

in which position L is equivalent to position 0. The distance between two genes g and g' such that $P_C(g) < P_C(g')$ is given by:

$$\Delta_C(g, g') = \min \begin{cases} P_C(g') - P_C(g) \\ P_C(g) + L - P_C(g'). \end{cases}$$

The permutation $\pi_C = (g_1 \ldots g_n)$ is still well defined for circular chromosomes, but so are the permutations, for $1 < m \le n$:

$$\pi_C^{(m)} = (g_m \ldots g_n g_1 \ldots g_{m-1}).$$

A δ-chain in a circular chromosome is any δ-chain of at least one of these permutations. A *circular* δ-chain is a δ-chain $(g_1 \ldots g_k)$ such that $\Delta_C(g_k, g_1) \le \delta$: it goes all around the chromosome. All other definitions of Section 2 apply without modifications.

Adapting algorithm FastFindTeams to circular chromosomes requires a special case for the treatment of circular δ-chains. Indeed, in Section 3.2, the beginning and end of a chromosome provided obvious starting places to detect leagues. In the case of circular chromosomes, assume that S is a league of chromosomes C and D, and that the genes of S are respectively ordered in C and D, from arbitrary starting points, as:

$$(c_1 \ldots c_n) \text{ and } (d_1 \ldots d_n).$$

If none of these sequence is a circular δ-chain, then there is a gap of length greater than δ on each chromosome, and the problem is reduced to a problem of linear chromosomes. If both are circular δ-chains, then S is a δ-team. Thus, the only special case is when one is a circular δ-chain, and the other, say $(c_1 \ldots c_n)$ has a gap greater than δ between two consecutive elements, or between the last one and the first one. Without loss of generality, we can assume that the gap is between c_n and c_1. Then, if S is not a team, there exists a value $p \le n/2$ such that one of the following sequence is a league:

$$(c_1 \ldots c_p)$$
$$(c_{n-p+1} \ldots c_n.)$$

The extraction procedure is similar to the one in Section 3.2, but both extracted leagues can again be circular δ-chains.

Circularity can be detected in $\mathcal{O}(p)$ steps, since the property is destroyed if and only if an extracted gene creates a gap of length greater than δ between its two neighbors.

4.3   Teams with a Designated Member

A particular case of the team problem is to find, for various values of δ, all δ-teams that contain a designated gene g. Clearly, the output of algorithm FastFindTeams can be filtered for the designated gene, but it is possible to do better. In lines 13 and 14 of the algorithm of Fig. 1, the original problem is split in two subproblems. Consider the first case, in which the sub-league $(c_1 \ldots c_p)$ is identified:

1. If gene g belongs to $(c_1 \ldots c_p)$, then the second recursive call is unnecessary.

2. If gene g does not belong to $(c_1 \ldots c_p)$, then the extraction of (d'_1, d'_p), and the first recursive call, are not necessary.

These observation lead to a simpler recurrence for the time complexity of this problem, since roughly half of the work can be skipped at each iteration. With arguments similar to those in Section 3.3, we get that the number of operations is bounded by a function of the form:

$$T(n) = \alpha(n/2)log(n/2) + T(n/2),$$

where $T(1) = 1$, and whose solution is: $T(n) = \alpha n \log n - 2\alpha n + 2\alpha + 1$.

5   Conclusions and Perspectives

We defined the unifying notion of *gene teams* and we constructed two distinct identification algorithms for n genes belonging to two or more chromosomes, the faster one achieving $O(mn \log^2(n))$ time for m linear or circular chromosomes. Both algorithms require only linear space. The implementation of the faster algorithm required some care about the data structures in order to achieve the announced complexity. Indeed, the asymptotic behavior of Equation 1 relies heavily on the fact that the work needed to split a problem in two parts depends only on the size of the smaller part. Implementations for all variants are available for arbitrary values of m.

We intend to extend our work in two directions that will further clarify and simplify the concepts and algorithms used in comparative genomics. The first is to relax some aspect of the definition of gene teams. For large values of m, the constraint that a set S be a δ-chain in all m chromosomes might be too strong. Sets that are δ-chains in a quorum of the m chromosomes could have biological significance as well. We also assumed, in this paper, that each gene in the set Σ had a unique position in each chromosome. Biological reality can be more complex. Genes can go missing in a certain species – their function being taken over by others, and genes can have duplicates.

In a second phase, we plan to extend our notions and algorithms to combine distance with other relations between genes. For example, interactions between proteins are often studied through metabolic or regulatory pathways, and these graphs impose further constraints on teams.

Acknowledgement

We would like to thank Marie-France Sagot for interesting discussions, Laure Vescovo and Luc Nicolas for a careful reading, and Myles Tierney for suggestions on the terminology. A special thanks goes to Laurent Labarre for his bibliography work.

References

1. Arvind K. Bansal *An automated comparative analysis of 17 complete microbial genomes*. Bioinformatics, Vol 15, No. 11, p 900-908, (1999).
2. Tristan Colombo, Alain Guénoche and Yves Quentin, *Inférence fonctionnelle par l'analyse du contexte génétique: une application aux transporteurs ABC*. Oral presentation, Journées ALBIO, Montpellier, March 2002.
3. Thomas Cormen, Charles Leiserson and Ronald Rivest. Introduction to Algorithms. The MIT Press, Cambridge Mass., Eighth Printing, 1028 pages (1992).
4. Thomas Dandekar, Berend Snel, Martijn Huynen M and Peer Bork. *Conservation of gene order: a fingerprint of proteins that physically interact*. Trends Biochem Sci, Vol 23, No. 9, p 324-328, (1998).
5. Wataru Fujibuchi, Hiroyuki Ogata, Hideo Matsuda and Minoru Kanahisa *Automatic detection of conserved gene clusters in multiple genomes by graph comparison and P-quasi grouping*. Nucleic Acids Research, Vol 28, No. 20, p 4029-4036, (2000).
6. Steffen Heber and Jens Stoye. *Finding all common intervals of k permutations*, CPM 01 Proceedings, LNCS 2089, Springer-Verlag, Berlin: 207-218 (2001).
7. Martijn Huynen, Berend Snel, Warren Lathe III, and Peer Bork *Predicting Protein Function by Genomic Context: Quantitative Evaluation and Qualitative Inferences*. Genome Research, Vol 10, p 1024-1210, (2000).
8. Anne Morgat. *Synténies bactériennes*. Oral presentation, Entretiens Jacques Cartier on Comparative Genomics, Lyon, December 2001.
9. Hiroyuki Ogata, Wataru Fujibuchi, Susumu Goto and Minoru Kanehisa *A heuristic graph comparison algorithm and its application to detect functionally related enzyme clusters*. Nucleic Acids Research, Vol 28, No. 20, p 4021-4028, (2000).
10. Ross Overbeek, Michael Fonstein, Mark D'Souza, Gordon D. Push and Natalia Maltsev *The use of gene clusters to infer functional coupling*. Proc. Natl. Acad. Sci. USA, Vol 96, p 2896-2901, (1999).
11. T. Uno and M. Yagiura. *Fast algorithms to enumerate all common intervals of two permutations*. Algorithmica 26(2), 290-309, (2000).
12. Roman L. Tatusov, Michael Y. Galperin, Darren A. Natale and Eugene V. Koonin *The COG database: a tool for genome-sacle analysis of protein functions and evolution*. Nucleic Acids Research, Vol 28, No. 1, p 33-36, (2000).

Combinatorial Use of Short Probes
for Differential Gene Expression Profiling

Liling L. Warren[1,2] and Ben Hui Liu[1,2]

[1] Bioinformatics Program, North Carolina State University
Raleigh, NC 27695, USA
[2] Bio-informatics Group Inc., Cary, NC, 27511, USA

Abstract. Motivation: Gene chips or microarrays have made it possible to perform large-scale gene-expression experiments at the whole-genome level. To reduce the cost and identify genes that are differentially expressed between samples from hybridization signal intensity, we are proposing a suite of methods to use a relatively small number of short oligo probes for whole-genome or tissue-specific gene-expression studies. Results: We present methods to choose a set of short oligos to design a genome or tissue-specific biochip and then to solve a set of equations for gene-expression levels to determine genes that are differentially expressed between samples. The methods have been tested to define a set of 4,000 8mers as probes to identify genes that have fold changes for more than 6,000 identified yeast ORFs. These methods can also be expanded to design genome-specific or tissue-specific biochips for other organisms with full gene-sequence information.

1 Introduction

DNA microarrays are powerful tools to monitor expression levels for many genes in parallel. Probes deposited on the chips can either be cDNAs or oligos. Targets are often mRNA samples from various sources such as disease vs. normal, different time points from the same source, or different cancer cell lines. Upon hybridization, expression levels for many genes can be measured and these values are then analyzed for studies such as disease gene identification, drug target finding, cancer type classification [1,2,3,4].

Depending on the array format, a single gene-specific probe or multiple tiling probes are used to measure the expression level for a particular gene. With the Affymetrix GeneChips^{TM}, each gene is identified by a series of ~20 nucleotide tiling probes [5,6,7,8]. With the Stanford microarrays, cDNAs are serving as gene-specific probes to measure gene expression level [9]. With the Agilent arrays, gene-specific oligos of 60–80 bases long are used as probes for expression study [10,11,12].

In our study, the goal is to use a relatively small set of short oligo probes to identify genes that are differentially expressed between samples. The sets of genes can be all expressed genes in a genome or those in a specific tissue. In order to determine such a set of short probes, sequencing information for all genes either

R. Guigó and D. Gusfield (Eds.): WABI 2002, LNCS 2452, pp. 477–490, 2002.
© Springer-Verlag Berlin Heidelberg 2002

in a certain genome or in a specific tissue must be known. Various genome-sequencing projects have provided rich resources for such information. As more genomes being sequenced and more genes being annotated, the application areas of our methods will expand quickly.

2   Methods

When using a set of short oligos as probes in microarray experiments, one probe can hybridize to one or more genes having the complementary probe sequence. For each probe, the fluorescent signal we observe after hybridization reactions is a sum of all genes that have hybridized to this particular probe. Let g_j be the number of transcripts of gene j in a sample and x_{ij} be the number of occurrences of probe i in gene j; then we have

$$z_i = \sum_{j=1}^{n} g_j x_{ij} \qquad (1)$$

where z_i is the expected number of copies of the ith oligo, which corresponds to the total number of copies of all genes that hybridize to this probe. Once a set of probes are chosen, we would know x_{ij}, and when the experiment is done, we would observe y_i, the probe hybridization signal intensity. Due to variations in hybridization efficiency among probes of the same length, we need to adjust y_i in order to get z_i. Let h_i denote the hybridization efficiency for probe i; we can adjust the observed probe signal intensity based on $z_i = y_i/h_i$. In the following calculations, we will use the expected oligo copy numbers. With known x_{ij} and z_i, the goal here is to estimate g_j, the number of transcripts for gene j. For the whole microarray experiment, define X as the abundance distribution of m probes in n genes,

$$X = \begin{bmatrix} x_{11} & x_{12} & \cdots & x_{1n} \\ x_{21} & x_{22} & \cdots & x_{2n} \\ \cdots & \cdots & x_{ij} & \cdots \\ x_{m1} & x_{m2} & \cdots & x_{mn} \end{bmatrix} \qquad (2)$$

Z as the expected fluorescent signal intensity values for m probes, and G as the transcript numbers of n genes,

$$G = \begin{bmatrix} g_1 \\ \cdots \\ g_j \\ \cdots \\ g_n \end{bmatrix} \quad \text{and} \quad Z = \begin{bmatrix} z_1 \\ \cdots \\ z_i \\ \cdots \\ z_m \end{bmatrix} \qquad (3)$$

then the task is to estimate vector G based on $Z = XG$. When matrix X is of full rank, we can solve the set of equations with a conventional least-square method,

$$\hat{G} = (X'X)^{-1}X'Z \qquad (4)$$

Since the purpose here is to use a smaller set of probes to measure quantitative expression levels for a larger set of genes, we have more parameters (n) to be estimated than the total number of equations (m). The X matrix is thus not of full rank and consequently, the set of equations cannot be solved by the least-square method. Instead, we are proposing an EM-like algorithm to obtain estimates for the gene transcript numbers as follows,

Step 0 (Initialization): Initialize the vector G to the overall average of fluorescence signal intensity values,

$$\hat{g}'_j = \frac{\sum_i z_i}{\sum \sum_{ij} x_{ij}} \tag{5}$$

Step 1 (Expectation): Use $\hat{g}'_j$ to estimate the number of copies of each of the m oligos:

$$\begin{cases} w_j = \dfrac{\hat{g}'_j}{\sum \hat{g}'_j / n} \\[2ex] c_{ij} = \dfrac{z_i}{w_j x_{ij}} \end{cases} \tag{6}$$

Step 2 (Maximization): Obtain the new estimate of the gene transcript copy numbers:

$$\hat{g}'_j = \frac{\sum_{i=1}^{m} c_{ij} x_{ij}}{\sum_{i=1}^{m} x_{ij}} \tag{7}$$

Step 3 (Iteration): Repeat Steps 1 & 2 until the estimates converge. The convergence is reached for $|\hat{g}^t_j - \hat{g}^{t-1}_j| < \tau$, where τ is set to be 0.0001.

Upon convergence, we may or may not obtain the correct answers. It depends on how many probes we use, how many genes we have, and the structure of the X matrix. When the right set of oligos are identified, we can identify genes that are differentially expressed in different samples. In the next section, we describe the results of this approach on a simulated data set and on the yeast genome to choose a set of short oligomers (8mers) as probes for differential gene-expression profiling studies.

3   Implementation

We first implemented our method in a simulation-based study. Then we applied the method to the yeast (*Saccharomyces cerevisiae*) genome to see if we could identify a relatively small set of probes to determine genes that are differentially expressed between two samples for all identified yeast ORFs. The key idea here is that the combination of different short oligos for each gene is unique, even though each oligo can simultaneously hybridize to more than one gene. Instead of identifying a gene with one or multiple probe, our methods take advantage of the combinatorial use of short probes to obtain gene expression profiles.

In the simulation-based study, we need to simulate the X matrix (the oligo distribution among genes) and the G vector (the gene transcript numbers). To simulate X, we have chosen the number of genes as 1,000 and used a number of oligos from 100 to 1,000 in increments of 100. We assign each oligo randomly to three genes. To simulate G, we have assumed that g_j follows a χ^2 distribution with $df = 1$. This choice was suggested by real data from various microarray experiments [13,14,15,16]. Having simulated X and G, the product of X and G gives us Z, the expected signal intensity value for all probes. Then based on Z and X, we apply our algorithm to obtain the estimator for G, $\hat{G}$. If $\hat{G}$ and G are very close, then we have found a good solution. To determine the accuracy of our estimates, we used the correlation coefficient between G and $\hat{G}$ and the mean square error (MSE) between G and $\hat{G}$ as our two judging criteria. Correlation measures the overall similarity between the estimated and true values, while MSE indicates how far the estimates are from the true values. MSE is the expected value of the square of the difference between an estimator and the true value of a parameter. If the estimator is unbiased then MSE is simply the variance of the estimator. It is calculated as:

$$\sum_{j=1}^{n}(g_j - \hat{g}_j)^2 \tag{8}$$

A very high correlation and a small MSE indicate that the estimates are very close to the true values.

To implement this approach for the yeast genome, we used 6,329 yeast coding ORF sequences downloaded from the *Saccharomyces* Genome Database (SGD) [17]. We first looked at all possible 65,536 8mers' distribution in the genome. We then sorted 8mer frequency counts in ascending order. Ordered 8mers were grouped into sections to form a series of X matrices. To generate the Y vector (the fluorescence signal intensity values for all probes), we used both simulated gene copy numbers and real signal intensity values from yeast microarray experiments [13,14,15,16]. For each set of 8mers or the corresponding X matrix, we applied our algorithm to test whether all yeast ORFs transcript numbers can be accurately estimated. We then combined the top four sets of 8mers giving the best estimates into one large set. Random samples of 8mers were further drawn from the combined set to create a number of X matrices for estimating gene copy numbers. The set of 8mers that produced most accurate results could be used as probes for yeast gene-expression microarray experiments. As a final step, the chosen set of 8mers were checked and modified to ensure that all genes could be estimated accurately. Figure 1 shows the sequential steps involved in this process.

Once the set of short probes are chosen, we then randomly select a portion of the genes to be differentially expressed between two samples. Using our methods, we want to see if genes we have chosen to be differentially expressed can be identified. We are also interested in studying such issues as sensitivity and specificity.

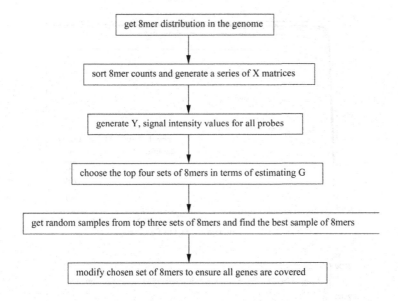

Fig. 1. Flow chart of the steps involved in choosing a set of 8mers for yeast genome microarray experiments.

4   Results

4.1   Simulation Results

For the simulated data set, we tried to estimate transcript numbers for 1,000 genes with different numbers of oligos, ranging from 100 to 1,000. The results are shown in Figure 2.

From Figure 2a, we can see that, as the number of iterations increases, the correlation between estimates and true values increase, indicating that the estimates are getting closer to the true values. Overall, convergence happens quickly. After 100 iterations, estimates for all six probes sets have converged. Upon convergence, when only 100 or 200 oligo probes are used, the estimates are not good, while when 300 and more probes are used, the estimates are quite accurate. When 400 probes are used, the estimates are close to 100% correlated to the true values. Figure 2b shows similar results. Instead of using correlation, MSE between estimates and true values are used to see how accurate the estimates are after some number of iterations for all six probe sets. (A smaller MSE indicates a more accurate estimate.) When 400 probes are used, MSE is about 1,000—that is, on average, for 1,000 genes, the difference between estimate and true value is about one copy of gene transcript. Combining both correlation and MSE results, we can see for this set of 1,000 genes, a set of 400 probes has sufficient accuracy to be used to measure gene transcript numbers. In Figure 2c, both correlation and MSE are plotted against the number of probes. As more probes are used, curves for both correlation and MSE show that the estimates are converging towards true values.

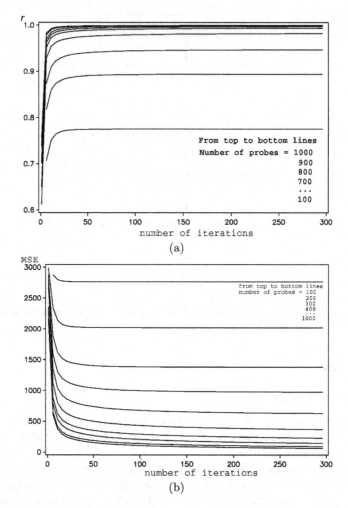

Fig. 2. Simulated data: (a) correlations between estimates and true values against the number of iterations for six probe sets with the number of probes ranging from 100 to 1,000; (b) MSE between estimates and true values against the number of iterations for six probe sets with the number of probes ranging from 100 to 1,000; and (c) correlation and MSE against the number of probes.

4.2   Yeast Genome Results

In order to pick a relatively small set of 8mers as probes to measure all yeast ORFs expression levels, we first looked at all possible 4^8 (65,536) 8mers' distribution in more than 6,000 yeast coding ORFs. Counts of 8mer occurrence in the yeast genome were sorted and plotted in a histogram, as shown in Figure 3(a).

Due to reasons such as codon bias and GC content variation, different 8mers are not evenly distributed in the genome. Instead, they follow a Poisson distribution [18]. From Figure 4, we can see that most of the 8mers have a relatively

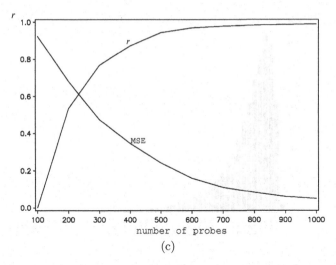

(c)

Fig. 2. (Continued)

low abundance in the genomes while a small number of probes have high abundance. The mean of the 8mer abundance in 6,329 yeast ORFs is about 136 and the median is around 100.

Knowing the 8mers' distribution in the genome, the next question is which set of 8mers to use as probes for expression experiments. To help determine which 8mers to pick, we grouped sorted 8mers based on their abundance counts in yeast ORFs in 22 (65,536/3,000) sets of 3,000 8mers. For each set of 3,000 8mers, we formed a corresponding X matrix where each row represents an 8mer, each column refers to a yeast ORF, and each element in the matrix is the count of the occurrences of a particular 8mer in the corresponding ORF. To generate Y, we used both simulated data and real data from yeast microarray experiments [13]. In terms of solving the system of equations, it did not seem to matter whether we used simulated gene expression data or real experimental data. In Figure 3(b), we show the results from this analysis with data from real microarray experiments. Each point in the plot corresponds to the correlation between estimates and true values using one of the 22 sets of 8mers.

The overall trend from the plot shows that using a more sparse X matrix gives more accurate results. Based on this observation, we then pooled the top four sets of 8mers to form a refined set for further analysis. To see how different numbers of probes affect the accuracy of estimation, we have ranged the probe number from 1,000 to 6,000 in an increment of 1,000. The results for this analysis is presented in Figure 4. The results from yeast genome data are similar to those from simulated data. We can see that, as the probe number increases, the solutions are getting more and more accurate. At the convergence point, estimates from 3,000 probes can have a correlation of 0.96 and those from 4,000 probes can have a correlation of 0.98.

Another question we want to address is how much variation there is between different sets of 8mers for the same number of probes. In other words, we want

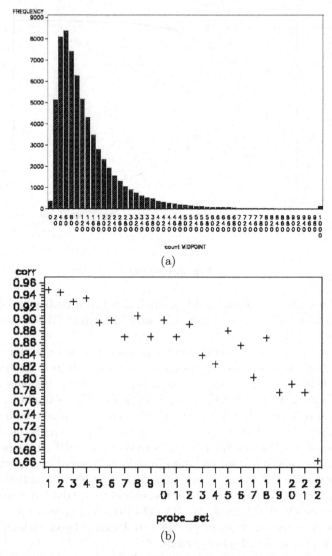

Fig. 3. (a) 8mer distribution in yeast coding ORFs; (b) sequential analysis results of using ordered 8mers based on their frequency counts in yeast coding ORFs.

to see when X matrices are of the same dimension and how the structure of the X matrix affects the estimation accuracy. To do this, we first fixed the probe number to be 3,000, then we sampled different sets of 8mers from the top four sets of 8mers identified from the previous analysis. Although the number of probe choices has been significantly reduced (from 65,536 down to 12,000), the number of choices left, $\binom{12000}{3000}$, remains much too large. Instead of generating all possible 8mer sets of size 3,000, we chose to do random sampling. Each sample is

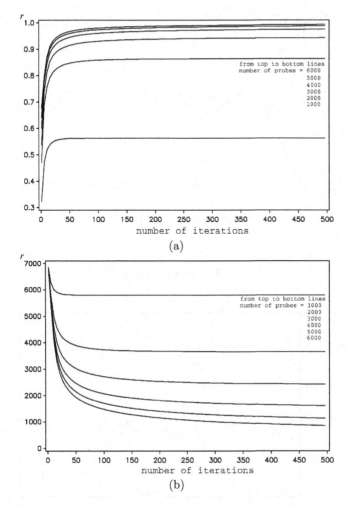

Fig. 4. Yeast genome data: (a) correlations between estimates and true values against number of iterations for six probe sets with the number of probes ranging from 1,000 to 6,000; (b) MSE between estimates and true values against number of iterations for six probe sets with the number of probes ranging from 1,000 to 6,000; and (c) correlation and MSE against number of probes.

generated by picking 3,000 8mers from the 12,000 8mer set uniformly at random. Results of this analysis are shown in Figure 5.

We can see that results from different samples of 8mers vary: for instance, the correlation coefficients from different random samples range from 0.8 to 0.98. Thus we need to do further sampling for the same number of probes.

We should note that our method is not good at estimating gene transcript number for genes with relatively low copy numbers, as can be observed from Figure 6, the scatter plot between true value and estimates. As the gene copy

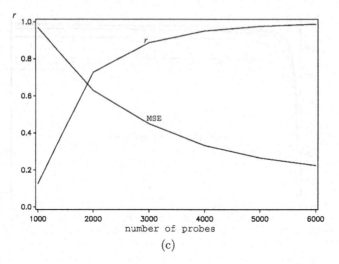

(c)

Fig. 4. (Continued)

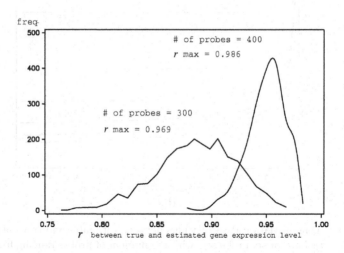

Fig. 5. Correlation results between gene copy number estimates and true values for random samples of 3,000 8mers.

number increases, the correlation between true values and estimates increases. When applying our methods for real experiments, we would advise using a cutoff value to discard genes with expression levels lower than the cutoff value.

To apply our methods to identify genes that are differentially expressed between two samples, we simulated a portion of the genes to have fold changes between 2 and 5. The portion of the genes are set to be 2%, 4%, 6%, 8% and 10% of all genes. The results are summarized in terms of false positive rate and power. False positive rate refers to the proportion of genes that do not have fold changes and yet are identified as differentially expressed genes, while power

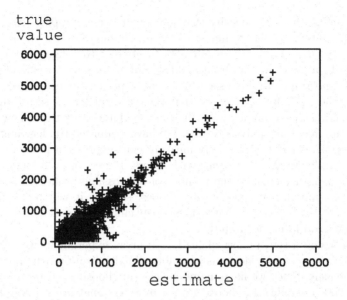

Fig. 6. Scatter plot between true gene copy numbers and estimates.

Table 1. False positive and power for identifying differentially expressed genes using a cutoff value.

Genes with fold change	False positive	Power
2 percent	0.002	0.880
4 percent	0.005	0.906
6 percent	0.007	0.922
8 percent	0.010	0.953
10 percent	0.018	0.923

refers to proportion of genes identified as differentially expressed genes that are actually fold-changing genes. We used a cutoff value of 50 for both estimates and true values—this choice of cutoff lies roughly at the 1 percentile in the range of overall gene copy number. The results are summarized in Table 1.

5   Discussion

In applying our method to design genome-specific biochips with a relatively small set of short oligos, we want to identify differentially expressed genes between samples based on probe hybridization signal intensity values. Since before the experiment, we can always tell whether the selected probes can give good estimates, we have the flexibility to add or replace 8mers to ensure precise estimates of gene transcript copy numbers. For example, when the correlation between estimates and simulated gene copy numbers is below 100%, we can see which genes are not estimated accurately. Consequently, we can either add additional 8mers that can identify that particular gene or replace some 8mers in that gene column in the X matrix to rerun the analysis till the solutions are precise estimates of the true values. When none of the oligos in the probe set can identify a gene, we can easily add one or more gene-specific oligos to the probe set and run analysis to see whether all genes can be estimated accurately.

From these results, we can see that the art of picking a set of 8mers is the key to making this approach work. It would be interesting to see how the probe length can affect our analysis results. We have evaluated the feasibility of using a set of 3,000 8mers to accurately measure gene-expression levels for the yeast genome; similar work can be done with 9mers, 10mers or even longer oligos. As the probe length increases, the number of all possible probes goes up quickly, as does the number of gene-specific probes, so that it will be challenging to find a small set of probes that can have specific combinations to measure all gene-expression levels accurately.

When using a set of short oligos for an array experiment, we need to optimize the experimental conditions to enable all hybridization reactions to happen simultaneously. Since we have many choices (in choosing 3,000 or 4,000 probes from 12,000 possibilities), choosing a set of 8mers that are suitable for experiments is possible. In reality, we would first identify all probes that are suitable for the same experimental condition, then apply our algorithm to choose an appropriate set of probes. The process can be optimized for specific genome.

A great advantage of using a relatively small set of short oligos as probes for microarray experiment is that we can also assess individual 8mer hybridization kinetics. Hybridization efficiency between probes of the same length can vary significantly [19,20]. Once hybridization efficiency for a probe set is quantitatively measured in an array experiment, we can incorporate it as a weight factor in our algorithm. The Y vector, representing probe hybridization signal intensity, can be easily modified to incorporate variation in hybridization efficiency among probes. If we define h_i as the coefficient of hybridization efficiency, then each y_i can be adjusted accordingly by h_i and we can then use the adjusted Y vector in the set of equations to estimate G, the gene transcript numbers.

Our method can be easily extended to design short probes for microarray experiments in other organisms. Although organisms such as human and mouse have much larger genomes, their gene expression is highly tissue-specific and only a subset of all genes in the genome are expressed in a specific tissue. For a certain tissue, once all the genes expressed in it are identified, we can apply our probe selection process to design probes for tissue-specific microarray experiments. NCBI's Unigene set have provided such tissue-specific gene expression information [21].

6   Conclusions

We have described a method to mine and pick a set of short oligomers that can be used as probes for large-scale genome-specific or tissue-specific microarray gene-expression experiments. The major advantages of using our method is to significantly reduce the overall cost in array fabrication and oligo synthesis. The process of mining probe set depends on knowing the gene-sequence information in a specific genome or tissue. As more genomes being sequenced, more genes annotated, and more expression libraries constructed, our method offers promise for more accurate and less expensive microarray experiments.

Acknowledgements

We thank Dr. Ross Whetten (molecular geneticist) for talking with us about the experimental feasibility of the study and for his input in helping design experimental strategies for future experiments using our methods.

References

1. Zhang, L. Zhou, W., Velculescu, V.E., Kern, S.E., Hruban, R.H., Hamilton, S.R., Vogelstein, B. and Kinzler, K.W. (1997) Gene expression profiles in normal and cancer cells. *Science* **276**, 1268–1272.
2. Golub, T.R., Slonim, D.K., Tamayo, P., Huard, C., Gaasenbeek, M., Mesirov, J.P., Coller, H., Loh, M.L., Downing, J.R., Caligiuri, M.A., Bloomfield, C.D., Lander, E.S. (1999) Molecular classification of cancer: class discovery and class prediction by gene expression monitoring. *Science* **286**(5439), 531–537
3. Perou, C.M., Sorlie, T., Eisen, M.B., van de Rijn, M., Jeffrey, S.S., Rees, C.A., Pollack, J.R., Ross, D.T., Johnsen, H., Akslen, L.A., Fluge, O., Pergamenschikov, A., Williams, C., Zhu, S.X., Lonning, P.E., Borresen-Dale, A.L., Brown, P.O. and Botstein, D. (2000) Molecular portraits of human breast tumours. *Nature.* **406**(6797), 747–752.
4. Scherf, U., Ross, D.T, Waltham, M., Smith, L.H., Lee, J.K., Tanabe, L., Kohn, K.W., Reinhold, W.C., Myers, T.G., Andrews, D.T., Scudiero, D.A., Eisen, M.B., Sausville, E.A., Pommier, Y., Botstein, D., Brown, P.O. and Weinstein, J.N. (2000) A gene expression database for the molecular pharmacology of cancer. *Nat. Genetics* **24**(3), 236–244.
5. Fodor, S.P., Read, J.L., Pirrung, M.C., Stryer, L., Lu, A.T., Solas, D. (1991) Light-directed, spatially addressable parallel chemical synthesis. *Science*, 251(4995), 767–773.
6. Lockhart, D.J. *et al.* (1996) Expression monitoring by hybridization to high-density oligonucleotide arrays. *Nat. Biotechnol.* **14**, 1675–1680.
7. Chee, M.S., Yang, R., Hubbell, E., Berno, A., Huang, X.C., Stern, D., Winkler, J., Lockhart, D.J., Morris, M.S. and Fodor, S.P.A. (1996) Accessing Genetic Information with High-Density DNA Arrays. *Science* **274**, 610–614.
8. Lipshutz, R.J., Fodor, S., Gingeras, T., and Lockhart, D. (1999) High-density synthetic oligonucleotide arrays. *Nat. Genet.* **21** (Suppl.), 20–24.
9. Schena, M., Shalon, D., Davis, R.W. and Brown, P.O. (1995) Quantitative monitoring of gene expression patterns with a complementary DNA microarray. *Science*, **270**(5235), 467–470.
10. Blanchard, A.P. (1998) Synthetic DNA Arrays. *Genetic Engineering* **20**, 111–123.
11. Blanchard, A.P. and Friend, S.H. (1999) Cheap DNA arrays—it's not all smoke and mirrors. *Nat. Biotechnol.* 17(10), 953.

12. Hughes, T.R., Mao, M., Jones, A.R., Burchard, J., Marton, M.J., Shannon, K.W., Lefkowitz, S.M., Ziman, M., Schelter, J.M., Meyer, M.R., Kobayashi, S., Davis, C., Dai, H., He, Y.D., Stephaniants, S.B., Cavet, G., Walker, W.L., West, A., Coffey, E., Shoemaker, D.D., Stoughton, R., Blanchard, A.P., Friend, S.H. and Linsley, P.S. (2001) Expression profiling using microarrays fabricated by an ink-jet oligonucleotide synthesizer. *Nat Biotechnol.* **19**(4), 342–347.

13. DeRisi, J.L., Iyer, V.R. and Brown, P.O. (1997) Exploring the Metabolic and Genetic Control of Gene Expression on a Genomic Scale, *Science* **278**, 680–686

14. Cho, R.J., Campbell, M.J., Winzeler, E.A., Steinmetz, L., Conway, A., Wodicka, L., Wolfsberg TG, Gabrielian AE, Landsman D, Lockhart DJ, Davis RW. (1998) A genome-wide transcriptional analysis of the mitotic cell cycle. *Mol Cell.* **2**(1), 65–73.

15. Hughes, T.R., Marton, M.J., Jones, A.R., Roberts, C.J., Stoughton, R., Armour, C.D., Bennett, H.A., Coffey, E., Dai, H., He, Y.D., Kidd, M.J., King, A.M., Meyer, M.R., Slade, D., Lum, P.Y., Stepaniants, S.B., Shoemaker, D.D., Gachotte, D., Chakraburtty, K., Simon, J., Bard, M. and Friend, S.H. (2000) Functional discovery via a compendium of expression profiles. *Cell.* **102**(1), 109–126.

16. Epstein, C.B., Waddle, J.A., Hale, W. IV, Dave, V., Thornton, J., Macatee, T.L., Garner, H.R., Butow, R.A., (2001) Genome-Wide Responses to Mitochondrial Dysfunction, *Mol. Biol. Cell* **12**, 297–308.

17. Cherry, J.M., Adler, C., Ball, C., Chervitz, S.A, Dwight, S.S., Hester, E.T., Jia, Y., Juvik, G., Roe, T., Schroeder, M., Weng, S. and Botstein, D. (1998) SGD: Saccharomyces Genome Database. *Nucleic Acids Res.*, **26**(1), 73–79.

18. van Dam, R.M. and Quake, S.R. (2002) Gene expression analysis with universal n-mer arrays. *Genome Res.* **12**(1), 145–152.

19. Maskos, U. and Southern, E.M. (1993) A study of oligonucleotide reassociation using large arrays of oligonucleotides synthesized on a glass support. *Nucleic Acids Res.,* **21**, 4663–4669.

20. Southern, E.M., Mir, K. and Shchepinov, M. (1999) Molecular interactions on microarrays. *Nat. Genet.* **21**, 5–9.

21. Schuler, G.D. (1997) Pieces of the puzzle: expressed sequence tags and the catalog of human genes. *J Mol Med* **75**(10), 694–698.

Designing Specific Oligonucleotide Probes for the Entire *S. cerevisiae* Transcriptome

Doron Lipson[1], Peter Webb[2], and Zohar Yakhini[1,2]

[1] Technion, Haifa 32000, Israel
dlipson@cs.technion.ac.il
[2] Agilent Laboratories
{peter_webb,zohar_yakhini}@agilent.com

Abstract. Probe specificity plays a central role in designing accurate microarray hybridization assays. Current literature on specific probe design studies algorithmic approaches and their relationship with hybridization thermodynamics. In this work we address probe specificity properties under a stochastic model assumption and compare the results to actual behavior in genomic data. We develop efficient specificity search algorithms. Our methods incorporate existing transcript expression level data and handle a variety of cross-hybridization models. We analyze the performance of our methods. Applying our algorithm to the entire *S. cerevisiae* transcriptome we provide probe specificity maps for all yeast ORFs that may be used as the basis for selection of sensitive probes.

1 Introduction

A *probe*, in the context of the current study is a (nucleic acid) molecule that strongly interacts with a specific target in a detectable and quantifiable manner. Oligonucleotides are used as probes in an increasing number of molecular biology techniques. They are central in Southern and northern blotting, in *in-situ* hybridization assays, in quantitative PCR, and in array based hybridization assays (chips) where they are immobilized on a surface [1,2]. Some applications and protocols require probe labeling while in others (e.g. arrays) the target is being labeled.

There are, roughly speaking, two parameters by which to evaluate candidate probes for a given application. *Sensitivity* - is it really going to strongly interact with its target, under the assay's conditions, and how much target is needed for the reaction to be detectable or quantifiable; and *specificity* - how well does the probe discriminate between its intended target and other messages that it might cross hybridize to. This study addresses specificity design questions.

A particular application in which design and specificity issues arise is gene expression profiling, in which the particular subset of genes expressed at a given stage and its quantitative composition are queried. Such information can help in characterizing sequence to function relationships, in determining effects (and side effects) of experimental treatments, and in understanding other molecular

R. Guigó and D. Gusfield (Eds.): WABI 2002, LNCS 2452, pp. 491–505, 2002.

biological processes [3,4], many with clinical implications [5,6]. Gene expression profiling is typically performed using array based hybridization assays. The actual probes that populate a custom designed expression profiling array are specifically designed and chosen to measure the expression levels of a defined set of genes. Given the state of the human genome sequence draft and the extent of gene hunting efforts currently invested by the scientific community, it is reasonable to approach gene expression profiling assuming complete knowledge of the sequences of the genes of interest, as well as those of many of the genes expressed in the background message.

Let $\mathcal{N} = A, C, G, T$ be the alphabet representing the four different nucleotides. Our general design question is as follows. We are given a *target gene*, g - a sequence over N, of length 500-10000; and a *background message, D* - a large set of sequences (with lengths in the same range), representing all possible mRNA molecules that might be active in our sample[1]. In g we seek substrings that represent Watson-Crick (WC) complements to molecules that have a high WC mismatch to the background message. These substrings are presumably good probe binding site candidates, in terms of specificity. Since they don't have a close WC match in the background message, they do not have a high cross-hybridization potential. An equivalent computational task is to find, in g, the substrings that are far, in Hamming distance, from the background message (as a set). We seek many symbolically specific probe candidates since specificity screening is only one stage in the probe selection process. Specificity issues arise in probe design as described above (and in other forms), as well as in the design of PCR primers [7] and the design and development of anti-sense drugs.

We re-emphasize that distance (as above, in the Watson-Crick or Hamming sense) is not the only parameter that determines the specificity of a probe candidate. The issues of multiplicity (how many times do mismatches occur in the background), abundance (how strongly expressed are genes with close matches), hybridization potential (how competitive is the background, in terms of thermodynamics and not only homology), and others, play an important if not well understood and not easily quantified role.

Previous work [8] presents an algorithm for selecting specific probes using suffix trees and a model of hybridization thermodynamic. The authors apply the algorithm on several genomes providing few candidate probes per gene. Other methods are described in [9]. In this study we address additional issues:

- A process for specific probe design, whose output is specificity map of all possible probes for any specific gene, based on Hamming distance from the background message or other possible thermodynamic models. This map can be used as a basis for selecting sensitive probes.
- Analysis of the efficiency and effectiveness of the proposed algorithm.
- Consideration of the relative abundance of background transcripts by applying different filters to different relative expression ratios. Abundance, for this purpose, is obtained from pre-existing data. Such data is expected to grow

---

[1] Typically, we shall use some database that stores much of (what is known of) this organism/tissue/stage specific information

as more expression measurements are performed and as databases storing this information continue to evolve.
- Rigorous stochastic analysis of the statistical protection provided by probe length when pre-existing background sequence data is not sufficient.

In Section 2 we study a stochastic model that indicates how long we should expect probes with a given threshold specificity to be. In Section 3 we present algorithmic approaches to assessing probe candidate specificity against available data. We analyze some efficient and effective heuristics and modifications that address transcript abundance and thermodynamic models. Finally, in Section 4, we discuss implementations and the design of specific probes for the entire *S. cerevisiae* transcriptome.

2   Statistical Properties of Symbolic Specificity

In this section we analyze the expected behavior of a probe candidate mismatch to the background message, under uniform stochastic model assumptions. Understanding this behavior is instrumental in determining practical probe lengths that provide statistical defense against cross hybridization. Probe length decisions are central to any array development and manufacturing process. Probe lengths influence pricing, probe fidelity[2] and other parameters. Clearly - the uniform distribution assumed in the model is certainly not the observed reality. We therefore test the quality of the stochastic model predictions and their validity for genomic data.

We denote the Hamming distance between two strings $s, t \in \mathcal{N}^k$ by $H_k(s,t)$. More precisely, if $s = s_1 \ldots s_k$ and $t = t_1 \ldots t_k$ then $H_k(s,t) = \sum_{i=1}^{k} 1_{[s_i \neq t_i]}$.

When considering a probe candidate and a given background message, we are interested in the distance between the probe and the entire message. For a string $s \in \mathcal{N}^k$ and a string $B \in \mathcal{N}^N$, where $\mathcal{N} >> k$, we set

$$d(s, B) = \min_{1 \leq i \leq N-k+1} H_k(s, B_i^{i+k-1}).$$

This is analogous to treating B as a set of k-long strings and measuring the Hamming distance from s to the said set.

In our stochastic model we assume that both g, the gene of interest, and B, the background message, were drawn uniformly and independently over $\mathcal{N}^m$ and $\mathcal{N}^N$, respectively. Given k and N as set forth consider the random variable $d_{N,k} = d(p, B)$, where p is a probe candidate, a k-substring of g. The distribution of $d_{N,k}$ gives us starting points for answering the probe design and length determination questions described in the introduction.

2.1   Poisson Approximation

Our study of the distribution of $d_{N,k}$ utilizes approximation methods that are part of a general approximation theory developed by Stein and others [10,11]. For completeness we state the results that are directly applicable in our context.

---

[2] Fraction of full-length probe in any nominal feature.

Let Γ be an index set. Let $I_i : i \in \Gamma$ be a collection of indicator random variables on some probability space. Let $W = \sum_{i \in \Gamma} I_i$. Let $\lambda = E(W)$. Assume that for every i there is a set of indices Γ_i such that I_i is independent of $I_j : j \notin \Gamma_i$. Γ_i is called I_i's *dependence neighborhood*.

Theorem: *Let W as above. Let Z ˜$Poisson(\lambda)$.*
Set $b_1 = \sum_{i \in \Gamma} \sum_{j \in \Gamma_i} E(I_i)E(I_j)$ and $b_2 = \sum_{i \in \Gamma} E(I_i) \sum_{j \in \Gamma_i} E(I_j | I_i = 1)$.
Then $\left| Pr(W = 0) - e^{-\lambda} \right| \leq (1 \wedge \lambda^{-1})(b_1 + b_2)$.

Remark: The independence assumption can, indeed, be relaxed. An additional error term is then added.
For $r = 0, 1, 2 \ldots k$, let $\mu_r = \dfrac{1}{4^k} \displaystyle\sum_{\rho=0}^{r} \binom{k}{\rho} \cdot 3^\rho$ and $\lambda(N, r) = (N - k + 1)\mu_r$.

Theorem: *Fix a word s_0 of length k. Consider the random variable $d_N(s_0)$ as above. For $r = 0, 1, 2, \ldots, k$, let $F(r) = Pr(d_N \leq r)$.*
Set $b_1(r) = (N - k + 1)(2k - 1)\mu_r^2$.
Then $\left| F(r) - (1 - e^{-\lambda(N,r)}) \right| \leq (1 \wedge \lambda(N, r)^{-1})b_1(r)$.

Proof: For $r = 1, 2, \ldots, k$ define random variables

$$V_N(r) = \sum_{i=1}^{N-k+1} 1_{B_r(s_0)}(B_i^{i+k-1}),$$

Where $B_r(s_0)$ is the Hamming ball of radius r around s_0, in $\mathcal{N}^k$. We then have $1 - F(r) = Pr(V_N(r) = 0)$. $V_N(r)$ is a sum of indicator random variables. Clearly $E(V_N(r)) = \lambda(N, r)$. A Poisson distribution is not always a good approximation for that of $V_N(r)$. This can be seen by considering the case $r = 0$ and a word s_0 with small periods. In this case $V_N(r)$ will have compound Poisson distribution. However, we are only interested in estimating the mass at 0.
For $1 \leq i \leq N - k + 1$ set

$$J_i = 1_{B_r(s_0)}(B_i^{i+k-1}) \cdot \left(\prod_{j=i+1}^{i+k-1} (1 - (1_{B_r(s_0)}(B_j^{j+k-1}))) \right) \text{ and } U_N(r) = \sum_{i=1}^{N-k+1} J_i$$

$U_N(r)$ counts the left-most visits to the Hamming ball around s_0, encountered in the background sequence B. We will assess a Poisson approximation to $U_N(r)$. Since $Pr(V_N(r) = 0) = Pr(U_N(r) = 0)$ we will have an estimate of the quantity of interest.

To apply the Poisson approximation theorem stated above we use the dependence neighborhoods $\Gamma_i = 1 \leq j \leq N = k + 1 : |j - 1| \leq k$. Estimating the bounds we get $b_1 = (N - k + 1)(2k - 1)\mu_r^2$.
By stationarity: $b_2 = 2(N - k + 1)\mu_r \sum_{j=2}^{k} E(I_j | I_1 = 1)$.

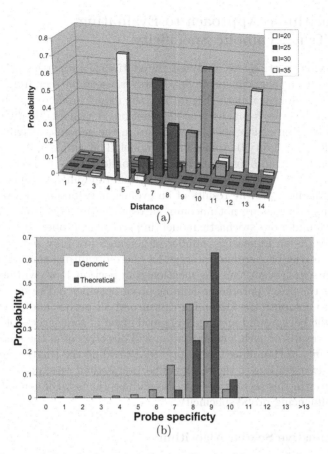

Fig. 1. a) Expected distribution of probe specificities (given in distance from background message of length N=9Mb - the size of the yeast transcriptome), for probes of different lengths l. b) Distribution of probe specificity according to theoretical computation in comparison to the specificity distribution retrieved by a search over a genomic database (150 ORFs of *S. cerevisiae*).

The summands here are all 0 since the presence of s_0 at a given position means a left-most visit can not occur in the next $k - 1$ positions. Thus $b_2 = 0$.

The above approximation result enables us to compute the distribution of $d_{N,k}$ for various assumed transcriptome sizes and various probe lengths, as shown in Figure 1a. In Figure 1b we compare the results obtained for the stochastic model to theses actually observed when measuring distances of candidate probes to the rest of the yeast transcriptome. The reasons for the observed bias are discussed in the next section. Despite the bias, it is clear that the model calculation provides us with information on the length needed for desired statistical protection against cross hybridization. Model results should be taken as over-estimates of the actual distances.

3   Algorithmic Approach to Evaluating the True Symbolic Specificity

In the previous section, we analyzed the statistical properties of an unknown background, using a stochastic model. However, as can be seen from the comparison to actual biological data (Figure 1b), the model predicted probe specificity distribution deviates from the true distribution. A remarkable difference is the noticeable fraction of low specificity probes in the genomic distribution. These are virtually probability 0 events, in the model. The simple and straightforward explanation for this difference is that the genomic data is structured and therefore is expected to contain more similarities than random data. Specifically, since genomic sequences are the results of an evolutionary process, total and partial duplications of genes are not uncommon, resulting in repetitions that are very improbable, under our stochastic model (in particular, probes with 0 specificity).

However, as an increasing number of genomes are fully sequenced and annotated, the true background message for these organisms can be evaluated by means of a search algorithm. As mentioned in Section 1, we will use Hamming distance as the proxy parameter defining probe specificity. The specificity search attempts to compute the Hamming distance of all probe-candidates of a given length to the background message (assumed given in some accessible data structure). The results of such a computation will directly entail the identification of good (specific, in Hamming terms) probes. Providing the length of probe candidates was set according to the results of Section 2, then we expect plenty of good probes. Later in this section we will also address the issue of gene abundance as a factor influencing probe specificity.

3.1   Exhaustive Search Algorithm

We now state the computational problem in more formal terms. We address the case in which the set of candidate probes is the full set of substrings of g, of a given length.
Full Distance Search (Statement of the problem)
Input: The target gene g (of length m), a background message D (of length N)[3], a probe length l.
Output: A set of Hamming distances $\{d(g_i^{i+l-1}, D) : 1 \leq i \leq m - l + 1\}$.
In practice, $N \approx 1 - 100Mb$ and $m \approx 1 - 10Kb$.

The naive algorithm performs an exhaustive search of all candidate probes against the background message, computing all Hamming distances as above.

3.2   Probe Classification Algorithm

Recall, that we want to design specific probes for g. This means that we are not really interested in computing the set of exact distances $\{d(g_i^{i+l-1}, D) : 1 \leq i \leq m - l + 1\}$, but only in classifying the probe candidates to those of small

---

[3] D may be given as a set of sequences with a total length N, or as a concatenated sequence

Let I be an integer lists, with 4^ξ entries, addressable as words in $\mathcal{N}^\xi$.

Pre-Processing	Scanning

`for (i=0; i<m-ξ+1; i++)`
    `Insert(i,I($g_i^{i+\xi+1}$))`
	`for (j=0; j<N-ξ+1; j++)`
	    `for (i∈I($D_j^{j+\xi-1}$))`
	        `ld=d($g_i^{i+l-1}, D_j^{j+l-1}$)`

Initializing d
`for (i=0; i<m-ξ+1; i++)`
    `d(i)=1`
	`        for (a=0; a<l-ξ; a++)`
	`            d(i-a)=min(d(i-a),ld)`
	`            ld=ld-$1_{(g_{i+l-1-a} \neq D_{j+l-1-a})} + 1_{(g_{i+l-a} \neq D_{j+l-a})}$`

Fig. 2. IClaS: Computing some (but possibly not all) distances and classifying probe candidates, using an indexed search.

distance (bad probes) and those of large distance (good probes). Formally, the computational problem reduces to:

Classification of Probe Candidates (Statement of the problem)

Input: The target gene g (of length m), a background message D (of length N), a probe length l, a quality threshold r.

Output: A partition of the set $\{1 \leq i \leq m - l + 1\}$ into $B = \{i : d(g_i^{i+l-1}, D) < r\}$ and $G = \{i : d(g_i^{i+l-1}, D) \geq r\}$

This formulation of the problem brings about a simple but useful observation:

Observation: If two strings have $H_l(s_1^l, t_1^l) \leq d$, then there must be at least one substring of s, of length $\geq \lceil (l - d)/(d + 1) \rceil$, which is a perfect match to the corresponding substring of t. This observation comes from the fact that the worst distribution of d mismatches (i.e. the distribution that produces the shortest maximal perfect match) is the one in which the mismatches are distributed evenly along the string - between intervals of matches that are $\lceil (l - d)/(d + 1) \rceil$ long.

Therefore, to classify probes we can perform an indexed search. A version that is applicable here is formally described in **IClaS** (Indexed Classified Search - Figure 2). Here is an informal description: Generate and index of all words of a given length, to be called the *seed length*, and denoted ξ. In each entry of this index keep a list of the occurrences of the corresponding word in the target gene g. Scan all the background message with a window of size ξ. For each such location, go to all the occurrences in g of the encountered word, grow l-strings around the two short words (in g and in D) in parallel, compute the distances and update the distance vector when necessary (that is, whenever the currently interrogated l-strings have a match closer than anything previously encountered).

Performance. The performance of **IClaS** can be analyzed by counting the number of probe-length comparisons performed by the algorithm. Clearly, the exhaustive search performs $O(mN)$ such comparisons[4]. The number of comparisons

---

[4] The performance of both **IClaS** and the exhaustive search can be improved slightly by using various implementation "shortcuts". These do not affect the running time by more than a constant factor.

made by **IClaS** cannot be determined directly, since it depends on the actual input and more specifically on the total number of index entries each one of the seed ξ-mers. We analyze the order of the expected number of such comparisons assuming that g and D are uniformly and independently drawn: Given a seed s, set $\gamma(s)$ to be the number of occurrences of s in g (i.e. the number of index entries for s). Set $\delta(s)$ to be the number of occurrences of s in D. For this s, for each of its appearances in D, $l - \xi + 1$ comparisons are performed for each index entry of s. The total number of probe-length comparisons performed by the algorithm is therefore in the order of $\sum_{s \in N^\xi} l \cdot \gamma(s) \cdot \delta(s)$. Employing our uniformity and independence assumptions we calculate the expected number of such comparisons to be: $E\left(\sum_{s \in N^\xi} l \cdot \gamma(s) \cdot \delta(s)\right) = l \cdot \sum_{s \in N^\xi} E\left(\gamma(s)\right) \cdot E\left(\delta(s)\right) = l \cdot 4^\xi \frac{m}{4^\xi} \frac{N}{4^\xi} = \frac{lmN}{4^\xi}$

Thus, the relative efficiency of **IClaS** over the exhaustive search may be expected to be in the order of $l/4^\xi$, which becomes very significant as ξ grows.

Testing of the actual performance of **IClaS** for various values of ξ, in comparison to the actual performance of the exhaustive search, was performed on a Pentium III 650MHz. For an an arbitrary *S. cerevisiae* transcript of length $m = 300$ against the entire transcriptome wit $l = 30$ the following running times (in seconds) were measured for $\xi = 9, 8, 7, 6, 5, 4$ and an exhaustive search, respectively: 73, 79, 116, 283, 781, 2750 (46 mins), and 23011 (6.4 hrs).

Reliability. In the previous section we observed that working with a small seed size, ξ, will guarantee catching all bad (not specific enough) probes in a target gene. In the previous section we saw, however, that the time complexity of the indexed search strongly depends on the size of the seed. In this section we examine the connection between ξ and the reliability of the corresponding **IClaS** indexed search. The reliability is measured in terms of the probability of false positives: bad probes that were not caught by **IClaS**.

We start by stating a combinatorial result, a special case of the problems studied in [12,13]:

Definition: Given a string s over an alphabet Σ and $\sigma \in \Sigma$, a substring s_i^j is a *leftmost run* of σ's if $s_i^j \in \{\sigma\}^*$ and either $s_{i-1} \neq \sigma$ or $i = 1$. For example, the string $AABBBBBBBAA$ contains only one leftmost run of 3 B's while the string $AABBBABBBAA$ contains two leftmost runs of 3 B's.

Consider all words of length l over the binary alphabet $\{A, B\}$. Consider all such words that have exactly $l(A)$ occurrences of A, and exactly $l(B) = l - l(A)$ occurrences of B. The number of such words that have at least j leftmost runs of at least σ B's can be computed by taking all words of $l(A)$ A's and $(l(B) - \sigma j)$ B's, and for each such word taking all combinations of insertions of the j runs between the $l(A)$ A's. In total there are $\binom{l(A)+1}{j}\binom{l - \sigma j}{l(A)}$ such words.

However, the same word may be overcounted in the above expression. For example, for $j = 1$, a sequence containing exactly two leftmost runs of at least σ B's will be counted twice - once for containing at least each of the two runs. Using the inclusion and exclusion principle, the number of such words that have at least one run of at least σ B's is: $\sum_{j \geq 1}(-1)^{j+1}\binom{l(A)+1}{j}\binom{l - \sigma j}{l(A)}$.

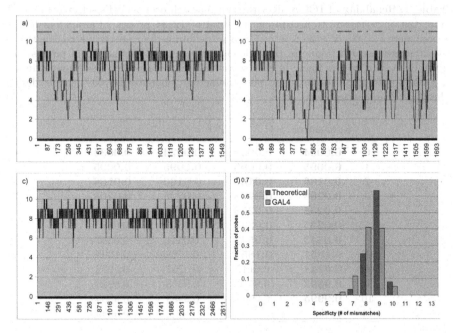

Fig. 3. Probe distance maps for three genes of the GAL family. For each gene, the specificity of all possible probes (in Hamming distance from the entire transcriptome) are plotted against their position along the sequence. Areas of specific probes (above a threshold $r = 7$) are marked above: (a) the sequence of GAL1 contains an area where probes are less specific (this is an area of homology to GAL3), (b) the sequence of GAL2 contains only small stretches of specific probes (because of high similarity to genes from the HXT family), (c) the sequence of GAL4 is almost totally specific, with a probe specificity distribution similar to theory, as shown in (d).

Consequently, for a probe p of length l and a seed length ξ, the probability that a background sequence b of length l uniformly drawn from the Hamming sphere of radius d from p (i.e. $H(p, b) = d$) will contain a ξ long match is:

$$\alpha(l, d, \xi) = \frac{\sum_{j \geq 1} (-1)^{j+1} \binom{d+1}{j} \binom{l - \xi j}{d}}{\binom{l}{d}},$$

which represents a lower bound for the reliability of the indexed search.

Given a probe design problem, it is therefore possible to choose a maximal seed length ξ that ensures the desired reliability α. As we already noted, biological data does not usually conform to statistical results derived over a uniform distribution of sequences. However, since in this analysis the worst case scenario is the one in which the entropy is maximal, structured biological data may be expected to give better results (higher reliability). Moreover, as indicated in Table 1, even when choosing a seed for $\alpha = 95\%$, the reliability of detecting a bad probe is virtually 100%.

Table 1. Reliability of **IClaS** algorithm for a search over 7 ORFs selected randomly from the *S. cerevisiae* transcriptome (a total of 12184 probes). Parameters for the search were $l = 30$, $\alpha = 95\%$, $r = 7/5/3$ (*e* denoting the respective seed length), or an exhaustive search. Value in table denotes the cumulative fraction of probes for a given distance d or less, found in each search. Since the algorithm never overestimates a probe's specificity, results show that the actual reliability is 100% at the desired threshold (emphasized cells).

d	Exhaustive	$r = 7, e = 5$	$r = 5, e = 7$	$r = 3, e = 9$
0	**0.0034**	**0.0034**	**0.0034**	**0.0034**
1	**0.0095**	**0.0095**	**0.0095**	**0.0095**
2	**0.0166**	**0.0166**	**0.0166**	**0.0166**
3	**0.0208**	**0.0208**	**0.0208**	**0.0208**
4	**0.0258**	**0.0258**	**0.0258**	0.0257
5	**0.0335**	**0.0335**	**0.0335**	0.0334
6	**0.0543**	**0.0543**	0.0541	0.0502
7	**0.1857**	**0.1857**	0.1788	0.1399
8	**0.6148**	0.6147	0.5767	0.4272
9	**0.9592**	0.9590	0.9358	0.8285
10	**0.9995**	0.9995	0.9988	0.9880
11	1	1	1	0.9999

3.3   Addressing Gene Abundance

Choosing the mismatch threshold r (or equivalently, the seed parameter ξ) depends on the degree of interference that may be expected due to cross-hybridization. Gene abundance is therefore significant in evaluating candidate probes. When comparing probes of a gene g to a background gene b, compare the situation in which g is of high abundance whilst b is rare, to the opposite situation in which g is rare whilst b is abundant. In the first scenario, cross hybridization between b and the probe designed for g is unlikely to significantly interfere with the correct hybridization between the probe and g itself. Accordingly, a choice of a low mismatch threshold r (or a long seed length ξ) is appropriate for this situation. In the second scenario b is fierce competition for g over the probe and a high mismatch threshold r (a short seed length ξ) is required.

Since gene expression profiling experiments are specifically intended to measure the abundance of gene transcripts it is clearly impossible to assume full a-priori knowledge of this information. However, for some organisms (e.g. *S. cerevisiae*) there exists substantial experimental information about the typical abundance of gene transcripts. In such cases, rough relative abundances of the different genes may be used for a wiser choice of parameters r and ξ. In addition, abundance information may be used in an iterative design process. Fortunately, **IClaS** easily lends itself to such modification. An important observation is that although gene abundance may be regarded as a continuous parameter, the parameters r and ξ themselves are discrete, and usually fall within a short range (e.g. for $l = 30$ r is usually confined to the range 4-8). Therefore, rather than use some continuous function that calculates the optimal r for the abundance levels of g and each $b \in D$, it is equivalent and simpler to make use of a direct lookup table for determining r.

Classification of Probe Candidates by Abundance Weighting. (Statement of the problem)
Input: The target gene g (of length m), the background message D partitioned into $\{D_1, D_2, \cdots, D_k\}$ where each D_i is an abundance category, a probe length l, a vector $(r_1, r_2, \cdots, r_k)$ where r_i is the quality threshold for abundance group D_i.
Output: A partition of the set $\{1 \le i \le m-l+1\}$ into $B = \{i : \exists j\ d(g_i^{i+l-1}, D_j) < r_j\}$ and $G = \{i : \forall j\ d(g_i^{i+l-1}, D_j) < r_j\}$.

The algorithm **IClaSA** is a loop of runs of **IClaS**, where in each iteration the complete set of probes of g is compared to D_i using the threshold r_i to produce the appropriate vector of approximate distances.

3.4   Classifying Probes by Melting Temperature of Potential Cross-Hybridization Nucleation Complexes

As shown, the algorithm for classifying probes according to a mismatch threshold r is analogous to classifying the same probes according to a seed length ξ that may be calculated from r. For many practical applications ξ is even more appropriate as a classification parameter since cross hybridization to a probe (of a non-specific message) is most probably the result of hybridization initiated around a perfect match nucleus (of length ξ)[5]. In this approach we directly set the seed parameter ξ than calculate it using r. A different, perhaps more realistic, method of classifying probes is according to the melting temperature (T_m) of a nucleation complex that may initiate cross-hybridization. Following this reason, we define the proximity of a probe p from the background B to be the maximal melting temperature of a subsequence of p that is a perfect match to some background subsequence. Formally:

$$f(p, D) = \max_{i,k} \left\{ T_m(p_i^{i+k-1}) : \exists d \in D, j \in \mathbf{N} \text{ s.t. } p_i^{i+k-1} = d_j^{j+k-1} \right\}.$$

Probe Nucleation Complex Algorithm. (Statement of the problem)
Input: The target gene g (of length m), the background message D (of length N), a probe length l, a melting temperature threshold t.
Output: A set of proximities of the candidate probes $\{f(g_i^{i+l-1}, D) : 1 \le i \le m - l + 1\}$.

T_m of short sequences may be approximated using different thermodynamic models (2-4 rule [15] nearest-neighbor [16] are examples). It should be noted that the expected size of the longest perfect match in a probe of length l is of order $\log_4(lN)$ which, for $l = 30$, $N = 3 \cdot 10^6$ is $\log_4(9 \cdot 10^7) > 13$. As a consequence, an efficient indexed search with a large seed size may be used to accurately locate

---

[5] "...The process begins by the formation of a transient nucleation complex from the interaction of very few base pairs. Duplex formation proceeds, one base pair at a time, through a zippering process." [14]

(a)

(b)

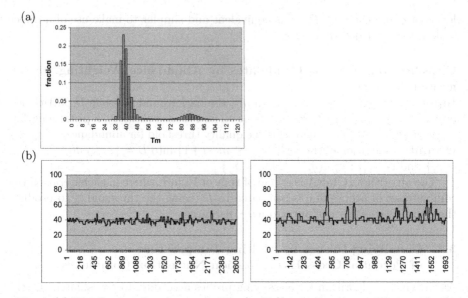

Fig. 4. (a) Distribution of probe proximities (in T_m) according to an arbitrary selection over a genomic database (200 ORFs of *S. cerevisiae*), for probe of length $l = 30$. (b) Distribution of probe proximities (in T_m) for two different ORFs: GAL4 - a gene with no specificity issues and GAL2 - a gene that is highly homologous to genes from the HXT family. X axis denotes position of probe along the sequence. Compare to specificity maps of the same genes depicted in Figure 3.

all possible locations of nucleation complex formation, followed by a rigorous calculation of T_m for these locations using the desired thermodynamic model.

Figure 4a depicts the distribution the proximities of candidate probes, in T_m, to the entire yeast transcriptome using the simple 2-4 model for T_m estimation. It is observed that this distribution is bimodal, composed of a distribution around $T_m \approx 40$, accounting for the expected probe proximities (13 bases with an average T_m of 3), and a second wider distribution around $T_m \approx 90$, for totally unspecific probes. From this distribution we may deduce that requiring a proximity threshold of $T_m > 50$ should allow identification of unacceptably unspecific probes. Figure 4b illustrates probe proximity maps for two different genes.

4   Application to the Yeast Transcriptome

In Section 3 an algorithm for efficiently mapping the specificity of probes against a background with known content and distribution was presented. In practice, complete information of this kind is still rare although it is rapidly accumulating. For the yeast *S. cerevisiae* the entire theoretical transcriptome has been extensively mapped and studied. Publicly accessible databases contain the complete set of theoretical transcripts [17] as well as substantial experimental information about their distribution [18]. We used this existing information to create a full set of probe specificity maps for the entire *S. cerevisiae* transcriptome.

As mentioned earlier, specificity is not the sole parameter for probe selection and therefore, for each transcript, rather than selecting the "best" probe, we calculated a map that quantifies the specificity of each candidate probe. For each transcript, this map can then be used as the basis for selecting highly specific probes, in conjunction with sensitivity parameters.

Figure 3 shows the specificity maps of three different genes, for seed value $\xi = 5$. Two typical specificity patterns are encountered: either a distribution of probe specificities that is similar to the theoretical distribution (e.g. GAL4) or a distribution of much lower values that arises from similarity between genes (e.g. GAL2). Many transcripts combine domains of both types - some parts of the sequence show low specificity against the background, while others are unique (e.g. GAL1). For these latter transcripts the process of mapping specificity is of extreme importance.

Probe specificity mapping for *S. cerevisiae* was performed using the following criteria:

- The abundance weighted threshold values used determined according to the abundance of the target a_t and that of the background transcript a_b. Specifically, we used (for the indicated a_t/a_b pairs): $h/h - 6$, $h/m - 4$, $h/l - 3$, $m/h - 7$, $m/m - 5$, $m/l - 3$, $l/h - 7$, $l/m - 6$ and $l/l - 4$, where h stands for high abundance (> 10 transcripts per cell on average), m - medium abundance (> 1 transcript per cell) and l - low abundance (all other transcripts).
- Since they are intended for use in gene expression profiling, the map for each transcript was created only for its 500 long 3'-terminus (for transcripts shorter than 500 bases, the entire transcript was considered).

Results. The complete set of specificity maps for the entire *S. cerevisiae* transcriptome will be available at our website. We present here a brief summary of the results of the mapping, and highlight some specificity issues.

Of the 6310 ORFs (theoretical transcripts) that appear in the *S. cerevisiae* genome databank [17] 71% (4475 ORFs) contain 450 or more valid probes, satisfying the aforementioned criteria. This bulk of the transcripts are those that have no significant specificity problems. The distance distribution for these transcripts is similar to the theoretical distribution described in Section 2. 24% of the ORFs (1497 ORFs) contain between 1 and 449 valid probes. These contain certain domains that are highly similar to other ORFs. Specificity mapping is important in this case, allowing the discrimination between specific and non-specific probes.

The remaining 5% of the ORFs (338 ORFs) are ones for which no satisfactory probe was found. Investigation of these ORFs revealed some explanations:

- A significant portion of the entirely non-specific ORFs are high-abundance transcripts (25% of the non-specific ORFs compared to 5% in the entire ORF population, see Figure 6). The non-specificity originated from gene duplicates that are either completely similar (e.g. TEF1/2 translation elongation factor, CUP1-1/2 copper-binding metallothionein) or almost completely similar

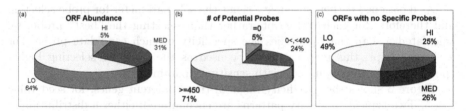

Fig. 5. Distribution of probe specificities for the 6310 ORFs of *S. cerevisiae*: (a) The overall distribution of ORF abundances, (b) the distribution of ORFs according to the number of their potential probes, (c) the distribution of probeless ORFs according to ORF abundance. Notice the large fraction of highly abundant ORFs that have no probes (25%) in relation to their fraction in the ORF population (5%).

(e.g. HHF1/2, HHT1/2 histone proteins, and many RPL and RPS ribosomal proteins). Although expression differences between such transcripts may be biologically significant (for example, if their regulation differs), the non-specificity of these transcripts is inherent to any hybridization-based assay.
- There are several groups of low-medium abundance transcripts that are very similar and therefore difficult to discriminate (e.g. HXT13/15/16/17 high-affinity hexose transporters, PAU1-7 seripauperin family). Functional difference between proteins in these groups is not always clear so the fact that they are indistinguishable may be problematic. Interrogating the expression levels of members of these sets requires further case-specific design.
- High homology between an abundant and scarce transcript may cause the latter's specificity to drop below the required criteria. A large percentage of non-specific ORFs appear in the group of unidentified ORFs, also referred to as hypothetical transcripts (7% non-specific, as opposed to 4% of the identified ORFs). Some of these may be dead genes, i.e. silent remnants of some partial duplication event. The hypothetical transcript YCL068C, for example, is very similar to a region within the abundant RAS guanyl-nucleotide exchange factor BUD5. In these cases non-specificity is may or may not an issue as the expression of the unidentified ORF may not be detected due to this similarity.

5   Summary

Specificity issues that arise in the process of designing hybridization probes are discussed in this work. We start by providing rigorous approximation formulae for the distribution of the Hamming distance of an arbitrary word of length k (representing a candidate probe) to a uniformly drawn much longer sequence over the same alphabet. We then provide an algorithmic approach to efficient specificity mapping. The approach is then applied to the yeast transcriptome and results are compared to the stochastic model. A complete specificity map of the yeast transcriptome was computed. For each ORF and each position we obtain an estimate of the potential cross-hybridization to the set of all other ORFs.

When computing this potential we consider symbolic Hamming distance as well as relative abundance. Measurements for low copy messages are very sensitive to cross hybridization to high abundance messages. The majority of yeast ORFs follow the stochastic model. Message homologies result in specificity issues for many other ORFs. We expect such issues will need to be addressed case by case.

References

1. A.P. Blanchard, and L. Hood, Sequence to array: Probing the Genome's Secrets, Nature Biotechnology 14:1649, 1996.
2. Y. Lysov, A. Chernyi, A. Balaev, F. Gnuchev, K. Beattie, and A. Mirzabekov, DNA Sequencing by Contiguous Stacking Hybridization on Modified Oligonucleotide Matrices, Molecular Biology 29(1):62-66, 1995.
3. E.S. Lander, Array of Hope, Nature Genetics 21:3-4, 1999.
4. P.T. Spellman, G. Sherlock, M.Q. Zhang, V.R Iyer, K. Anders, M.B. Eisen, P.O. Brown, D. Botstein, and B. Futcher, Comprehensive Identification of Cell Cycle-regulated Genes of the Yeast Saccharomyces cerevisiae by Microarray Hybridization, Molecular Biology of the Cell 9(12):3273-97, 1998.
5. M. Bittner et al, Molecular Classification of Cutaneous Malignant Melanoma by Gene Expression Profiling, Nature, 406(6795):536-40, 2000.
6. T. R. Golub et al, Molecular Classification of Cancer: Class Discovery and Class Prediction by Gene Expression Monitoring, Science, 286(5439):531-7, 1999.
7. M. Mitsuhashi, A. Cooper, M. Ogura, T. Shinagawa, K. Yano and T. Hosokawa, Oligonucleotide Probe Design - a New Approach, Nature 367:759-761, 1994.
8. F. Li, G.D. Stormo, Selection of Optimal DNA Oligos for Gene Expression Arrays, Bioinformatics 17(11):1067-1076, 2001.
9. N. Hosaka N, K. Kurata, H. Nakamura, Comparison of Methods for Probe Design, Genome Informatics 12: 449-450, 2001.
10. A.D. Barbour, L. Holst, and S. Janson, Poisson Approximation, Clarendon Press, Oxford, 1992.
11. C. Stein, Approximate Computation of Expectations, Institute of Mathematical Statistics Monograph Series, Vol. 7, 1996.
12. M. Morris, G. Schachtel, and S. Karlin, Exact Formulas for Multitype Run Statistics in a Random Ordering, SIAM J. Disc. Math., 6(1):70-86, 1993.
13. A.M. Mood, The Distribution Theory of Runs, Ann. Math. Stat. 11:367-392, 1940.
14. E. Southern, K. Mir, and M. Shchepinov, Molecular Interactions on Microarrays, Nature Genetics, 21(1):5-9, 1999.
15. T. Strachan and A.P. Read, Human Molecular Genetics, John Wiley & Sons, New York, 1997.
16. J. SantaLucia, A unified view of Polymer, Dumbbell, and Oligonucleotide DNA Nearest-neighbor Thermodynamics, PNAS USA 95, 1460-1465, 1998.
17. Cherry, J. M., Ball, C., Dolinski, K., Dwight, S., Harris, M., Matese, J. C., Sherlock, G., Binkley, G., Jin, H., Weng, S., and Botstein, D., Saccharomyces Genome Database, ftp://genome-ftp.stanford.edu/pub/yeast/SacchDB/
18. F.C.P. Holstege, E.G. Jennings, J.J. Wyrick, T.I. Lee, C.J. Hengartner, M.R. Green, T.R. Golub, E.S. Lander, and R.A. Young, Dissecting the Regulatory Circuitry of a Eukaryotic Genome, Cell, 95:717-728, 1998. http://web.wi.mit.edu/young/expression/transcriptome.html

K-ary Clustering with Optimal Leaf Ordering for Gene Expression Data

Ziv Bar-Joseph[1], Erik D. Demaine[1], David K. Gifford[1], Angèle M. Hamel[3], Tommi S. Jaakkola[2], and Nathan Srebro[1]

[1] MIT Laboratory for Computer Science
{zivbj,gifford,nati}@mit.edu
[2] MIT Artificial Intelligence Laboratory
tommi@ai.mit.edu
[3] Wilfrid Laurier University, Dept. of Physics and Computing
ahamel@wlu.ca

Abstract. A major challenge in gene expression analysis is effective data organization and visualization. One of the most popular tools for this task is hierarchical clustering. Hierarchical clustering allows a user to view relationships in scales ranging from single genes to large sets of genes, while at the same time providing a global view of the expression data. However, hierarchical clustering is very sensitive to noise, it usually lacks of a method to actually identify distinct clusters, and produces a large number of possible leaf orderings of the hierarchical clustering tree. In this paper we propose a new hierarchical clustering algorithm which reduces susceptibility to noise, permits up to k siblings to be directly related, and provides a single optimal order for the resulting tree. Our algorithm constructs a k-ary tree, where each node can have up to k children, and then optimally orders the leaves of that tree. By combining k clusters at each step our algorithm becomes more robust against noise. By optimally ordering the leaves of the tree we maintain the pairwise relationships that appear in the original method. Our k-ary construction algorithm runs in $O(n^3)$ regardless of k and our ordering algorithm runs in $O(4^{k+o(k)}n^3)$. We present several examples that show that our k-ary clustering algorithm achieves results that are superior to the binary tree results.

1 Introduction

Hierarchical clustering is one of the most popular methods for clustering gene expression data. Hierarchical clustering assembles input elements into a single tree, and subtrees represent different clusters. Thus, using hierarchical clustering one can analyze and visualize relationships in scales that range from large groups (clusters) to single genes. However, hierarchical clustering is very sensitive to noise, since in a typical implementation if two clusters (or genes) are combined they cannot be separated even if farther evidence suggests otherwise [11]. In addition, hierarchical clustering does not specify the clusters, making it hard to distinguish between internal nodes that are roots of a cluster and nodes which

R. Guigó and D. Gusfield (Eds.): WABI 2002, LNCS 2452, pp. 506–520, 2002.

only hold subsets of a cluster. Finally, the ordering of the leaves, which plays an important role in analyzing and visualizing hierarchical clustering results, is not defined by the algorithm. Thus, for a binary tree, any one of the 2^{n-1} orderings is a possible outcome.

In this paper we propose a new hierarchical clustering algorithm which reduces susceptibility to noise, permits up to k siblings to be directly related, and provides a single optimal order for the resulting tree, without sacrificing the pairwise relationships between neighboring genes and clusters in the result. Our solution replaces the binary tree of the hierarchical clustering algorithm with a k-ary tree. A k-ary tree is a tree in which each internal node has at most k children. When grouping k clusters (or genes) together, we require that all the clusters that are grouped together will be similar to one another. It has been shown (e.g. in CLICK [11]) that relying on similarities among large groups of genes helps reduce the noise effects that are inherent in expression data. Our algorithm utilizes this idea for the hierarchical case.

The number of children of each internal node is not fixed to k. Rather, k is an upper bound on this number, and if the data suggests otherwise this number can be reduced. An advantage of such a method is that it allows us to highlight some of the actual clusters since nodes with less than k children represent a set of genes that are similar, yet significantly different from the rest of the genes. Such a distinction is not available when using a binary tree.

Finally, our algorithm re-orders the resulting tree so that an optimally ordered tree is presented. This ordering maximizes the sum of the similarity of adjacent leaves in the tree, allowing us to obtain the best pairwise relationships between genes and clusters, even when $k > 2$.

The running time of our tree construction algorithm (for small values of k) is $O(n^3)$, which is similar to the running time of currently used hierarchical clustering algorithms. In order to obtain this performance we use a heuristic algorithm for constructing k-ary trees. Our ordering algorithm runs in $O(n^3)$ for binary trees, while for $k > 2$ our algorithm runs in $O(4^{k+o(k)}n^3)$ time and $O(kn^2)$ space which is feasible even for a large n (when k is small).

The rest of the paper is organized as follows. In Section 2 we present an algorithm for constructing k-ary trees from gene expression data. In Section 3 we present our ordering algorithm. In Section 4 we present some experimental results, and Section 5 summarizes the paper and suggests directions for future work.

1.1 Related Work

The application of hierarchical clustering to gene expression data was first discussed by Eisen in [5]. Hierarchical clustering has become the tool of choice for many biologists, and it has been used to both analyze and present gene expression data [5,9]. A number of different clustering algorithms, which are more global in nature, where suggested and applied to gene expression data. Examples of such algorithms are K-means, Self organizing maps [12] and the graph based algorithms Click [11] and CAST [2]. These algorithms generate clusters which are all assumed to be on the same level and thus lack the ability to represent the relationships between genes and sub clusters as hierarchical clustering does. In addition, they are usually less suitable for large scale visualization tasks, since they do not generate a global ordering of the input data. In this work we try to combine the robustness of these clustering algorithms with the presentation and flexible groupings capabilities of hierarchical clustering.

Recently, Segal and Koller [10] suggested a probabilistic hierarchical clustering algorithm, to address the robustness problem. Their algorithm assumes a specific model for gene expression data. In contrast, our algorithm does not assume any model for its input data, and works with any similarity/distance measure. In addition, in this paper we present a method that allows not only to generate the clusters but also to view the relationships between different clusters, by optimally ordering the resulting tree.

The problem of ordering the leaves of a binary hierarchical clustering tree dates back to 1972 [8]. Due to the large number of applications that construct trees for analyzing datasets, over the years, many different heuristics have been suggested for solving this problem (c.f. [8,7,5]). These heuristics either use a 'local' method, where decisions are made based on local observations, or a 'global' method, where an external criteria is used to order the leaves. In [1] Bar-Joseph *et al* presented an $O(n^4)$ algorithm that maximizes the sum of the similarities of adjacent elements in the orderings for binary trees. This algorithm maximizes a global function with respect to the tree ordering, thus achieving both good local ordering and global ordering. In this paper we extend and improve this algorithm by constructing a time and space efficient algorithm for ordering k-ary trees.

Recently it has come to our attention that the optimal leaf ordering problem was also addressed by Burkard *et al* [3]. In that paper the authors present an $O(2^k n^3)$ time, $O(2^k n^2)$ space algorithm for optimal leaf ordering of PQ-trees. For binary trees, their algorithm is essentially identical to the basic algorithm we present in section 3.2, except that we propose a number of heuristic improvements. Although these do not affect the asymptotic running time, we experimentally observed that they reduce the running time by 50-90%. For k-trees, the algorithms differ in their search strategies over the children of a node. Burkard *et al.* suggest a dynamic programming approach which is more computationally efficient ($O(2^k n^3)$ vs. $O(4^{k+o(k)} n^3)$), while we propose a divide and conquer approach which is more space-efficient ($O(kn^2)$ vs. $O(2^k n^2)$). The number of genes (n) in a expression data set is typically very large, making the memory requirements very important. In our experience, the lower space requirement, despite the price in running time, enables using larger ks.

2   Constructing K-ary Trees

In this section we present an algorithm for constructing k-ary trees. We first formalize the k-ary tree problem, and show that finding an optimal solution is hard (under standard complexity assumptions). Next, we present a heuristic algorithm for constructing k-ary trees for a fixed k, and extend this algorithm to allow for nodes with at most k children.

2.1   Problem Statement

As is the case in hierarchical clustering, we assume that we are given a gene similarity matrix S, which is initially of dimensions n by n. Unlike binary tree clustering, we are interested in joining together groups of size k, where $k > 2$. In this paper we focus on the average linkage method, for which the problem can be formalized as follows. Given n clusters denote by C the set of all subsets of n of size k. Our goal is to find a subset $b \in C$ s.t. $V(b) = max\{V(b')|b' \in C\}$ where V is defined in the following way:

$$V(b) = \sum_{i,j \in b, i < j} S(i,j)$$

That is, $V(b)$ is the sum of the pairwise similarities in b. After finding b, we merge all the clusters in b to one cluster and compute a revised similarity matrix in the following way. Denote by i a cluster which is not a part of b, and let the cluster formed by the merging all the clusters of b be denoted by j. For a cluster m, let $|m|$ denote the number of genes in m, then:

$$S(i,j) = \frac{\sum_{m \in b} |m| S(m,i)}{\sum_{m \in b} |m|}$$

which is similar to the way the similarity matrix is updated in the binary case. This process is repeated $(n-1)/(k-1)$ times until we arrive at a single root cluster, and the tree is obtained.

   Finding b in each step is the most expensive part of the above problem, as we show in the next lemma. In this lemma we use the notion of $W[1]$ hardness. Under reasonable assumptions, a $W[1]$ hard problem is assumed to be *fixed parameter* intractable, i.e. the dependence on k cannot be separated from the dependence on n (see [4] for more details).

Lemma 1. *Denote by $MaxSim(k)$ the problem of finding the first b set for a given k. Then $MaxSim$ is NP-hard for arbitrary k, and $W[1]$ hard in terms of k.*

proof outline: We reduce MAX-CLIQUE to $MaxSim(k)$ by constructing a similarity matrix S_G and setting $S_G(i,j) = 1$ iff there is an edge between i and j in G. Since MAX-CLIQUE is NP and $W[1]$ complete, $MaxSim(k)$ is NP and $W[1]$ hard.

2.2   A Heuristic Algorithm for Constructing k-ary Trees

As shown in the previous section, any optimal solution for the k-ary tree construction problem might be prohibitive even for small values of k, since n is very large. In this section we present a heuristic algorithm, which has a running time of $O(n^3)$ for any k, and reduces to the standard average linkage clustering algorithm when $k = 2$. The algorithm is presented in Figure 1 and works in the following way. Starting with a set of n clusters (initially each gene is assigned to

```
KTree(n, S) {
    C = {1 . . . n}
    for all j ∈ C  // preprocessing step
        L_j = ordered linked list of genes based on similarity to j
        b_j = j ⋃ first k − 1 genes of L_j
    for i = 1 : (n − 1)/(k − 1) {  // main loop
        b = max_{j∈C}{V(b_j)}
        C = C \ b
        Let p = min{m ∈ b}
        for all clusters j ∈ C
```
$$S(p,j) = \frac{\sum_{m \in b} |m| S(m,j)}{\sum_{m \in b} |m|}$$
```
        remove all clusters in b from L_j
        insert p into L_j
        b_j = j ⋃ first k − 1 cluster of L_j
        C = C ⋃ p
        generate L_p from all the clusters in C and find b_p
    }
    return C  // C is a singleton which is the root of the tree
}
```

Fig. 1. Constructing k-trees from expression data.

a different cluster), we generate a linked list of clusters for each cluster i, ordered by their similarity to i in ascending order. For each cluster i we compute $V(b_i)$ where b_i consists of i and the clusters that appear in the first $k - 1$ places on L_i. Next, we find $b = max_i\{V(b_i)\}$, merge all the clusters in b to a single cluster denoted by p and recompute the similarity matrix. After finding b and recomputing S we go over each linked list and delete all the clusters that are a subset of b from all the lists, insert p to each list and recompute b_j for all clusters. In addition, we generate a new linked list for p, and compute b_p.

Note that using this algorithm, it could be that even though j and i are the most similar clusters, j and i will not end up in the same k group. If there is a cluster t s.t. b_t includes i but does not include j, and if $V(b_t) > V(b_j)$, it could be that j and i will not be in the same k cluster. This allows us to use this algorithm to overcome noise and missing values since, even when using this heuristic we still need a strong evidence from other clusters in order to combine two clusters together.

The running time of this algorithm is $O(n^3)$. Generating the L_js lists can be done in $O(n^2 \log n)$, and finding b_j for all genes j can be done in kn time. Thus the preprocessing step takes $O(n^2 \log n)$.

For each iteration of the main loop, it takes $O(n)$ to find b, and $O(nk)$ to recompute S. It takes $O(kn^2 + n^2 + kn)$ to delete all the members of b from all the L_js, insert p into all the L_js and recompute b_j. We need another $O(n \log n + k)$ time to generate L_p and compute b_p. Thus, the total running time of each iteration is $O(kn^2)$. Since the main loop is iterated $(n - 1)/(k - 1)$ time, the total

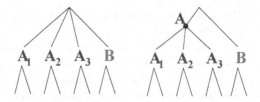

Fig. 2. In the left hand tree k is fixed at 4. This results in a cluster (A) which does not have any internal node associated with it. On the right hand side k is at most 4. Thus, the three subtrees that form cluster A can be grouped together and then combined with cluster B at the root. This results in an internal node that is associated with A.

running time of the main loop is $O(k(n-1)n^2/(k-1)) = O(n^3)$ which is also the running time of the algorithm. The running time of the above algorithm can be improved to $O(n^2 \log n)$. However, since our ordering algorithm operates in $O(n^3)$, this will not reduce the asymptotic running time of our algorithm, and thus we left the details out.

2.3   Reducing the Number of Children

Using a fixed k can lead to clusters which do not have a single node associated with them. Consider for example a dataset in which we are left with four internal nodes after some main loop iterations. Assume $k = 4$ and that the input data is composed of two real clusters, A and B such that three of the subtrees belong to cluster A, while the fourth belongs to cluster B (see Figure 2). If k was fixed, we would have grouped all the subtrees together, which results in a cluster (A) that is not associated with any internal node. However, if we allow a smaller number of children than k we could have first grouped the three subtrees of A and later combine them with B at the root. Note that is this case, nodes with less than k children represent a set of genes that are similar, yet significantly different than the rest of the genes.

We now present a permutation based test for deciding on how many clusters to combine in each iteration of the main loop. There are two possible approaches one can take in order to perform this task. The first is to join as many clusters as possible (up to k) unless the data clearly suggests otherwise. The second is to combine as few clusters as possible unless the data clearly suggests otherwise. Since we believe that in most cases more than 2 genes are co-expressed, in this paper we use the first approach, trying to combine as many clusters as possible unless the data clearly suggests otherwise.

Let $k = 3$ and let $b_c = argmax_i\{V(b_i)\}$ where i goes over all the clusters we currently have. Let d and e be the first and second clusters on L_c respectively. We now wish to decide weather to combine the first two clusters (c and d) or combine all three clusters. Let $max_e = max\{S(c,e), S(d,e)\}$ that is S_e is the maximal similarity between e and one of the two clusters we will combine in any case.. In order to test the relationship between max_e and $S(c,d)$, we perform the following test. In our case, each cluster c is associated with a profile (the average expression value of the genes in c). Assume our dataset contains m experiments, and let p_c, p_d and p_e be the three clusters profiles. Let p be the 3 by m matrix, where every row of p is a profile of one of the clusters. We permute each column of p uniformly and independently at random, and for the permuted p we compute the best (s_1) and second best (s_2) similarities among its rows. We repeat this procedure r times, and in each case test if s_2 is bigger than max_e or smaller. If $s_2 > max_e$ at least αr times (where α is a user defined value between 0 and 1) we combine c and d without e, otherwise we combine all three clusters. Note that if c and d are significantly different from e then it is unlikely that any permutation will yield an s_2 that is lower than max_e, and thus the above test will cause us to separate c and d from e. If c and d are identical to e, then all permutations will yield an s_2 that is equal to max_e, causing us to merge all three clusters. As for the values of α, if we set α to be close to 1 then unless e is very different from c and d we will combine all three clusters. Thus, the closer α is to 1, the more likely our algorithm is to combine all three clusters.

For $k > 3$ we repeat the above test for each $k' = 3 \ldots k$. That is, we first test if we should separate the first two clusters from the third cluster, as described above. If the answer is yes, we combine the first two clusters and move to the next iteration. If the answer is no we apply the same procedure, to test weather we should separate the first three clusters from the fourth and so on. The complexity of these steps is rk^2 for each k' (since we need to compute the pairwise similarities in each permutations), and at most rk^3 for the entire iteration. For a fixed r, and $k << n$ this permutation test does not increase the asymptotic complexity of our algorithm. Note that if we combine $m < k$ clusters, the number of main loop iteration increases. However, since in this case each the iteration takes $O(n^2 m)$ the total running time remains $O(n^3)$.

3   Optimal Leaf Ordering

In this section we discuss how we preserve the pairwise similarity property of the binary tree clustering in our k-ary tree algorithm. This is done by performing optimal leaf ordering on the resulting tree. After formally defining the optimal leaf ordering problem, we present an algorithm that optimally orders the leaves of a binary tree in $O(n^3)$. We discuss a few improvements to this algorithm which further reduces its running time. Next, we extend this algorithm and show how it can be applied to order k-ary trees.

3.1   Problem Statement

First, we formalize the optimal leaf ordering problem, using the following notations. For a tree T with n leaves, denote by $z_1, \cdots, z_n$ the leaves of T and by $v_1 \cdots v_{n-1}$ the $n-1$ internal nodes of T. A **linear ordering consistent with** T is defined to be an ordering of the leaves of T generated by flipping internal nodes in T (that is, changing the order between the two subtrees rooted at v_i, for any $v_i \in T$). See Figure 3 for an example of node flipping.

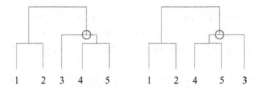

Fig. 3. When flipping the two subtrees rooted at the red circled node we obtain different orderings while maintaining the same tree structure. Since there are $n-1$ internal nodes there are 2^{n-1} possible orderings of the tree leaves.

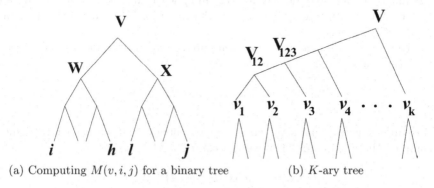

(a) Computing $M(v,i,j)$ for a binary tree   (b) K-ary tree

Fig. 4. (a) A binary tree rooted at V. (b) Computing $M(v,i,j)$ for the subtrees order $1 \ldots k$. For each possible ordering of $1 \ldots k$ we can compute this quantity by adding internal nodes and using the binary tree algorithm.

Since there are $n-1$ internal nodes, there are 2^{n-1} possible linear orderings of the leaves of a binary tree. Our goal is to find an ordering of the tree leaves, that maximizes the sum of the similarities of adjacent leaves in the ordering. This could be stated mathematically in the following way. Denote by Φ the space of the 2^{n-1} possible orderings of the tree leaves. For $\phi \in \Phi$ we define $D^\phi(T)$ to be:

$$D^\phi(T) = \sum_{i=1}^{n-1} S(z_{\phi_i}, z_{\phi_{i+1}})$$

where $S(u,w)$ is the similarity between two leaves of the tree. Thus, our goal is to find an ordering ϕ that maximize $D^\phi(T)$. For such an ordering ϕ, we say that $D(T) = D^\phi(T)$.

3.2   An $O(n^3)$ Algorithm for Binary Trees

Assume that a hierarchical clustering in form of a tree T has been fixed. The basic idea is to create a table M with the following meaning. For any node v of T, and any two genes i and j that are at leaves in the subtree defined by v (denoted $T(v)$), define $M(v,i,j)$ to be the cost of the best linear order of the leaves in $T(v)$ that begins with i and ends with j. $M(v,i,j)$ is defined only if node v is the least common ancestor of leaves i and j; otherwise no such ordering is possible. If v is a leaf, then $M(v, v, v) = 0$. Otherwise, $M(v, i, j)$ can be computed as follows, where w is the left child and x is the right child of v (see Figure 4 (a)):

$$M(v, i, j) = \max_{h \in T(w), l \in T(x)} M(w, i, h) + S(h, l) + M(x, l, j). \tag{1}$$

Let $F(n)$ be the time needed to compute all defined entries in table $(M(v, i, j))$ for a tree with n leaves. We analyze the time to compute Equation 1 as follows: Assume that there are r leaves in $T(w)$ and p leaves in $T(x)$, $r + p = n$. We must first compute recursively all values in the table for $T(w)$ and $T(x)$; this takes $F(r) + F(p)$ time.

To compute the maximum, we compute a temporary table $Temp(i, l)$ for all $i \in T(w)$ and $l \in T(x)$ with the formula

$$Temp(i, l) = \max_{h \in T(w)} M(w, i, h) + S(h, l); \tag{2}$$

this takes $O(r^2 p)$ time since there are rp entries, and we need $O(r)$ time to compute the maximum. Then we can compute $M(v, i, j)$ as

$$M(v, i, j) = \max_{l \in T(x)} Temp(i, l) + M(x, l, j). \tag{3}$$

This takes $O(rp^2)$ time, since there are rp entries, and we need $O(p)$ time to compute the maximum.

Thus the total running time obeys the recursion $F(n) = F(r) + F(p) + O(r^2 p) + O(rp^2)$ which can be shown easily (by induction) to be $O(n^3)$, since $r^3 + p^3 + r^2 p + rp^2 \leq (r + p)^3 = n^3$.

The required memory is $O(n^2)$, since we only need to store $M(v, i, j)$ once per pair of leaves i and j.

For a balanced binary tree with n leaves we need $\Omega(n^3)$ time to compute Equation 1; hence the algorithm has running time $\Theta(n^3)$.

We can further improve the running time of the algorithm in practice by using the following techniques:

Early Termination of the Search. We can improve the computation time for Equation 2 (and similarly Equation 3) by pruning the search for maximum whenever no further improvement is possible. To this end, set $s_{\max}(l) = \max_{h \in T(w)} S(h, l)$. Sort the leaves of $T(w)$ by decreasing value of $M(w, i, h)$, and compute the maximum for Equation 2 processing leaves in this order. Note that if we find a leaf h for which $M(w, i, h) + s_{\max}(l)$ is bigger than the maximum that we have found so far, then we can terminate the computation of the maximum, since all other leaves cannot increase it.

Top-Level Improvement. The second improvement concerns the computation of Equations 2 and 3 when v is the root of the tree. Let w and x be the left and right children of v. Unlike in the other cases, we do not need to compute $M(v, i, j)$

for all combinations of $i \in T(w)$ and $j \in T(x)$. Rather, we just need to find the maximum of all these values $M_{\max} = \max_{i,j} M(v, i, j)$.

Define $\text{Max}(v, i) = \max_{h \in T(v)} M(v, i, h)$. From Equation 1 we have

$$
\begin{aligned}
M_{\max} &= \max_{i \in T(w), j \in T(x)} \ \max_{h \in T(w), l \in T(x)} M(w, i, h) + S(h, l) + M(x, l, j) = \\
&= \max_{h \in T(w), l \in T(x)} \text{Max}(w, h) + S(h, l) + \text{Max}(x, l).
\end{aligned}
$$

Therefore we can first precompute values $\text{Max}(w, h)$ for all $h \in T(w)$ and $\text{Max}(x, l)$ for all $l \in T(x)$ in $O(n^2)$ time, and then find $M_{\max}$ in $O(n^2)$ time. This is in contrast to the $O(n^3)$ time needed for this computation normally.

While the above two improvements do not improve the theoretical running time (and in fact, the first one increases it because of the sorting step), we found in experiments that on real-life data this variant of the algorithm is on average 50–90% faster.

3.3   Ordering *k*-ary Trees

For a *k*-ary tree, denote by $v_1 \ldots v_k$ the *k* subtrees of *v*. Assume $i \in v_1$ and $j \in v_k$, then any ordering of $v_2 \ldots v_{k-2}$ is a possibility we should examine. For a specified ordering of the subtrees of *v*, $M(v, i, j)$ can be computed in the same way we computed *M* for binary trees by inserting $k - 2$ internal nodes that agree with this order (see figure 4 (b)).

Thus, we first compute $M(v_{1,2}, h, l)$ for all *h* and *l* leaves of v_1 and v_2. Next we compute $M(v_{1,2,3}, *, *)$ and so on until we compute $M(v, i, j)$ for this order. This results in the optimal ordering of the leaves when the subtrees order is $v_1 \ldots v_k$. Since there are *k*! possible ordering of the subtrees, going over all *k*! orderings of the subtrees in the manner described above gives rise to a simple algorithm for finding the optimal leaf ordering of a *k*-ary tree. Denote by $p_1 \ldots p_k$ the number of leaves in $v_1 \ldots v_k$ respectfully. Denote by Θ the set of *k*! possible orderings of $1 \ldots k$. The running time of this algorithm is $O(k!n^3)$ as can be seen using induction from the following recursion:

$$
\begin{aligned}
F(n) &= \sum_i F(p_i) + \sum_{\theta \in \Theta} \sum_{i=1}^{k-1} \left(\left(\sum_{j=1}^i p_{\theta(j)} \right)^2 p_{\theta(i+1)} + \left(\sum_{j=1}^i p_{\theta(j)} \right) p_{\theta(i+1)}^2 \right) \\
&\le k! \sum_i p_i^3 + \sum_{\theta \in \Theta} \left(\left(\sum_i p_i \right)^3 - \sum_i p_i^3 \right) \\
&= k!(p_1 + p_2 + \ldots p_k)^3 = k!n^3
\end{aligned}
$$

Where the inequality uses the induction hypothesis. As for space complexity, for two leaves, $i \in v_1$ and $j \in v_k$ we need to store $M(v, i, j)$. In addition, it might be that for two other leaves $m \in v_2$ and $l \in v_{k-1}$ *i* and *j* are two boundary leaves in the internal ordering of the subtrees of *v*, and thus we need to store the distance between them for this case as well. The total number of subdistances we need to store for each pair is at most $k - 2$, since there are only $k - 2$ subtrees between the two leftmost and rightmost nodes, and thus by deleting all subpaths which we do not use we only need $O(kn^2)$ memory for this algorithm.

Though $O(k!n^3)$ is a feasible running time for small ks, we can improve upon this algorithm using the following observation. If we partition the subtrees of v into two groups, v' and v'', then we can compute $M(v)$ for this partition (i.e. when the subtrees of v' are on the right side and the subtrees of v'' on the left) by first computing the optimal ordering on v' and v'' separately, and then combining the result in the same way discussed in Section 3.2. This gives rise to the following divide and conquer algorithm. Assume $k = 2^m$, recursively compute the optimal ordering for all the $\binom{k}{k/2}$ possible partitions of the subtrees of v to two groups of equal size, and merge them to find the optimal ordering of v. In order to compute the running time of this algorithm we introduce the following notations: We denote by $\Gamma(v)$ the set of all the possible partitions of the subtrees of v to two subsets of equal size. For $\gamma \in \Gamma(v)$ let $v_{\gamma(1)}$ and $v_{\gamma(2)}$ be the two subsets and let $p_{\gamma(1)}$ and $p_{\gamma(2)}$ be the number of leaves in each of these subsets. The running time of this algorithm can be computed by solving the following recursion:

$$F(n) = F(p_1 + p_2 \ldots + p_k) = \sum_{i=1}^{k} F(p_i) + D(v)$$

Where

$$D(v) = \sum_{\gamma \in \Gamma(v)} p_{\gamma(1)}^2 p_{\gamma(2)} + p_{\gamma(1)} p_{\gamma(2)}^2 + D(v_{\gamma(1)}) + D(v_{\gamma(2)})$$

and $D(i) = 0$ if i is a singleton containing only one subtree. The following lemma discusses the maximal running time of the divide and conquer approach.

Lemma 2. *Assume $k = 2^m$ then*

$$D(v) \leq \frac{(2^m)!}{\prod_{i=0}^{m-1} (2^i)!} (\sum_{j=1}^{k} p_j)^3 - \sum_{j=1}^{k} p_j^3$$

Proof. By induction on m.
For $m = 1$ we have $k = 2$ and we have already shown in Section 3.2 how to construct an algorithm that achieves this running time for binary trees. So $D(v) = p_1^2 p_2 + p_1 p_2^2 < 2(p_1 + p_2)^3 - p_1^3 - p_2^3$.
Assume correctness for $m - 1$. Let $k = 2^m$. Then for every $\gamma \in \Gamma(v)$ we have

$$p_{\gamma(1)}^2 p_{\gamma(2)} + p_{\gamma(1)} p_{\gamma(2)}^2 + D(v_{\gamma(1)}) + D(v_{\gamma(2)}) \leq$$

$$p_{\gamma(1)}^2 p_{\gamma(2)} + p_{\gamma(1)} p_{\gamma(2)}^2 + \frac{(2^{m-1})!}{\prod_{i=0}^{m-2} (2^i)!} (\sum_{j \in \gamma(1)} p_j)^3 - \sum_{j \in \gamma(1)} p_j^3 +$$

$$+ \frac{(2^{m-1})!}{\prod_{i=0}^{m-2} (2^i)!} (\sum_{j \in \gamma(2)} p_j)^3 - \sum_{j \in \gamma(2)} p_j^3 \leq \frac{(2^{m-1})!}{\prod_{i=0}^{m-2} (2^i!)} (\sum_{j=1}^{k} p_j)^3 - \sum_{j=1}^{k} p_j^3$$

The first inequality comes from the induction hypothesis, and the second inequality arises from the fact that $\gamma(1)$ does not intersect $\gamma(2)$ and $\gamma(1) \bigcup \gamma(2) = \{1 \ldots k\}$. Since $|\Gamma(v)| = \binom{k}{k/2} = \frac{(2^m)!}{(2^{m-1})!(2^{m-1})!}$, summing up on all $\gamma \in \Gamma$ proves the lemma.

It is now easy to see (by using a simple induction proof, as we did for the binary case) that the total running time of this algorithm is: $O(\frac{k!}{\prod_{i=0}^{m-1}(2^i)!}n^3)$ which is faster than the direct extension discussed above. If $2^m < k < 2^{m+1}$ then the same algorithm can be used by dividing the subtrees of v to two groups of size 2^m and $k - 2^m$. A similar analysis shows that in this case the running time is $O(\frac{k!}{\prod_{i=0}^{\lfloor \log k \rfloor -1}(2^i)!}n^3)$, which (using the Sterling approximation) is $O(4^{k+o(k)}n^3)$.

4   Experimental Results

In this section we compare the binary and k-ary clustering using synthetic and real datasets, and show that in all cases we looked at we only gain from using the k-ary clustering algorithm.

We report results that where generated with $k = 4$. We found that the results do not change much when using higher values of k. Due to the fact that the running time increases as a function of k, we concentrate on $k = 4$.

Generated Data: To test the effect of k-ary clustering and ordering on the presentation of the data, we generated a structured input data set. This set represents 30 temporally related genes, each one with 30 time points. In order to reduce the effect of pairwise relationships, we chose 6 of these genes, and manually removed for each of them 6 time points, making these time points missing values. Next we permuted the genes, and clustered the resulting dataset with the three methods discussed above. The results are presented in Figure 5. As can be seen, using optimal leaf ordering (o.l.o.) with binary hierarchical clustering improved the presentation of the dataset, however o.l.o. was unable to overcome missing values, and combined pairs of genes which where similar due to the missing values, but where not otherwise similar to a larger set of genes. Using the more robust k-ary tree algorithm, we where able to overcome the missing values problem. This resulted in the correct structure as can be seen in Figure 5. Note that by simply fusing together binary tree internal nodes one cannot arrive at the correct order achieved by the 4-ary algorithm. The correct order only arises from the fact that we allow the algorithm to combine 4 leaves simultaneously.

Visualizing Biological Datasets: In [9] the authors looked at the chicken immune system during normal embryonic B-cell development and in response to the overexpression of the *myc* gene. This dataset consists of 13 samples of transformed bursal follicle (TF) and metastatic tumors (MT). These samples where organized in decreasing order based on the *myc* overexpression in each of them. 800 genes showing 3 fold change where clustered using Eisen's Cluster [5], and

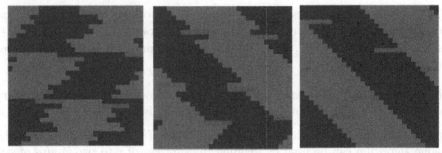

Hierarchical clustering   Binary clustering with o.l.o. 4-tree clustering with o.l.o.

Fig. 5. Color comparison between the three different clustering methods on a manually generated dataset. Green corresponds to decrease in value (-1) and red to increase (1). Gray represents missing values.

based on manual inspection, 5 different clusters where identified. The 5 clusters where separated into two groups. The first contains four of the clusters (A,B,E and D) which (in this order) contain genes that are decreasingly sensitive to *myc* overexpression. The second is the cluster C which contains genes that are not correlated with *myc* overexpression. We compared the results of the three clustering algorithms for this dataset (results not shown, see web supplement at: http://psrg.lcs.mit.edu/ zivbj/WABI02/chick.html). We found that when using the k-ary clustering algorithm these clusters were displayed in their correct order. Furthermore, each cluster is organized (from bottom to top) based on the required level of *myc* overexpression. This allows for an easier inspection and analysis of the data. On the other hand, using binary tree clustering with o.l.o. does not yield the same result and the A and E clusters were broken into two parts. In addition, from looking at the 4-ary tree our algorithm generated we observed that some of its internal nodes that contain less than 4 children correspond to clusters that where identified in the original paper. Had we used a fixed $k = 4$, these clusters might not have had a single node associated with them.

Clustering Biological Datasets: The second biological dataset we looked at is a collection of 79 expression experiments that where performed under different conditions, from [5]. In order to compare our k-ary clustering to the binary clustering we used the MIPS complexes categories (from http://www.mips.biochem.mpg.de). We focused on the 979 genes that appeared in both the dataset and the MIPS database. For each complex family, and each clustering algorithm (binary and 4-ary), we determined the cluster (internal node in the tree) that holds at least half of the genes in that family, and has the highest ratio of genes from that family to the total number of genes in that cluster. We report the results for the 8 families having more than 30 genes in Table 1. For three families (Proteasome, Respiration chain, and Translation) our 4-ary tree contains clusters in which these families are significantly overrepresented. The binary tree does not contain a significant cluster for the Respiration chain complex. In addition, for the Translation complex our 4-ary tree contains

Table 1. Comparison between the binary tree and 4-ary clustering algorithms using the MIPS complexes database. See text for complete details and analysis.

Complex	# genes	binary tree			4-ary tree		
		#∈ complex	cluster size	ratio	#∈ complex	cluster size	ratio
Cytoskeleton	48	48	979	.05	27	438	.06
Intracellular	72	49	644	.08	44	541	.08
Proteasome	35	28	28	1	28	28	1
Replication	48	36	236	.15	31	235	.13
Respiration chain	31	16	111	.14	26	32	.81
RNA processing	107	74	644	.12	107	979	.11
Translation	208	105	106	.99	111	111	1
Transcription	156	156	979	.16	87	438	.20

a cluster in which all 111 genes belong to this family, while the binary tree contains a cluster with less genes (106) out of which one gene does not belong to the Translation complex. These results indicate that our k-ary clustering algorithm is helpful when compared to the binary hierarchical clustering algorithm, and that it does not generate a tree which simply fuses binary tree nodes. Note that for some of the families both the binary and 4-ary algorithms do not contain an internal node that is significantly associated with the complex. This is not surprising since we have only used the top level categorization of MIPS, and thus some of the complexes should not cluster together. However, those that do, cluster better when using the 4-ary algorithm.

5   Summary and Future Work

We have presented an algorithm for generating k-ary clusters, and ordering the leaves of these clusters so that the pairwise relationships are preserved despite the increase in the number of children. Our k-ary clustering algorithm runs in $O(n^3)$ and our ordering algorithm has a running time of $O(4^{k+o(k)}n^3)$ which is on the order of $O(n^3)$ for a constant k. We presented several examples in which the results of our algorithm are superior to the results obtained using binary tree clustering, both with and without ordering.

An interesting open problem is if we can further improve the asymptotic running time of the optimal leaf ordering algorithm. A trivial lower bound for this problem is n^2 which leaves a gap of order n between the lower bound and the algorithm presented in this paper. Note that hierarchical clustering can be performed in $O(n^2)$ (see [6]), thus reducing the running time of the o.l.o. solution would be useful in practice as well.

Acknowledgements

We thank Therese Biedl, Brona Brejová and Tomáš Vinař who made important contributions to Section 3.2. Z.B.J was supported by the Program in Mathematics and Molecular Biology at the Florida State University with funding from the

Burroughs Wellcome Fund. A.M.H. was supported by the Leverhulme foundation, UK and NSERC.

References

1. Z. Bar-Joseph, D. Gifford, and T. Jaakkola. Fast optimal leaf ordering for hierarchical clustering. In *ISMB01*, 2001.
2. A. Ben-Dor, R. Shamir, and Z. Yakhini. Clustering gene expression patterns. *Journal of Computational Biology*, 6:281–297, 1999.
3. R. E. Burkard, Deineko V. G., and G. J. Woeginger. The travelling salesman and the pq-tree. *Mathematics of Operations Research*, 24:262–272, 1999.
4. R. G. Downey and M. R. Fellows. *Parameterized Complexity*. Springer, New-York, NY, 1999.
5. M.B. Eisen, P.T. Spellman, P.O. Brown, and D. Botstein. Cluster analysis and display of genome-wide expression patterns. *PNAS*, 95:14863–14868, 1998.
6. D. Eppstein. Fast hierarchical clustering and other applications of dynamic closest pairs. In *Proceedings of the 9th ACM-SIAM Symp. on Discrete Algorithms*, pages 619–628, 1998.
7. N. Gale, W. C. Halperin, and C.M. Costanzo. Unclassed matrix shading and optimal ordering in hierarchical cluster analysis. *Journal of Classification*, 1:75–92, 1984.
8. G. Gruvaeus and H. Wainer. Two additions to hierarchical cluster analysis. *British Journal of Mathematical and Statistical Psychology*, 25:200–206, 1972.
9. P.E. Neiman and et al. Analysis of gene expression during myc oncogene-induced lymphomagenesis in the bursa of fabricius. *PNAS*, 98:6378–6383, 2001.
10. E. Segal and D. Koller. Probabilistic hierarchical clustering for biological data. In *Recomb02*, 2002.
11. R. Sharan, R. Elkon, and R. Shamir. Cluster analysis and its applications to gene expression data. *Ernst Schering workshop on Bioinformatics and Genome Analysis*, 2001.
12. P. Tamayo and et al. Interpreting patterns of gene expression with self organizing maps: Methods and applications to hematopoietic differentiation. *PNAS*, 96:2907–2912, 1999.

Inversion Medians Outperform Breakpoint Medians in Phylogeny Reconstruction from Gene-Order Data

Bernard M.E. Moret[1], Adam C. Siepel[2], Jijun Tang[1], and Tao Liu[1]

[1] Department of Computer Science
University of New Mexico
Albuquerque, NM 87131, USA
{moret,jtang,tigerliu}@cs.unm.edu
[2] Department of Computer Science and Engineering
University of California at Santa Cruz
Santa Cruz, CA 95064
acs@cse.ucsc.edu

Abstract. Phylogeny reconstruction from gene-order data has attracted much attention over the last few years. The two software packages used for that purpose, **BPAnalysis** and **GRAPPA**, both use so-called breakpoint medians in their computations. Some of our past results indicate that using inversion scores rather than breakpoint scores in evaluating trees leads to the selection of better trees. On that basis, we conjectured that phylogeny reconstructions could be improved by using inversion medians, which minimize evolutionary distance under an inversions-only model of genome rearrangement. Recent algorithmic developments have made it possible to compute inversion medians for problems of realistic size.

Our experimental studies unequivocally show that inversion medians are strongly preferable to breakpoint medians in the context of phylogenetic reconstruction from gene-order data. Improvements are most pronounced in the reconstruction of ancestral genomes, but are also evident in the topological accuracy of the reconstruction as well as, surprisingly, in the overall running time. Improvements are strongest for small average distances along tree edges and for evolutionary scenarios with a preponderance of inversion events, but occur in all cases, including evolutionary scenarios with high proportions of transpositions.

All of our tests were run using our **GRAPPA** package, available (under GPL) at www.cs.unm.edu/~moret/GRAPPA; the next release will include the inversion median software we used in this study. The software used includes **RevMed**, developed by the authors and available at www.cs.unm.edu/~acs, and A. Caprara's inversion median code, generously made available for testing.

Keywords: breakpoint, genome rearrangement, genomic distance, inversion, reversal

R. Guigó and D. Gusfield (Eds.): WABI 2002, LNCS 2452, pp. 521–536, 2002.
© Springer-Verlag Berlin Heidelberg 2002

1   Introduction

Biologists can infer the ordering and strandedness of genes on a chromosome, and thus represent each chromosome by an ordering of signed genes (where the sign indicates the strand). These gene orders can be rearranged by evolutionary events such as inversions and transpositions and, because they evolve slowly, give biologists an important new source of data for phylogeny reconstruction (see, e.g., [7,16,17,19]). Appropriate tools for analyzing such data may help resolve some difficult phylogenetic reconstruction problems. Developing such tools is thus an important area of research—indeed, the recent DCAF symposium was devoted to this topic, as was a workshop at DIMACS.

A natural optimization problem for phylogeny reconstruction from gene-order data is to reconstruct an evolutionary scenario with a minimum number of the permitted evolutionary events on the tree. This problem is NP-hard for most criteria—even the very simple problem of computing the median[1] of *three* genomes under such models is NP-hard [4,18]. All approaches to phylogeny reconstruction for such data must therefore find ways of handling significant computational difficulties. Moreover, because suboptimal solutions can yield very different evolutionary reconstructions, exact solutions are strongly preferred over approximate solutions in all phylogenetic work (see [25]).

For some datasets (e.g., chloroplast genomes of land plants), biologists conjecture that rearrangement events are predominantly *inversions.* In other datasets, transpositions and inverted transpositions are viewed as possible, but their relative preponderance with respect to inversions is unknown, so that it is difficult to define a suitable distance measure based on these three events. Sankoff proposed using the *breakpoint* distance (the number of pairwise gene adjacencies present in one genome but absent in the other), a measure of distance between genomes that is independent of any particular mechanism of rearrangement, to reconstruct phylogenies; the *breakpoint phylogeny*, introduced by Blanchette *et al.* [2], is the most parsimonious tree with respect to breakpoint distances.

The two software packages for reconstructing the breakpoint phylogeny, the original BPAnalysis of Sankoff and Blanchette [21] and the more recent and much faster GRAPPA [14], both use as their basic optimization tool an algorithm for computing the breakpoint median of three genomes. Recent work, however, based on the elegant theory of Hannenhalli and Pevzner [9], has shown that inversion distance can be computed in linear time [1], a development that has allowed us to base the reconstruction process on the inversion score of a tree rather than on its breakpoint score, with significant resulting improvements in the accuracy of reconstructions [13]. Other recent results have shown that the inversion median of three genomes can be obtained quickly for a reasonable range of instances [5,23,24] (in spite of the NP-hardness of the problem). These developments have enabled us to extend GRAPPA by replacing the breakpoint median routine with an inversion median routine—which we conjectured

---

[1] The median of k genomes is a genome that minimizes the sum of the pairwise distances between itself and each of the k given genomes.

would yield better phylogenetic reconstructions (at least when inversions are the dominant mechanism of rearrangement), because inversion medians score significantly better in terms of inversion distance, and are much closer to being unique, than breakpoint medians [24]. Note that it would be even more desirable to use medians based on a general measure of distance that considers transpositions (and inverted transpositions) as well as inversions. Currently, however, efficient algorithms are not available either to find transposition distance or to compute medians that consider transpositions.

In this paper, we present the results of a series of experiments designed to compare the quality of reconstructions obtained by GRAPPA running with breakpoint medians and with inversion medians. To our knowledge, no such comparison has previously been conducted, despite much speculation about the value of inversion medians in phylogeny reconstruction. In brief, we found that inversion medians are strongly preferable to breakpoint medians in all but the most extreme cases, in terms of quality of tree topologies and of ancestral genome reconstruction, and also (surprisingly) preferable in many cases in terms of speed of execution.

The rest of the paper is organized as follows. We begin by reviewing the pertinent prior work. We then introduce the required terminology, present our experimental design, and briefly highlight the main attributes of the median-finding routines described in [5,23,24]. Next we present a cross-section of our results and discuss their implications. Finally, we conclude with suggestions for further work.

2   Prior Results

BPAnalysis. Blanchette *et al.* [2] proposed the breakpoint phylogeny (finding the tree with the fewest breakpoints) and developed a reconstruction method, BPAnalysis [21], for that purpose. Their method examines every possible tree topology in turn and for each topology, it generates a set of ancestral genomes so as to minimize the total breakpoint distance in the tree. This method returns good results, but takes exponential time: the number of topologies is exponential and generating a set of ancestral genomes is achieved through an unbounded iterative process that must solve an instance of the Travelling Salesperson Problem (TSP) for each internal node at each iteration, so that the total running time is exponential in both the number of genes and the number of genomes.

GRAPPA. We reimplemented BPAnalysis in order to analyze our larger datasets and also to experiment with alternative approaches. Our program, called GRAPPA [14], includes all of the features of BPAnalysis, but is more flexible and runs up to six orders of magnitude faster [12]. It also allows the user to base the optimization on a tree's inversion score or on its breakpoint score (although medians are still computed with the TSP methods of Sankoff and Blanchette, which is optimal only for breakpoint medians).

Inversion distance and inversion medians. Inversion distance can be computed in linear time [1]; an efficient implementation of this algorithm is provided in GRAPPA. As mentioned earlier, computing the inversion (or breakpoint) median

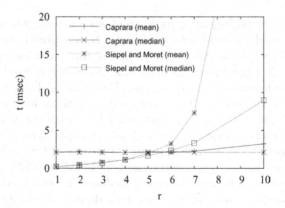

Fig. 1. Mean and median running times (in ms) of the two algorithms for three genomes of 50 genes as a function of the evolutionary rate r (expected number of events per edge) under an inversion-only scenario. Values shown are based on 100 experiments. When r is large, mean times significantly exceed median times, because the algorithms occasionally take much longer than usual to find a median.

of three signed permutations is an NP-hard problem. Breakpoint medians can be computed very efficiently for reasonable problem sizes [14]. Two algorithms have recently been proposed for computing inversion medians (also called "reversal medians") and shown to run efficiently (although much more slowly than the algorithm for breakpoint medians) for a range of parameters that covers most organellar genomes; one by Caprara [5] and one by Siepel and Moret [24] (subsequently refined in [22,23]). Both algorithms use branch-and-bound strategies, but the algorithm of Siepel and Moret directly searches the space of genome rearrangements for an optimal solution, using the metric property of inversion distance for bounding, while the algorithm of Caprara uses edge contraction on a "multibreakpoint graph" (a version of the breakpoint graph generalized to accommodate more than two signed permutations). The edge-contraction operation modifies a multibreakpoint graph by removing an edge and its adjoining nodes, and making appropriate adjustments to *matchings* associated with the graph. Caprara's branch-and-bound algorithm uses the property that the best solution to an instance of the problem can be expressed in terms of the best solutions to subproblems in which edges have been contracted. (The algorithms of Caprara and of Siepel and Moret are both moderately complicated, and fuller descriptions are beyond the scope of this paper; we refer readers to [5] and [22] for details). The algorithm of Caprara generally runs faster than the other, although the "sorting median" refinement of Siepel is faster when edge lengths are small between permutations, as seen in Figure 1. Both algorithms are sensitive to the distances separating input permutations (although Caprara's much less so); hence their use in phylogeny reconstruction depends on relatively short edge lengths between nodes. Timings in this study use Caprara's algorithm; note, however, that a dynamic switch to the algorithm of Siepel and Moret for small distances (which are common within most phylogenies) would considerably improve execution time.

3   Background and Terminology

3.1   Evolutionary Events

When each genome has the same set of genes and each gene appears exactly once, a genome can be described by an ordering (circular or linear) of these genes, each gene given with an orientation that is either positive (g_i) or negative $(-g_i)$. Let G be the genome with signed ordering $g_1, g_2, \ldots, g_n$. An *inversion* between indices i and j, $i \leq j$, produces the genome with linear ordering

$$g_1, g_2, \ldots, g_{i-1}, -g_j, -g_{j-1}, \ldots, -g_i, g_{j+1}, \ldots, g_n$$

A *transposition* on the ordering G acts on three indices, i, j, k, with $i \leq j$ and $k \notin [i, j]$, picking up the interval $g_i, g_{i+1}, \ldots, g_j$ and inserting it immediately after g_k. Thus the genome G above (for $k > j$) is replaced by

$$g_1, \ldots, g_{i-1}, g_{j+1}, \ldots, g_k, g_i, g_{i+1}, \ldots, g_j, g_{k+1}, \ldots, g_n$$

An *inverted transposition* is a transposition followed by an inversion of the transposed piece. The *distance* between two gene orders is the minimum number of inversions, transpositions, and inverted transpositions needed to transform one gene order into the other. When only one type of event occurs in the model, we speak of *inversion distance* or *transposition distance*.

Given two genomes G and G' on the same set of genes, a *breakpoint* in G is defined as an ordered pair of genes (g_i, g_j) such that g_i and g_j appear consecutively in that order in G, but neither (g_i, g_j) nor $(-g_j, -g_i)$ appear consecutively in that order in G'. The number of breakpoints in G relative to G' is the *breakpoint distance* between G and G'.

The *Nadeau-Taylor* model [15] of genome evolution, as generalized by Wang and Warnow [27], uses only genome rearrangement events, so that all genomes have equal gene content. The model assumes that each of the three types of events obeys a Poisson distribution on each edge—with the three means for the three types of events in some fixed ratio.

3.2   Model Trees: Simulating Evolution

A model tree is a rooted binary tree in which each edge e has an associated non-negative real number, λ_e, denoting the expected number of events on e. The model tree also has a weight parameter, a triple of values which defines the probability that a rearrangement event is an inversion, transposition, or inverted transposition. We denote this triple by $(1 - a - b, a, b)$.

Given a model tree with N leaves, a set of N "contemporary" gene orderings can be generated as follows. First, the root is labeled with the identity gene ordering $g_1, g_2, \ldots, g_n$; then the tree is traversed recursively, and each node is assigned a label that is derived by applying random rearrangements to the label of its parent. Suppose a node u is separated from its parent p by an edge having expected number of events λ_e, and suppose the parent is labeled with gene ordering π_p. The label π_u for u is determined by drawing numbers of inversions, transpositions, and inverted transpositions from the appropriate Poisson distributions (having expected values $\lambda_e(1 - a - b)$, $\lambda_e a$, and $\lambda_e b$, respectively) and applying those events to π_p in random order, at randomly-selected indices. If an effective reconstruction algorithm is applied to the labels at the leaves of such a tree, it should infer an evolutionary history resembling the model tree.

3.3   Labeling Internal Nodes

One of the major advantages of median-based phylogeny reconstruction over alternative methods (such as the encoding methods of [13,26]) is that it estimates the configurations of ancestral genomes as well as the topology of the tree. The approach proposed by Sankoff and Blanchette to estimate ancestral genomes is iterative, using a local optimization strategy. It is applied in turn to each plausible tree topology; in the end, the topologies are selected that require the fewest total breakpoints to explain. After initial labels have been assigned in some way to internal (ancestral) nodes in a given topology, the procedure repeatedly traverses the tree, computing for each node the breakpoint median of its three neighbors and using the median as the new label if this change improves the overall breakpoint score—the entire process is also known as "Steinerization." The median-of-three subproblems are transformed into instances of the Travelling Salesperson Problem (TSP) and solved optimally. The overall procedure is a heuristic without any approximation guarantee, but does well in practice on datasets with a small number of genomes.

GRAPPA uses the same overall iterative strategy and also solves the median-of-three problem in its TSP formulation to obtain potential labels for internal nodes. GRAPPA, however, has the option of accepting a relabeling of an internal node based on either the breakpoint score (as in BPAnalysis) or the inversion score of the tree.

3.4   Performance Criteria

Let T be a tree leaf-labelled by the set S. Deleting some edge e from T produces a bipartition π_e of S into two sets. Let T be the true tree and let T' be an estimate of T. Then the *false negatives* of T' with respect to T are those non-trivial bipartitions[2] that appear in T, but not in T'; conversely, the *false positives* of T' with respect to T are those non-trivial bipartitions that appear in T', but not in T. The numbers of false positives and false negatives are normalized by dividing by the number of non-trivial bipartitions of T, to obtain the *fraction of false positives* and the *fraction of false negatives*, respectively. If

---

[2] An edge of zero length is said to produce a *trivial* bipartition. If an edge e' in T' has zero length but a corresponding edge e in T (one producing the same bipartition) does not have zero length, e will count as a false negative; but if e' and e both have zero length, no penalty will occur. The converse holds for false positives.

the fraction of false negatives is 0, then T' equals T or refines it; if the fraction of false positives is also 0, then T' equals T (assuming differences in zero-length edges are not considered important).

4 Our Experimental Design

We designed a set of experiments to assess the impact of replacing breakpoint medians with inversion medians on three critical aspects of phylogenetic reconstruction: speed of execution, topological accuracy of reconstructed trees, and accuracy of estimated ancestral labels. As mentioned, we sought to test the conjecture that exact inversion medians—which score better than breakpoint medians in terms of inversion distance and are more unique [24]—would lead to more accurate reconstructions (both in terms of topology and ancestral labels); at the same time, we sought to characterize the running-time penalty for using the considerably slower inversion median computations. We mostly used simulated datasets (where we know the true tree and the true ancestral labels and thus can directly assess accuracy), but also used two datasets of chloroplast genomes from a number of land plants and algae (where we can compare tree scores and running times). A simulated dataset is determined by four parameters: (i) the number of genes n; (ii) the number of genomes N; (iii) the triple $(1 - a - b, a, b)$ describing the proportions of inversions, transpositions, and inverted transpositions used in the simulated evolution; and (iv) the amount of evolution taking place, represented by r, the expected number of evolutionary events occurring along a tree edge. A simulated dataset is obtained by selecting a tree topology uniformly at random, labeling its edges uniformly with an expected number r of rearrangement events (that is, for simplicity, we use the same λ_e value everywhere), and simulating evolution down the tree, as explained earlier. Because we know the true tree and the true ancestral labels, we can compare the reconstructions to the true tree in terms of topology as well as in terms of ancestral labels. In order to assess variability, we repeated all simulation experiments for at least 10 datasets generated using the same parameters.

Because computing the exact inversion median of three genomes is much more expensive than computing their exact breakpoint median (while both tasks are NP-hard, the breakpoint version has a much simpler structure), replacing breakpoint medians by inversion medians was expected to cause considerable slowdown. However, if inversion medians lead quickly to better tree scores, some performance gain can accrue from more effective pruning of tree topologies. The net effect on running time will thus be the result of a trade-off between more expensive median computations and better pruning. We evaluated this net effect by running the fastest version of GRAPPA, with the improved bounding and layered search described in [12]; we used both breakpoint medians (evaluated with inversion scores, as previously supported) and true inversion medians (supported by new extensions). We compared both total running times and pruning rates, the latter being of interest as an implementation-independent measure.

To assess improvements in accuracy, we ran experiments with fixed evolutionary rate and increasing numbers of genomes. We repeated experiments for several different numbers of genes—a parameter that affects computational cost more than it affects accuracy, but that does have an impact on accuracy through discretization (with a small number of genes, the range of possible values is limited and any error is therefore magnified). To assess topological accuracy, we ran GRAPPA with breakpoint medians and with inversion medians and compared the best reconstructed topologies with the true tree. While the trees returned by GRAPPA are ostensibly binary, they often contain edges of zero length and thus are not fully resolved. (Such lack of resolution is much more pronounced with breakpoint medians than with inversion medians.) We thus measured both false positive and false negative rates (a reconstruction can avoid introducing false positives by leaving relationships unresolved, but only at the cost of false negatives).

To assess the accuracy of reconstructed ancestral genomes, we ran GRAPPA on the leaf genomes as before, but this time with a single topology—that of the true tree. In effect, we used GRAPPA to score a single tree by computing ancestral genomes for the internal nodes of the tree. The reconstructed tree is thus always isomorphic to the true tree, and we can compare internal nodes directly, without having to adjust for inconsistent topologies or lack of branch resolution. We assessed overall accuracy by summing, over all internal nodes of the tree, the pairwise distances (in both inversion and breakpoint versions, with no significant changes in outcomes) between the true and the reconstructed ancestral genomes and normalizing this sum by the number of internal nodes in the tree (for independence from tree size). This simple measure describes how far are the ancestral nodes of the reconstructed tree (in the aggregate) from being precisely correct. (Notice that this measure has the potential quickly to become saturated when reconstructions are poor, because errors accumulate as one moves from leaves to root; we found, however, that reconstructed trees were good enough to avoid this problem).

To assess robustness of results, we ran experiments with many different evolutionary scenarios, including inversions only $(1,0,0)$, equally weighted inversions, transpositions, and inverted transpositions $(\frac{1}{3}, \frac{1}{3}, \frac{1}{3})$, a weighting of $(\frac{2}{3}, \frac{1}{6}, \frac{1}{6})$ that is believed to be more realistic (recommended by Sankoff), and the completely mismatched transposition-only scenario, $(0,1,0)$. We can expect inversion medians to be beneficial in inversion-only evolutionary scenarios, but need to evaluate their usefulness in more adverse scenarios, including the completely mismatched scenarios where we have $a + b = 1$. (One argument for breakpoints is that they provide a model-independent measure: our goal then must be to test whether using a model-dependent measure such as inversion distance provides sufficient benefits when the model is well matched and avoids excessive problems when the model is poorly matched.)

5   Results and Discussion

5.1   Speed

We used two real datasets of chloroplast genomes that we had previously analyzed using a variety of methods. Our first set contains 12 *Campanulaceae* plus a

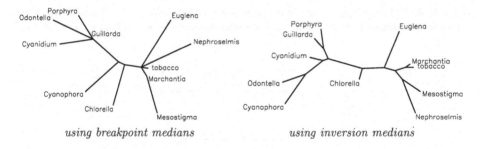

Fig. 2. Phylogenies of 11 plants based on chloroplast gene orders.

Tobacco outgroup, for a total of 13 taxa, each with 105 genes. Evolutionary rates on this set appear low and most evolutionary changes are believed to have occurred through inversions [6]. Earlier analyses reduced the running times from an original estimate of several centuries with BPAnalysis to 9 hours on a fast workstation [12], with the best-scoring tree having an inversion score of 67. Running the same code with the inversion median routine reduced the running time to just one hour on the same machine, while yielding many trees with an inversion score of 64—but with topologies identical to the previous best trees. (Bourque and Pevzner [3] had found one tree with a score of 65 using a somewhat different approach, while Larget *et al.* [11] just reported finding a number of trees that, like ours, required 64 inversions—again these trees have the same topologies as our previous best trees.) While the inversion median routine itself was much slower than the breakpoint median routine, the pruning rate more than made up for it: of the 13.7 billion trees to be examined, all but 100,000 were pruned away when using the inversion medians (a pruning rate of over 99.999%), while 8.7 million remained unpruned when using breakpoint medians (a pruning rate of 99.94%).

Our second set has 11 taxa (including land plants as well as red and green algae), each with 33 genes. This set was also analyzed with a variety of methods [20]; unlike the *Campanulaceae* set, it has high divergence, with intertaxa inversion distances averaging over 10. The analyses of [20] used a variety of techniques, including GRAPPA with breakpoint medians, which yielded, in about 14 hours of computation (with a pruning rate of 25%), trees with inversion scores of 95; using inversion medians brought the pruning rate up to 99.9%, the running time down to $\frac{1}{4}$ hour, and the inversion score down to a very low 82, with complete binary resolution, as illustrated in Figure 2. The new tree (on the right-hand side) has monophyletic greens, correctly groups together tobacco and Marchantia, and offers a resolution of red algae that, while not the accepted one (the placement of Cyanophora, in particular, is curious), is not entirely impossible; in contrast, the old tree has poor resolution and does not group tobacco and Marchantia. Of interest in this reconstruction is the fact that a method predicated on an inversion model did much better (both in biological terms and in terms of scoring) than one using the neutral model of breakpoints, which could be viewed as additional support for the conjecture that most of the rearrangements in the chloroplast genome have occurred through inversions.

Table 1. Ratios of the running time of GRAPPA with breakpoint medians to that with inversion medians for two genome sizes, under three different evolutionary scenarios.

	$(1,0,0)$		$(\frac{2}{3},\frac{1}{6},\frac{1}{6})$		$(\frac{1}{3},\frac{1}{3},\frac{1}{3})$	
For 10 taxa:	$r=2$	$r=4$	$r=2$	$r=4$	$r=2$	$r=4$
$n=50$	1.01	0.90	1.07	1.12	1.36	1.25
$n=100$	1.00	1.20	1.43	1.08	2.90	1.93

	$(1,0,0)$		$(\frac{2}{3},\frac{1}{6},\frac{1}{6})$		$(\frac{1}{3},\frac{1}{3},\frac{1}{3})$	
For 11 taxa:	$r=2$	$r=4$	$r=2$	$r=4$	$r=2$	$r=4$
$n=50$	1.02	0.93	0.99	0.99	1.22	1.13
$n=100$	1.01	1.00	1.01	1.02	1.72	1.43

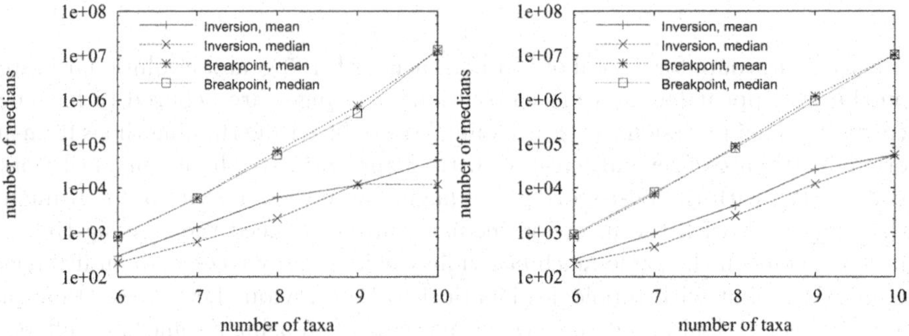

Fig. 3. The number of calls made to each median routine as a function of the number of taxa, for 30 genes and two values of r, plotted on a semilogarithmic scale.

We ran a large number of tests on simulated data as well. Table 1 shows the ratio of the running times with breakpoint medians to the running times with inversion medians. Note that the ratios are very close to 1, indicating that the cost of the inversion median computation is neatly compensated by the reduction in the number of passes necessary to score a tree and by the improved pruning rate. Whenever the new code runs faster, it is a consequence of significantly better pruning due to tighter bounds. (The bounds are computed with exactly the same formula, relying only on the triangle inequality, but the distance estimates based on inversion distances are evidently much better than those based on breakpoint medians.) Figure 3 shows the number of calls made to each of the median procedures. The use of inversion medians, by deriving better initial solutions and by tightening upper bounds quickly, considerably reduces the number of trees that must be examined and, for these trees, the number of passes required to score them.

5.2   Topological Accuracy

Figure 4 shows the fractions of false negative edges in the reconstructions for $r = 2$, $n = 30$, and three different proportions of rearrangement events. False posi-

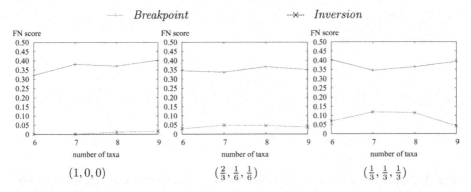

Fig. 4. False negative (FN) scores as a function of the number of taxa for 30 genes at evolutionary rate of $r = 2$.

tive fractions are negligible for both breakpoint and inversion medians—although even here, the values for inversion medians improve on those for breakpoint medians. These results are typical of what we observed with other parameter settings. The rate of false negatives is quite high for breakpoint medians, in good part because breakpoint medians are much more frequently "trivial" than inversion medians. That is, a valid breakpoint median for three gene orderings is frequently one of the orderings itself; a node in the tree and its parent are then assigned the same ordering and a zero-length edge results, which counts as a false negative as long as the corresponding edge in the true tree is nontrivial. Zero-length edges are undesirable for other reasons as well (see below), so that their avoidance constitutes an important advantage of inversion medians over breakpoint medians. Remarkably, the percentage of false negatives in reconstructions using the inversion median hovers at or below 10, for all scenarios and all tree sizes tested— demonstrating extraordinary accuracy in reconstruction even when only one third of the events are in fact inversions. In contrast, the reconstructions based on breakpoint medians fare poorly under all scenarios, although their performance is, as expected, insensitive to the proportions of the three types of events.

As suggested by these small fractions of false positive and negative edges, GRAPPA was highly effective at accurately reconstructing model trees. Indeed, in most cases, the program found a tree that was identical in topology to the true tree (it generally reports multiple trees of equal score, each of which has a different topology). Figure 5 shows the percentage of cases in which at least one exact reconstruction was obtained, under various conditions. With less favorable evolutionary scenarios (having a substantial percentage of transpositions), we see a slow decrease in performance, but the gap between breakpoint medians and inversions medians remains pronounced. Surprisingly, even in unfavorable cases, the percentage of perfect topologies obtained using inversion medians remains high (Figure 6). In the course of performing these experiments, we found that many more trees of equal score tend to be returned when using breakpoint medians, presumably because breakpoint medians are highly non-unique [24] and because trivial medians are common. We also

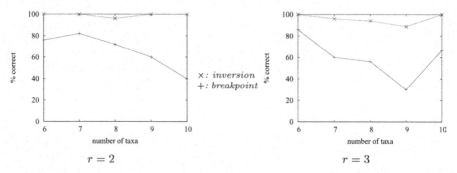

Fig. 5. Percentage of datasets, as a function of the number of taxa, for which GRAPPA returns at least one topology identical to the true tree; all datasets have 30 genes and were generated under an inversion-only scenario. Each point is based on 50 datasets.

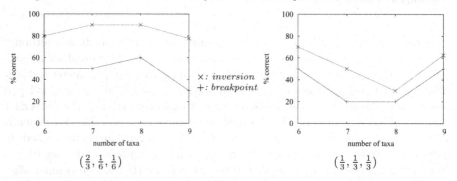

Fig. 6. Percentage of datasets, as a function of the number of taxa, for which GRAPPA returns at least one topology identical to the true tree; all datasets have 30 genes and were generated with an evolutionary rate of $r = 2$. Each point is based on 10 datasets.

found in many cases that GRAPPA fails to find an exact reconstruction with the breakpoint median because of false negatives: that is, it finds trees consistent with the true tree, except that they have many unresolved nodes. The main advantages of the inversion median with respect to topology, then, seem to be that it produces more fully-resolved trees and fewer equally scoring trees.

Also of interest, in regard to topological accuracy, is the effect on overall parsimony scores of using the inversion median. While parsimony scores (the sum of the minimum inversion distances along the edges of the tree) do not have direct topological significance, the goal of the GRAPPA optimization is to obtain a most parsimonious tree, so that any reduction in the score is an indication that the algorithm is behaving more effectively. Table 2 shows percent reductions for a variety of settings; note that reductions of 3% or more generally translate into significant changes in tree topology.

5.3   Accuracy of Ancestral Genomes

Figure 7 shows the average inversion distance between corresponding ancestral genomes in true and reconstructed trees, as a function of the number of

Table 2. The reduction (in percent) in the parsimony score of trees when using inversion medians, for two genome sizes, under three different evolutionary scenarios.

For 10 taxa:

	$(1,0,0)$		$(\frac{2}{3},\frac{1}{6},\frac{1}{6})$		$(\frac{1}{3},\frac{1}{3},\frac{1}{3})$	
	$r=2$	$r=4$	$r=2$	$r=4$	$r=2$	$r=4$
$n=50$	3.2%	5.8%	2.7%	5.1%	1.6%	6.2%
$n=100$	2.7%	2.4%	1.5%	2.1%	1.9%	2.0%

For 11 taxa:

	$(1,0,0)$		$(\frac{2}{3},\frac{1}{6},\frac{1}{6})$		$(\frac{1}{3},\frac{1}{3},\frac{1}{3})$	
	$r=2$	$r=4$	$r=2$	$r=4$	$r=2$	$r=4$
$n=50$	4.8%	9.9%	2.1%	3.9%	2.9%	5.4%
$n=100$	0.0%	2.8%	0.5%	1.6%	3.5%	2.7%

taxa for fixed evolutionary rates under different evolutionary scenarios. This measure was computed by fixing tree topologies and letting GRAPPA simply label ancestral nodes. Again, the high degree of accuracy of reconstructions based on inversion medians is striking and in sharp contrast with the poor reconstructions obtained with breakpoint medians. Even in the optimal case of inversion-only scenarios, the results exceeded our expectations, as almost all internal genomes are correctly reconstructed; also of note is that, in the case of transposition-only scenarios. the inversion-based reconstructions remain superior to the breakpoint-based ones, in spite of the complete mismatch of evolutionary models. We have shown elsewhere that inversion medians provide highly accurate estimates of actual intermediate genomes when $N = 3$ and evolution occurs by inversions only [24]. Our results here indicate that they remain accurate—and clearly superior to breakpoint medians—as N is increased (with multiple intermediate nodes determined by Sankoff and Blanchette's Steinerization method) and as other types of rearrangement events are introduced.

6   Conclusions and Future Work

Finding the inversion median of three genomes is a complex optimization task; finding the median of three genomes under a mix of evolutionary events (inversions, transpositions, and inverted transpositions) has not yet been addressed and seems considerably harder. Yet our experiments indicate that further research on these problems should prove very useful: the reconstructions we have obtained by using inversion medians are clearly better, in every respect, than those obtained with breakpoint medians. In particular, we obtained very accurate reconstructions of the ancestral genomes, a crucial step in the scoring of a particular tree topology as well as an important biological datum.

The results we have presented here, along with other results on handling transversions and gene duplication (see [8,10]) justify a cautious optimism: over the next few years, the reconstruction of phylogenies from gene-order data should become applicable to much larger datasets, extending the approach from organellar genomes to some nuclear genomes. We also expect that some of the lessons learned in the process will yield distinct improvement to the

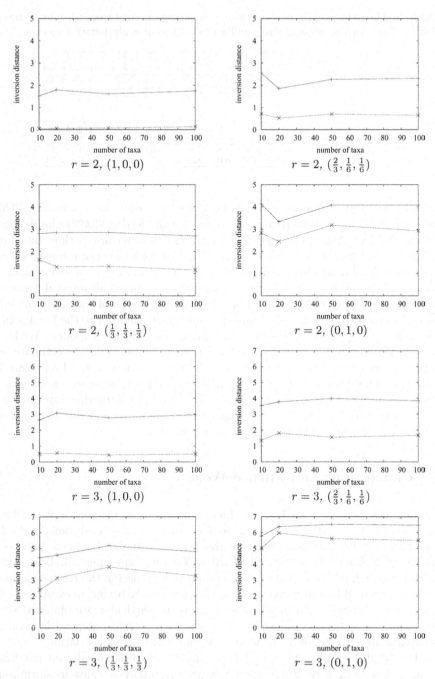

Fig. 7. Average inversion distance between corresponding ancestral genomes in the true tree and the reconstructed tree as a function of the number of taxa, under various evolutionary scenarios, for 30 genes. Upper curve corresponds to reconstructions based on breakpoint medians, lower curve to reconstructions based on inversion medians.

reconstruction of phylogenies from simpler molecular data (such as RNA, DNA, or amino-acid sequences).

Acknowledgements

We thank Alberto Caprara from the Università di Bologna for letting us use his code for inversion medians. Bernard Moret's work is supported by the National Science Foundation under grants ACI 00-81404, EIA 01-13095, EIA 01-23177, and DEB 01-20709.

References

1. D. Bader, B. Moret, and M. Yan. A fast linear-time algorithm for inversion distance with an experimental comparison. *J. Comput. Biol.*, 8(5):483–491, 2001.
2. M. Blanchette, G. Bourque, and D. Sankoff. Breakpoint phylogenies. In S. Miyano and T. Takagi, editors, *Genome Informatics 1997*, pages 25–34. Univ. Academy Press, 1997.
3. G. Bourque and P. Pevzner. Genome-scale evolution: reconstructing gene orders in the ancestral species. *Genome Research*, 12:26–36, 2002.
4. A. Caprara. Formulations and hardness of multiple sorting by reversals. In *Proc. 3rd Int'l Conf. on Comput. Mol. Biol. RECOMB99*, pages 84–93. ACM Press, 1999.
5. A. Caprara. On the practical solution of the reversal median problem. In *Proc. 1st Workshop on Algs. in Bioinformatics WABI 2001*, volume 2149 of *Lecture Notes in Computer Science*, pages 238–251. Springer-Verlag, 2001.
6. M. Cosner, R. Jansen, B. Moret, L. Raubeson, L. Wang, T. Warnow, and S. Wyman. A new fast heuristic for computing the breakpoint phylogeny and experimental phylogenetic analyses of real and synthetic data. In *Proc. 8th Int'l Conf. on Intelligent Systems for Mol. Biol. ISMB-2000*, pages 104–115, 2000.
7. S. Downie and J. Palmer. Use of chloroplast DNA rearrangements in reconstructing plant phylogeny. In P. Soltis, D. Soltis, and J. Doyle, editors, *Plant Molecular Systematics*, pages 14–35. Chapman and Hall, 1992.
8. N. El-Mabrouk. Genome rearrangement by reversals and insertions/deletions of contiguous segments. In *Proc. 11th Ann. Symp. Combin. Pattern Matching CPM 00*, volume 1848 of *Lecture Notes in Computer Science*, pages 222–234. Springer-Verlag, 2000.
9. S. Hannenhalli and P. Pevzner. Transforming cabbage into turnip (polynomial algorithm for sorting signed permutations by reversals). In *Proc. 27th Ann. Symp. Theory of Computing STOC 95*, pages 178–189. ACM Press, 1995.
10. S. Hannenhalli and P. Pevzner. Transforming mice into men (polynomial algorithm for genomic distance problems). In *Proc. 36th Ann. IEEE Symp. Foundations of Comput. Sci. FOCS 95*, pages 581–592. IEEE Press, 1995.
11. B. Larget, J. Kadane, and D. Simon. A Markov chain Monte Carlo approach to reconstructing ancestral genome rearrangements. Technical Report, Carnegie Mellon University, Pittsburgh, PA, 2002. Available at www.stat.cmu.edu/tr/tr765/.
12. B. Moret, J. Tang, L.-S. Wang, and T. Warnow. Steps toward accurate reconstructions of phylogenies from gene-order data. *J. Comput. Syst. Sci.*, 2002. in press.

13. B. Moret, L.-S. Wang, T. Warnow, and S. Wyman. New approaches for reconstructing phylogenies from gene-order data. In *Proc. 9th Int'l Conf. on Intelligent Systems for Mol. Biol. ISMB 2001*, volume 17 of *Bioinformatics*, pages S165–S173, 2001.

14. B. Moret, S. Wyman, D. Bader, T. Warnow, and M. Yan. A new implementation and detailed study of breakpoint analysis. In *Proc. 6th Pacific Symp. Biocomputing PSB 2001*, pages 583–594. World Scientific Pub., 2001.

15. J. Nadeau and B. Taylor. Lengths of chromosome segments conserved since divergence of man and mouse. *Proc. Nat'l Acad. Sci. USA*, 81:814–818, 1984.

16. R. Olmstead and J. Palmer. Chloroplast DNA systematics: a review of methods and data analysis. *Amer. J. Bot.*, 81:1205–1224, 1994.

17. J. Palmer. Chloroplast and mitochondrial genome evolution in land plants. In R. Herrmann, editor, *Cell Organelles*, pages 99–133. Springer Verlag, 1992.

18. I. Pe'er and R. Shamir. The median problems for breakpoints are NP-complete. *Elec. Colloq. on Comput. Complexity*, 71, 1998.

19. L. Raubeson and R. Jansen. Chloroplast DNA evidence on the ancient evolutionary split in vascular land plants. *Science*, 255:1697–1699, 1992.

20. L. A. Raubeson, B. M. Moret, J. Tang, S. K. Wyman, and T. Warnow. Inferring phylogenetic relationships using whole genome data: A case study of photosynthetic organelles and chloroplast genomes. Technical Report TR-CS-2001-19, U. of New Mexico, Albuquerque, New Mexico, 2001.

21. D. Sankoff and M. Blanchette. Multiple genome rearrangement and breakpoint phylogeny. *J. Comp. Biol.*, 5:555–570, 1998.

22. A. Siepel. Exact algorithms for the reversal median problem. Master's thesis, U. New Mexico, Albuquerque, NM, 2001. Available at www.cs.unm.edu/~acs/thesis.html.

23. A. Siepel. An algorithm to find all sorting reversals. In *Proc. 6th Int'l Conf. on Comput. Mol. Biol. RECOMB02*. ACM Press, 2002. to appear.

24. A. Siepel and B. Moret. Finding an optimal inversion median: experimental results. In *Proc. 1st Workshop on Algs. in Bioinformatics WABI 2001*, volume 2149 of *Lecture Notes in Computer Science*, pages 189–203. Springer-Verlag, 2001.

25. D. Swofford, G. Olson, P. Waddell, and D. Hillis. Phylogenetic inference. In D. Hillis, C. Moritz, and B. Mable, editors, *Molecular Systematics, 2nd ed.*, chapter 11. Sinauer Associates, 1996.

26. L.-S. Wang, R. Jansen, B. Moret, L. Raubeson, and T. Warnow. Fast phylogenetic methods for the analysis of genome rearrangement data: an empirical study. In *Proc. 7th Pacific Symp. Biocomputing PSB 2002*, pages 524–535. World Scientific Pub., 2002.

27. L.-S. Wang and T. Warnow. Estimating true evolutionary distances between genomes. In *Proc. 33rd Symp. on Theory of Comp. STOC01*, pages 637–646. ACM Press, 2001.

Modified Mincut Supertrees

Roderic D.M. Page*

DEEB, IBLS, University of Glasgow
Glasgow G12 8QQ, UK
r.page@bio.gla.ac.uk

Abstract. A polynomial time supertree algorithm could play a key role in a divide-and-conquer strategy for assembling the tree of life. To date only a single such method capable of accommodate conflicting input trees has been proposed, the MINCUTSUPERTREE algorithm of Semple and Steel. This paper describes this algorithm and its implementation, then illustrates some weaknesses of the method. A modification to the algorithm that avoids some of these problems is proposed. The paper concludes by discussing some practical problems in supertree construction.

1 Introduction

A central goal of systematics is the construction of a "tree of life," a tree representing the relationships among all living things. One approach to this goal is to assemble a "supertree" from many smaller trees, inferred for different sets of taxa using a variety of data types and (in many cases) a range of phylogenetic methods [1]. Steel et al. [2] list five desirable properties of a supertree method:

1. Changing the order of the trees in the set of input trees $\mathcal{T}$ does not change the resulting supertree.
2. Relabeling the input species results in the corresponding relabeling of species in the supertree.
3. If there are one or more parent trees with which every tree in $\mathcal{T}$ is compatible then the supertree is one of those parent trees.
4. Any leaf that occurs in at least one input tree occurs in the supertree.
5. The supertree can be computed in polynomial time.

Currently the most widely used supertree algorithm in phylogenetics is Matrix Representation using Parsimony (MRP) [3,4]. MRP encodes the input trees into binary characters, and the supertree is constructed from the resulting data matrix using a parsimony tree building method. The MRP supertree method

* I thank David Bryant, James Cotton, Mike Steel, Joe Thorley, and three anonymous referees for their helpful comments on the manuscript. Computational resources were provided by a grant from the Wolfson Foundation.

R. Guigó and D. Gusfield (Eds.): WABI 2002, LNCS 2452, pp. 537–551, 2002.

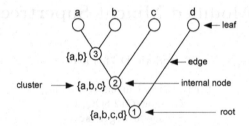

Fig. 1. The tree terminology used in this paper

does not satisfy property 5 above because the problem of finding the most parsimonious tree is NP-complete [5]. Hence the task of constructing an MRP supertree is of the same order of complexity as inferring individual phylogenies. Similarly, supertrees by flipping [6] is NP-complete.

To date the only supertree method that satisfies all five desirable properties is the mincut supertree algorithm of Semple and Steel [7]. This algorithm is very attractive because it can potentially scale to handle large problems. This is particularly important if supertrees are going to be an effective tool for synthesizing very large trees, such as those contained in TreeBASE [8] (`http://www.treebase.org`). A polynomial time supertree algorithm could play a key role in a divide-and-conquer strategy for assembling the tree of life.

In this paper I start by outlining the mincut supertree method, and discussing its implementation. I then note some limitations of the method, and propose modifications that increase its practical utility.

2   Supertrees

2.1   Tree Terminology

The terminology used in this paper is illustrated in figure 1. I will consider only rooted trees. The *leaves* of the tree represent species. Let $\mathcal{L}(T)$ be the set of leaves for tree T. The set of leaves that descend from an internal node in a tree comprise that node's *cluster*. In the tree shown in figure 1, node 3 has the cluster $\{a, b\}$. The most recent common ancestor (MRCA) of a set of nodes in a tree is the node furthest from the root of the tree whose cluster contains all those nodes. For example, in figure 1 MRCA(b, c) is node 2. Given a tree T, if A and B are subsets of $\mathcal{L}(T)$ and MRCA(A) is a descendant of $MRCA(B)$, then $A <_T B$, that is, A nests in B [9]. Examples of nestings in figure 1 include $\{a, b\} <_T \{a, b, c\}$ and $\{b, c\} <_T \{a, b, c, d\}$.

An *edge* in the tree is *internal* it connects two internal nodes (one of which can be the root). Given two trees T and T', T' is said to obtained from T *by contraction* if T' can be obtained from T by contracting one or more internal edges. Let A be a subset of $\mathcal{L}(T)$, then the subtree of T with the leaf set A is the *subtree of T induced by A*, represented as $T|A$. A tree T *displays* a tree t if t is an induced subtree of T, or can be obtained from an induced subtree of T

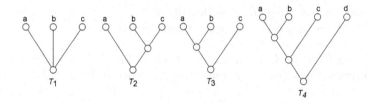

Fig. 2. Examples of tree compatibility. Tree T_1 can be obtained by contraction from either T_2 or T_3. Tree T_3 is the subtree of T_4 induced by the set $\{a, b, c\}$ (i.e., $T_4|\{a, b, c\}$). Hence T_4 displays trees T_1 and T_3. Trees T_2 and T_3 are incompatible with each other, as are trees T_2 and T_4

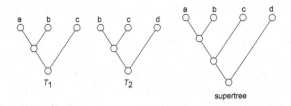

Fig. 3. Two input trees T_1 and T_2 and a supertree

by contraction. A pair of trees is *compatible* if there is a tree that displays them (see fig. 2).

Suppose we have a set of k trees $\mathcal{T}$ with different, overlapping leaf sets. Let $S = \cup_{i=1}^{k} \mathcal{L}(T_i)$ (i.e., the set of all species which are in at least one of the trees in $\mathcal{T}$). A *supertree method* takes $\mathcal{T}$ as input and returns a tree with the leaf set S, the *supertree* (fig. 3).

2.2   ONETREE Algorithm

Given a set of rooted trees, the question of whether they are compatible – that is, whether there is a supertree with which each input tree is compatible – can be answered in polynomial time using algorithms that owe their origin to [10]. To appreciate how Semple and Steel's algorithm works, let us first consider Bryant and Steel's [11] version of algorithm ONETREE [12]. Given three species a, b, c, let the triplet $ab|c$ be the rooted tree where a and b are more closely related to each other than either is to c. A binary tree can be decomposed into a set of triplets. Bryant and Steel construct the graph $[R, S]$ where the nodes S correspond to the species, and there is an edge between nodes a and b if there are any triplets of the form $ab|c$ in the set of input trees. The algorithm ONETREE takes as input a set of triplets and a set of species $S = \{x_1, \ldots, x_n\}$.

procedure OneTree(R, S)

1. **if** $n = 1$ **then return** a single node labelled by x_1.
2. **if** $n = 2$ **then return** a tree with two leaves labelled by x_1 and x_2.

3. Otherwise, construct $[R, S]$ as described.
4. **if** $[R, S]$ has only one component **then return** 'no tree'.
5. **for** each component S_i of $[R, S]$ **do**
   if ONETREE(R, S_i) returns a tree **then** call it T_i **else** return 'no tree'.
6. Construct a new tree T by connecting the roots of the trees T_i to a new root
   r.
7. **return** T

end

The key step in this recursive algorithm is step 4. Bryant and Steel [11] showed that a set of rooted triplets R is compatible if and only if, for any subset S' of S of size at least 2 the graph $[R, S']$ is disconnected. If at any point in the execution of ONETREE the graph $[R, S]$ is connected then the algorithm exits without returning a tree. Hence ONETREE offers a polynomial time test for tree compatibility.

If there is a supertree with which all the input trees are compatible, then ONETREE will return that tree, otherwise the algorithm returns the result that no such tree exists. This severely limits its application to the supertree problem in phylogenetics. Any non-trivial set of phylogenetic trees is likely to be incompatible for a range of reasons, including sampling error, inaccuracies, or biases in tree building algorithms. What is needed is an algorithm that will find a supertree even if the input trees are incompatible. Given that ONETREE exits if $[R, S]$ is connected, one approach to extending the algorithm would be to disconnect $[R, S]$ whenever it is connected. This is the rationale behind Semple and Steel's [7] mincut algorithm.

2.3 MINCUTSUPERTREE

Semple and Steel's MINCUTSUPERTREE algorithm modifies ONETREE so that it always returns a tree. Instead of constructing $[R, S]$ from a set of triplets, MINCUTSUPERTREE constructs a graph $S_{\mathcal{T}}$, where the nodes are species, and nodes a and b are connected if a and b are in a proper cluster in at least one of the input trees (i.e., if there is a tree in which MRCA(a, b) is not the root of the tree). Semple and Steel's algorithm takes a set of k rooted trees $\mathcal{T}$ and a set of species $S = \cup_{i=1}^{k} \mathcal{L}(T_i) = \{x_1, \ldots, x_n\}$.

procedure MinCutSupertree($\mathcal{T}$)

1. **if** $n = 1$ **then return** a single node labelled by x_1.
2. **if** $n = 2$ **then return** a tree with two leaves labelled by x_1 and x_2.
3. Otherwise, construct $S_{\mathcal{T}}$ as described.
4. **if** $S_{\mathcal{T}}$ is disconnected **then**
   Let S_i be the components of $S_{\mathcal{T}}$.
   else Create graph $S_{\mathcal{T}}/E_{\mathcal{T}}^{\max}$ and delete all edges in $S_{\mathcal{T}}/E_{\mathcal{T}}^{\max}$ that are in a
      minimum cut set of $S_{\mathcal{T}}$. Let S_i be the resulting components of $S_{\mathcal{T}}/E_{\mathcal{T}}^{\max}$.
5. **for** each component S_i **do**

$T_i = MinCutSupertree(\mathcal{T}|S_i)$, where $\mathcal{T}|S_i$ is the set of input trees with any species not in S_i pruned.

6. Construct a new tree $\mathcal{T}$ by connecting the roots of the trees T_i to a new root r.

7. **return** T

end

In step 4 Semple and Steel ensure that $S_{\mathcal{T}}$ yields more than one component by using minimum cuts. Given a connected graph G, a set of edges whose removal disconnects the graph is a *cut set*. If each edge in G has a weight assigned to it, then a cut set with the smallest sum of weights is a *minimum cut* of the graph. Semple and Steel do not find minimum cuts of $S_{\mathcal{T}}$, but rather of an associated graph $S_{\mathcal{T}}/E_{\mathcal{T}}^{\max}$ which is constructed as follows::

1. Weight each edge (a, b) in $S_{\mathcal{T}}$ by the number of trees in $\mathcal{T}$ in which a and b are in the same proper cluster.
2. Let $E_{\mathcal{T}}^{\max}$ be the set of edges that have weight k, where k is the number of trees in $\mathcal{T}$.
3. Merge any nodes in $S_{\mathcal{T}}$ that are connected by edges in $E_{\mathcal{T}}^{\max}$.

For example, given the two input trees in figure 4, the edge (a, b) in $S_{\mathcal{T}}$ has a weight of 2, and hence the nodes a and b are merged. This procedure ensures that any nesting found in all of the input trees $\mathcal{T}$ will be in the supertree returned by MINCUTSUPERTREE[1].

2.4 Implementation

Because the mincut problem can be solved in polynomial time, so can MIN-CUTSUPERTREE. The crucial computational step in the MINCUTSUPERTREE algorithm is determining for each edge in $S_{\mathcal{T}}/E_{\mathcal{T}}^{\max}$ whether that edge is in a minimum cut set of $S_{\mathcal{T}}/E_{\mathcal{T}}^{\max}$. Semple and Steel suggest an algorithm for doing this that requires computing the minimum cut for all the subgraphs of $S_{\mathcal{T}}/E_{\mathcal{T}}^{\max}$ that result from deleting a single edge of that graph. Doing this for all the edges of $S_{\mathcal{T}}/E_{\mathcal{T}}^{\max}$ rapidly becomes tedious for all but fairly small examples. We can improve on this by using Picard and Queyranne's [13] algorithm to find all minimum cuts of the graph. Firstly all (s, t) minimum cuts of $S_{\mathcal{T}}/E_{\mathcal{T}}^{\max}$ are found [14], where s and t are nodes in $S_{\mathcal{T}}/E_{\mathcal{T}}^{\max}$ that are disconnected by the cut. For each (s, t) cut, the graph $S_{\mathcal{T}}/E_{\mathcal{T}}^{\max}$ is transformed into a directed graph, and the maximum $s - t$ flow computed. The strongly connected components of the residual graph R for this flow are then computed. Any edge in R which connects

---

[1] Note that in Semple and Steel's original algorithm each input tree in $\mathcal{T}$ can have a weight $w(T)$ assigned to it, and in step 1 above they weight each edge by the sum of the weights of the trees in which a and b are in the same proper cluster. The set of edges $E^{\max}$ comprises those edges for which $w_{sum} = \sum_{T \in \mathcal{T}} w(T)$. For simplicity here I consider only the case where all trees have the same unit weight.

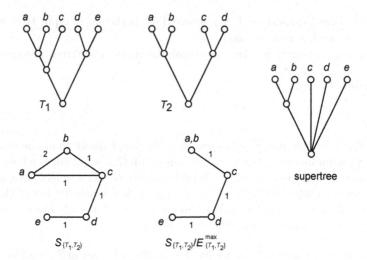

Fig. 4. An example of the MINCUTSUPERTREE algorithm showing the two input trees T_1 and T_2, and the graphs S_T and S_T/E_T^{max}. The graph S_T/E_T^{max} has three minimum cut sets, which yield the components $\{a,b\}, \{c\}, \{d\}$, and $\{e\}$, which in turn yield the supertree (from [7])

nodes belonging to different strongly connected components is a member of a minimum cut set of S_T/E_T^{max} [13].

Another shortcut concerns the recursion in MINCUTSUPERTREE. As originally described the algorithm keeps calling itself recursively until $\mathcal{T}|S$ contains less than three leaves (steps 1 and 2 above), at which point the algorithm returns either a single-leaf or a two-leaf tree. However, if at any point there is only a single tree T in $\mathcal{T}|S_i$ with any leaves, then that tree can be grafted directly onto the growing supertree.

3   Properties of MINCUTSUPERTREE

MINCUTSUPERTREE has at least two attractive properties: it is quick to compute, and any nesting found in all the input tree will appear in the supertree [7]. However, when applied to some examples it can yield slightly disconcerting results. Fig. 5 shows two trees that share only three leaves (a, b, and c), and disagree on the relationships among those leaves. Both trees have unique sets of leaves with fully resolved relationships. In the tree produced by MINCUTSUPERTREE for these two trees the relationships among $x_1 - x_3$ are unresolved, although there is no information in the input trees that contradicts this grouping. In contrast, relationships among $y_1 - y_4$ are fully resolved. Furthermore, even though the two trees disagree about the relations between a, b, and c, the supertree contains the triplet $ab|c$.

These features of the supertree are due, at least in part, to the different sizes of the two input trees. Fig. 6 shows the graph S_T/E_T^{max} for these two

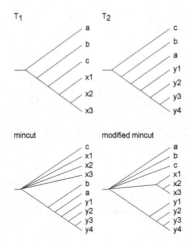

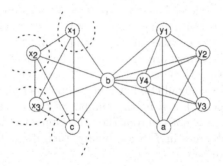

Fig. 5. Two source trees that share the leaves a, b, and c, and their mincut and modified mincut supertrees

Fig. 6. The graph $S_{\mathcal{T}}/E_{\mathcal{T}}^{\max}$ for the trees shown in figure 5. The four minimum weight cuts of the graph are indicated by dashed lines

trees. Because there is no triplet common to the two subtrees, no edge has a weight greater than one. In this case, minimum-weight cuts sets will involve edges connected to nodes with the lowest degree. These are all nodes in the smaller of the two source trees (T_1).

This example shows that MINCUTSUPERTREE can be sensitive to the size of the input trees, and can fail to include information that is not contradicted in the set of input trees. Indeed, the original illustration of Semple and Steel's method (fig. 4) provides an example of the latter problem. The edge connecting nodes d and e in the graph $S_{\mathcal{T}}/E_{\mathcal{T}}^{\max}$ represents the uncontradicted nesting $\{d, e\} <_{T_1} \{a, b, c, d\}$. This nesting is not present in the supertree, and as a result the supertree has less information than the input trees. Although there is no consensus method that can always display all the uncontradicted information in a set of trees [2], it would be desirable to maximize in some sense the amount of uncontradicted information a supertree displays.

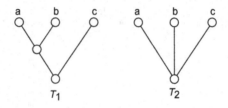

Fig. 7. A triplet (T_1) and a fan (T_2)

If mincut supertrees are to be useful then we need ways to avoid these problems. The next section explores one way of modifying the way the graph S_T/E_T^{max} is created so as to minimize the impact of these problems.

3.1   Modifications

We can refine Semple and Steel's approach by distinguishing between "unanimous" and "uncontradicted" nestings in $T|S$. A unanimous nesting occurs in every tree in $T|S$. These nestings correspond to edges with weight $w(e) = k$, where k is the number of trees in $T|S$. Semple and Steel's algorithm ensures that all unanimous nestings are preserved in the mincut supertree. I define an uncontradicted nesting to be a nesting found in a subset of trees in $T|S$, but which no tree in $T|S$ contradicts. Let c be the number of trees in $T|S$ in which leaves i and j co-occur. Let n be the number of trees containing both i and j for which $\mathrm{MRCA}(i,j)$ is not the root (i.e., the number of trees in which i and j nest in $\mathcal{L}(T)$). If all trees in $T|S$ are binary, then if $c > 0$ and $n = c$ then the nesting is uncontradicted.

Fans. For nonbinary trees the situation is a little more complicated. Consider the two trees shown in figure 7. Tree T_1 contains the nesting $(a,b) <_T (a,b,c)$, whereas T_2 other does not. Whether we regard T_2 as disagreeing with T_1 depends on our interpretation of the polytomy in T_2. Polytomies can be true ("hard") or apparent ("soft") [15,16]. A true polytomy results when three or more lineages diverge from a single ancestor at the same time. In a tree derived from a single data set an apparent polytomy represents a lack of information, such as no substitutions occurring along a branch on the tree [16]. If the tree is a consensus tree that summarizes a set of input trees, then an apparent polytomy is used to depict conflict between the original trees. If the polytomy in tree T_2 is true, then T_1 and T_2 conflict, if the polytomy is only apparent then T_2 is consistent with any of the three rooted trees for a,b, and c, and hence T_1 and T_2 are consistent.

The distinction between true and apparent polytomies corresponds to different uses of the term "compatible" [17,12]. Ng and Wormald [12] regard two trees as compatible only if one is a subtree of the other, whereas for Steel [17] two trees will also be considered compatible if one tree can be obtained by contraction of the other tree. For example, in figure 7 T_2 is not a subtree of T_1, but by removing an internal edge from T_1 we can collapse T_1 into the fan T_2. In Steel's terminology – which I follow here – T_1 and T_2 are compatible.

Given that true polytomies are likely to be rare [16], in most cases any polytomies in the trees in $\mathcal{T}|S$ will be apparent (soft), and hence we need to take this into account when deciding whether a nesting is contradicted. Specifically, let f be the number of trees in $\mathcal{T}|S$ for which MRCA(i,j) is the root of that tree, and the root has degree > 2. A nesting between two taxa is uncontradicted if $c - n - f = 0$, that is, in every tree in which i and j co-occur, $\{i,j\}$ is nested in the tree, or is part of a fan.

Contradicted Edges. We can modify Semple and Steel's algorithm by assigning to each edge in $S_\mathcal{T}$ one of three colors: unanimous, uncontradicted, or contradicted. Our goal is to modify S in such a way as to minimize the number of uncontradicted edges that are cut. One way to do this is to extend Semple and Steel's approach of collapsing nodes linked by unanimous edges to include nodes linked by uncontradicted edges:

1. Construct $S' = S_\mathcal{T}/E_\mathcal{T}^{\max}$ as described in MINCUTSUPERTREE.
2. Let C be the set of contradicted edges in S'.
3. Remove all contradicted edges from S' to yield the graph S'/C.
4. **if** S'/C is disconnected **then**
     S' can be cut without cutting an uncontradicted edge. Find the components of S'/C and merge the nodes in each component.
5. **else**
     Let AC be the set of all edges in S' adjacent to a contradicted edge. Remove all edges in AC to yield $S'/(C \cup AC)$.
     if $S'/(C \cup AC)$ is disconnected **then** find the components of $S'/(C \cup AC)$ and merge the nodes in each component.
6. Restore all deleted edges (removing any parallel edges resulting from nodes merged in step 4 and/or step 5). The graph S' now comprises nodes representing both unanimous and uncontradicted nestings.

When then find all edges in at least one cutset of S' and proceed as in MINCUTSUPERTREE.

Step 4 above tests whether we can disconnect the graph by cutting one or more contradicted edge. If this is the case, then we can preserve all uncontradicted edges at this step. If this test fails, then at least one uncontradicted edge must be cut to disconnect $S' = S_\mathcal{T}/E_\mathcal{T}^{\max}$. Given that we would like the cuts to include contradicted edges wherever possible, in Step 5 we try to achieve this by removing all edges that are adjacent to a contradicted edge (fig.8a). If this disconnects the graph, then we have identified at least one cut that contains a contradicted edge.

Example. Applying the modified algorithm to the two trees in figure 5 yields the $S_\mathcal{T}/E_\mathcal{T}^{\max}$ graph shown in fig.8. This graph has two contradicted edges. Removing

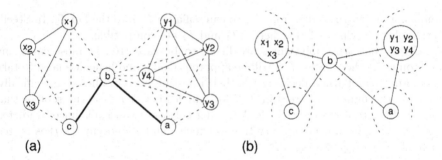

Fig. 8. Modified mincut algorithm. (a) The graph $S_T/E_T^{\max}$ for the trees shown in figure 5. The two bold edges $\{a, b\}$ and $\{b, c\}$ are contradicted, the edges drawn as dashed lines are uncontradicted but adjacent to a contradicted edge. Deleting the contradicted and the adjacent to contradicted edges disconnects the graph. Merging the nodes in each component, then restoring the deleted edges results in the graph S' (b). The six minimum weight cuts of S' are indicated by dashed lines

these does not disconnect the graph. However, removing all edges adjacent to the contradicted edges does disconnect the graph. The resulting supertree (fig. 5) is more resolved than that found by Semple and Steel's original algorithm, and includes the resolved subtrees for $x_1 - x_3$ from T_1 and $y_1 - y_4$ from T_2. The relationships between the shared leaves a, b, and c are represented as a basal polytomy. It should be noted here that polytomies in mincut supertrees are interpreted in a similar fashion to polytomies in Adams consensus trees [9]. This means that groupings in the supertree tree are not necessarily clusters (i.e., monophyletic groups of species)^2.

3.2 Empirical Example

Figure 9 shows the modified mincut supertree for the two trees used by [1] to illustrate the MRP supertree method. For comparison, figure 9 also shows the strict consensus tree of the 8 equally parsimonious trees obtained for the MRP matrix for the two input trees. The two input trees share only four taxa (shown in **boldface** in fig. 9), and disagree about their relationships for three of the species: one tree has the triplet *Astragalus alpinus* | (*Oxytropis deflexa, Cilianthus puniceus*), the other tree has the triplet (*Astragalus alpinus, Oxytropis deflexa*) | *Cilianthus puniceus*. Given we have only two sources of information about the relationships among these three species, and that these sources conflict, it is a little disconcerting that the MRP tree resolves this triplet in favor of the second tree. In contrast, the mincut supertree displays a fan for these three species.

---

^{2} Although the notion that Adams consensus trees are "difficult" to interpret has become part of taxonomic folklore, some biologists have argued that this way of interpreting polytomies "merges with, or is implicit within, routine taxonomic practice" [18].

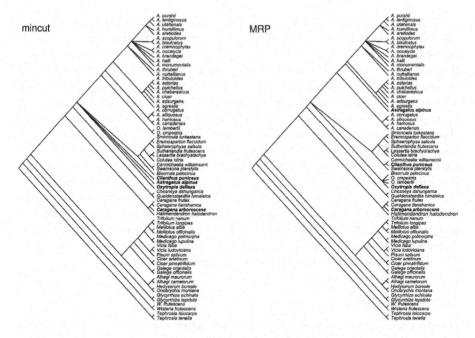

Fig. 9. Comparison of the modified mincut supertree and the MRP supertree for the two trees shown in [1, fig. 2]. The four species in common to the two input trees are shown in **boldface**

4   Discussion

The modified mincut supertree algorithm presented here retains the desirable property of being able to be computed in polynomial time, as well as retaining more of the information shared among the input trees. However, its properties require further exploration. MRP is a global optimization method that seeks to find one or more optimal supertrees. The mincut supertree method does not have an explicit global optimality criterion, instead it uses the locally optimal criterion of minimum cuts. Semple and Steel [7] show that it preserves any nesting found in all input trees, but its other properties are unclear. In addition to theoretical work on the properties of modified mincut supertrees, there are practical issues that are relevant to supertrees in general.

4.1   Labels

A supertree method requires that the same taxon be given the same name in all of the input trees containing that taxon. This requirement is obvious, but not always met. Proper application of the rules of taxonomic nomenclature should ensure that the same taxon is called the same thing in different studies but, sadly, this is not always the case. Even in very well known groups such as birds and mammals, there is disagreement about the name given to some taxa.

A further source of difficulty concerns combining trees containing taxa of different taxonomic rank. In phylogenies constructed from sequence data the leaves of the tree are typically individual organisms, such as *Mus musculus*. However, trees constructed from morphological data sets may have higher taxa as terminals, such as "Rodentia". In the absence of any additional information, a supertree algorithm has no way of knowing that *Mus musculus* in one tree is equivalent to Rodentia in another. One solution would be to include information about higher taxon names in the input trees. TreeBASE [8] has a facility to do this, for example. With this information in hand we could, for example substitute a tree containing *Mus* and *Rattus* species for a leaf labelled Rodentia in another tree. However, this method of "taxon substitution" [19], which has parallels with substitution and adjunction operations in tree adjoining grammars [20], requires that higher taxon names are applied consistently. Given current controversy about the meaning and application of higher taxon names [21], this may be difficult to guarantee. Consequently, any mapping between taxonomic names at different levels would need to be tested for internal consistency.

4.2   Constraints

A potentially serious problem affecting supertree construction is poor overlap between different studies. Sanderson et al. [1] introduced the concept of a "tree-graph," where each node represents an input tree, and nodes are connected by edges weighted by the number of taxa shared by the corresponding trees. For a supertree to be constructed from two trees, those trees must share a minimum of two taxa in common [1]. If the degree of overlap between input trees is poor then this may compromise the resulting supertree. This is a particular problem if input trees use different "exemplars" of higher taxa. For example, if one study on mammalian phylogeny used a mouse to represent rodents, and another study used a rat, then a supertree for those two studies would not group mouse and rat together. This same problem can be encountered when assembling species trees from trees for gene families [22]. One solution would be to impose a constraint on the set of possible solutions, for example by specifying a constraint tree [23] which must be displayed by the supertree. In the previous example, we could require that any solution grouped mouse and rat. Constraints could be readily implemented in the mincut supertree method by collapsing the corresponding nodes in the graph $S_T/E_T^{\max}$ prior to finding the minimum cuts.

4.3   How Good Is a Supertree?

In addition to computational complexity, another criterion for evaluating the efficacy of a supertree method is the degree to which the supertree agrees with the input trees $\mathcal{T}$ [24]. This can be measured using a tree comparison metric. For example, we can measure the agreement between an input tree T and a supertree by the size of the maximum agreement subtree (MAST) for the two trees [6]. Although efficient algorithms for MAST are available [25,26], it has limitations as a measure of tree similarity, particularly if the input trees are not fully resolved [27]. These limitations could be overcome by computing the size of the maximum compatible tree [28]. Instead, here I use triplet measures that are the rooted equivalent of the quartet measures described by [29]. For each input tree T we compare that tree with the subtree of the supertree that results when any taxa not in T are pruned. By analogy with quartets [29], for any pair of triplets (one from each tree) there are five possible outcomes: the triplet is resolved in both trees is identical (d), or different (s); the triplet is resolved in one tree ($r1$) or the other ($r2$) but not both, or the triplet is unresolved in both trees (x). For our purposes, only d, s, and $r2$ are relevant. Triplets that are resolved only in the supertree ($r1$) reflect information present in other input trees rather than T, and triplets not resolved in either T or the supertree (x) are not counted as "agreement" because they reflect a shared lack of information. Hence, for each tree T the fit to the supertree is given by:

$$1 - \frac{d + r2}{d + s + r2}$$

The average of this value over all input trees is a measure of fit between $\mathcal{T}$ and the supertree.

4.4 Software

The algorithms described here have been implemented in the C++ programming language and compiled using the GNU gcc compiler, and commercial compilers from Borland and Metrowerks. Paul Lewis' NEXUS Class Library (http://lewis.eeb.uconn.edu/lewishome/ncl2/NexusClassLibrary.html) is used to read NEXUS format tree files. The code makes extensive use of the Graph Template Library (GTL) (http://www.infosun.fmi.uni-passau.de/GTL/) which provides a Standard Template Library (STL) based library of graph classes and algorithms. Triplet tree comparison measures are computed using the quartet algorithm developed by Douchette [30]. The program reads trees in either NEXUS or Newick format, and outputs a supertree. The program can also output the graphs S_T and $S_T/E_T^{\max}$ in GML (http://www.infosun.fmi.uni-passau.de/Graphlet/)and dot formats (http://www.research.att.com/sw/tools/graphviz/) so the progress of the algorithm can be viewed. The supertree software is available at (http://darwin.zoology.gla.ac.uk/~rpage/supertree/).

References

1. Sanderson, M. J., Purvis, A., Henze, C.: Phylogenetic supertrees: assembling the trees of life Trends Ecol. Evol. **13** (1998) 105–109
2. Steel, M., Dress, A., Böcker, S.: Simple but fundamental limitations on supertree and consensus tree methods. Syst. Biol **49** (2000) 363–368
3. Baum, B.R.: Combining trees as a way of combining data sets for phylogenetic inference, and the desirability of combining gene trees. Taxon **41** (1992) 3–10
4. Ragan,M.A.: Phylogenetic inference based on matrix representation of trees. Mol. Phylogen. Evol. **1** (1992) 53–58

5. Graham, R.L., Foulds, L.R.: Unlikelihood that minimal phylogenies for a realistic biological study can be constructed in reasonable computational time. Math. Biosci. **60** (1982) 133–142

6. Chen, D., Eulenstein, O., Fernández-Baca, D., Sanderson, M.: Supertrees by flipping. Technical Report TR02-01, Department of Computer Science, Iowa State University (2001)

7. Semple, C., Steel, M.: A supertree method for rooted trees. Disc. Appl. Math. **105** (2000) 147–158

8. Piel, W.: Phyloinformatics and tree networks In: Wu, C.H., Wang, P., Wang, J.T.L. (eds.): Computational Biology and Genome Informatics. World Scientific Press (2001)

9. Adams, E.N.: N-trees as nestings: complexity, similarity, and consensus. J. Classif. **3** (1986) 299–317

10. Aho, A.V., Sagiv, Y., Szymanski, T.G., Ullman, J.D.: Inferring a tree from lowest common ancestors with an application to the optimization of relational expressions. SIAM J. Comput. **10** (1981) 405–421

11. Bryant, D., Steel, M.: Extension operations on sets of leaf-labelled trees. Adv. Appl. Math. **16** (1995) 425–453

12. Ng, M. P., Wormald, N. C.: Reconstruction of rooted trees from subtrees. Disc. Appl. Math. **69** (1996) 19–31

13. Picard, J.-C., Queyranne, M.: On the structure of all minimum cuts in a network and applications. Math. Prog. Study **13** (1980) 8–16

14. Stoer, M., Wagner, F.: A simple min cut algorithm. J. ACM **44** (1997) 585–591

15. Maddison, W. P.: Reconstructing character evolution on polytomous cladograms. Cladistics **5** (1989) 365–377

16. Slowinski, J.B. Molecular polytomies. Mol. Phylogen.Evol. **19** (2001) 114–120

17. Steel, M.: The complexity of reconstructing trees from qualitative characters and subtrees. J. Classif. **9** (1992) 91–116

18. Nelson, G., Platnick, N.I. Multiple branching in cladograms: two interpretations. Syst. Zool. **29** (1980) 86–91

19. Wilkinson, M., Thorley, J.L., Littlewood, D.T.J., Bray, R.A.: Towards a phylogenetic supertree of Platyhelminthes? In: Littlewood, D.T.J., Bray, R.A. (eds.): Interrelationships of the Platyhelminthes. Taylor and Francis, London (2001) 292–301

20. Joshi, A.K.: An introduction to tree adjoining grammars In: Manaster-Ramer, A. (ed.): Mathematics of Language. John Benjamins Publishing Co., Amsterdam (1987) 87–115

21. de Queiroz, K., Gauthier, J.: Phylogenetic taxonomy. Ann. Rev. Ecol. Syst. **23** (1992) 449–480

22. Page, R.D.M.: Extracting species trees from complex gene trees: reconciled trees and vertebrate phylogeny. Mol. Phylogen. Evol. **14** (2000) 89–106

23. Constantinescu, M., Sankoff, D.: Tree enumeration modulo a consensus. J. Classif. **3** (1986) 349–356

24. Day, W.H.E.: The role of complexity in comparing classifications. Math. Biosci. **66** (1983) 97–114

25. Steel, M.A., Warnow, T.: Kaikoura tree theorems: Computing the maximum agreement subtree. Info. Proc. Letters **48** (1993) 77–82

26. Farach, M., Przytycka, T.M. and Thorup, M.: On the agreement of many trees. Info. Proc. Letters **55** (1995) 297–301

27. Swofford, D.L.: When are phylogeny estimates from molecular and morphological data incongruent? In: Miyamoto, M. M., Cracraft, J. (eds.): Phylogenetic analysis of DNA sequences. Oxford University Press, New York (1991) 295–333

28. Ganapathysaravanabavan, G. and Warnow, T.: Finding a maximum compatible tree for a bounded number of trees with bounded degree is solvable in polynomial time. In: Gascuel, O., and Moret, B.M.E.Moret (eds.): Proceedings of the First International Workshop on Algorithms in Bioinformatics (WABI 2001) (Lecture Notes in Computer Science **2149**) (2001) 156–163

29. Day, W.H.E.: Analysis of quartet dissimilarity measures between undirected phylogenetic trees. Syst. Zool. **35** (1986) 325–333

30. Douchette, C.R.: An efficient algorithm to compute quartet dissimilarity measures. BSc(Hons) dissertation. Department of Computer Science, Memorial University, Newfoundland, Canada (1985)

Author Index

Abouelhoda, M.I., 449
Adebiyi, E.F., 140
Ajana, Y., 300

Bafna, V., 29
Bar-Joseph, Z., 506
Bergeron, A., 464
Berman, P., 419
Bertone, P., 419
Bhattacharyya, C., 1
Bilu, Y., 263
Bryant, D., 375

Chen, Z.-Z., 82
Clark, A.G., 44
Corteel, S., 464
Csűrös, M., 10

DasGupta, B., 419
Demaine, E.D., 506
Desper, R., 357
Duhovny, D., 185
Durbin, R., 331

Edwards, N., 68
El-Mabrouk, N., 300
Eriksen, N., 316

Fiziev, P., 60

Gascuel, O., 357
Gerstein, M., 419
Gifford, D.K., 506
Goldwasser, M.H., 157
Grate, L.R., 1

Halpern, A.L., 126
Hamel, A.M., 506
Hatzigeorgiou, A., 60
Huson, D.H., 126

Istrail, S., 29, 44

Jaakkola, T.S., 506
Jiang, T., 82
Jordan, M.I., 1

Kahng, A.B., 435
Kao, M.-Y., 157, 419
Kaufmann, M., 140
Kurtz, S., 449

Lancia, G., 29
Lefebvre, J.-F., 300
Leong, H.-W., 392
Lin, G., 82
Linial, M., 263
Lippert, R., 68
Lipson, D., 491
Liu, B.H., 477
Liu, T., 521
Lu, H.-I., 157

Măndoiu, I.I., 435
Ma, B., 97
Mian, I.S., 1
Milik, M., 172
Milosavljevic, A., 10
Moret, B.M.E., 343, 521
Moulton, V., 375
Myers, G., 331

Nakhleh, L., 287
Nussinov, R., 185, 235

Ohlebusch, E., 449
Olszewski, K.A., 172

Page, R.D.M., 537
Pedersen, C.N.S., 220
Pertea, M., 210
Pevzner, P.A., 435
Preparata, F.P., 392

Raffinot, M., 464
Rahmann, S., 434
Reczko, M., 60
Reda, S., 435
Reinert, K., 126
Rizzi, R., 29
Rocke, E., 251
Roshan, U., 287, 343

Salzberg, S.L., 210
Scharling, T., 220
Schwartz, R., 44
Shapiro, E., 418
Shatsky, M., 235
Siepel, A.C., 521
Skiena, S., 404
Snir, S., 404
Snyder, M., 419
Srebro, N., 506
Staub, E., 60
Sung, W.-K., 392
Szalma, S., 172

Tang, J., 521
Tillier, E.R.M., 300

Ukkonen, E., 277

Vawter, L., 287

Wang, L., 97, 112
Warnow, T., 287, 343
Warren, L.L., 477
Webb, P., 491
Wen, J., 82
Wheeler, R., 201
Willy, H., 392
Wolfson, H.J., 185, 235

Xu, D., 82
Xu, Y., 82

Yakhini, Z., 491

Zelikovsky, A., 435
Zhang, L., 97

Lecture Notes in Computer Science

For information about Vols. 1–2380
please contact your bookseller or Springer-Verlag

Vol. 2382: A. Halevy, A. Gal (Eds.), Next Generation Information Technologies and Systems. Proceedings, 2002. VIII, 169 pages. 2002.

Vol. 2383: M.S. Lew, N. Sebe, J.P. Eakins (Eds.), Image and Video Retrieval. Proceedings, 2002. XII, 388 pages. 2002.

Vol. 2384: L. Batten, J. Seberry (Eds.), Information Security and Privacy. Proceedings, 2002. XII, 514 pages. 2002.

Vol. 2385: J. Calmet, B. Benhamou, O. Caprotti, L. Henocque, V. Sorge (Eds.), Artificial Intelligence, Automated Reasoning, and Symbolic Computation. Proceedings, 2002. XI, 343 pages. 2002. (Subseries LNAI).

Vol. 2386: E.A. Boiten, B. Möller (Eds.), Mathematics of Program Construction. Proceedings, 2002. X, 263 pages. 2002.

Vol. 2387: O.H. Ibarra, L. Zhang (Eds.), Computing and Combinatorics. Proceedings, 2002. XIII, 606 pages. 2002.

Vol. 2388: S.-W. Lee, A. Verri (Eds.), Pattern Recognition with Support Vector Machines. Proceedings, 2002. XI, 420 pages. 2002.

Vol. 2389: E. Ranchhod, N.J. Mamede (Eds.), Advances in Natural Language Processing. Proceedings, 2002. XII, 275 pages. 2002. (Subseries LNAI).

Vol. 2390: D. Blostein, Y.-B. Kwon (Eds.), Graphics Recognition. Proceedings, 2001. XI, 367 pages. 2002.

Vol. 2391: L.-H. Eriksson, P.A. Lindsay (Eds.), FME 2002: Formal Methods – Getting IT Right. Proceedings, 2002. XI, 625 pages. 2002.

Vol. 2392: A. Voronkov (Ed.), Automated Deduction – CADE-18. Proceedings, 2002. XII, 534 pages. 2002. (Subseries LNAI).

Vol. 2393: U. Priss, D. Corbett, G. Angelova (Eds.), Conceptual Structures: Integration and Interfaces. Proceedings, 2002. XI, 397 pages. 2002. (Subseries LNAI).

Vol. 2394: P. Perner (Ed.), Advances in Data Mining. VII, 109 pages. 2002. (Subseries LNAI).

Vol. 2395: G. Barthe, P. Dybjer, L. Pinto, J. Saraiva (Eds.), Applied Semantics. IX, 537 pages. 2002.

Vol. 2396: T. Caelli, A. Amin, R.P.W. Duin, M. Kamel, D. de Ridder (Eds.), Structural, Syntactic, and Statistical Pattern Recognition. Proceedings, 2002. XVI, 863 pages. 2002.

Vol. 2397: G. Grahne, G. Ghelli (Eds.), Database Programming Languages. Proceedings, 2001. X, 343 pages. 2002.

Vol. 2398: K. Miesenberger, J. Klaus, W. Zagler (Eds.), Computers Helping People with Special Needs. Proceedings, 2002. XXII, 794 pages. 2002.

Vol. 2399: H. Hermanns, R. Segala (Eds.), Process Algebra and Probabilistic Methods. Proceedings, 2002. X, 215 pages. 2002.

Vol. 2400: B. Monien, R. Feldmann (Eds.), Euro-Par 2002 – Parallel Processing. Proceedings, 2002. XXIX, 993 pages. 2002.

Vol. 2401: P.J. Stuckey (Ed.), Logic Programming. Proceedings, 2002. XI, 486 pages. 2002.

Vol. 2402: W. Chang (Ed.), Advanced Internet Services and Applications. Proceedings, 2002. XI, 307 pages. 2002.

Vol. 2403: Mark d'Inverno, M. Luck, M. Fisher, C. Preist (Eds.), Foundations and Applications of Multi-Agent Systems. Proceedings, 1996-2000. X, 261 pages. 2002. (Subseries LNAI).

Vol. 2404: E. Brinksma, K.G. Larsen (Eds.), Computer Aided Verification. Proceedings, 2002. XIII, 626 pages. 2002.

Vol. 2405: B. Eaglestone, S. North, A. Poulovassilis (Eds.), Advances in Databases. Proceedings, 2002. XII, 199 pages. 2002.

Vol. 2406: C. Peters, M. Braschler, J. Gonzalo, M. Kluck (Eds.), Evaluation of Cross-Language Information Retrieval Systems. Proceedings, 2001. X, 601 pages. 2002.

Vol. 2407: A.C. Kakas, F. Sadri (Eds.), Computational Logic: Logic Programming and Beyond. Part I. XII, 678 pages. 2002. (Subseries LNAI).

Vol. 2408: A.C. Kakas, F. Sadri (Eds.), Computational Logic: Logic Programming and Beyond. Part II. XII, 628 pages. 2002. (Subseries LNAI).

Vol. 2409: D.M. Mount, C. Stein (Eds.), Algorithm Engineering and Experiments. Proceedings, 2002. VIII, 207 pages. 2002.

Vol. 2410: V.A. Carreño, C.A. Muñoz, S. Tahar (Eds.), Theorem Proving in Higher Order Logics. Proceedings, 2002. X, 349 pages. 2002.

Vol. 2412: H. Yin, N. Allinson, R. Freeman, J. Keane, S. Hubbard (Eds.), Intelligent Data Engineering and Automated Learning – IDEAL 2002. Proceedings, 2002. XV, 597 pages. 2002.

Vol. 2413: K. Kuwabara, J. Lee (Eds.), Intelligent Agents and Multi-Agent Systems. Proceedings, 2002. X, 221 pages. 2002. (Subseries LNAI).

Vol. 2414: F. Mattern, M. Naghshineh (Eds.), Pervasive Computing. Proceedings, 2002. XI, 298 pages. 2002.

Vol. 2415: J.R. Dorronsoro (Ed.), Artificial Neural Networks – ICANN 2002. Proceedings, 2002. XXVIII, 1382 pages. 2002.

Vol. 2416: S. Craw, A. Preece (Eds.), Advances in Case-Based Reasoning. Proceedings, 2002. XII, 656 pages. 2002. (Subseries LNAI).

Vol. 2417: M. Ishizuka, A. Sattar (Eds.), PRICAI 2002: Trends in Artificial Intelligence. Proceedings, 2002. XX, 623 pages. 2002. (Subseries LNAI).

Vol. 2418: D. Wells, L. Williams (Eds.), Extreme Programming and Agile Methods – XP/Agile Universe 2002. Proceedings, 2002. XII, 292 pages. 2002.

Vol. 2419: X. Meng, J. Su, Y. Wang (Eds.), Advances in Web-Age Information Management. Proceedings, 2002. XV, 446 pages. 2002.

Vol. 2420: K. Diks, W. Rytter (Eds.), Mathematical Foundations of Computer Science 2002. Proceedings, 2002. XII, 652 pages. 2002.

Vol. 2421: L. Brim, P. Jančar, M. Křetínský, A. Kučera (Eds.), CONCUR 2002 – Concurrency Theory. Proceedings, 2002. XII, 611 pages. 2002.

Vol. 2422: H. Kirchner, Ch. Ringeissen (Eds.), Algebraic Methodology and Software Technology. Proceedings, 2002. XI, 503 pages. 2002.

Vol. 2423: D. Lopresti, J. Hu, R. Kashi (Eds.), Document Analysis Systems V. Proceedings, 2002. XIII, 570 pages. 2002.

Vol. 2425: Z. Bellahsène, D. Patel, C. Rolland (Eds.), Object-Oriented Information Systems. Proceedings, 2002. XIII, 550 pages. 2002.

Vol. 2426: J.-M. Bruel, Z. Bellahsène (Eds.), Advances in Object-Oriented Information Systems.Proceedings, 2002. IX, 314 pages. 2002.

Vol. 2430: T. Elomaa, H. Mannila, H. Toivonen (Eds.), Machine Learning: ECML 2002. Proceedings, 2002. XIII, 532 pages. 2002. (Subseries LNAI).

Vol. 2431: T. Elomaa, H. Mannila, H. Toivonen (Eds.), Principles of Data Mining and Knowledge Discovery. Proceedings, 2002. XIV, 514 pages. 2002. (Subseries LNAI).

Vol. 2432: R. Bergmann, Experience Management. XXI, 393 pages. 2002. (Subseries LNAI).

Vol. 2434: S. Anderson, S. Bologna, M. Felici (Eds.), Computer Safety, Reliability and Security. Proceedings, 2002. XX, 347 pages. 2002.

Vol. 2435: Y. Manolopoulos, P. Návrat (Eds.), Advances in Databases and Information Systems. Proceedings, 2002. XIII, 415 pages. 2002.

Vol. 2436: J. Fong, C.T. Cheung, H.V. Leong, Q. Li (Eds.), Advances in Web-Based Learning. Proceedings, 2002. XIII, 434 pages. 2002.

Vol. 2438: M. Glesner, P. Zipf, M. Renovell (Eds.), Field-Programmable Logic and Applications. Proceedings, 2002. XXII, 1187 pages. 2002.

Vol. 2439: J.J. Merelo Guervós, P. Adamidis, H.-G. Beyer, J.-L. Fernández-Villacañas, H.-P. Schwefel (Eds.), Parallel Problem Solving from Nature – PPSN VII. Proceedings, 2002. XXII, 947 pages. 2002.

Vol. 2440: J.M. Haake, J.A. Pino (Eds.), Groupware: Design, Implementation and Use. Proceedings, 2002. XII, 285 pages. 2002.

Vol. 2442: M. Yung (Ed.), Advances in Cryptology – CRYPTO 2002. Proceedings, 2002. XIV, 627 pages. 2002.

Vol. 2443: D. Scott (Ed.), Artificial Intelligence: Methodology, Systems, and Applications. Proceedings, 2002. X, 279 pages. 2002. (Subseries LNAI).

Vol. 2444: A. Buchmann, F. Casati, L. Fiege, M.-C. Hsu, M.-C. Shan (Eds.), Technologies for E-Services. Proceedings, 2002. X, 171 pages. 2002.

Vol. 2445: C. Anagnostopoulou, M. Ferrand, A. Smaill (Eds.), Music and Artificial Intelligence. Proceedings, 2002. VIII, 207 pages. 2002. (Subseries LNAI).

Vol. 2446: M. Klusch, S. Ossowski, O. Shehory (Eds.), Cooperative Information Agents VI. Proceedings, 2002. XI, 321 pages. 2002. (Subseries LNAI).

Vol. 2447: D.J. Hand, N.M. Adams, R.J. Bolton (Eds.), Pattern Detection and Discovery. Proceedings, 2002. XII, 227 pages. 2002. (Subseries LNAI).

Vol. 2448: P. Sojka, I. Kopeček, K. Pala (Eds.), Text, Speech and Dialogue. Proceedings, 2002. XII, 481 pages. 2002. (Subseries LNAI).

Vol. 2449: L. Van Gool (Ed.), Pattern Recognotion. Proceedings, 2002. XVI, 628 pages. 2002.

Vol. 2451: B. Hochet, A.J. Acosta, M.J. Bellido (Eds.), Integrated Circuit Design. Proceedings, 2002. XVI, 496 pages. 2002.

Vol. 2452: R. Guigó, D. Gusfield (Eds.), Algorithms in Bioinformatics. Proceedings, 2002. X, 554 pages. 2002.

Vol. 2453: A. Hameurlain, R. Cicchetti, R. Traunmüller (Eds.), Database and Expert Systems Applications. Proceedings, 2002. XVIII, 951 pages. 2002.

Vol. 2454: Y. Kambayashi, W. Winiwarter, M. Arikawa (Eds.), Data Warehousing and Knowledge Discovery. Proceedings, 2002. XIII, 339 pages. 2002.

Vol. 2455: K. Bauknecht, A M. Tjoa, G. Quirchmayr (Eds.), E-Commerce and Web Technologies. Proceedings, 2002. XIV, 414 pages. 2002.

Vol. 2456: R. Traunmüller, K. Lenk (Eds.), Electronic Government. Proceedings, 2002. XIII, 486 pages. 2002.

Vol. 2458: M. Agosti, C. Thanos (Eds.), Research and Advanced Technology for Digital Libraries. Proceedings, 2002. XVI, 664 pages. 2002.

Vol. 2462: K. Jansen, S. Leonardi, V. Vazirani (Eds.), Approximation Algorithms for Combinatorial Optimization. Proceedings, 2002. VIII, 271 pages. 2002.

Vol. 2463: M. Dorigo, G. Di Caro, M. Sampels (Eds.), Ant Algorithms. Proceedings, 2002. XIII, 305 pages. 2002.

Vol. 2464: M. O'Neill, R.F.E. Sutcliffe, C. Ryan, M. Eaton, N. Griffith (Eds.), Artificial Intelligence and Cognitive Science. Proceedings, 2002. XI, 247 pages. 2002. (Subseries LNAI).

Vol. 2469: W. Damm, E.-R. Olderog (Eds.), Formal Techniques in Real-Time and Fault-Tolerant Systems. Proceedings, 2002. X, 455 pages. 2002.

Vol. 2470: P. Van Hentenryck (Ed.), Principles and Practice of Constraint Programming – CP 2002. Proceedings, 2002. XVI, 794 pages. 2002.

Vol. 2479: M. Jarke, J. Koehler, G. Lakemeyer (Eds.), KI 2002: Advances in Artificial Intelligence. Proceedings, 2002. XIII, 327 pages. (Subseries LNAI).

Vol. 2483: J.D.P. Rolim, S. Vadhan (Eds.), Randomization and Approximation Techniques in Computer Science. Proceedings, 2002. VIII, 275 pages. 2002.